Advanced Control Systems for Electric Drives

Advanced Control Systems for Electric Drives

Editor

Adel Merabet

MDPI • Basel • Beijing • Wuhan • Barcelona • Belgrade • Manchester • Tokyo • Cluj • Tianjin

Editor
Adel Merabet
Saint Mary's University
Canada

Editorial Office
MDPI
St. Alban-Anlage 66
4052 Basel, Switzerland

This is a reprint of articles from the Special Issue published online in the open access journal *Electronics* (ISSN 2079-9292) (available at: https://www.mdpi.com/journal/electronics/special_issues/control_electric_drives).

For citation purposes, cite each article independently as indicated on the article page online and as indicated below:

LastName, A.A.; LastName, B.B.; LastName, C.C. Article Title. *Journal Name* **Year**, *Volume Number*, Page Range.

ISBN 978-3-03943-699-6 (Hbk)
ISBN 978-3-03943-700-9 (PDF)

Contents

About the Editor

Adel Merabet received his Ph.D. in Engineering from the University of Québec at Chicoutimi, Chicoutimi, Canada, in 2007. In 2007–2008, he held a Postdoctoral Fellow position at the Department of Electrical and Computer Engineering at Dalhousie University, Halifax, Canada. He joined Saint Mary's University, Halifax, Canada, in 2009 and currently is an Associate Professor at the Division of Engineering. In 2016–2017, he was a Visiting Academic at the Department of Sustainable and Renewable Energy Engineering, University of Sharjah, Sharjah, United Arab Emirates. In 2019, he was awarded the Tan Chin Tuan Exchange Fellowship in Engineering from Nanyang Technological University (NTU) in Singapore and spent 3 months, as a Research Fellow, in the School of Electrical and Electronic Engineering at NTU. He has authored or coauthored over 90 research articles and 2 book chapters. His research interests include renewable (wind–solar) energy conversion systems, energy management, advanced control, electric drives, artificial intelligence, and smart grids.

Editorial

Advanced Control for Electric Drives:
Current Challenges and Future Perspectives

Adel Merabet

Division of Engineering, Saint Mary's University, Halifax, NS B3H 3C3, Canada;
adel.merabet@smu.ca; Tel.: +1-902-420-5712

Received: 2 October 2020; Accepted: 21 October 2020; Published: 23 October 2020

Abstract: In the Special Issue "Advanced Control for Electric Drives", the objective is to address a variety of issues related to advances in control techniques for electric drives, implementation challenges, and applications in emerging fields such as electric vehicles, unmanned aerial vehicles, maglev trains and motion applications. This issue includes 15 selected and peer-reviewed articles discussing a wide range of topics, where intelligent control, estimation and observation schemes were applied to electric drives for various applications. Different drives were studied such as induction motors, permanent magnet synchronous motors and brushless direct current motors.

Keywords: motor drives; advanced control; power converters; estimation; sensor; artificial intelligence

---

1. Introduction

In recent years, electric drives have attracted the attention of researchers in emerging fields such as electric vehicles (EV), renewable energy systems (wind, tidal, ocean, etc.) and high precision motion applications [1–3]. Electric drives are composed of electrical machines, power electronic converters and control systems. Their efficient operation, for position and speed regulation, is determined by the control system. Furthermore, electric drives are a highly nonlinear, multivariable, time-varying system, depending on the type of the electrical machine, and require more complex methods of control. Therefore, they constitute a theoretical and practical challenge in control, estimation and efficient operation in different applications [4–7].

2. The Special Issue

This Special Issue covers a wide range of topics in the field of electric drives. It contains 15 articles studying advanced control schemes, estimation, disturbance rejection and energy management in different applications of electric drives. This section summarizes the content of these articles.

In [8], a predictive controller was developed for tracking a reference current provided by the speed control loop in permanent magnet synchronous motors (PMSM). The control scheme includes a disturbance observer, based on an equivalent input-disturbance approach combined with the deadbeat predictive controller, to improve its robustness. The provided results demonstrate that the proposed control method, compared to the conventional proportional-integral (PI) control, offers smaller fluctuation and shorter settling time for the starting current of the motor. In [9], the bird swarm algorithm, a bio-inspired evaluation approach, was improved and implemented to develop an energy management of extended-range electric vehicles. In this optimization technique, the spatial distance from the center of the bird swarm, instead of fitness function value, was used to stand for their intimacy of relationship. In [10], a multi-sensor data fusion method was developed to detect a safe distance between the unmanned aerial vehicle (UAV) and the transmission lines. This approach takes into consideration the physical model of the UAV in the complex electromagnetic field and the main factors affecting the UAV to develop an adaptive weighted fusion algorithm to conduct analysis on the sensor data. In [11], a cascade second order sliding mode control (SMC) scheme was applied to a PMSM for speed and current control loops. This SMC technique includes an integral action and takes into consideration the nonlinear dynamics and uncertainties. The control robustness, against parametric variations, external disturbance and unknown load, was improved using this control scheme. Another predictive control technique for PMSM current was proposed in [12]. The method uses a two-step-ahead prediction to compensate the delay between the measurements and actuation that causes current ripples. Furthermore, the deadbeat principle and the inverse model of PMSM were used to obtain the reference stator voltage, and an identification system was developed to deal with parametric inaccuracies of the predictive model. In [13], an adaptive SMC scheme was proposed to obtain the desired torque trajectory of the clutch transmission in the mode transition of single-shaft parallel hybrid electric vehicles. Furthermore, a proportional-integral (PI) observer was designed to estimate the actual transmission torque of the clutch. Finally, a fractional order proportional-integral-differential (FOPID) controller, with optimized control parameters by particle swarm optimization (PSO), was employed to realize the accurate position tracking of the direct current (DC) motor clutch required to ensure clutch transmission torque tracking. In [14], an efficient proportional derivative (PD) position controller was developed for three-phase motor drives. The effect of the load disturbance was compensated by a feed forward term. The proposed method was validated on the two different drives (induction motor, PMSM) and demonstrated good performance with respect to parametric uncertainties, unknown load disturbance and measurement noise in the position and current loops. In [15], an estimator and an intelligent controller were proposed for sensorless control of the PMSM drive. The estimator, for position and rotor speed estimation, was constructed using a sliding mode observer and a phase-locked loop. The intelligent controller was used for the velocity control loop. This includes a radial basis function neural network (RBFNN) for self-tuning of the proportional-integral-derivative (PID) controller. In [16], a low-order finite impulse response (FIR) filter, for a high-frequency signal injection method in the permanent magnet linear synchronous motor (PMLSM), was designed for position observation. The filter coefficients were obtained using constraint equations based on the amplitude–frequency characteristics of the FIR filter. In [17], state-of-the art position sensor technologies were investigated for use in PMSM drives applied to automotive electric powertrain systems. Multiple sensor systems were analyzed based on different influencing factors and performance indicators in the automotive field. In [18], the speed control for brushless direct current (BLDC) motors, in steel rolling applications, was investigated using a cascade control loop. The load torque and the speed reference were calculated from the rolling process. In [19], the influence of the harmonic torque on the performance of PMSM drive, in a pure electric vehicle, was investigated by considering the dead-time, the voltage drop effects and the nonlinear characteristics of the transmission system. In [20], a feedback controller was designed for controlling a three-phase linear induction motor. The system modeling was carried out using a neural network identifier to deal with uncertainties due to disturbances, unmodeled quantities, sensors and actuators. In [21], a fuzzy logic control was used for speed tracking in PMSM drives. The intelligent controller was adjusted by a radial basis function neural network. Furthermore, sensorless control was granted by a sliding mode observer for the rotor position and the speed estimation. In [22], the SMC method was combined with a fractional order synergetic technique for speed and current control to deal with nonlinearities and uncertainties. Furthermore, a sliding mode observer was used to estimate the rotor position and speed, based on the motor voltages and currents, for sensorless operation. In order to ensure the high performance of the sensorless control system, a fault detection system was implemented for the current sensors.

3. Future Perspectives

The use of electric drives in sophisticated systems, such as electric and unmanned vehicles, renewable energy systems and robots, is growing with complicated motion and high precision requirements. Furthermore, the evolution and involvement of multi-phase electric machines and power converters require precise, fast and efficient control for the speed and the position, which cannot be achieved by conventional control systems. Smart control systems, which incorporate artificial intelligence tools, sensorless and wireless concepts, and data analysis for adaptive modeling, are emerging to enhance the operation of electric drives. Therefore, the research in this field will remain very active [23–26].

Funding: The guest editing of this Special Issue and publication of this editorial was funded by the Natural Sciences and Engineering Research Council of Canada (NSERC), discovery grant RGPIN-2018-05381.

Acknowledgments: The author would like to thank the editorial board of *Electronics* for the opportunity to serve as the Guest Editor of this Special Issue. Many thanks to all researchers who submitted articles to this Special Issue for their excellent contributions. Gratitude and thank you to all reviewers for their comments and suggestions to improve the quality of contributions. Finally, congratulations for the *Electronics* editorial office team for the success in managing this Special Issue.

Conflicts of Interest: The author declares no conflict of interest.

References

1. Shao, L.; Karci, A.E.H.; Tavernini, D.; Sorniotti, A.; Cheng, M. Design approaches and control strategies for energy-efficient electric machines for electric vehicles—A review. *IEEE Access* **2020**, *8*, 116900–116913. [CrossRef]
2. Nair, R.; Gopalaratnam, N. Emulation of wind turbine system using vector controlled induction motor drive. *IEEE Trans. Ind. Appl.* **2020**, *56*, 4124–4133. [CrossRef]
3. Khatri, P.; Wang, X. Comprehensive review of a linear electrical generator for ocean wave energy conversion. *IET Renew. Power Gener.* **2020**, *14*, 949–958. [CrossRef]
4. Salem, A.; Narimani, M. A review on multiphase drives for automotive traction applications. *IEEE Trans. Transport. Electrif.* **2019**, *5*, 1329–1348. [CrossRef]
5. Li, L.; Liu, Q. Research on IPMSM drive system control technology for electric vehicle energy consumption. *IEEE Access* **2019**, *7*, 186201–186210. [CrossRef]
6. Rafaq, M.S.; Jung, J.-W. A comprehensive review of state-of-the-art parameter estimation techniques for permanent magnet synchronous motors in wide speed range. *IEEE Trans. Ind. Inform.* **2020**, *16*, 4747–4758. [CrossRef]
7. Tanvir, A.A.; Merabet, A. Artificial neural network and Kalman filter for estimation and control in standalone induction generator wind energy DC microgrid. *Energies* **2020**, *13*, 1743. [CrossRef]
8. Liu, X.; Zhang, Q. Robust current predictive control-based equivalent input disturbance approach for PMSM drive. *Electronics* **2019**, *8*, 1034. [CrossRef]
9. Wu, D.; Feng, L. On-off control of range extender in extended-range electric vehicle using bird swarm intelligence. *Electronics* **2019**, *8*, 1223. [CrossRef]
10. Zhang, W.; Ning, Y.; Suo, C. A method based on multi-sensor data fusion for UAV safety distance diagnosis. *Electronics* **2019**, *8*, 1467. [CrossRef]
11. Merabet, A. Cascade second order sliding mode control for permanent magnet synchronous motor drive. *Electronics* **2019**, *8*, 1508. [CrossRef]
12. Tang, P.; Dai, Y.; Li, Z. Unified predictive current control of PMSMs with parameter uncertainty. *Electronics* **2019**, *8*, 1534. [CrossRef]
13. Ding, J.; Jiao, X. A novel control method of clutch during mode transition of single-shaft parallel hybrid electric vehicles. *Electronics* **2020**, *9*, 54. [CrossRef]
14. Alkorta, P.; Barambones, O.; Cortajarena, J.A.; Martija, I.; Maseda, F.J. Effective position control for a three-phase motor. *Electronics* **2020**, *9*, 241. [CrossRef]
15. Hoai, H.-K.; Chen, S.-C.; Than, H. Realization of the sensorless permanent magnet synchronous motor drive control system with an intelligent controller. *Electronics* **2020**, *9*, 365. [CrossRef]
16. Zhao, H.; Zhang, L.; Liu, J.; Cai, J.; Shen, L. Design of a low-order FIR filter for a high-frequency square-wave voltage injection method of the PMLSM used in maglev train. *Electronics* **2020**, *9*, 729. [CrossRef]
17. Datlinger, C.; Hirz, M. Benchmark of rotor position sensor technologies for application in automotive electric drive trains. *Electronics* **2020**, *9*, 1063. [CrossRef]

18. Nandakumar, M.; Ramalingam, S.; Nallusamy, S.; Rangarajan, S.S. Hall-sensor-based position detection for quick reversal of speed control in a BLDC motor drive system for industrial applications. *Electronics* **2020**, *9*, 1149. [CrossRef]

19. Hu, J.; Yang, Y.; Jia, M.; Guan, Y.; Fu, C.; Liao, S. Research on harmonic torque reduction strategy for integrated electric drive System in pure electric vehicle. *Electronics* **2020**, *9*, 1241. [CrossRef]

20. Alanis, A.Y.; Rios, J.D.; Gomez-Avila, J. Discrete-time neural control of quantized nonlinear systems with delays: Applied to a three-phase linear induction motor. *Electronics* **2020**, *9*, 1274. [CrossRef]

21. Hoai, H.-K.; Chen, S.-C.; Chang, C.-F. Realization of the neural fuzzy controller for the sensorless PMSM drive control system. *Electronics* **2020**, *9*, 1371. [CrossRef]

22. Nicola, M.; Nicola, C. Sensorless fractional order control of PMSM based on synergetic and sliding mode controllers. *Electronics* **2020**, *9*, 1494. [CrossRef]

23. Liu, D.; Song, X.; Deng, F.; Dong, J. Investigation into multi-phase armature windings for high-temperature superconducting wind turbine generators. *IEEE Trans. Appl. Supercond.* **2020**, *30*, 5201005. [CrossRef]

24. Bakir, H.; Merabet, A.; Dhar, R.K.; Kulaksiz, A.A. Bacteria foraging optimisation algorithm based optimal control for doubly-fed induction generator wind energy system. *IET Renew. Power Gener.* **2020**, *14*, 1850–1859. [CrossRef]

25. Schenke, M.; Kirchgässner, W.; Wallscheid, O. Controller design for electrical drives by deep reinforcement learning: A proof of concept. *IEEE Trans. Ind. Inform.* **2020**, *16*, 4650–4658. [CrossRef]

26. El-Sousy, F.F.M.; Abuhasel, K.A. Nonlinear robust optimal control via adaptive dynamic programming of permanent-magnet linear synchronous motor drive for uncertain two-axis motion control system. *IEEE Trans. Ind. Appl.* **2020**, *56*, 1940–1952. [CrossRef]

Publisher's Note: MDPI stays neutral with regard to jurisdictional claims in published maps and institutional affiliations.

 *electronics*

Article

Robust Current Predictive Control-Based Equivalent Input Disturbance Approach for PMSM Drive

Xudong Liu [1] and Qi Zhang [2,*]

[1] College of Automation, Qingdao University, Qingdao 266071, China; xudong19871982@163.com
[2] School of Control Science and Engineering, Shandong University, Jinan 250061, China
* Correspondence: zhangqi2013@sdu.edu.cn

Received: 11 August 2019; Accepted: 12 September 2019; Published: 15 September 2019

Abstract: The implementation and experimental validation of current control strategy based on predictive control and equivalent input disturbance approach is discussed for permanent magnet synchronous motor (PMSM) control system in the paper. First, to realize the current decoupling control, the deadbeat predictive current control technique is adopted in the current loop of PMSM. Indeed, it is well known that the traditional deadbeat current control cannot completely reject the disturbance and realize the zero error current tracking control. Then, according to the model uncertainties and the parameter variations in the motor, an equivalent input disturbance approach is introduced to estimate the lump disturbance in the system, which will be used in the feed-forward compensation. Thus, a compound current controller is designed, and the proposed algorithm reduces the tracking error caused by the disturbance; the robustness of the drive system is improved effectively. Finally, simulation and experiment are accomplished on the control prototype, and the results show the effectiveness of the proposed current control algorithm.

Keywords: PMSM drive; current control; deadbeat predictive control; equivalent input disturbance

1. Introduction

Owning to the multiple advantages of high efficiency, high power density, and exceptional reliability, permanent magnet synchronous motor (PMSM) has been widely used in different applications [1], such as electric vehicle drive system, rail traffic, and robot. However, the main weakness of PMSM is the complex controller for its nonlinear and strong coupling characteristics. Therefore, the vector control strategy is employed for the practical applications in general. Consequently, a double closed-loop control method with the inner current loop and the out speed loop is formed. This paper mainly concentrates on solving the current control problem of PMSM in the face of different disturbance.

In general, the proportional plus integral (PI) current control method is popular for industrial applications and is not designed based on the mathematical model. The good performance of zero steady-state error and fixed switching frequency has promoted extensive industrial application. However, it may not meet the requirement in some special occasions, and the transient response may be limited. Meanwhile, it is a challenge for engineers to select the PI parameters by trial and error because of the large parameter uncertainties [2]. Thus, many approaches have been proposed for the motor drive system, such as, sliding mode variable structure control [3], feedback linearization control [4], finite-time control [5], predictive control [6], and passive control [7]. These methods may improve the performance of motor drive system in different aspects, such as good transient response, strong robustness, or lower torque ripple in the steady state.

Electronics **2019**, *8*, 1034; doi:10.3390/electronics8091034 www.mdpi.com/journal/electronics

Among these methods, predictive control, as an advanced control strategy, has been widely applied to the power electronics and drives [8,9]. Predictive control can achieve good tracking control and can be completed easily. For the current control of PMSM, predictive control methods mainly includes deadbeat predictive control [10,11] and model predictive control (MPC) [12–15]. In comparison, the requirement for high computational resources is one of the main drawbacks in a drive with MPC. Deadbeat predictive control, as one of the simplest and best-known predictive control methods, can obtain good tracking performance with less computational burden. In this method, the discrete time model is used to compute the reference voltage, and the zero-error can be achieved within one sampling time. Then, the voltage is translated to the corresponding switching configurations through pulse width modulation or the space vector PWM [16]. Thus, the deadbeat predictive control is characterized with fixed switching frequency, fast current dynamic response, and less computational burden.

It is worth emphasizing that the deadbeat predictive control is a model-based control technique, and the exact model is required. The sensitivity of deadbeat predictive control against uncertainties is a well-known disadvantage. If the motor parameters are known, the tuning problem is reduced to the selection of one parameter only in deadbeat predictive control. However, the PMSM drive system faces inevitable model uncertainties and parameter variations, as the values of stator resistance, stator inductance, and the rotor flux may change, along with the changes of the operation conditions. Influenced by system uncertainty, the motor current cannot track the reference value, which will affect the field-oriented control of the motor, then the performance of the control system will be degraded. Thus, the uncertainties inside the motor and outside disturbances are the main factors to reduce the system performances. Confronted with the problem of system uncertainty, effective methods include the disturbance estimation and attenuation method [17]. Meanwhile, it is proven that this technology has a different but complementary mechanism to widely used robust control and adaptive control [18]. In the work by the authors of [19], a generalized proportional integral observer method is proposed to estimate the time-varying disturbance for the speed current control of PMSM. In the work by the authors of [20], a generalized predictive current control method is proposed for the current control of PMSM, and a sliding mode compensation controller is designed to eliminate the disturbance. Meanwhile, several papers focusing on improving the robustness of deadbeat controller have been published. In the work by the authors of [21], a robust current controller is designed by using an additional integrator term in deadbeat control. In the work by the authors of [22], a robust deadbeat current control method is studied through calculating the switch signals applied in the next sampling period. In the work by the authors of [23], the parameter identification method is used to estimate the motor parameters in real time, and the robustness is improved, but this method depends on the identification precision of the model parameters, and the model still contains uncertainties due to unmodeled dynamics and disturbances. In the work by the authors of [24], a high-order sliding mode observer is designed for the estimation of disturbance in the current loop for PMSM. In [25–27], the disturbance observer or extended state observer is designed in the deadbeat current control for PMSM. In these methods, the system disturbance is estimated through the observer, then they are used for the feed-forward compensation control to improve the robustness.

In this paper, to enhance the robustness of deadbeat predictive control and improve the consequent system performance degradation for the uncertainties, a novel disturbance attenuation method based on equivalent-input-disturbance (EID) is proposed. EID is a signal on the input voltage that produces the same effect on the current output as actual disturbance does [28–30]. To the best of our knowledge, this method has not been used in the current control of PMSM at present. In this paper, we applied EID into the current control of PMSM, and it is used to deal with the disturbance in the drive system. Thus, a composite current controller by combining the deadbeat predictive control and EID approach is designed. The decoupled current control is completed by the deadbeat control. Then, according to the parameter variations and model uncertainties, EID is designed to eliminate the influence triggered by the uncertainties. In the EID, the lumped disturbance that consists of the model uncertainties and the parameter variations is regarded, and it is used to the feed-forward compensation control. The main contribution of the paper is the idea that the EID is introduced to estimate the disturbance in the current control for PMSM. Only the nominal motor parameters are needed in the controller. Meanwhile, the designed current controller is not complex and the parameters are convenient to be adjusted. Finally, the simulation and experimental verification on a speed current closed-loop control system of PMSM is completed, and the effectiveness is verified in different conditions. Meanwhile, the designed controller can also be used to the torque control of PMSM.

The remainder of this paper is organized as follows. The dynamic model of PMSM considering the disturbance is derived. The robust predictive current controller based on equivalent-input-disturbance is studied in Section 3. The simulation and experiment are demonstrated in Section 4, and the final part states the conclusions.

2. Mathematical Model of PMSM

The electromagnetic model of PMSM in d-q axes can be expressed as [1]

$$\begin{cases} L_d \frac{di_d}{dt} = -R_s i_d + n_p \omega L_q i_q + u_d + \xi_d \\ L_q \frac{di_q}{dt} = -R_s i_q - n_p \omega L_d i_d - n_p \omega \Phi + u_q + \xi_q \end{cases} \tag{1}$$

where L_d and L_q represent d-axes and q-axes stator inductances, respectively; i_d and i_q are the stator currents; u_d and u_q are the stator input voltages in dq-axes reference frame; R_s is the per-phase stator resistance; n_p is the number of pole pairs; ω is the mechanical angular speed; Φ is the rotor flux; and ξ_d and ξ_q represent the disturbance caused by the parameter variations and model uncertainties as

$$\begin{cases} \xi_d = -(\Delta R_s i_d - \Delta L_q n_p \omega i_q + \Delta L_d \frac{di_d}{dt} + \eta_d) \\ \xi_q = -(\Delta R_s i_q - \Delta L_d n_p \omega i_d + \Delta L_q \frac{di_q}{dt} + \Delta \Phi n_p \omega + \eta_q) \end{cases} \tag{2}$$

where η_d and η_q represent the model uncertainties.

$$\Delta R_s = R_{st} - R_s, \Delta L_d = L_{dt} - L_d,$$
$$\Delta L_q = L_{qt} - L_q, \Delta \Phi = \Phi_t - \Phi, \tag{3}$$

in which $R_s, L_d, L_q,$ and Φ denote the nominal parameter values, and $R_{st}, L_{dt}, L_{qt},$ and Φ_t denote the actual parameter values.

3. Current Control Scheme for PMSM

The control structure of PMSM control system with the proposed current control method is shown in Figure 1. The controller uses a cascade control structure including an out speed loop and two inner current loops. Here, the PI controller is used in the speed loop to realize the tracking control of motor speed, thus the q-axes reference current $i_q^* = k_p(\omega_r - \omega) + k_i \int_0^t (\omega_r - \omega)dt$, where ω_r is the reference speed. The reference current i_d^* is set to be zero. In this paper, a novel deadbeat predictive controller with equivalent-input-disturbance is used to solve the current control problem.

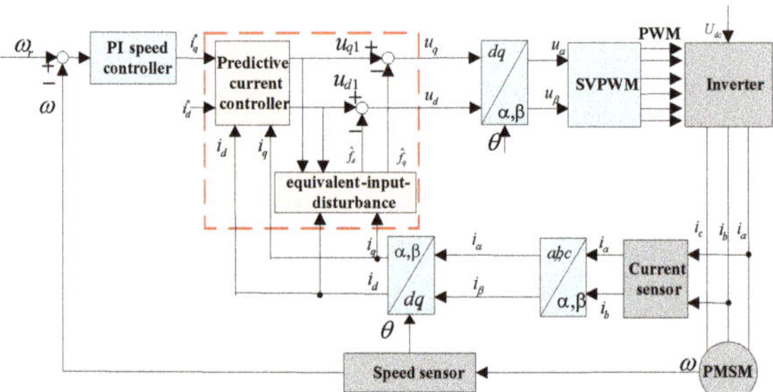

Figure 1. Block diagram of permanent magnet synchronous motor (PMSM) control system.

3.1. Deadbeat Predictive Current Controller for Pmsm

The model of the system (1) can be expressed as

$$
\begin{bmatrix} \frac{di_d}{dt} \\ \frac{di_q}{dt} \end{bmatrix} = \begin{bmatrix} -\frac{R_s}{L_d} & 0 \\ 0 & -\frac{R_s}{L_q} \end{bmatrix} \begin{bmatrix} i_d \\ i_q \end{bmatrix} + \begin{bmatrix} \frac{1}{L_d} & 0 \\ 0 & \frac{1}{L_q} \end{bmatrix} \begin{bmatrix} u_d \\ u_q \end{bmatrix} + \begin{bmatrix} \frac{1}{L_d} & 0 \\ 0 & \frac{1}{L_q} \end{bmatrix} \begin{bmatrix} f_d \\ f_q \end{bmatrix} \tag{4}
$$

where $f_d = n_p \omega L_q i_q + \zeta_d$ and $f_q = -n_p \omega L_d i_d - n_p \omega \Phi + \zeta_q$ are considered as the lump disturbance, which include the disturbance and the back-electromotive force.

In the condition of ignoring the lump disturbance, the predictive model of (4) is expressed as a discrete-time form:

$$
\begin{bmatrix} i_d(k+1) \\ i_q(k+1) \end{bmatrix} = \begin{bmatrix} 1 - \frac{R_s}{L_d}T_s & 0 \\ 0 & 1 - \frac{R_s}{L_q}T_s \end{bmatrix} \begin{bmatrix} i_d(k) \\ i_q(k) \end{bmatrix} + \begin{bmatrix} \frac{T_s}{L_d} & 0 \\ 0 & \frac{T_s}{L_q} \end{bmatrix} \begin{bmatrix} u_{d1}(k) \\ u_{q1}(k) \end{bmatrix} \tag{5}
$$

where T_s is the sample time. u_{d1} and u_{q1} are the control input without considering the lump disturbance.

Define the state variable as $x(k) = \begin{bmatrix} x_1(k) & x_2(k) \end{bmatrix}^T = \begin{bmatrix} i_d(k) & i_q(k) \end{bmatrix}^T$, the input variable as $u_1(k) = \begin{bmatrix} u_{d1}(k) & u_{q1}(k) \end{bmatrix}^T$, and the output variable as $y(k) = \begin{bmatrix} y_1(k) & y_2(k) \end{bmatrix}^T = \begin{bmatrix} i_d(k) & i_q(k) \end{bmatrix}^T$.

The premise of deadbeat predictive current control is that the actual current $I(k)$ is sampled at the beginning of the $k - th$ carrier cycle, and the predicted value of current deviation vector $\Delta I(k)$ is obtained, then the reference voltage output is calculated [22]. For achieving the current control, the required input voltage $u_1(k)$ at the current time can be calculated through the sample current $x(k)$ at kT and $x(k+1)$ at $(k+1)T$. However, in a practical control system, the reference voltage $u_1(k)$ is not added to the inverter immediately at kT, and it is carried out at $(k+1)T$. Therefore, the sample current $x(k+1)$ just reaches the reference current $x^*(k) = \begin{bmatrix} i_d^* & i_q^* \end{bmatrix}^T$ at kT.

Take the reference current $x^*(k)$ as the predictive current value at $(k+1)T$, thus $\begin{bmatrix} i_d(k+1) & i_q(k+1) \end{bmatrix} = \begin{bmatrix} i_d^* & i_q^* \end{bmatrix}$. According to Equation (5), the following can be calculated.

$$
\begin{bmatrix} u_{d1}(k) \\ u_{q1}(k) \end{bmatrix} = \begin{bmatrix} \frac{T_s}{L_d} & 0 \\ 0 & \frac{T_s}{L_q} \end{bmatrix}^{-1} \left(\begin{bmatrix} i_d^*(k) \\ i_q^*(k) \end{bmatrix} - \begin{bmatrix} 1 - \frac{R_s}{L_d}T_s & 0 \\ 0 & 1 - \frac{R_s}{L_q}T_s \end{bmatrix} \begin{bmatrix} i_d(k) \\ i_q(k) \end{bmatrix} \right) \tag{6}
$$

The control laws in (6) are the reference dq-axes voltage based on deadbeat predictive control. The designed controller can solve the deadbeat control problem in general without additional means of support, but the lump disturbance is ignored, and the current control results will be affected in the face of the strong disturbance.

3.2. Disturbance Observer Based on EID

To suppress the lump disturbance in the system and improve the robustness, an equivalent-input-disturbance approach combined with the deadbeat predictive controller is studied. EID approach is a simple disturbance attenuation method and it can be easily implemented in the digital controller because it does not require an inverse model of the plant or prior information on the disturbance. Usually, an EID-based control system includes a state observer, an EID estimator, and state feedback [29]. The structural diagram of EID is shown in Figure 2. Thus, the deadbeat predictive controller can be seen as the state feedback.

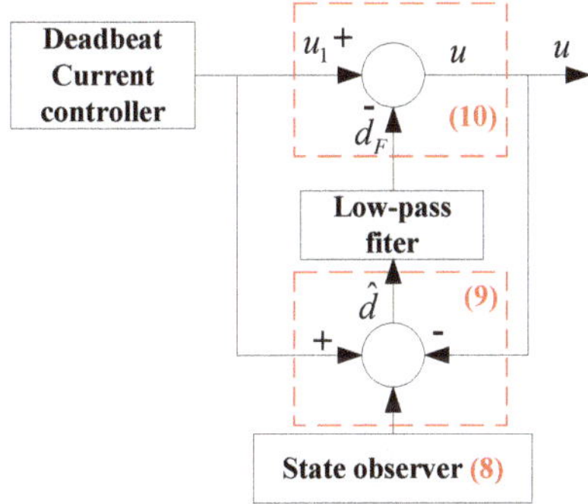

Figure 2. Structural diagram of equivalent-input-disturbance (EID).

Firstly, according to (4), the model can be described as

$$\begin{cases} x = Ax + Bu + Bd \\ y = cx \end{cases} \tag{7}$$

where $A = \begin{bmatrix} -\frac{R_s}{L_d} & 0 \\ 0 & -\frac{R_s}{L_q} \end{bmatrix}$, $B = \begin{bmatrix} \frac{1}{L_d} & 0 \\ 0 & \frac{1}{L_q} \end{bmatrix}$, $C = \begin{bmatrix} 1 & 0 \\ 0 & 1 \end{bmatrix}$, $u = \begin{bmatrix} u_d & u_q \end{bmatrix}^T$ and $d = \begin{bmatrix} f_d & f_q \end{bmatrix}^T$.

The system is controllable and observable, and it has no zeros on the imaginary axis, which guarantees the internal stability of the motor system and allows the speed to track the reference value [31]. Assume the distubance d satisfies $\|d\|_\infty < d_M$, where d_M is an unknown positive real number.

The EID is estimated by making the best use of the state observer of the system. A full-order observer is used to estimate the EID, then a state observer is defined as

$$\begin{cases} \dot{\hat{x}} = A\hat{x} + Bu_1 + L[y - \hat{y}] \\ \hat{y} = c\hat{x} \end{cases} \tag{8}$$

where $u_1 = \begin{bmatrix} u_{d1} & u_{q1} \end{bmatrix}^T$, L is the observer gain, and $\hat{x}$ is the reconstructed state of x.

According to the literatures [29,30], the estimator of EID is derived as

$$\hat{d} = B_1 LC[x - \hat{x}] + u_1 - u \tag{9}$$

where $B_1 = (B^T B)^{-1} B^T$.

As the output y contains a measurement noise, to depress the noise of the measured output current, a low-pass filter is used in the estimator of EID, thus $d_F = F(s)\hat{d}$, $F(s) = \frac{B_F}{s + A_F}$ is the low-pass filter.

Combing the deadbeat controller and the EID estimator, the final current controller of PMSM control system can be given as

$$u = u_1 - d_F \tag{10}$$

In the paper, the model of PMSM is represented as the linear model, which is shown in Equation (7). The stability of the motor drive system can be broken down into two independent parts: state feedback and the observer. First, the state feedback is designed by the deadbeat predictive control method. Then, according to the Theorem 1 in the work by the authors of [31], the conditions are satisfied for the motor system. Thus, the stability of the control system can be guaranteed.

Meanwhile, the following simulation and experiment in different conditions prove that the motor can run steadily.

4. Simulation and Experiment

The proposed current control method with deadbeat predictive control and equivalent input disturbance approach is implemented in simulation and dSPACE based experiment. The motor parameters used in the simulation and experiment are given in Table 1.

Table 1. Parameters of PMSM.

Description	Value	Unit
rated speed	3000	r/min
rated torque	2.3	N·m
resistance	4.8	Ω
d-axes inductance	19.5	mH
q-axes inductance	27.5	mH
rotor flux	0.15	Wb

4.1. Simulation and Analysis

The simulation is completed in the speed control system of PMSM. The double closed-loop vector control method is used for the speed and current control, the out loop is the speed loop, and the inner loop is the current loop. The motor reference speed is given as $\omega_r = 1000$ r/min, and the load torque 1 N·m is added on the motor. The PI controller is adopted in the speed loop. In order to verify the merit of the proposed method, in the current loop, both PI current controller and proposed current control method have been completed, respectively. The parameters of speed loop are consistent with each other in both methods. In addition, to validate the current tracking performance of the proposed control method while the motor has parameter perturbation, the values of rotor–flux linkage, armature resistance, and dq-axes inductance are changed to 80%, 200%, and 150% of the normal values at $t = 0.5$ s, respectively. Figure 3 shows the motor response waveform based on PI control. Figure 3a,b shows the dq-axes current waveform.

The motor speed response curve is shown in Figure 3c. The simulation results based on the proposed current control method are shown in Figure 4.

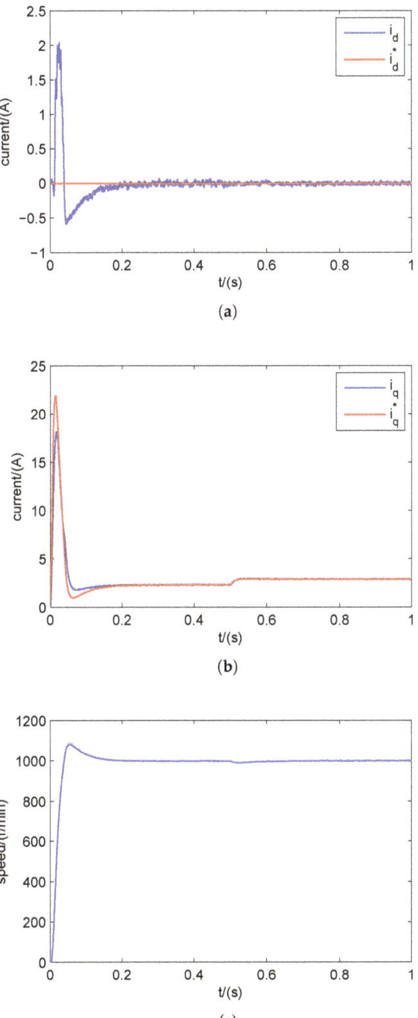

Figure 3. Simulation results with proportional plus integral (PI) control: (**a**) d-axes current; (**b**) q-axes current; and (**c**) speed response.

As shown in Figures 3 and 4, compared with the PI control, the proposed predictive current control method has better current control performance with a faster current response and smaller current fluctuation. When the motor parameters are changed, the results show that the parameter variations has little influence to the current control performance, and the proposed method in this paper still has precise current tracking and antidisturbance performance. Meanwhile, the control system has the good speed tracking performance.

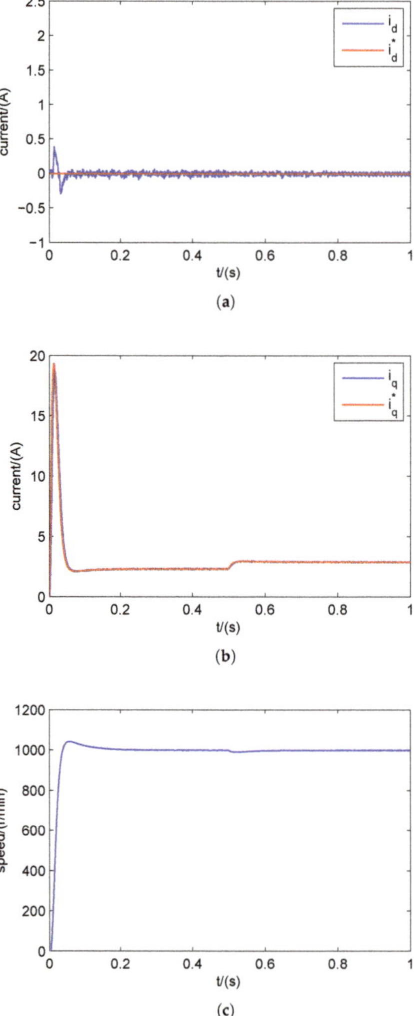

Figure 4. Simulation results with the proposed method: (**a**) d-axes current; (**b**) q-axes current; and (**c**) speed response.

To verify the effectiveness of the proposed method under the change of reference current, the d-axes reference current is changed from 0 to -1A at $t = 0.5$ s, and Figure 5 shows the current and speed response waveform. As shown in the figures, the dq-axes current varies with the change of the reference current, and the controller also good current tracking performance. Meanwhile, the motor speed has a small fluctuation, and it can be stable for a short period of time.

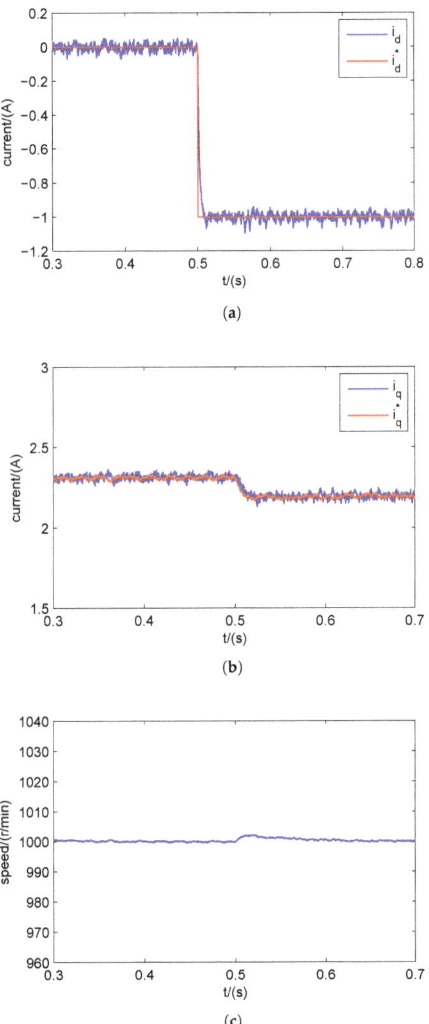

Figure 5. Simulation results with the proposed method when i_d^* is changed: (**a**) d-axes current, (**b**) q-axes current, and (**c**) speed response.

In addition, to test the performance of the controller with load disturbance, the reference speed is given as 1000 r/min, the load torque is changed from 1 N·m to 2 N·m at $t = 0.5$ s, and other parameters are fixed. The results are show in Figure 6. The figures reveal that with the increase of load torque, the q-axis current increases quickly to produce the same electromagnetic torque. Although the reference current has a variation with the change of load torque, the motor current can still track the reference value, and the robustness can be maintained in steady-state.

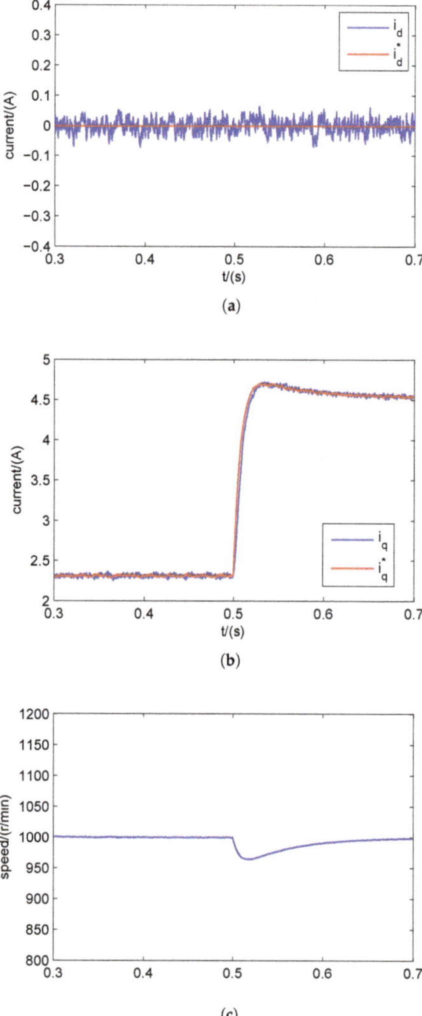

Figure 6. Simulation results with the proposed method when the load torque changed: (**a**) d-axes current, (**b**) q-axes current, and (**c**) speed response.

To better validate the advantage of the proposed current control method under parameter disturbance, three different comparative methods—PI current control, deadbeat predictive control without equivalent input disturbance approach, and the proposed method—have been used in the current loop controller of the drive system, respectively. The reference speed is also given as 1000 r/min, but the motor parameters are changed when the motor is started. The results based on three current control methods are shown in Figures 7–9. The speed loop has the same parameters as the PI control method.

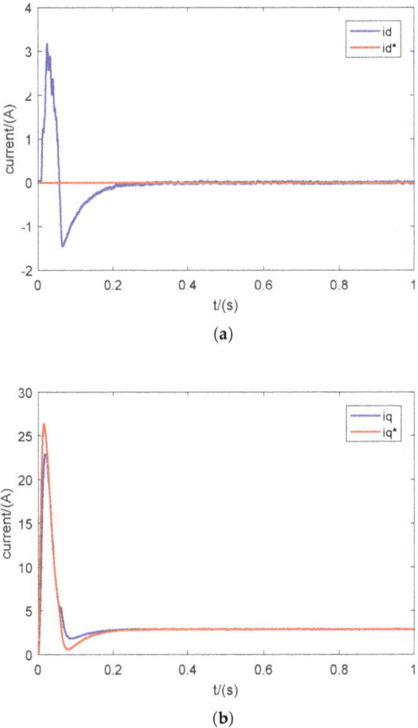

(a)

(b)

Figure 7. Simulation results with PI control method under the parameter disturbance: (**a**) d-axes current and (**b**) q-axes current.

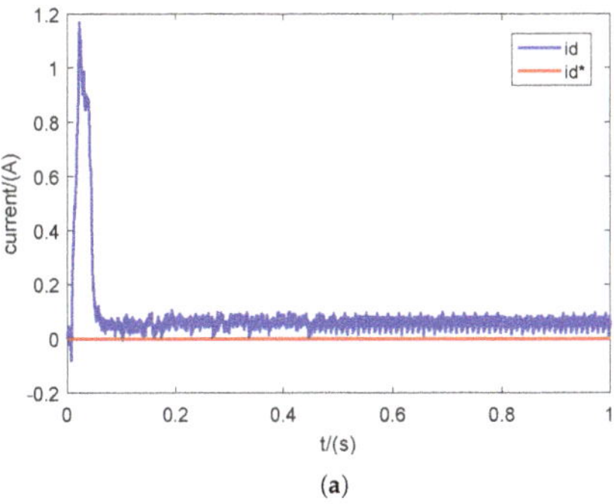

(a)

Figure 8. *Cont.*

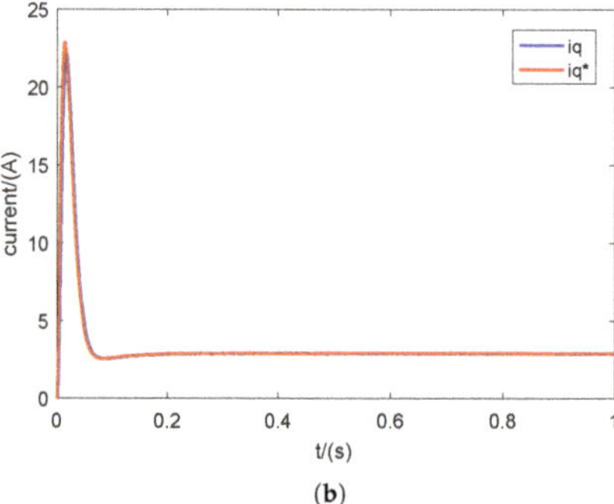

(b)

Figure 8. Simulation results with deadbeat control method under the parameter disturbance: (**a**) d-axes current and (**b**) q-axes current.

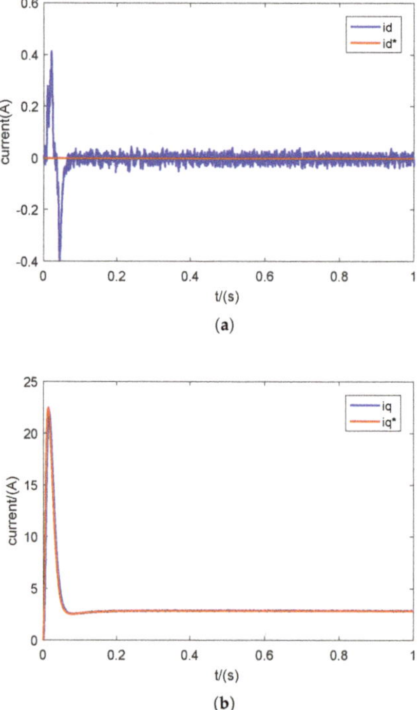

(a)

(b)

Figure 9. Simulation results with the proposed method under the parameter disturbance: (**a**) d-axes current and (**b**) q-axes current.

The results show that compared to the PI control, the proposed method has the smaller current fluctuation and the shorter settling time when the motor starts. Both methods have strong robustness for the parameter of disturbance. Although the deadbeat predictive current controller has the fast response, there is an obvious error between the output current and the reference values, and the dq-axes current cannot track the reference values accurately. The above results verify that the proposed method has better current control performance.

4.2. Experiment and Analysis

The experiment is implemented on a dSPACE-based PMSM control platform, as shown in Figure 10. The system includes an interior PMSM, a dynamometer, a converter based on IPM, and the dSPACE MicroAutoBox as the control board. MicroAutoBox is a rapid prototyping (RCP) system, and it is ideally suited as hardware for prototyping in the motor drive system. The current is measured by the Hall sensor and it is turned to the digital signal through ACMC module. The experiment is completed on a speed control system of PMSM based on a doubled closed-loop structure. The deadbeat predictive controller with EID is used in the current loop control. The parameters of the PMSM are given in Table 1. The sample time is chosen as $T_s = 0.1$ ms. The observer gain L is the main parameter to be adjusted in the controller. The larger the gain is, the faster the observer converges. However, in general, the gain, L, cannot be too large. Otherwise, the system will be too sensitive to the interference signal and the stability of the system will decrease. So the observer gain of $L = \begin{bmatrix} 100 & 0 \\ 0 & 100 \end{bmatrix}$ is chosen in the controller. The low-pass filter is chosen as $F(s) = \frac{200}{s+200}$.

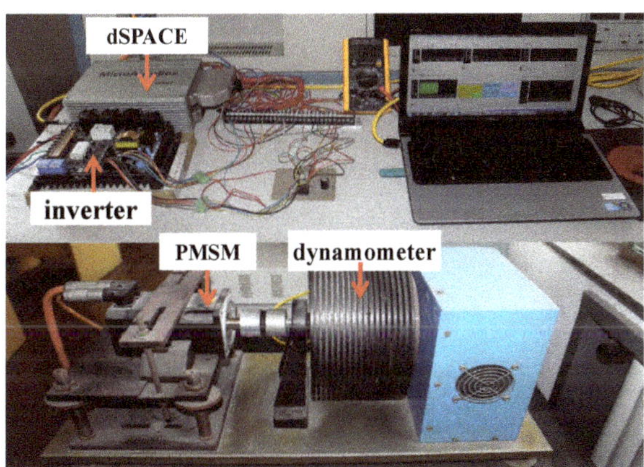

Figure 10. Experimental platform of PMSM drive system.

First, the motor starting speed is given as 1000 r/min. When the motor is stable, the reference speed is changed from 1000 r/min to 1500 r/min, and the experimental results are shown in Figures 11 and 12, which include the dq-axes current and speed response waveforms. As seen from Figures 11 and 12, when the motor is starting, the large starting current is produced, and the dq-axes current reaches to the reference current value soon. The designed current controller has the good tracking performance. The practical output current has a small fluctuation, which may be caused by the current harmonic, but the tracking performance is not affected. The motor speed can also arrive at the given value quickly.

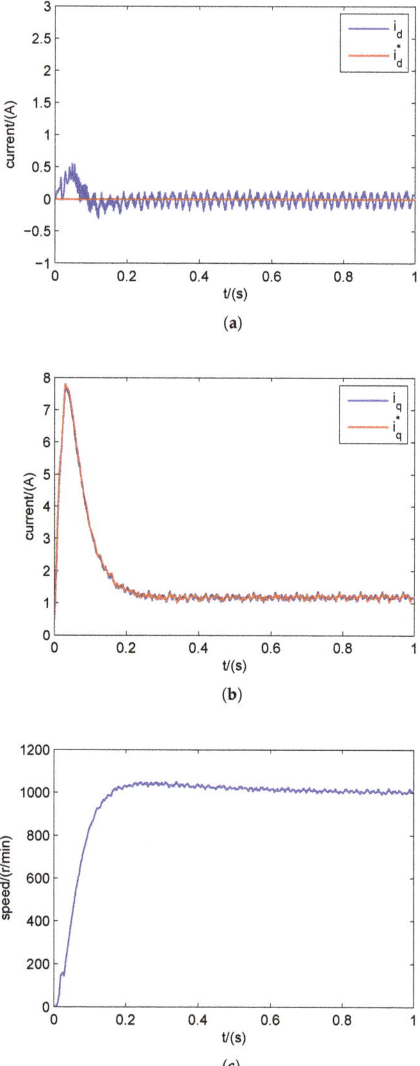

Figure 11. Experimental results with the proposed method: (**a**) d-axes current; (**b**) q-axes current; and (**c**) speed response.

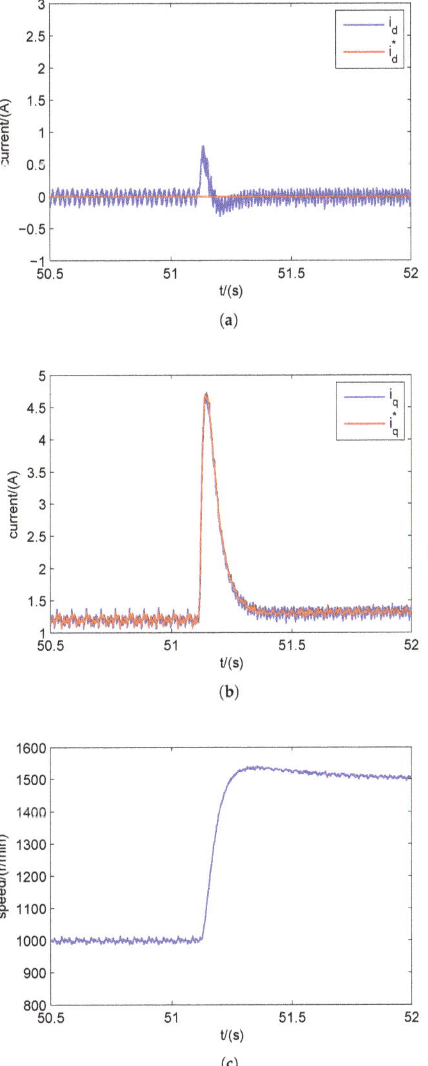

Figure 12. Experimental results with the increase of reference speed: (**a**) d-axes current; (**b**) q-axes current; and (**c**) speed response.

To further evaluate the performance of the proposed control method, the q-axe reference current is changed in the experiment. When the motor is working in 1000 r/min, the load torque disturbance is added to the motor at $t = 1$ s, and the experimental results are shown in Figure 13. We can see that the q-axes current changes with the increase of load torque, and it can converge to its reference value quickly. In this process, the d-axes current is not affected by the load disturbance, can stay at zero, and the robustness still can be maintained.

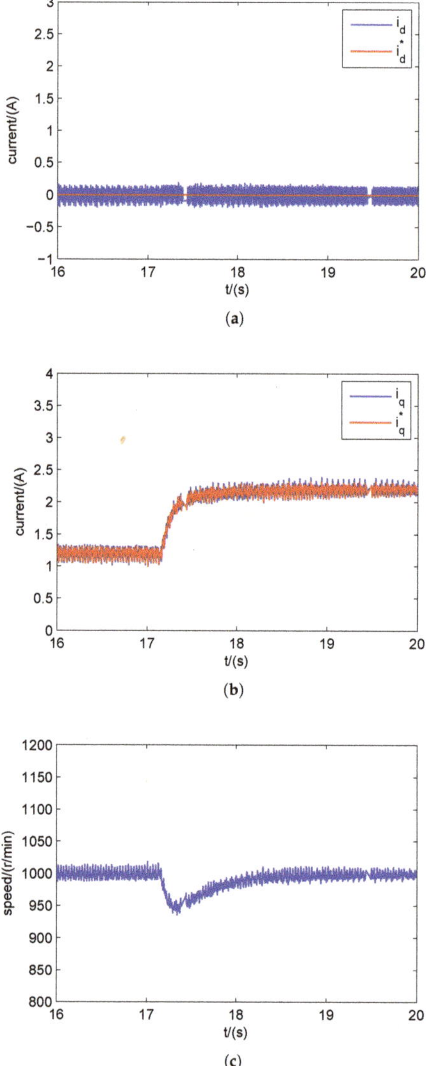

Figure 13. Experimental results with load disturbance: (**a**) d-axes current, (**b**) q-axes current, and (**c**) speed response.

To further verify the robustness of the proposed method with the parameters perturbation, the controller parameters are set mismatched with the real motor parameters. In comparison with the real values in the controller, the dq-axes inductances are set as two times of the rated values in the controller. The corresponding results are shown in Figure 14. Although the parameters are not consistent between the current controller and the motor, there is no large difference in the actual output current and the reference current. The experimental results prove the designed controller has excellent robustness for the disturbance.

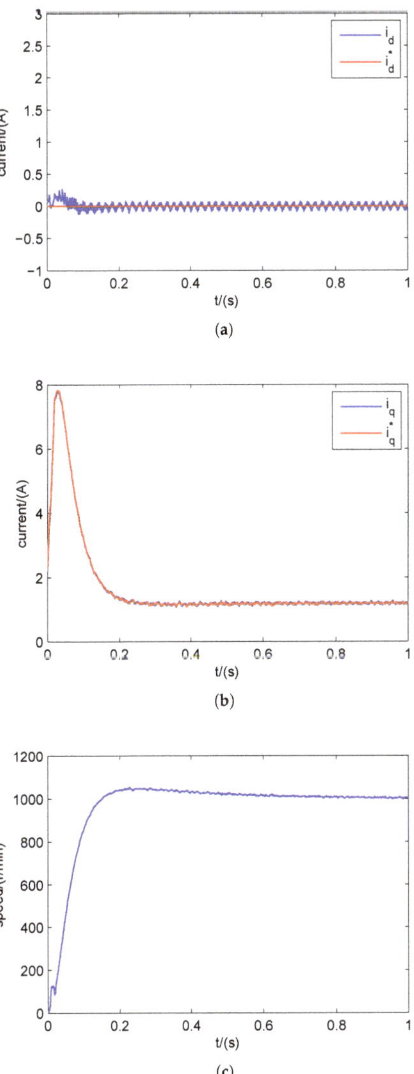

Figure 14. Experimental results with parameter variations: (**a**) d-axes current, (**b**) q-axes current, and (**c**) speed response.

5. Conclusions

In this paper, a novel current control method based on deadbeat predictive control and equivalent-input-disturbance theory has been proposed for PMSM drive. With the designed deadbeat controller, the current stabilizing control is achieved. According to the disturbance in the PMSM, we have designed an equivalent-input-disturbance estimator for the feed-forward compensation, and the robustness can be maintained under the disturbance. The simulation and experiment results have proved that the controller has good current tracking performance in different conditions. Meanwhile, the designed current controller is also suitable for torque control system of PMSM.

Author Contributions: Conceptualization, X.L.; methodology, X.L. and Q.Z.; software, X.L.; validation, X.L. and Q.Z.; formal analysis, Q.Z.; investigation, X.L.; resources, X.L.; data curation, Q.Z.; writing—original draft preparation, X.L.; writing—review and editing, X.L. and Q.Z.; visualization, X.L.; supervision, X.L.; project administration, X.L. and Q.Z.; funding acquisition, X.L. and Q.Z.

Funding: This research was funded by the National Natural Science Foundation of China under Grant 61703222, and in part by the China Postdoctoral Science Foundation under Grant 2018M632622.

Conflicts of Interest: The authors declare no conflict of interest.

References

1. Liu, X.; Yu, H.; Yu, J.; Zhao, L. Combined speed and current terminal sliding mode control with nonlinear disturbance observer for PMSM drive. *IEEE Access* **2018**, *6*, 29594–29601. [CrossRef]
2. Ren, J.; Ye. Y.; Xu, G.; Zhao, Q. Uncertainty-and-disturbance-estimator-based current control scheme for PMSM drives with a simple parameter tuning algorithm. *IEEE Trans. Ind. Electron.* **2017**, *32*, 5712–5722.
3. Qian, J.; Ji, C.; Pan, N.; Wu, J. Improved sliding mode control for permanent magnet synchronous motor speed regulation system. *Appl. Sci.* **2018**, *8*, 2491. [CrossRef]
4. Merabet, A.; Tanvir, A.; Beddek, K. Torque and state estimation for real-time implementation of multivariable control in sensorless induction motor drives. *IET Electr. Power Appl.* **2017**, *11*, 653–663. [CrossRef]
5. Yang, X.; Yu, J.; Wang, Q.; Zhao, L. Adaptive fuzzy finite-time command filtered tracking control for permanent magnet synchronous motors. *Neurocomputing* **2019**, *337*, 110–119. [CrossRef]
6. He, L.; Wang, F.; Wang, J.; Rodriguez, J. Zynq implemented lunenberger disturbance observer based predictive control scheme for PMSM drives. *IEEE Trans. Power Electron.* **2019**. [CrossRef]
7. Liu, X.; Yu, H.; Yu, J.; Zhao, Y. A novel speed control method based on port-controlled hamiltonian and disturbance observer for PMSM drives. *IEEE Access* **2019**, *7*, 111115–111123. [CrossRef]
8. Merabet, A.; Labib, L.; Ghias, A. Robust model predictive control for photovoltaic inverter system with grid fault ride-through capability. *IEEE Trans. Smart Grid* **2018**, *9*, 5699–5709. [CrossRef]
9. Kouro, S.; Cortes, P.; Vargas, C.; Ammann, U. Model predictive control—a simple and powerful method to control power converters. *IEEE Trans. Ind. Electron.* **2008**, *56*, 1826–1838. [CrossRef]
10. Turker, T.; Buyukkeles, U.; Bakan, A. A robust predictive current controller for PMSM drives. *IEEE Trans. Ind. Electron.* **2016**, *63*, 3906–3914. [CrossRef]
11. Guo, X.; Du, S.; Li, Z.; Chen, F. Analysis of current predictive control algorithm for permanent magnet synchronous motor based on three-level inverters. *IEEE Access* **2019**, *7*, 87750–87759. [CrossRef]
12. Chai, S.; Wang, L.; Rogers, E. A cascade MPC control structure for a PMSM with speed ripple minimization. *IEEE Trans. Ind. Electron.* **2013**, *60*, 2978–2987. [CrossRef]
13. Zhang, G.; Chen, C.; Gu, X.; Zhang, Z. An improved model predictive torque control for a two-level inverter fed interior permanent magnet synchronous motor. *Electronics* **2019**, *8*, 769. [CrossRef]
14. Zhang, X.; Zhang, L.; Zhang, Y. Model predictive current control for PMSM drives with parameter robustness improvement. *IEEE Trans. Power Electron.* **2019**, *34*, 1645–1657. [CrossRef]
15. Luo, Y.; Liu, C. A simplified model predictive control for a dual three-phase PMSM with reduced harmonic currents. *IEEE Trans. Ind. Electron.* **2018**, *65*, 9079–9089. [CrossRef]

16. Zhang, X.; Hou, B.; Mei, Y. Deadbeat predictive current control of permanent magnet synchronous motors with stator current and disturbance observer. *IEEE Trans. Power Electron.* **2016**, *32*, 3818–3834. [CrossRef]

17. Chen, W.; Yang, J.; Guo, L.; Li, S. Disturbance-observer-based control and related methods-an overriew. *IEEE Trans. Ind. Electron.* **2016**, *63*, 1083–1095. [CrossRef]

18. Yang, J.; Chen, W.; Li, S.; Guo, L. Disturbance/uncertainty estimation and attenuation techniques in PMSM drives—A survey. *IEEE Trans. Ind. Electron.* **2017**, *64*, 3273–3285. [CrossRef]

19. Sun, Z.; Guo, T.; Yan, Y.; Wang, X. A composite current-constrained control for permanent magnet synchronous motor with time-varying disturbance. *Adv. Mech. Eng.* **2017**, *9*, 1–13. [CrossRef]

20. Liu, X.; Zhang, C.; Li, K.; Zhang, Q. Speed control for permanent magnet synchronous motor based on an improved extended state observer. *ISA Trans.* **2017**, *71*, 542–552. [CrossRef]

21. Turker, T.; Yanik, G.; Buyukkeles, U.; Mese, E. A discrete-time nonlinear robust controller for current regulation in PMSM drives. *J. Electr. Eng. Technol.* **2017**, *12*, 1921–1931.

22. Niu, L.; Yang, M.; Wang, G.; Xu, D. Research on the robust current control algorithm of permanent magnet synchronous motor based on deadbeat control principle. *Proc. CSCC* **2012**, *33*, 78–85.

23. Wang, W.; Xiao, X. Research on predictive control for PMSM based on online parameter identification. In Proceedings of the 38th IEEE Industrial Electronics Society Annual Conference, Glendale, AZ, USA, 7–10 November 2010; pp. 1982–1986.

24. Jiang, Y.; Xu, W.; Mu, C.; Liu, Y. Improved deadbeat predictive current control combined sliding mode strategy for PMSM drive system. *IEEE Trans. Veh. Technol.* **2018**, *67*, 251–263. [CrossRef]

25. Mohamed, Y.; EI-Saadany, E. Robust high bandwidth discrete-time predictive current control with predictive internal model-a unified approach for voltage-source PWM converters. *IEEE Trans. Ind. Electron.* **2008**, *23*, 126–136. [CrossRef]

26. Yang, M.; Liang, X.; Long, J.; Xu, D. Flux immunity robust predictive current control with incremental model and extended state observer for PMSM drive. *IEEE Trans. Power Electron.* **2017**, *23*, 9267–9279. [CrossRef]

27. Yang, R.; Wang, M.; Li, L.; Zhang, C. Robust predictive current control with variable-gain adaptive disturbance observer for PMLSM. *IEEE Access* **2018**, *6*, 13158–13169. [CrossRef]

28. Du, Y.; Cao, W.; Wu, M.; She, J. Disturbance rejection and control system design using improved equivalent-input-disturbance approach. *IEEE Trans. Ind. Electron.* **2019**. [CrossRef]

29. Liu, R.; Liu, G.; Wu, M. Robust disturbance rejection based on equivalent-input-disturbance approach. *IET Control Theory Appl.* **2013**, *7*, 1261–1268. [CrossRef]

30. Gao, F.; Wu, M.; She, J. Disturbance rejection in nonlinear systems based on equivalent-input-disturbance approach. *Appl. Math. Comput.* **2016**, *282*, 244–253. [CrossRef]

31. She, J.; Fang, M.; Ohyama, Y. Improving disturbance-rejection performance based on an Equivalent-Input-Disturbance approach. *IEEE Trans. Ind. Electron.* **2008**, *55*, 380–389. [CrossRef]

Article

On-Off Control of Range Extender in Extended-Range Electric Vehicle using Bird Swarm Intelligence

Dongmei Wu * and Liang Feng

Jiangsu Engineering Lab for IOT Intelligent Robots, College of Automation & College of Artificial Intelligence, Nanjing University of Posts and Telecommunications, Nanjing 210023, China; 1218053626@njupt.edu.cn
* Correspondence: wudm@njupt.edu.cn

Received: 21 September 2019; Accepted: 22 October 2019; Published: 26 October 2019

Abstract: The bird swarm algorithm (BSA) is a bio-inspired evolution approach to solving optimization problems. It is derived from the foraging, defense, and flying behavior of bird swarm. This paper proposed a novel version of BSA, named as BSAII. In this version, the spatial distance from the center of the bird swarm instead of fitness function value is used to stand for their intimacy of relationship. We examined the performance of two different representations of defense behavior for BSA algorithms, and compared their experimental results with those of other bio-inspired algorithms. It is evident from the statistical and graphical results highlighted that the BSAII outperforms other algorithms on most of instances, in terms of convergence rate and accuracy of optimal solution. Besides the BSAII was applied to the energy management of extended-range electric vehicles (E-REV). The problem is modified as a constrained global optimal control problem, so as to reduce engine burden and exhaust emissions. According to the experimental results of two cases for the new European driving cycle (NEDC), it is found that turning off the engine ahead of time can effectively reduce its uptime on the premise of completing target distance. It also indicates that the BSAII is suitable for solving such constrained optimization problem.

Keywords: BSAII; Euclidean distance; energy management; E-REV

---

1. Introduction

Nowadays, society is facing the increasing depletion of petrochemical energy, the serious destruction of the ecological environment, and increasing car ownership. These factors promote the rapid development of new energy vehicles like the electric vehicle. However, the power battery of the pure electric vehicle has a series of problems, such as high cost, short range and over discharge, which is not conducive to long-distance driving.

As a transitional model of pure electric vehicle, the extended-range electric vehicle (E-REV) can effectively address the shortcomings above. The basic structure of a typical E-REV is shown in Figure 1. The auxiliary engine and power generation device has been added to the mechanism of the electric vehicle, which extends driving distance of electric vehicle. The integration of engine and generator constitute is called the range extender (RE), the main function of which is to charge the battery under the condition of insufficient power supply, for purpose of providing enough power to extend driving distance. Because of separation of the engine from the road load and the balance of the battery load, E-REV can keep the engine at the optimum working efficiency point (85%) and improve the fuel efficiency greatly. Additional E-REV has two energy sources: engine and power battery, so an efficient control strategy is essential to practice the coordination of the two devices, improve vehicle performance, e.g., fuel efficiency and exhaust pollution.

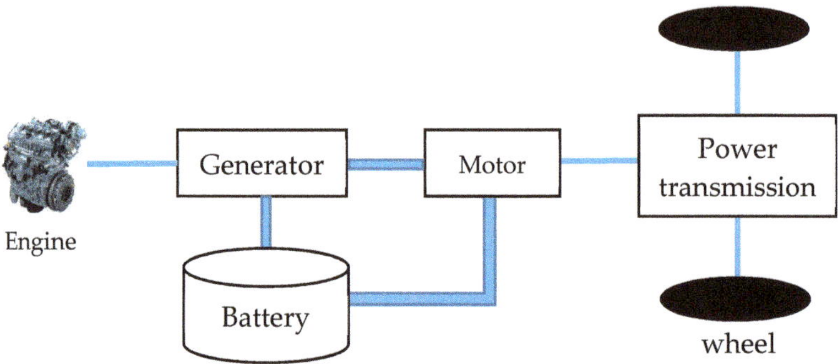

Figure 1. System structure of extended-range electric vehicle (E-REV).

The energy management of E-REV has always been a research hotspot [1–3]. Under the different driving conditions, the on-off time of RE for the E-REV is optimized with the target distance as the constraint condition. The main principle of the E-REV energy management strategy is that the use of engine is as little as possible as well as keeps the vehicle running in pure electric mode. The traditional control strategy is that when the battery power reaches the minimum threshold, the vehicle enters the extended range mode, the engine starts and drives the generator to produce electricity. Part of the generated electricity charges the battery, and the other part drives the vehicle to continue driving. When the battery power reaches the maximum threshold, the engine shuts down and vehicle enters the pure electric drive mode.

Control method like fuzzy control has been adopted in the energy management. It has been used for powering the battery, to keep the state of charge (SOC) in the designed threshold and avoid overcharge and over discharge [4]. As energy management can be considered an optimization problem, conventional planning methods were applied to the problem, such as dynamic programing, genetic algorithm (GA), and particle swarm optimization (PSO), etc. A hybrid genetic particle swarm optimization (GPSO) algorithm was proposed to optimize the parameters of energy management strategy [5]. In order to solve the problem of frequent start-stop of electric vehicle engines, a non-dominant sequencing genetic algorithm was used to optimize the start-stop interval of engines. The optimization effect of the running time of the extender under the two control modes of early opening and early closing is analyzed, in new European driving cycle (NEDC), urban dynamometer driving schedule (UDDS) cycles [6]. Energy management strategy of E-REV based on dynamic programming was designed, and optimal control rules of extender start-stop corresponding to SOC and motor power were established [7]. Driving behavior based on prediction of vehicle speeds was integrated into the energy management of the electric vehicle [8].

In recent years, with the unprecedented development of bionic optimization, a series of novel algorithms have emerged [9–17]. These include the teaching and learning optimization algorithm (TLBO, 2011) and its variants, the grey wolf optimizer (GWO, 2014) and its variants, the pigeon swarm algorithm (PSA, 2014), the whale optimization algorithm (WOA, 2016) and the bird swarm algorithm (BSA, 2016). Compared with GA, PSO and other mature algorithms, the optimization performance of these new bio-derived algorithms has been greatly improved. Therefore, the application of these algorithms in engineering attracted the attention of researchers.

The BSA as a novel algorithm, simulates the foraging behavior, defensive behavior and flight behavior of birds. It has the advantages of few parameters and it is easy to adjust. This paper extends the basic idea proposed in [17]. We propose a new method called BSAII with new coefficients for evaluating birds' ability to reach to the center. Then solve the optimization problem of engine on-off control in E-REV with the proposed algorithm.

The rest of paper will be organized as follows. Section 2 will outline the background to BSA, the formulation of motions for BSA, and its variants. We also propose another formulation of defense behavior of birds and explain the workflows of optimization with BSAII. Section 3 formulates the energy management in E-REV. Section 4 conducts extensive optimizing simulation, and analyzes the experiment results. Section 5 will present the experiment results of the application of BSAII on energy management of E-REV. Conclusions are drawn at the end of the paper.

2. Principle of Brid Swarm Intelligence

2.1. Bird Swarm Intelligence

Bird swarm foraging is easier to collect more information than individual foraging. It has survival advantages and good foraging efficiency. BSA is inspired by foraging behavior, defense behavior and flight behavior in the foraging process of birds. It is based on information sharing mechanism and search strategy in the foraging process of birds. The core of social behaviors and interactions in the bird swarm put forward a novel optimization algorithm BSA. Ideally, the basic principles of BSA can be elaborated as the following five rules [17].

(1) Each bird freely converts between defense and foraging behavior, which is a random behavior.
(2) In the process of foraging, each bird can record and update its own optimal information and global optimal information about the food. This information is used to find new sources of food. At the same time, the whole population share the social information.
(3) During the defense, each bird tries to move toward the center, but this behavior is influenced by competition among populations. Birds with high alertness are more likely to approach the center than low-alert birds.
(4) The swarm flies to another place each time. The identity of a bird converts between a producer and a beggar. That is, the most alert bird becomes a producer, while the lowest alert birds become a beggar. Birds with alertness between the two birds randomly become producers or beggars.
(5) Producers actively seek food, and beggars follow the producers at random.

The above five rules are described in mathematical terms as follows:

We suppose the size of the swarm is M, the number of dimensions is N. Foraging behavior in rule (1) is formulated;

$$x_i^{t+1} = x_i^t + c_1 r_1 (p_i - x_i^t) + c_2 r_2 (g - x_i^t) \tag{1}$$

where, x_i^t is the position of each bird, t represents the current number of iterations, $i = 1, 2 \ldots M$. c_1 and c_2 are non-negative constants which represent cognitive and social acceleration coefficients independently. r_1 and r_2 are the random numbers with uniform distribution in $[0,1]$. p_i and g record the historical optimal location of the ith bird and the historical optimal location of the whole swarm respectively.

According to the Rule (3), birds in the swarm are trying to get close to the central area, but there is a competitive relationship between birds. These behaviors can be expressed as follows;

$$x_{i,j}^{t+1} = x_{i,j}^t + A_1 r_3 (mean_j - x_{i,j}^t) + A_2 r_4 (p_{k,j} - x_{i,j}^t) \tag{2}$$

$$A_1 = a_1 \times e^{\left(- \frac{pFit_i}{sumFit+\varepsilon} \times M\right)} \tag{3}$$

$$A_2 = a_2 \times e^{\left(- \frac{pFit_i - pFit_k}{|pFit_k - pFit_i| + \varepsilon} \times \frac{M \times pFit_k}{sumFit+\varepsilon}\right)} \tag{4}$$

Among them, a_1 and a_2 are the constants of $[0,2]$, $pFit_i$ represents the optimal value of the ith bird, $sumFit$ represents the sum of the optimal value of the whole swarm. ε is the smallest real number in a computer. $mean_j$ is average value of positions in the jth dimension. r_3 is the random number between $(0, 1)$, r_4 is the random number between $(-1, 1)$. $k \neq i$. A_1 controls a bird approaching to center position of the whole swarm and $A_1 r_3 \in (0,1)$. A_2 represents the competitiveness of ith bird versus

kth bird. The greater A_2 means compared with ith bird, the kth bird is more likely to move to the center of the swarm.

According to the Rule (4), every once in a while FQ, birds may fly to another place for seeking food, some birds may become producers, others will become beggars, behavior of producers and beggars are regulated their new position according to;

$$x_{i,j}^{t+1} = x_{i,j}^t + r_5 x_{i,j}^t \tag{5}$$

$$x_{i,j}^{t+1} = x_{i,j}^t + FL r_6 (x_{k,j}^t - x_{i,j}^t) \tag{6}$$

r_5 is a Gaussian random number that satisfies the variance of 0 and the mean of 1. r_6 is the random number between (0, 1), and FL stands for the beggars getting food information from producers, $FL \in [0,2]$. The workflow of BSA for solving optimization problem is illustrated as Figure 2.

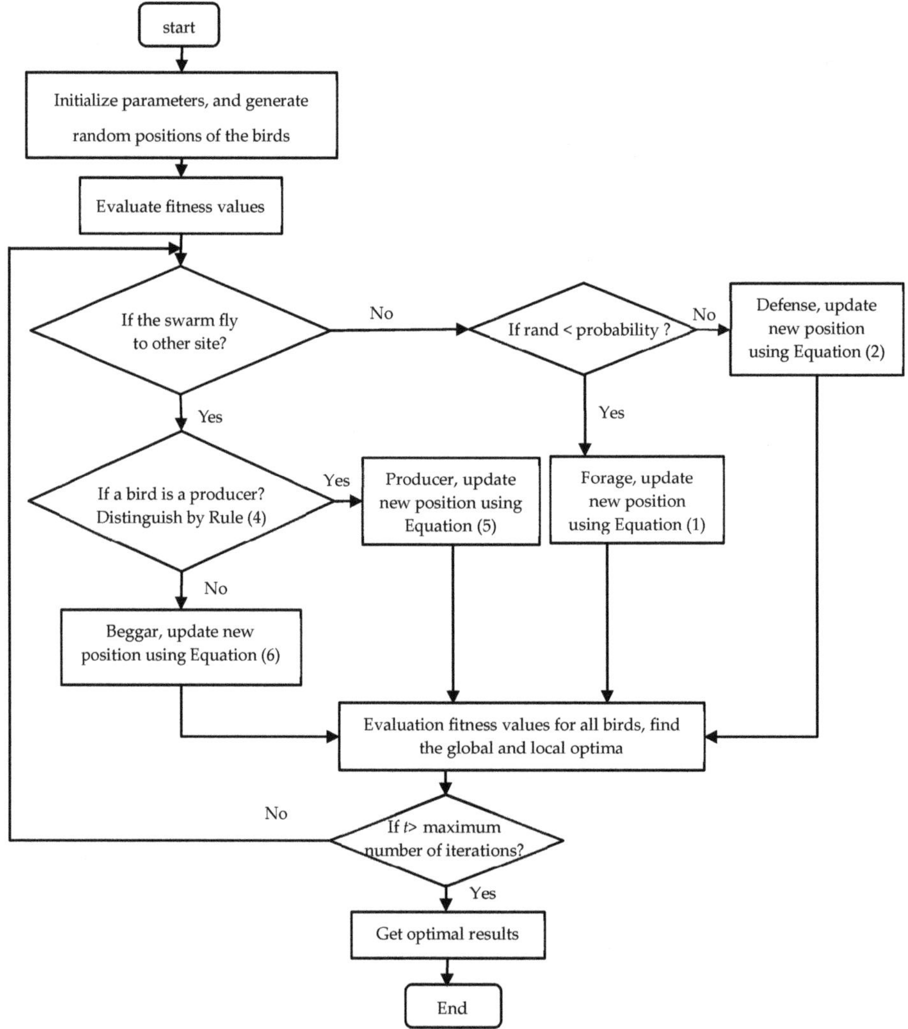

Figure 2. Workflow of bird swarm algorithm (BSA).

2.2. Related Improvement Methods

As a relatively new optimization algorithm, there is not much research on improvement of BSA. The algorithm is improved by defining inertia weight, with linear differential decline strategy, and linearly adjusting cognitive coefficient and social coefficient. Then different models are optimized [18]. Levy flight strategy is applied to position initialization or iteration of BSA [19–21]. In [20], the random walk mode of Levy flight strategy increased the diversity of population and conduced to jumping out of local optimum. Inertia weight modified by random uniform distribution improved the search ability of BSA, besides, linear adjustment of cognitive and social coefficients was used to improve the solution accuracy. Boundary constraints were adopted to modify candidate solutions outside or on the boundary in the iteration process, which improves the diversity of groups and avoids premature problems. On the other hand, accelerated foraging behavior by adjusting the sine-cosine coefficients of cognitive and social components was achieved in [22].

2.3. BSAII

In the defensive state, a bird should not only move to the center as far as possible, but also compete with other neighbors. The parameters A_1 and A_2 are two factors that reflect ability of a bird moving to the center and competition with its neighbor bird respectively. In the traditional version of BSA, the fitness function was used to evaluate the weight coefficients of birds flying towards the center and affected by other birds. The function is one-dimensional, based on which the central position of the bird swarm is not accurate.

In this paper, we use spatial coordinates of birds to formulate A_1 and A_2. Based on the position of the bird group's center coordinate, the European distance between a bird's position and the center is calculated separately, and the traction and competitiveness of a bird flying toward the center are judged. We used other representations as;

$$A_1 = a_1 \times e^{pd_i} \tag{7}$$

$$A_2 = a_2 \times e^{-\frac{pd_k - pd_i}{|pd_k - pd_i|}} \tag{8}$$

where pd_i is the normalized Euclidean distance between coordinates of a bird p_i and the center of the swarm *meanp*.

$$pd_i = norm(|p_i - meanp|) \tag{9}$$

An example of normalized Euclidean distance is shown in Figure 3. The red points are the four coordinate positions distributed in two-dimensional space, and the blue pentagonal star represents the central position determined by the average coordinate values of the four points, which is considered as the center point of the swarm. Pd_1-pd_4 in the figure mean the normalized Euclidean distance between four points and the center one, which range from 0 to 1.

Foraging and flying behaviour are formulized as Equations (1), (5) and (6), the same as in the previous version of BSA.

Initially, we set parameters, i.e., maximum number of iterations T, size of population M, flying interval FQ, c_1, c_2, a_1 and a_2, and created populations x randomly.

For each cycle, within each time interval, we only need to consider two behaviours of birds, foraging and defence. A bird behaviour is determined randomly, if the bird is looking for food, it would update position using Equation (1). Otherwise, the bird is on the defensive, and tries to move to the centre of the swarm. As each bird wants to fly to the centre, it is inevitable to compete with others. We used A_1 and A_2 related to normalized European distance to evaluate centralized flight of the bird, shown as Equations (7) and (8). Meanwhile the new position is regulated via Equation (2). If the swarm stays at one site for FQ, it needs to move to the next location as a whole. In the flying process, each bird plays a different role, i.e., beggar or producer. Birds move to new positions according to Equations (5) and (6) respectively. The outline of BSAII can be written as Algorithm 1.

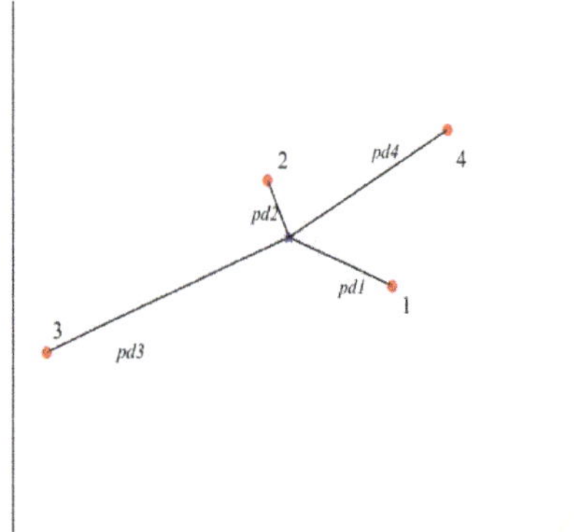

Figure 3. Normalized Euclidean distance in a two-dimensional (x-y) coordinate system.

Algorithm 1. BSAII.

Step 1: Set parameters:

    Population and Dimension of bird swarm [*M*, *N*].

    Iteration *T*, flying interval *FQ*, a_1, a_2, *FL*, Foraging probability *P*, etc.

Step 2: Initialize the original positions of birds, $x_i = [x_{i1}, x_{i2}, \cdots, x_{iN}]$, *i*=1,2,…*M*.

Step 3: Calculate the fitness function *f*(*xᵢ*), find local and global optimal solutions.

Step 4: For *r* = 1:*T*

1   **While** *r* is not an exact multiple of *FQ* **do**

        If (*Pᵣ*<*P*)

            Bird forages for food according to Equation (1).

            Otherwise

            Bird conducts defensive action based on Equations (2), (7), and (8).

        End if

    End while

2   Bird swarm is divided as producers and beggars, and flying to other site. The producers fly to new position of Equation (5), and the beggars follow the producers moving to Equation (6).

3   Calculate the fitness function *f*(*xᵢ*), update the local and global optimal solutions.

Step 5: Output the optimal results.

---

3. Energy Management of E-REV

Energy management of E-REV in this paper mainly refers to optimize the on-off timing of RE in E-REV so as to reduce the running time of engine. This problem can be mathematically summed up as a constrained objective optimization problem. In this paper, the penalty function method is used to solve this optimization problem.

3.1. Constrained Optimization Problem

Optimization problem with constraints is formulized as,

$$\begin{aligned}
&\min f(x), x \in R^n \\
&\text{s.t. } h_i(x) = 0, i = \{1,\ 2,\ \cdots,\ l\} \\
&\quad\ g_j(x) \geq 0, j = \{1,\ 2,\ \cdots,\ m\}
\end{aligned} \tag{10}$$

h_i and g_j are equality and inequality constraints respectively. The feasible region Ω of the problem is defined as $\Omega = \left\{ x \in R^n \middle| h_i(x) = 0, g_j(x) \geq 0 \right\}$. A popular method to solve constrained optimization problems is penalty function method. The penalty function [23] is constructed as;

$$\overline{P}(x) = \sum_{i=1}^{l} h_i^2(x) + \sum_{j=1}^{m} \left[min\left\{ 0, g_j(x) \right\} \right]^2 \tag{11}$$

Therefore, the objective function is transformed to;

$$P(x,\delta) = f(x) + \delta\overline{P}(x) \tag{12}$$

where $\delta > 0$ is penalty factor. The bigger δ is, the heavier the punishment will be. When $x \in \Omega$, x is a feasible point, $P(x,\delta) = f(x)$, the objective function is not subject to additional penalties. While $x \notin \Omega$, x is an infeasible point, $P(x,\delta) > f(x)$, the objective function is subject to additional penalties. When the penalty exists in the objective function, the penalty function should be sufficiently small to make $P(x,\delta)$ reach the minimum value, so that the minimum point of $P(x,\delta)$ approximates the feasible region Ω sufficiently, and its minimum value naturally approximates the minimum value of $f(x)$ on Ω sufficiently. The constrained optimization problem converts to unconstrained optimization problem, which is expressed as;

$$min\ P(x,\delta_k) \tag{13}$$

here δ_k is positive sequence, and $\delta_k \rightarrow +\infty$.

3.2. Problem Formulation

A certain type of electric vehicle is selected as research object, and basic parameters of vehicle are the same with that in [24]. Assuming that the SOC of battery power in electric vehicle should be kept between 20% and 80%. In order to reduce the uptime of engine in E-REV, and make full use of the power in the battery, this paper optimizes the uptime of the engine with constraint of distance. The objective function of the optimization problem is engine running time t, defined as Equation (14);

$$min\ t = \frac{L_{off} - L_{on}}{L} T_{cycle} \tag{14}$$

T_{cycle} is the time period, and L is the driving distance under one test condition, e.g., the NEDC cycle. We choose an E-REV as research object, whose main parameters are listed in Table 1.

Table 1. The main parameters for E-REV.

Parameters	Value
Curb weight (m/kg)	1700
Full mass (m/kg)	2100
Wheel radius (r/m)	0.334
Windward area (A/m^2)	1.97
Drag coefficients (CD)	0.32
Maximum speed (km/h)	>140
Total distance (km)	400
Climbing gradient (%)	>20%
Transmission efficiency	0.95
State of charge (SOC) range (%)	20–80
Maximum pure electric driving range (km)	>50

The vehicle parameters and component matching parameters (motor, battery, and RE) of the E-REV were entered into the advanced vehicle simulator (ADVISOR 2002) simulator, then the NEDC cycle condition was selected. The simulation results including SOC change curve under the NEDC cycle were obtained. The results showed an approximate linear relationship between the driving distance of electric vehicles and the SOC of battery.

$$y = kx + b \tag{15}$$

x is SOC, %. y is driving distance of E-REV, km. k and b are constant coefficients, in relation to battery working mode. When driving distance can be accomplished with one charge and discharge cycle, the distance when engine starts and shuts down are calculated based on Equations (16) and (17).

$$L_{on} = k_1(100 - t_{on}) \tag{16}$$

$$L_{off} = k_1(100 - t_{on}) + k_2(t_{off} - t_{on}) \tag{17}$$

t_{on} and t_{off} are timing of engine start-up and shutdown independently, which are represented as SOC. The equality constraint is the requirement of trip range distance D,

$$D = k_1(100 - t_{on}) + k_2(t_{off} - t_{on}) + k_3(t_{off} - 20) \tag{18}$$

k_i, $i = 1,2,3$, are driving distance per unit charge under different conditions, the specific values are shown in Table 2.

Table 2. Driving distance per unit charge.

Distance of Unit Charge	Driving State	Value [km/%]
k_1	Initial pure electric	0.7310
k_2	Charging	0.3667
k_3	Pure electric after charging	0.7210

Inequality constraints satisfy,

$$t_{off} - t_{on} > 0$$
$$20 < t_{on}, t_{off} < 80$$

When the target distance exceeds one charge and discharge cycle, the engine repeatedly starts and closes. If there are n charge and discharge cycles in the whole trip. Evaluation is processed based on total uptime of the engine.

$$min\ t = \frac{k_2\big[60(n-1) + (L_{off} - L_{on})\big]}{L}T_{cycle} \tag{19}$$

The target trip distance is calculated as Equation (20).

$$D = k_1(100 - t_{on}) + 60(n-1)(k_2 + k_3) + k_2(t_{off} - t_{on}) + k_3(t_{off} - 20) \tag{20}$$

4. Computational Experiment

In order to verify the effectiveness of BSAII algorithm, 20 benchmark functions were used in computational experiments, including unimodal and multimodal examples [25,26]. BSA, particle swarm optimization (PSO), artificial bee colony (ABC) and differential evolution (DE) were used as algorithms for comparison. In general, two aspects were taken into account to evaluate performance of algorithms: (1) proximity to the real optima in single operation, and (2) stability and accuracy of optimal results using different algorithms in multiple operations. Tables 3 and 4 illustrate concrete content of benchmark functions. In all cases, the population size was 50. The dimensions had two different types, dimension of population in f_4, f_6–f_8, f_{11}–$f_{13}, f_{15}, f_{16} f_{18}$–$f_{20}$ were set to two, and in other cases were 20. The number of iterations was set to 1000. Other parameters used in simulation were tuned according to Table 5. *FQ* is the flying interval, and this was set to three. *FL* is the following coefficient, and this value was a random number between 0.5 and 0.9. c_3 and c_4 are acceleration constants. w is inertia weight linearly decreasing from 0.9 to 0.5 [27]. *CR* is crossover probability, and F is the mutation rate [28]. *foodnumber* is the number of the food sources which is equal to the number of employed bees. *limit* is a predetermined number of cycles [29]. All algorithms were programmed with MATLAB 2018a. The simulation environment was on a computer with Intel ® core ™ i5-8400 CPU @ 2.80 GHz.

Table 3. Benchmark functions.

Id.	Name	Dimension	Boundary	Optima
f_1	Sphere	20	$[-100,100]$	0
f_2	Sum of Different Powers	20	$[-1,1]$	0
f_3	Sum Squares	20	$[-5.12,5.12]$	0
f_4	Trid	2	$[-4,4]$	-2
f_5	Rotated Hyper-Ellipsoid	20	$[-65.536,65.536]$	0
f_6	Easom	2	$[-100,100]$	-1
f_7	Matyas	2	$[-10,10]$	0
f_8	Booth	2	$[-10,10]$	0
f_9	Griewank	20	$[-600,600]$	
f_{10}	Rastrigin	20	$[-5.12,5.12]$	0
f_{11}	Schwefel	2	$[-500,500]$	0
f_{12}	Shubert	2	$[-5.12,5.12]$	-186.7309
f_{13}	Schaffer Function No. 2	2	$[-100,100]$	0
f_{14}	Rosenbrock	20	$[-2.048,2.048]$	0
f_{15}	Beale	2	$[-4.5,4.5]$	0
f_{16}	Needle in a Haystack	2	$[-5.12,5.12]$	-3600
f_{17}	Zakharov	20	$[-5,10]$	0
f_{18}	Drop-Wave	2	$[-5.12,5.12]$	-1
f_{19}	Bukin Function No. 6	2	$x_1 \in [-15,5], x_2 \in [-3,3]$	0
f_{20}	Three-Hump Camel	2	$[-5,5]$	0

Table 4. Benchmark functions.

Id.	Function
f_1	$f_1(x) = \sum_{i=1}^{N} x_i^2$
f_2	$f_2(x) = \sum_{i=1}^{N} |x|^{i+1}$
f_3	$f_3(x) = \sum_{i=1}^{N} i x_i^2$
f_4	$f_4(x) = \sum_{i=1}^{N} (x_i - 1)^2 - \sum_{i=2}^{N} x_i x_{i-1}$
f_5	$f_5(x) = \sum_{i=1}^{N} \sum_{j=1}^{i} x_j^2$
f_6	$f_6(x) = -\cos(x_1)\cos(x_2)\exp(-(x_1-\pi)^2-(x_2-\pi)^2)$
f_7	$f_7(x) = 0.26(x_1^2 + x_2^2) - 0.48 x_1 x_2$
f_8	$f_8(x) = (x_1 + 2x_2 - 7)^2 + (2x_1 + x_2 - 5)^2$
f_9	$f_9(x) = \sum_{i=1}^{N} \frac{x_i^2}{4000} - \prod_{i=1}^{N} \cos(\frac{x_i}{\sqrt{i}}) + 1$
f_{10}	$f_{10}(x) = 10N + \sum_{i=1}^{N} [x_i^2 - 10\cos(2\pi x_i)]$
f_{11}	$f_{11}(x) = 418.9829N - \sum_{i=1}^{N} x_i \sin(\sqrt{x_i})$
f_{12}	$f_{12}(x) = (\sum_{i=1}^{5} i\cos((i+1)x_1 + i))(\sum_{i=1}^{5} i\cos((i+1)x_2 + i))$
f_{13}	$f_{13}(x) = 0.5 + \frac{\sin^2(x_1^2 - x_2^2) - 0.5}{[1+0.001(x_1^2 + x_2^2)]^2}$
f_{14}	$f_{14}(x) = \sum_{i=1}^{N-1} [100(x_{i+1} - x_i^2)^2 + (x_i - 1)^2]$
f_{15}	$f_{15}(x) = (1.5 - x_1 + x_1 x_2)^2 + (2.25 - x_1 + x_1 x_2^2)^2 + (2.625 - x_1 + x_1 x_2^3)^2$
f_{16}	$f_{16}(x) = -(\frac{3}{0.05+(x_1^2 + x_2^2)})^2 - (x_1^2 + x_2^2)^2$
f_{17}	$f_{17}(x) = \sum_{i=1}^{N} x_i^2 + (\sum_{i=1}^{N} 0.5 i x_i)^2 + (\sum_{i=1}^{N} 0.5 i x_i)^4$
f_{18}	$f_{18}(x) = -\frac{1+\cos(12\sqrt{x_1^2 + x_2^2})}{0.5(x_1^2 + x_2^2) + 2}$
f_{19}	$f_{19}(x) = 100\sqrt{|x_2 - 0.01 x_1^2|} + 0.01|x_1 + 10|$
f_{20}	$f_{20}(x) = 2x_1^2 - 1.05 x_1^4 + \frac{x_1^6}{6} + x_1 x_2 + x_2^2$

Table 5. Parameter setting.

Algorithm	Parameters
BSAII, BSA	$a_1 = a_2 = 1$, $c_1 = c_2 = 1.5$, $P \in [0.8,1]$, $FL \in [0.5,0.9]$, $FQ = 3$
PSO	$c_3 = c_4 = 2.0$, $w \in [0.5,0.9]$
DE	$CR = 0.9$, $F = 0.5$
ABC	$foodnumber = \frac{M}{2}$, $limit = 10$

First of all, each algorithm ran once with 1000 iteration independently. Convergence performance on 20 benchmark functions is shown in Figure 4. Based on the graphical results, except for f_{11}, f_{14}, and f_{19}, BSAII could reach the real optima in other 17 functions. In addition, better results can be obtained by BSAII, compared with other four algorithms. Therefore, it can be shown that the performance of BSAII algorithm outperforms other algorithms.

Additionally, each algorithm was independently performed for 50 times on 20 test instances. Due to randomness of initial solutions for all algorithms, multiple performance indexes were used for comparing the performance of BSAII with different algorithms. Table 6 gives the minimum (MIN), maximum (MAX), mean (MEAN) and standard variation (SD) of 50 trials on each case. We can make the following remarks from results of Table 6:

1. On instances of $f_1 \sim f_7, f_{12}, f_{17}$, and f_{20}, BSAII performed better than other algorithms, in terms of convergence rate and accuracy of optimal solution. Since the four indexes had the smallest values compared with these obtained by other algorithms.
2. On instances of f_8, it was evident that both BSAII and DE could get the real optima (0 in Table 3), better than the other algorithms.
3. On instances of f_9 and f_{10}, BSAII and BSA were better than PSO, ABC, and DE in terms of performance indexes.
4. On instances of f_{13}, f_{15} and f_{18}, BSAII, BSA and DE converged to the real optimal value, with the same accuracy. But the convergence rate of BSAII and BSA in the early stage was faster than DE.
5. Only on instance of f_{11}, DE acquired the best performance over other algorithms. Solutions found by BSAII and BSA occasionally fell into local optimum.
6. Only on instance of f_{16}, BSA converged to the real optimal value (–3600) every time. It was obvious that BSA was better than the other algorithm on this example. While BSAII fell into local optimum at times and was not convergent to –3600. But the average value of 50 trials was closer to the optimum than that obtained by other algorithms.
7. The statistical results of f_{14} and f_{19} were somewhat complicated. The minimum of optimal fitness on f_{19} was found by BSAII, but the other three performance indexes of *MAX*, *MEAN*, and *SD* were slightly worse than in DE. BSAII would fall into local optimum when solving benchmark function of f_{14}.

Overall, we claimed that BSAII produced better convergence and more stable performance than BSA, PSO, ABC, and DE, in most cases.

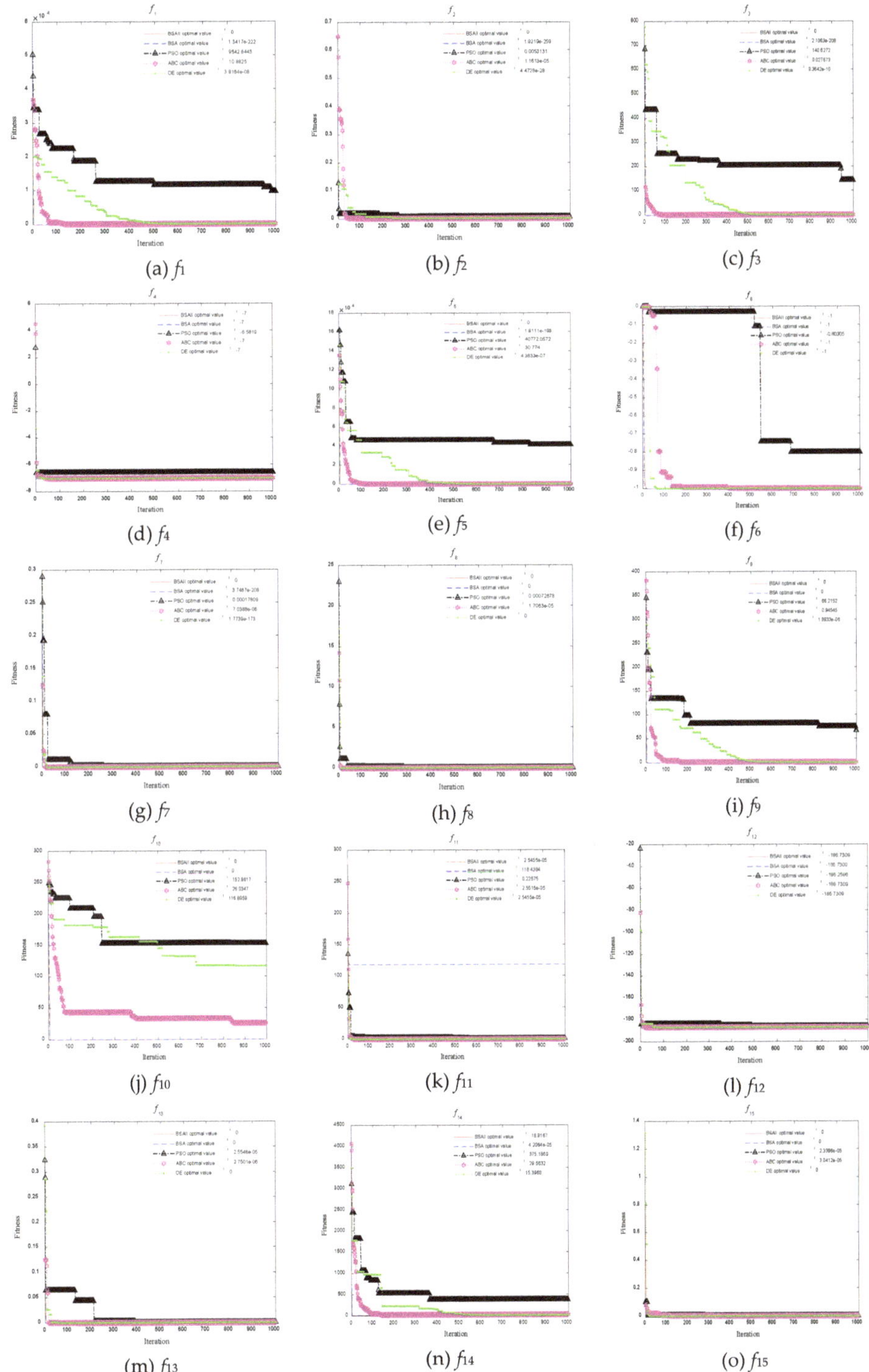

(a) f_1

(b) f_2

(c) f_3

(d) f_4

(e) f_5

(f) f_6

(g) f_7

(h) f_8

(i) f_9

(j) f_{10}

(k) f_{11}

(l) f_{12}

(m) f_{13}

(n) f_{14}

(o) f_{15}

Figure 4. *Cont.*

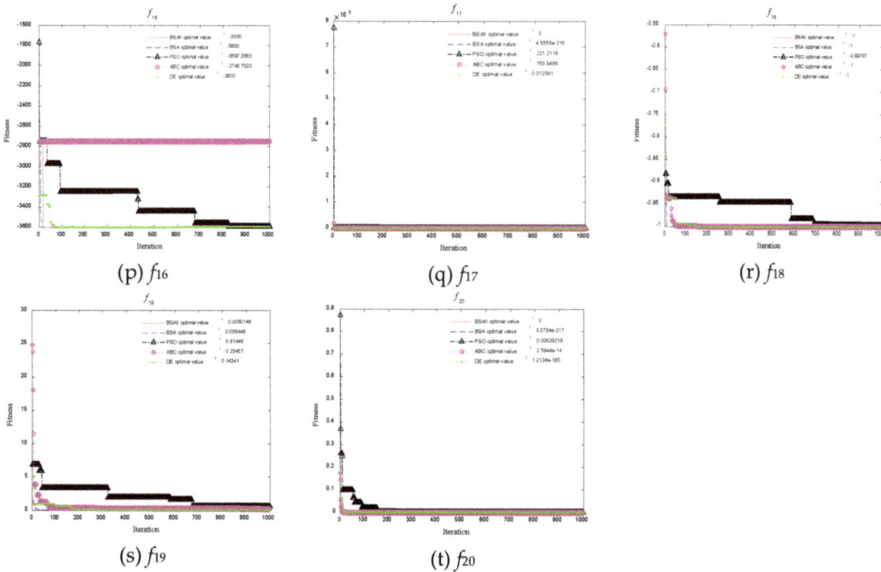

(p) f_{16}　　　　(q) f_{17}　　　　(r) f_{18}

(s) f_{19}　　　　(t) f_{20}

Figure 4. Convergence of benchmark functions using BSAII, BSA, particle swarm optimization (PSO), artificial bee colony (ABC) and differential evolution (DE).

Table 6. Statistical results obtained by BSAII, BSA, PSO, ABC, and DE in 50 trails (the best results are in bold).

Id.	Algorithm	MIN	MAX	MEAN	SD
	BSAII	**0**	**0**	**0**	**0**
	BSA	6.15×10^{-266}	9.98×10^{-176}	2.40×10^{-177}	0
f_1	PSO	3150.775	11924.62	7944.131	1938.963
	ABC	0.1627	12.08388	3.873986	2.878341
	DE	5.88×10^{-09}	8.31×10^{-07}	9.04×10^{-08}	1.23×10^{-07}
	BSAII	**0**	**0**	**0**	**0**
	BSA	5.43×10^{-256}	2.53×10^{-63}	5.07×10^{-65}	3.55×10^{-64}
f_2	PSO	0.000443	0.032429	0.006521	0.00539
	ABC	1.35×10^{-08}	1.64×10^{-05}	2.37×10^{-06}	3.26×10^{-06}
	DE	2.17×10^{-32}	1.02×10^{-20}	2.05×10^{-22}	1.43×10^{-21}
	BSAII	**0**	**0**	**0**	**0**
	BSA	1.13×10^{-242}	1.06×10^{-184}	3.27×10^{-186}	0
f_3	PSO	69.29799	315.1082	196.9478	56.54325
	ABC	0.001651	0.036862	0.016376	0.008962
	DE	2.98×10^{-10}	6.99×10^{-09}	1.51×10^{-09}	1.27×10^{-09}
	BSAII	**−7**	**−7**	**−7**	**2.66×10^{-15}**
	BSA	−7	−7	−7	2.29×10^{-14}
f_4	PSO	−6.99272	−6.45886	−6.90539	0.104025
	ABC	−7	−6.99997	−6.99999	6.26×10^{-06}
	DE	−7	−7	−7	2.66×10^{-15}
	BSAII	**0**	**0**	**0**	**0**
	BSA	1.32×10^{-237}	1.62×10^{-177}	3.27×10^{-179}	0
f_5	PSO	8375.27	48805.58	30703.65	7607.746
	ABC	1.394579	48.57402	14.07298	9.398897
	DE	2.91×10^{-08}	5.71×10^{-07}	1.59×10^{-07}	1.12×10^{-07}
	BSAII	**−1**	**−1**	**−1**	**0**
	BSA	−1	−1	−1	3.94×10^{-14}
f_6	PSO	−0.99599	$−1.18 \times 10^{-69}$	−0.29917	0.334968
	ABC	−1	$−8.11 \times 10^{-05}$	−0.9555	0.197304
	DE	**−1**	**−1**	**−1**	**0**
	BSAII	**0**	**0**	**0**	**0**
	BSA	3.79×10^{-251}	2.53×10^{-179}	5.07×10^{-181}	0
f_7	PSO	1.75×10^{-05}	0.002047	0.000348	0.000447
	ABC	2.20×10^{-07}	0.000157	3.22×10^{-05}	3.99×10^{-05}
	DE	1.42×10^{-178}	2.10×10^{-170}	6.36×10^{-172}	0

Table 6. *Cont.*

Id.	Algorithm	MIN	MAX	MEAN	SD
f_8	**BSAII**	**0**	**0**	**0**	**0**
	BSA	0	1.36×10^{-14}	2.72×10^{-16}	1.90×10^{-15}
	PSO	0.000117	0.054104	0.011096	0.012605
	ABC	2.91×10^{-08}	0.000174	3.09×10^{-05}	3.93×10^{-05}
	DE	**0**	**0**	**0**	**0**
f_9	**BSAII**	**0**	**0**	**0**	**0**
	BSA	**0**	**0**	**0**	**0**
	PSO	36.51811	113.051	77.5095	17.99438
	ABC	0.669883	1.17361	1.027435	0.086522
	DE	5.34×10^{-08}	0.027025	0.003197	0.006494
f_{10}	**BSAII**	**0**	**0**	**0**	**0**
	BSA	**0**	**0**	**0**	**0**
	PSO	118.5101	174.2391	150.3891	13.19794
	ABC	16.57457	41.5169	28.29116	5.126784
	DE	83.19845	134.5668	110.4572	12.10043
f_{11}	BSAII	2.55×10^{-05}	118.4384	40.26906	56.10528
	BSA	2.55×10^{-05}	118.4384	30.79399	51.95111
	PSO	0.004037	21.11573	2.520817	4.011557
	ABC	2.55×10^{-05}	6.29×10^{-05}	2.97×10^{-05}	6.14×10^{-06}
	DE	$\mathbf{2.55 \times 10^{-05}}$	$\mathbf{2.55 \times 10^{-05}}$	$\mathbf{2.55 \times 10^{-05}}$	**0**
f_{12}	**BSAII**	**−186.731**	**−186.731**	**−186.731**	$\mathbf{9.59 \times 10^{-14}}$
	BSA	−186.731	−186.731	−186.731	9.80×10^{-07}
	PSO	−186.717	−184.601	−186.243	0.436279
	ABC	−186.731	−186.731	−186.731	3.28×10^{-06}
	DE	−186.731	−186.731	−186.731	9.99×10^{-14}
f_{13}	**BSAII**	**0**	**0**	**0**	**0**
	BSA	**0**	**0**	**0**	**0**
	PSO	2.70×10^{-06}	0.017705	0.005071	0.004228
	ABC	8.24×10^{-10}	1.85×10^{-05}	2.82×10^{-06}	3.96×10^{-06}
	DE	**0**	**0**	**0**	**0**
f_{14}	BSAII	18.83609	18.97269	18.90242	**0.029108**
	BSA	$\mathbf{9.70 \times 10^{-13}}$	18.80166	**1.304165**	4.264392
	PSO	151.1679	709.0956	455.849	117.039
	ABC	19.55623	51.37853	28.38958	7.592035
	DE	12.17635	**16.29873**	14.56445	0.859034
f_{15}	**BSAII**	**0**	**0**	**0**	**0**
	BSA	**0**	**0**	**0**	**0**
	PSO	2.86×10^{-05}	0.012789	0.001782	0.002403
	ABC	2.63×10^{-08}	0.000207	5.00×10^{-05}	5.51×10^{-05}
	DE	**0**	**0**	**0**	**0**
f_{16}	BSAII	**−3600**	−2748.78	-3565.95	166.8039
	BSA	**−3600**	**−3600**	**−3600**	**0**
	PSO	−3599.74	−3024.95	−3504.13	103.0209
	ABC	−3598.49	−2748.78	−3307.88	297.6948
	DE	−3600	−2748.78	−2919.03	340.4871
f_{17}	**BSAII**	**0**	**0**	**0**	**0**
	BSA	1.21×10^{-240}	3.65×10^{-54}	7.34×10^{-56}	5.11×10^{-55}
	PSO	60.66649	438.8272	203.5698	81.22512
	ABC	101.2244	178.586	140.5003	16.80537
	DE	0.003087	0.190539	0.042672	0.037769
f_{18}	**BSAII**	**−1**	**−1**	**−1**	**0**
	BSA	**−1**	**−1**	**−1**	**0**
	PSO	−0.99988	−0.93622	−0.97521	0.020419
	ABC	−1	−0.99996	−0.99999	7.72×10^{-06}
	DE	**−1**	**−1**	**−1**	**0**
f_{19}	BSAII	$\mathbf{7.97 \times 10^{-05}}$	0.14995	0.066484	0.041502
	BSA	0.002607	0.147479	0.070706	0.039185
	PSO	0.169037	1.416421	0.729224	0.272299
	ABC	0.041325	0.298631	0.166209	0.043545
	DE	0.00036	**0.126775**	**0.063711**	**0.038273**
f_{20}	**BSAII**	**0**	**0**	**0**	**0**
	BSA	8.88×10^{-249}	6.76×10^{-170}	1.35×10^{-171}	0
	PSO	1.16×10^{-05}	0.003825	0.000797	0.000874
	ABC	1.00×10^{-16}	3.08×10^{-11}	1.96×10^{-12}	5.64×10^{-12}
	DE	2.31×10^{-187}	1.66×10^{-178}	3.68×10^{-180}	0

5. Optimization of On-Off Control Mode

This paper selects NEDC condition for research. The NEDC working condition test consists of two parts: the urban operation cycles and the suburban operation cycle (Figure 5). The urban operation cycle consisted of four urban operation cycle units. The test time of each cycle unit was 195 s, including idling, starting, accelerating and decelerating parking stages. The driving distance of the whole NEDC was 10.93 km and the testing time was 1184 s. The maximum velocity was 120 km/h, and average velocity was 33 km/h.

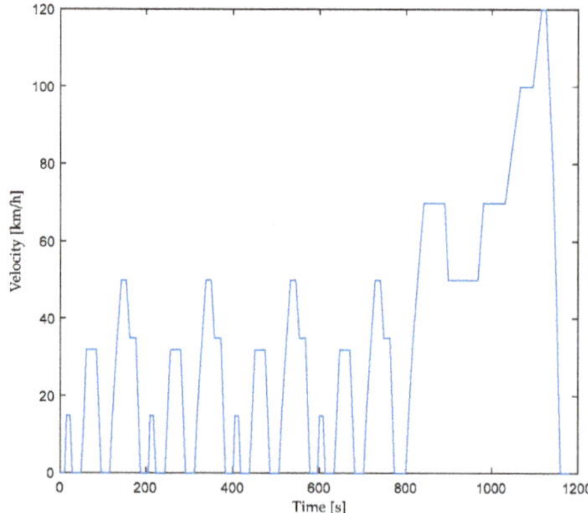

Figure 5. New European driving cycle (NEDC).

In this paper, two instances with different target distances for E-REV are analyzed. One is short-distance driving, in which the battery can complete the trip with one charge-discharge cycle. Another case is long-distance driving, and the battery needs multiple charge-discharge processes.

5.1. Traditional On-Off Control Mode

5.1.1. Distance = 100 km

The control strategy was to start the engine and recharge the battery when the power reduced to 20%, and shut down the engine when the battery was charged to 80%. According to the established model, the driving distance of the vehicle at engine start-up was 58 km. The driving distance of the vehicle was 80km when engine was shut down (Table 7). The engine was working in a full charge–discharge cycle, the uptime of the engine in E-REV was 2383 s. Figure 6 shows the variation of SOC. It can be observed from the figure, that when the battery dropped to 52% after charging, the vehicle's distance reached 100 km.

Table 7. On-off control mode.

Mode	Time [%]	Distance [km]	Working time [s]
t_{on}	20	58	2383 s
t_{off}	80	80	

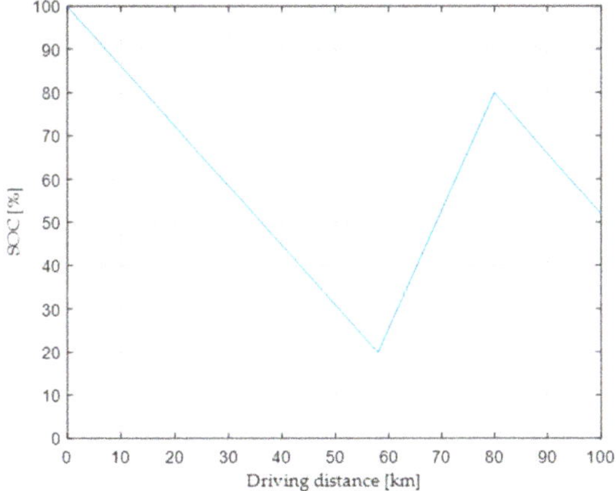

Figure 6. Battery volume of the electric vehicle, distance = 100 km.

5.1.2. Distance = 200 km

The distance exceeded 124 km, so more than one charge and discharge cycles would be repeated in the process of driving motion. According to Equation (20), there were three incomplete charge-discharge cycles. Working time of engine during vehicle in motion was 5959 s (Table 8). The SOC curve is shown as Figure 7. From the figure, it is evident that vehicle completed the target distance in the charging mode, while the battery SOC reached 50%.

Table 8. On-off control mode in the last cycle.

Mode	Time [%]	Distance [km]	Working time [s]
t_{on}	20	189	5959 s
t_{off}	50	200	

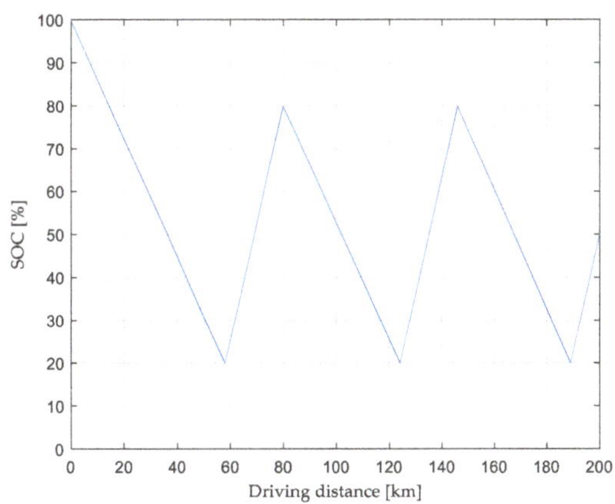

Figure 7. Battery volume of the electric vehicle, distance = 200 km.

5.2. On-Off Control Mode After Optimization

5.2.1. Distance = 100 km

BSAII was adopted to minimize the working time of engine so as to reduce fuel consumption and gas pollution. The trip distance was set to 100 km. The optimization problem and the constraints satisfied Equations (14)–(17). The number of iterations was set to 1000. Start-up and shut-down time of engine can be obtained by simulation, which are shown in Table 9. According to Table 9, it was obviously found that the early shutdown of the engine could help reduce fuel consumption and exhaust emissions of engine. The trade-off relation between the working time t with iterations is shown in Figure 8. The variation of SOC in the battery can be observed in Figure 9. The working time reduced to 36.4%, compared with the traditional control strategy.

Table 9. Optimal on-off control mode.

Mode	Time [%]	Distance [km]	Working time [s]
t_{on}	20	58	1515 s
t_{off}	58	80	

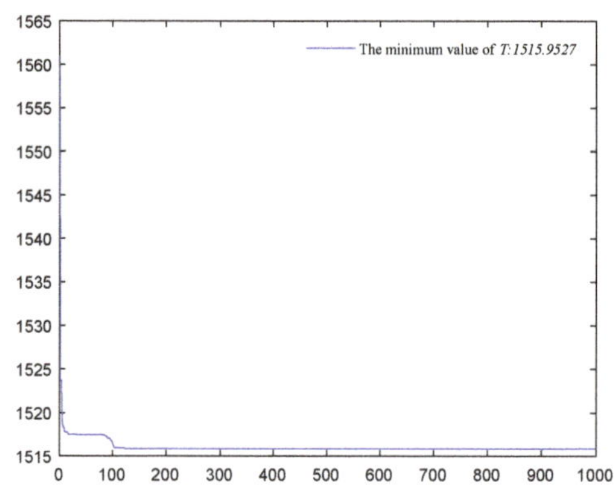

Figure 8. Convergence curve of engine uptime optimized via BSAII.

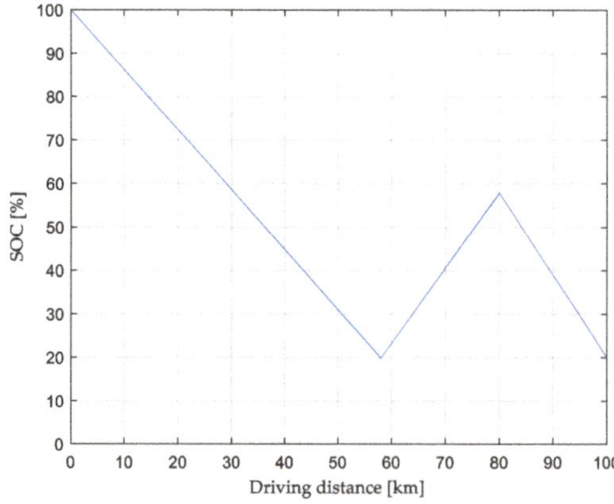

Figure 9. SOC of battery in electric vehicle after optimization, distance = 100 km.

5.2.2. Distance = 200 km

When the target distance was set to 200 km, it meant that the engine needed to be turned on and off repeatedly. The working hours of engine was calculated using Equation (19), and the distance constraint was formulized as Equation (20). The optimal on-off mode found by BSAII is illustrated in Table 10. According to the results, there were three charge–discharge cycles during driving process. In the last cycle, the engine turned off when SOC in battery reached 30%, and the electric power which was just enough to complete the entire distance (200 km). The convergence curve in the optimization problem is shown as Figure 10. After optimization, the total running time of engine was 5168 s, a 13% reduction in contrast to 5959 s under the traditional control strategy. The SOC of the battery power for electric vehicles while the car was in motion is shown in Figure 11.

Table 10. Optimal on-off control mode in the last cycle.

Mode	Time [%]	Distance [km]	Working time [s]
t_{on}	20	189	5168 s
t_{off}	30	193	

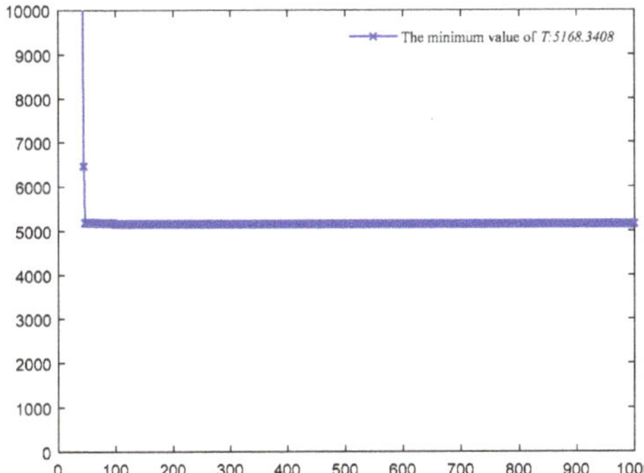

Figure 10. Convergence curve of t optimized with BSAII.

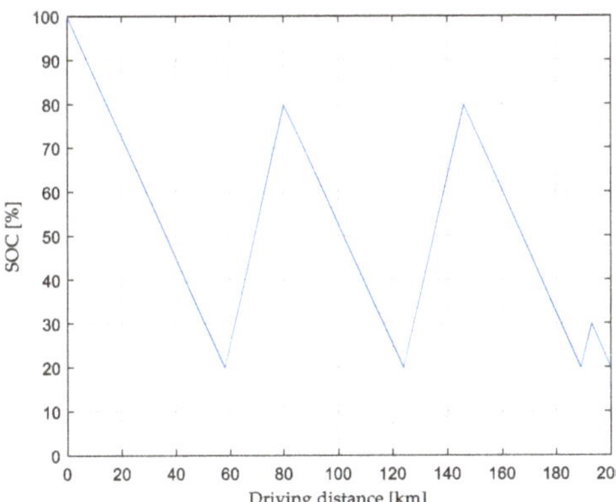

Figure 11. SOC of battery in electric vehicle after optimization, distance = 200 km.

6. Conclusions

This paper proposed another version of BSA, named as BSAII. In the version of BSAII algorithm, the spatial coordinates of birds in solution-space instead of the fitness function were used to determine the distance from the center of the whole bird group. Based on this method, the coefficients A_1 and A_2

were more accurate than that in BSA. We examined the performance of two different representations of defense behavior for BSA algorithms, and compared their experimental results with other bio-inspired algorithms, including PSO, ABC, and DE. It was evident from the statistical and graphical results highlighted that the BSAII outperformed other algorithms on most of the instances, in terms of convergence rate and accuracy of the optimal solution. Besides this, it was the first time that BSAII has been applied to the energy management of electric vehicles, which helps to reduce engine fuel consumption and exhaust emissions. Based on the analysis in the previous section, it is clear that:

- The BSAII performed statistically superior to or equal to the BSA on 16 benchmark problems. On these problems, it obtained the real optimal solution. However, in the case of Rosenbrock function (f_{14}), it was prone to falling into local optimum.
- The energy management of electric vehicles in this paper referred to minimization uptime of RE. The uptime was determined by the time interval between engine on and off. Two instances with different target driving distances were optimized with BSAII. Results indicated that the uptime of engine could be reduced by 36.4% with a target distance of 100 km, and 13% with a target distance of 200 km, respectively, in the NEDC condition. Hence we can draw a conclusion that based on the optimization strategy of BSAII, the on/off timing of the engine can be accurately controlled, which can effectively shorten the uptime of the engine, reduce fuel consumption and exhaust emissions, and also facilitate the next external charging.

It is hoped that in future work, more approaches will be designed to avoid the problem of local optimum, without sacrificing the convergence rate in BSA.

Author Contributions: Conceptualization, D.W. and L.F.; methodology, D.W.; software, D.W. and L.F.; validation, D.W.; formal analysis, L.F.; investigation, L.F.; writing—original draft, D.W.; writing—review and editing, L.F.; project administration, D.W.

Funding: This research was funded by Natural Science Foundation of Jiangsu Province, grant number BK20130873.

Conflicts of Interest: The authors of the manuscript declare no conflicts of interest with any of the commercial identities mentioned in the manuscript.

References

1. Amp, N.J.; Su, Z. On/off timing optimization for the range-extender in extended-range electric vehicles. *Automot. Eng.* **2013**, *35*, 418–423.
2. Zhao, J.; Han, B.; Bei, S. Start-stop moment optimization of range extender and control strategy design for extended-range electric vehicle. In Proceedings of the 5th Asia Conference on Materials Science and Engineering, Tokyo, Japan, 9–11 June 2017. [CrossRef]
3. Jiang, Y.C.; Zhou, H.; Cheng, L.; Wang, L. A study on energy management control strategy for an extended-range electric vehicle. In Proceedings of the 2015 International Conference on Energy and Mechanical Engineering, Wuhan, China, 17–18 October 2015; pp. 159–169.
4. Pan, C.F.; Han, F.Q.; Chen, L.; Xu, X.; Chen, L. Power distribution strategy of extend range electric vehicle based on fuzzy control. *J. ChongQing Jiaotong Univ. Nat. Sci.* **2017**, *33*, 140–144.
5. Niu, J.G.; Pei, F.L.; Zhou, S.; Zhang, T. Multi-objective optimization study of energy management strategy for extended-range electric vehicle. *Adv. Mater. Res.* **2013**, *694–697*, 2704–2709. [CrossRef]
6. Abdelgadir, A.A.; Alsawalhi, J.Y. Energy management optimization for an extended range electric vehicle. In Proceedings of the IEEE International Conference on Modeling, Sharjah, UAE, 4–6 April 2017.
7. Xi, L.H.; Zhang, X.; Geng, C.; Xue, Q.C. Energy management strategy optimization of extended-range electric vehicle based on dynamic programming. *J. Traffic Transp. Eng.* **2018**, *18*, 148–156.
8. Vatanparvar, K.; Faezi, S.; Burago, L.; Levorato, M.; Faruque, M.A.A. Extended range electric vehicle with driving behavior estimation in energy management. *IEEE Trans. Smart Grid* **2019**, *10*, 2959–2968. [CrossRef]
9. Rao, R.V.; Savsani, V.J.; Vakharia, D.P. Teaching-learning-based optimization: A novel method for constrained mechanical design optimization problems. *Comput. Aided Des.* **2001**, *43*, 303–315. [CrossRef]
10. García-Monzó, A.; Migallón, H.; Jimeno-Morenilla, A.; Sanchez-Romero, J.L.; Rico, H.; Rao, R.V. Efficient Subpopulation Based Parallel TLBO Optimization Algorithms. *Electronics* **2018**, *8*, 19. [CrossRef]

11. Mirjalili, S.; Mirjalili, S.M.; Lewis, A. Grey wolf optimizer. *Adv. Eng. Softw.* **2014**, *69*, 46–61. [CrossRef]

12. Al-Moalmi, A.; Luo, J.; Salah, A.; Li, K. Optimal virtual machine placement based on grey wolf optimization. *Electronics* **2019**, *8*, 283. [CrossRef]

13. Fatima, A.; Javaid, N.; Anjum Butt, A.; Sultana, T.; Hussain, W.; Bilal, M.; ur Rehman Hashmi, M.A.; Akbar, M.; Llahi, M. An enhanced multi-objective gray wolf optimization for virtual machine placement in cloud data centers. *Electronics* **2019**, *8*, 218. [CrossRef]

14. Goel, S. Pigeon optimization algorithm: A novel approach for solving optimization problems. In Proceedings of the International Conference on Data Mining & Intelligent Computing, New Delhi, India, 5–6 September 2014.

15. Duan, H.; Qiao, P.X. Pigeon-inspired optimization: A new swarm intelligence optimizer for air robot path planning. *Intern. J. Intell. Comput. Cybern.* **2014**, *7*, 24–37. [CrossRef]

16. Mirjalili, S.; Lewis, A. The whale optimization algorithm. *Adv. Eng. Softw.* **2016**, *95*, 51–67. [CrossRef]

17. Meng, X.B. A new bio-inspired optimisation algorithm: Bird swarm algorithm. *J. Exp. Theor. Artif. Intell.* **2015**, *28*, 673–687. [CrossRef]

18. Yang, W.R.; Ma, X.Y.; Xu, M.L.; Bian, X.L. Research on scheduling optimization of grid-connected micro-grid based on improved bird swarm algorithm. *Adv. Technol. Electr. Eng. Energy* **2018**, *2*, 53–60.

19. Yang, W.Y.; Ma, X.Y.; Bian, X.L. Optimization of Micro-grid operation using improved bird swarm algorithm under time-of-use price mechanism. *Renew. Energy Resour.* **2018**, *7*, 1046–1054.

20. Yang, W.Y.; Ma, X.Y.; Bian, X.L. Adaptive improved bird swarm algorithm based on levy flight strategy. *J. Hebei Univ. Technol.* **2017**, *5*, 10–16, 22.

21. Zhang, L.; Bao, Q.; Fan, W.; Cui, K.; Xu, H.; Du, Y. An improved particle filter based on bird swarm algorithm. In Proceedings of the 2017 10th International Symposium on Computational Intelligence and Design (ISCID), Hangzhou, China, 9–10 December 2017.

22. Wang, Y.; Wan, Z.; Peng, Z. A novel improved bird swarm algorithm for solving bound constrained optimization problems. *Wuhan Univ. J. Nat. Sci.* **2019**, *24*, 349–359. [CrossRef]

23. Zahara, E.; Hu, C.H. Solving constrained optimization problems with hybrid particle swarm optimization. *Eng. Optim.* **2008**, *40*, 1031–1049. [CrossRef]

24. Wang, Z.X.; Bei, S.Y.; Wang, W. The optimization control of on and off timing for the engine in extended-range electric vehicle. *Mach. Des. Manuf.* **2017**, *4*, 100–103.

25. Yang, X.S.; Deb, S. Cuckoo search: Recent advances and applications. *Neural Comput. Appl.* **2014**, *24*, 169–174. [CrossRef]

26. Virtual Library of Simulation Experiments: Test Functions and Datasets. Available online: http://www.sfu.ca/~{}ssurjano (accessed on 30 July 2019).

27. Eberhart, R.C.; Shi, Y.H. Particle swarm optimization: Development, application, and resources. In Proceedings of the 2001 Congress on Evolutionary Computation, Seoul, Korea, 27–30 May 2001. [CrossRef]

28. Storn, R.; Price, K. Differential evolution—A simple and efficient heuristic for global optimization over continuous spaces. *J. Glob. Optim.* **1997**, *11*, 341–359. [CrossRef]

29. Karaboga, D.; Basturk, B. A powerful and efficient algorithm for numerical function optimization: Artificial bee colony (ABC) algorithm. *J. Glob. Optim.* **2007**, *39*, 459–471. [CrossRef]

Article

A Method Based on Multi-Sensor Data Fusion for UAV Safety Distance Diagnosis

Wenbin Zhang [1,*], **Youhuan Ning** [1] and **Chunguang Suo** [2]

[1]   College of Mechanical and Electrical Engineering, Kunming University of Science and Technology, Kunming 650504, China; ningyouhuan@stu.kust.edu.cn
[2]   College of Science, Kunming University of Science and Technology, Kunming 650504, China; scgzhb@stu.kust.edu.cn
*   Correspondence: zwbscg@kust.edu.cn; Tel.: +86-1530-885-9327

Received: 5 November 2019; Accepted: 28 November 2019; Published: 3 December 2019

Abstract: With the increasing application of unmanned aerial vehicles (UAVs) to the inspection of high-voltage overhead transmission lines, the study of the safety distance between drones and wires has received extensive attention. The determination of the safety distance between the UAV and the transmission line is of great significance to improve the reliability of the inspection operation and ensure the safe and stable operation of the power grid and inspection equipment. Since there is no quantitative data support for the safety distance of overhead transmission lines in UAV patrol, it is impossible to provide accurate navigation information for UAV safe obstacle avoidance. This paper proposes a mathematical model based on a multi-sensor data fusion algorithm. The safety distance of the line drone is diagnosed. In these tasks, firstly, the physical model of the UAV in the complex electromagnetic field is established to determine the influence law of the UAV on the electric field distortion and analyze the maximum electric and magnetic field strength that the UAV can withstand. Then, based on the main factors affecting the UAV such as the maximum wind speed, inspection speed, positioning error, and the size of the drone, the adaptive weighted fusion algorithm is used to perform first-level data fusion on the homogeneous sensor data. Then, based on the improved evidence, the theory performs secondary fusion on the combined heterogeneous sensor data. According to the final processing result and the type of proposition set, we diagnose the current safety status of the drone to achieve an adaptive adjustment of the safety distance threshold. Lastly, actual measurement data is used to verify the mathematical model. The experimental results show that the mathematical model can accurately identify the safety status of the drone and adaptively adjust the safety distance according to the diagnosis result and surrounding environment information.

Keywords: overhead transmission line; UAV inspection; safe distance; multi-source data fusion; adaptive threshold

---

1. Introduction

With the continuous development of power systems, in order to ensure the safe and stable operation of a power system, the power industry needs to conduct regular inspections on overhead transmission lines, grasp the environment around the transmission line and its operation in a timely manner, and deal with potential problems accordingly [1–3]. The traditional transmission line inspections are mostly manual. In this way, the inspection line is very hard, and the operation period is very long. It is difficult to reach some blind spots with complicated or even dangerous terrain [4]. Due to its strong maneuverability, simple operation, various forms, high efficiency, and low cost, the unmanned aerial vehicle (UAV) inspection line has been widely used by domestic and foreign power companies for the inspection of transmission lines [5,6].

In order to accurately and clearly collect the operation and fault information of the power line, the UAV needs to be inspected close to the tower and the transmission line. However, the complex electromagnetic field environment around the high-voltage transmission line has a great influence on the inspection system, the navigation, and the communication system of the UAV. If the distance between the transmission line and the drone is too close, the complex electromagnetic environment will cause interference to the electronic components that make up the control system, resulting in data transmission interruption and equipment failure, affecting the flight performance of the drone, and even causing the drone to crash into the power line or other safety accidents. In such cases, the UAV will not complete the line inspection task. However, if this distance is too far, the drone's inspection system will not have the best inspection effect [7]. Accurate measurement of the safety distance between the UAV and the transmission line is a key factor restricting the development of the UAV transmission line inspection system. Accurately measuring the safe distance of the drone relative to the overhead transmission line is a challenge in the UAV line-of-sight system. An accurate measurement of the safety distance can not only ensure the safety of the power grid, the equipment, and the drones, but also improve the quality and reliability of the drone's inspection of long-distance outdoor high-voltage transmission lines.

In order to improve the mobility of the UAV inspection, to enable the UAV to perform inspections under a variety of working conditions, and to improve the safety and reliability of the UAV line inspection operations, accurately determine the safety between the drone and the transmission line distance is very important. Therefore, it is necessary to quantify the safe distance between the UAV and the transmission line. Some researchers have judged the safety distance through the detection and identification of power lines and achieved good research results. There are some more representative methods. Some researchers have used spectrum-space methods to detect power lines in UAV remote sensing images. Using k-means and the expectation maximization (EM) algorithm for spectral clustering, the pixels are divided into power lines and non-power lines. Thereby, the autonomous inspection of the power line by the drone can be further achieved [8]. Other researchers have used lidar, vision sensors, and infrared cameras to detect transmission lines for navigation and obstacle avoidance [9–11]. However, these methods have certain limitations. Although the laser ranging has high accuracy and the laser scanning method is used to realize the transmission line detection, it has a significant effect on the inspection of the drone. The load of the aircraft reduces the flying efficiency of the drone. At the same time, the processing of data is complicated because the information of power transmission lines and towers extracted from the pictures taken by laser point clouds and vision sensors is affected by the complex environment. The data collected by these methods mainly depend on manual processing, and the work efficiency is low.

In order to solve the detection errors problem of infrared cameras and lidars due to environmental factors such as temperature differences and the atmosphere, some researchers have used the edge detection method to effectively extract power lines and suppress non-power line features [12]. However this method is easily affected by straight roads and rivers and cannot extract the characteristics of transmission lines well. In [13], Matikainen et al. provided an extensive overview of the possibilities offered by modern remote sensing sensors in power line corridor surveys and discussed the potential and limitations of different approaches. The monitoring of power line components and their surrounding vegetation uses visual, near-infrared, and multispectral images. In [14], the researchers proposed a method to process continuous images with telemetry data sent by the autopilot, identify online vegetation, trees, and building areas near the power line, and calculate the relationship between power lines and vegetation, trees, and buildings the distance. At the same time, the system processes the images captured by the infrared camera to detect the poor conductivity and hot spots of power lines, transformers, and substations. In [15], a method for determining the safe flight area of a UAV patrol inspection based on an electromagnetic field calculation was proposed. The field strength distribution around the transmission line and the maximum electromagnetic field strength of the electronic equipment carried by the UAV were analyzed through simulation to realize the control of the prediction of a safe flight distance for UAV inspection. In [16], a method for avoiding obstacles of overhead transmission lines based on multi-sensor information fusion and cylindrical space inspection was proposed. These methods can better realize the detection and identification of power lines to a certain extent. They can also better solve the inspection of transmission lines by line inspection drones and achieve autonomous obstacle avoidance. However, these methods need to extract the transmission lines from a messy background due to its characteristics, it is difficult to process the data, the real-time performance is poor, and the efficiency of UAV inspection cannot be met.

The accurate measurement of safety distance is a key technology. At present, in the practical application of UAV inspection power line engineering, there is still no stable and reliable method to achieve autonomous line inspection and autonomous obstacle avoidance navigation. Operators still need to judge by subjective experience. Under the premise of ensuring that the drone is in a controllable range, the operator can use the control handle to make the drone is as close to the transmission line as possible to complete the inspection work [17]. This method can effectively avoid the occurrence of line inspection safety accidents; however, when the drone is far away from the control personnel, it is difficult to accurately control it, which limits the performance of the drone operation to a great extent. Relying on the stable electric field and magnetic field information around the transmission line to navigate the UAV can effectively solve the limitations of traditional sensors and the difficulty of data processing. The accurate measurement of the safety distance between the UAV and the transmission line has become a key factor affecting the development of the UAV transmission line inspection system [18]. Therefore, the study of the safety distance threshold can provide data support for the navigation of the UAV inspection operation, and it is of great significance for the navigation of the UAV autonomous obstacle avoidance navigation and the improvement of the safety and reliability of the inspection operation.

Based on the status of the research, this paper proposes an adaptive multi-sensor data fusion algorithm for multi-sensor data fusion on the electromagnetic field strength of the UAV's safe flight, the performance of the UAV control system, and the size of the UAV. According to the result and the type of proposition set, we judge the current safety status of the drone to realize the UAV's adaptive adjustment of the safety distance threshold. Finally, the mathematical model was verified by using certain inspection data. The experimental results show that the mathematical model can accurately identify the safety status of drones. UAVs that use this mathematical model for line inspection can be adaptively adjusted according to the surrounding environmental factors' safe distance. The multi-sensor adaptive fusion has a clear distinction of credibility, which better solves the problem that a single safety threshold adapts to a single condition, which brings about a large cruise error due to the field source positioning error.

The rest of the paper is organized as follows. Section 2 establishes the physical model of the UAV patrol overhead transmission line, analyzes the influence of the UAV on the electric field distribution and the variation law of the maximum distorted electric field on the surface of the UAV, and analyzes the anti-electromagnetic interference performance of the UAV. In Section 3, based on the multi-sensor data fusion algorithm, the mathematical model of autonomous obstacle avoidance navigation is established. The main influencing factors affecting the safety distance of the UAV are considered comprehensively, and the decision of the UAV inspection status based on the fusion result and the set proposition is realized. In Section 4, according to the actual situation of the UAV patrol overhead transmission line, the patrol data is collected to verify the mathematical model in the text, the results are analyzed, and finally, the scientific, practical, limitations, and considerations of the proposed algorithm and mathematical model are discussed in Section 5.

2. Materials and Methods

2.1. Analysis of the Influence of Distorted Electric Field on UAV

When a patrol drone is placed as an electric conductor in an electric field, an induced electric charge is generated on its surface to form an induced electric field. The superposition of the induced electric field with the original electric field changes the distribution of the original electric field to form a "distorted electric field" [19]. When the UAV is inspected near the overhead transmission line, the influence of the "distortion field" on the drone is much greater than that of the original field on the drone, the distortion is not linear, and it is difficult to completely eliminate it. Therefore, the analysis of the distortion generated by the UAV in the electric field is of great significance for the accurate identification of the safety distance threshold.

During the inspection of the overhead transmission line with a power frequency voltage, when the UAV is located under the transmission line for inspection, a capacitor series circuit is formed between the UAV, the overhead transmission line, and the ground. Its equivalent physical model is shown in Figure 1.

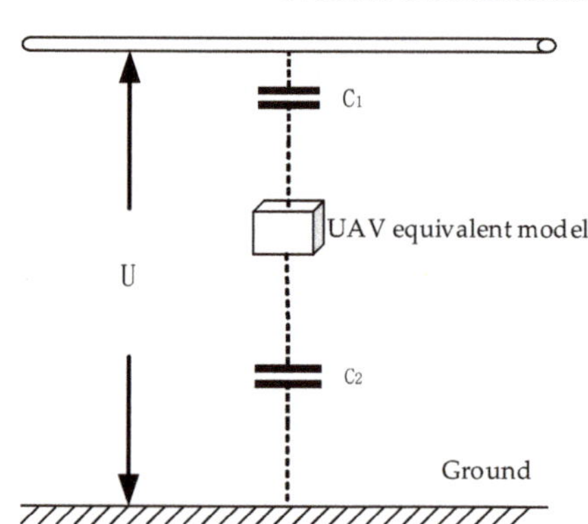

Figure 1. Unmanned aerial vehicle (UAV) patrol transmission line equivalent physical model.

As shown in Figure 1, where U is the phase voltage of the overhead transmission line, C_1 is the equivalent capacitance between the UAV and the transmission line, and C_2 is the equivalent capacitance between the UAV and the ground.

It can be seen from the analysis in Figure 1 that when the UAV is near the transmission line for patrol inspection, a capacitor series circuit is formed among the UAV, the overhead transmission line, and the ground, and there will be impedance among the UAV, the transmission line, and the ground.

Here, the impedance between the UAV and the transmission wire can be expressed as:

$$Z_1 = kD_1 \tag{1}$$

where k is the rate of impedance change, which is related to external environmental factors, and D_1 is the distance between the UAV and overhead transmission wire.

Therefore, the physical model of the field strength between the drone and the wire during the inspection process can be expressed as:

$$E_1 = \frac{Uk}{Z_1 + Z_2} \tag{2}$$

where U is the voltage between the conductor and the ground, Z_1 is the impedance between the UAV and the transmission conductor, and Z_2 is the impedance between the UAV and the ground.

Since the distance between the UAV and the conductor is far less than the distance between the UAV and the ground, the impedance change between the UAV and the ground in the inspection process is much smaller than that between the UAV and the transmission conductor, so the impedance between the UAV and the ground can be regarded as a fixed value [20]. Therefore, the analysis in Formula (2) shows that with the voltage U remaining unchanged, as the distance D_1 between the drone and the wire decreases, the field strength E_1 between them gradually increases, and if D_1 is extremely small, the air between the drone and the transmission line is broken down, causing the phenomenon of induced electrification. The field strength E_1 enhancement will interfere with the normal operation of the drone. In order to study the distortion of the electric field by the drone, the drone line is established. The simulation model is shown in Figure 2.

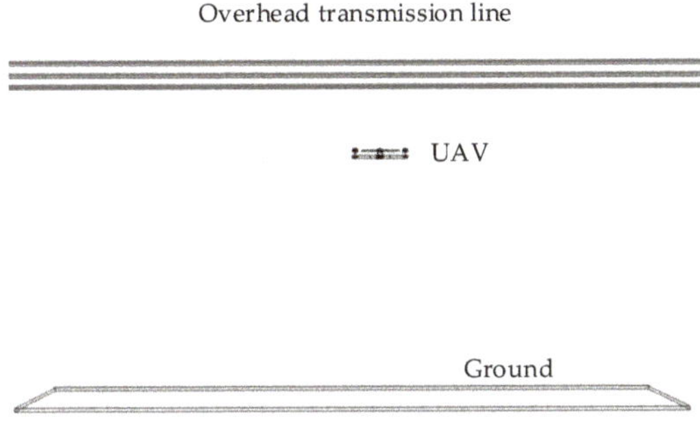

Figure 2. Modeling of a UAV in an electric field.

The main components of the patrol drone are modeled, and the finite element analysis method is used to simulate the UAV model placed in the electric field in order to study the effect of electric fields on drones through the inspection of power lines by drones. As shown in Figure 3, the electric field distortion caused by the drone's patrol operation is on the same horizontal plane as that of the edge conductor.

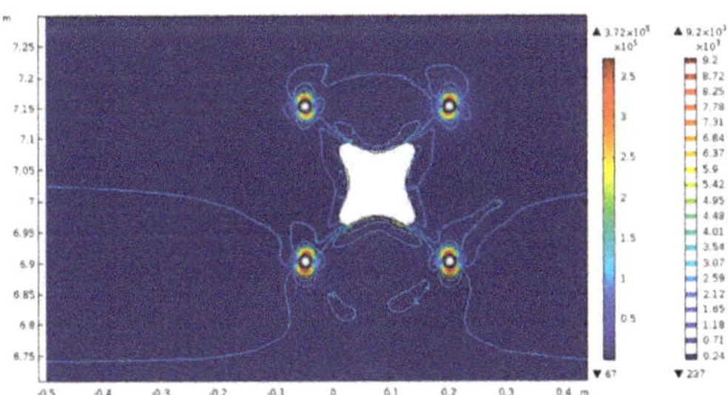

Figure 3. Effect of UAV on electric field distortion.

It can be seen from the analysis in Figure 3 that when the patrol drone is close to the transmission line for patrol inspection, the rotor, the tripod, and the internal electronic components of the fuselage will have distortion effects on the spatial electric field distribution around the line, especially near the motor. At the tip of the wing or at the stand, charge accumulation and field distortion are prone to occur, especially in the electric field distortion around the motor. This kind of distorted electric field will have a great impact on the navigation system and communication system of the drone, even causing the drone to get out of control and cause a safety accident, leading to an inability to complete the safety inspection. The analysis of the distortion field under actual working conditions has an extremely important influence on the accurate identification of the safety distance. Therefore, this paper simulates and analyzes the influence of a distorted electric field on a UAV inspection system in 500 kV and 220 kV common voltage transmission lines.

According to the relevant standards of China's GB 50545-2010 "110–750 kV overhead transmission line design" and "National Power Transmission and Transformation Engineering Universal Design (2011 Edition)", the related parameters such as wire type, ground height, and wire diameter of the overhead transmission line are set. Our experiments consider a wire to be a smooth cylinder with infinitely long straight parallel ground and the surface of the wire as an equipotential surface. Ignoring the end effect of the wire and the sag, the earth is approximated as an equipotential body with a potential of 0. After simplifying the high-voltage overhead transmission line, the AC/DC module of the finite element analysis software COMSOL Multiphysics 5.3 multiphysics simulation software is used to analyze the transient transmission lines of 500 kV and 220 kV AC voltage lines. The three-dimensional wire model uses a horizontally distributed four-split wire, the material is taken to be copper, and the wire is added to a three-phase alternating current. The simulation results are shown in Figures 4 and 5.

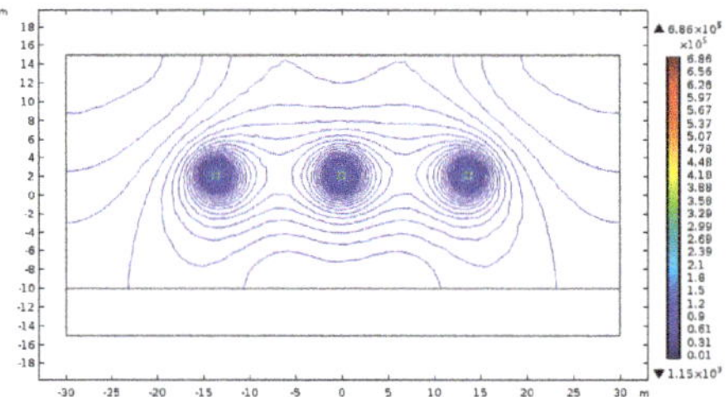

Figure 4. 500 kV vertical section electric field strength contour.

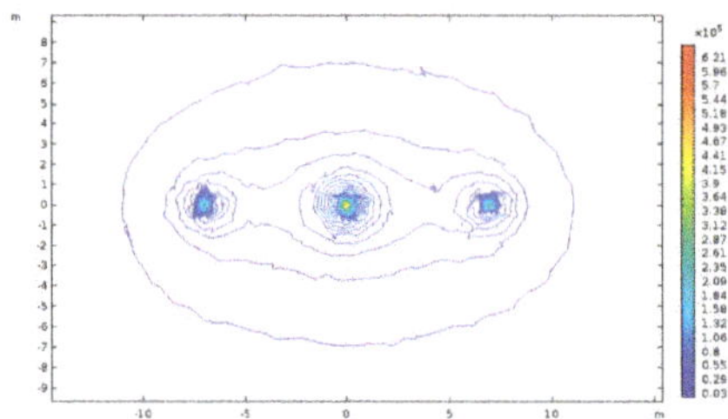

Figure 5. 220 KV vertical section electric field strength contour.

From the simulation results of Figures 4 and 5, the distribution law of the electric field intensity on the y–z plane perpendicular to the three-phase transmission line is as follows. At the 500 kV voltage level, the highest field strength is on the three transmission lines, and the maximum electric field strength is 6.87×10^5 V/m. At the voltage level of 220 kV, the highest field strength is on the three transmission lines, and the maximum electric field strength is 6.21×10^5 V/m. The maximum electric field strength appears on the wire, and the electric field strength contour intersects between the two transmission lines. The field strength of the electric field is gradually reduced away from the transmission line, and the field strength decreases faster as it moves away from the wire.

Then, we need to further analyze the influence of the UAV on the field strength distortion during the operation of the UAV patrol overhead transmission line and the variation of the distortion field with the distance between the transmission lines. In order to do so, we establish a four-rotor UAV inspection 500 kV, 220 kV high-voltage overhead transmission line simulation model and carry out simulation experiments. In the experiment, the distance from the wire is 0.5–12 m, and the maximum field strength of the surface of the patrol UAV is scanned. The scanning result is shown in Figure 6.

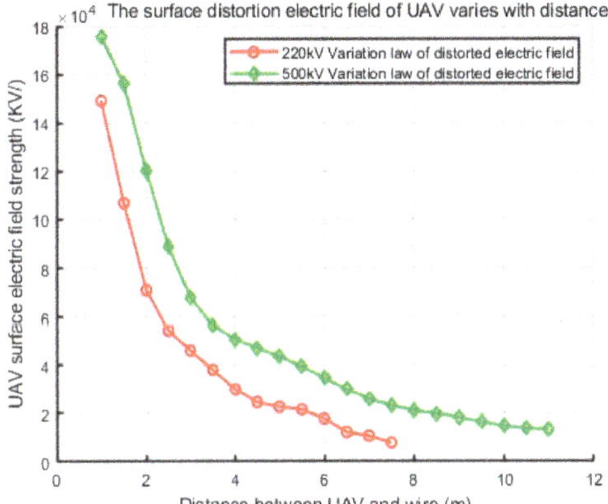

Figure 6. The relationship between the maximum electric field strength of the surface of the UAV and the distance.

It can be seen from the analysis in Figure 6 that the maximum field strength of the surface of the 220 kV voltage class overhead transmission line is reduced with the increase of the distance from the conductor, and the field strength and distance are roughly inversely proportional. The farther the distance, the smaller the rate of change. The variation of the distortion field strength in relation to the distance between the outer side and the middle line of the edge conductor is similar. The maximum field strength at a distance of 1 m is about 150 kV/m, and when the distance is 3.5 m, the field strength is attenuated to 40 kV/m. When the UAV patrols the 500 kV voltage class overhead transmission line, the maximum field strength and the field strength change rate of the UAV surface gradually decrease with the increase of the distance.

When the UAV is 1 m away from the side conductor, the maximum field strength is 180 kV/m, and when the distance is 5 m, the field strength is attenuated to 50 kV/m. The electric field distortion of the 500 kV voltage level has a greater influence on the surface of the UAV than the voltage level of 220 kV, but the variation of the field strength with the distance and the rate of change of the field strength are similar under the two voltage levels. For the electronic components that make up the UAV flight control system and the inspection system, as well as the conditioning circuit and protection circuit of the sensor, the maximum electric field strength that can resist electromagnetic interference under normal operation is generally 50 kV/m [21,22]. Therefore, in order to operate the UAV and the normal collection of environmental information, the maximum electric field strength of the UAV surface is generally less than 50 kV/m. In this paper, this electric field strength value is used as the safety field strength threshold of the normal inspection operation of the unmanned aerial vehicle.

2.2. Analysis of the Influence of Magnetic Field on UAV

The complex electromagnetic environment during the inspection of the unmanned aerial vehicle will cause interference to the electronic components that constitute the UAV patrol system and communication system, affecting the flight performance of the UAV. If the distance between the inspection drone and the transmission line is too close, the electromagnetic field will have a greater impact on the flight performance of the drone. When the electromagnetic interference is serious, the effect is macroscopic: due to the occurrence of clutter, the data transmission is not accurate, causing the equipment to malfunction and the drone to hit the power line. At the same time, according to Biot–Savart Law, the magnetic field strength at a point around the wire through which the current is passed is proportional to the current in the wire and inversely proportional to the distance from the point to the wire. When the UAV moves in a complex magnetic field, it will be attracted by the wire and hit the line; subsequently, it will be unable to complete the line inspection task. A current-carrying wire generates a magnetic field around the space as shown in Equation (3):

$$\vec{B} = \int_L \frac{\mu_0 I}{4\pi} \frac{dl \times \vec{e_r}}{r^2} \tag{3}$$

where $\mu_0 = 4\pi \times 10^{-7}$ is the vacuum permeability, I is the source current, L is the integral road force, dl is the microfluidic source, and r is the distance between a point in the space and the current-carrying wire.

The patrol UAV is generally composed of a magnetic field sensor, a GPS positioning system, an accelerometer gyroscope, a barometer, a motor, a flight control system, and a digital transmission module [23]. The magnetometer provides patrol navigation for the UAV, which is resistant to a magnetic interference that is three to four times the magnetic field strength. Therefore, during the inspection operation, if the magnetic induction generated by the current in the transmission line is greater than 220 μT, the magnetometer of the unmanned aerial vehicle will be disturbed, which will affect the normal operation.

The GPS receiver contains circuits of various frequencies, and the electronic components that make up the GPS internal circuit system are also subject to interference from electromagnetic fields. The national standard GB/T18314-2001 "Global Positioning System (GPS) Measurement Specifications" stipulates that when GPS observation is performed, the observation instrument should be 50m away from the high-voltage transmission line, affecting the inspection and monitoring of transmission lines using drones that carry GPS. In the literature [24], Silva and Whitney et al. carried out related research on the influence of the power carrier wave on the accuracy of NGPS (National GPS) received signals in the United States. The quality has a certain impact, because NGPRS uses a mid-band modulation signal. However, in [25], Xingfa and Hui et al. analyzed the impact of high-voltage transmission lines on GPS received signals by comparing positioning errors at different distances from the high-voltage transmission lines. Field tests were conducted under AC high-voltage transmission lines. The test results showed that the GPS positioning coordinate accuracy did not significantly decrease at different distances from the transmission lines, and there were no obvious rules for GPS positioning accuracy, distance, or whether the line was live. It proves that the error of the high-voltage transmission line on the GPS receiving signals of nearby operations is not enough to affect the navigation of the line-drone UAV. At the same time, research analysis shows that the carrier signal L_1 frequency of the GPS satellite is 1575.42 MHz, and the L_2 frequency is 1227.6 MHZ. The frequency used by overhead transmission lines in China is 50 Hz. The radio interference in this frequency band is close to background noise. The impact on GPS is within the range of the UAV positioning and navigation error. Therefore, under normal circumstances, the electromagnetic field radiation around the transmission line will not affect the operation of the GPS positioning system;

The servo motor is an important part of the UAV, and the internal reluctance is easily interfered by the electromagnetic field. The magnetic field required for normal operation is about 600 μT. The UAV used to analyze the distorted electric field was placed in the magnetic field for simulation analysis by finite element analysis. Taking the unmanned aerial vehicle in the same horizontal plane as the side conductor, the simulation results are shown in Figure 7.

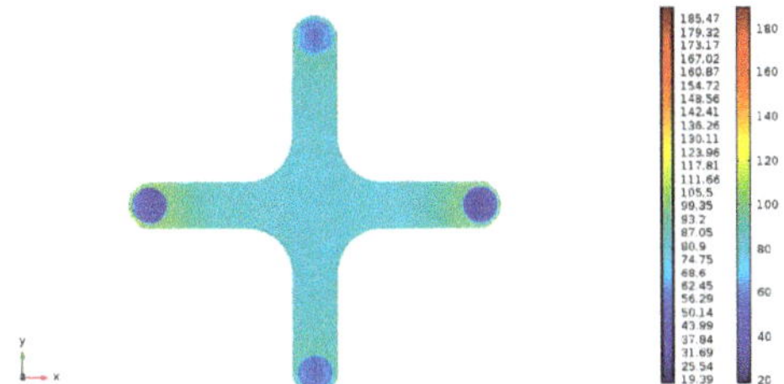

Figure 7. Magnetic field distribution on the surface of the drone.

From the analysis in Figure 7, it can be seen that in the simulation model, the UAV has a small effect on the spatial magnetic field distribution around the wire; this is mainly because the constituent materials of the UAV inspection system are mainly carbon fibers, which have a small magnetic permeability and are not easy magnetized.

In order to further analyze the relationship between the influence of the magnetic field on the UAV and the distance of the transmission line during the operation of the UAV patrol overhead transmission line, the four-rotor UAV patrols two 500 kV and 220 kV high-voltage overhead transmission lines as examples to carry out simulation experiments. The finite element analysis method is used to simulate the influence of the magnetic field distribution around the high-voltage transmission line on the patrol UAV. The transmission current of China's 220 kV overhead transmission line is 0.6–1.2 kA. The simulation model established in the paper applies 1.2 kA, the transmission current of the 500 kV transmission line is 2–4 kA, and the current applied to the two poles of this simulation line is 3 kA. The variation law of the maximum magnetic induction intensity on the surface of the drone inspection operation and the distance between the transmission wires is shown in Figure 8.

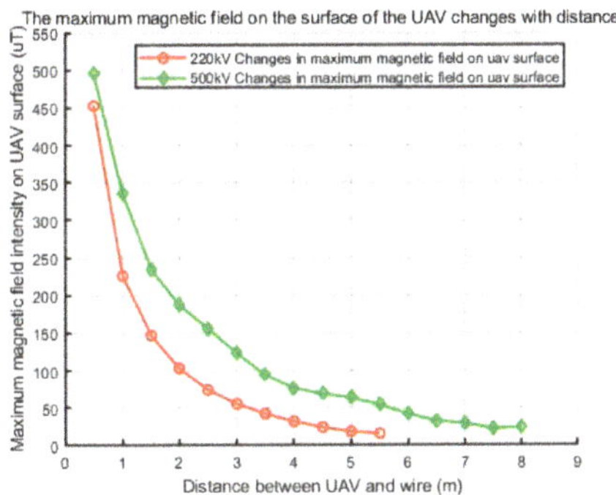

Figure 8. The variation of the surface magnetic field of the UAV with the distance.

It can be seen from the analysis of Figures 7 and 8 that the surface magnetic field distortion law of the UAV patrol overhead transmission line is similar to the electric field distribution. Taking the unmanned aerial vehicle in the same horizontal plane as the side-phase conductor as an example, the inspection model of the UAV in the complex electromagnetic field is established, and the rate of change of the magnetic field strength of the UAV surface is simulated and analyzed. The rate of change of the magnetic induction intensity on the surface of the UAV and the distance between the drone and the transmission line changes are shown in Figure 8. When the UAV patrols a 220 kV high-voltage overhead transmission line with a current intensity of 1.2 kA and a 500 kV voltage-grade high-voltage overhead transmission line with a current of 3 kA, the surface magnetic field changes with distance similar to the electric field. The magnetic field at the 500 kV voltage level has a greater impact on the drone, but its rate of change is similar. Generally, the overhead transmission line inspection has polarization effects and performance differences related to its components. In order to ensure the normal operation of the UAV, the magnetic field strength should be less than 220 µT. As can be seen from the analysis of Figure 8, the relationship between the minimum safe distance d and the transmission current I is [17]:

$$d = I. \tag{4}$$

2.3. Electromagnetic Field Experiment Results and Analysis

In the laboratory environment, a high-voltage transmission line simulation experiment platform was built to simulate and analyze the minimum safe distance of the electromagnetic interference caused by the inspection of the UAV. In the high-voltage laboratory, a 100 kV experiment transformer without local power-off frequency was built to generate a power-frequency electromagnetic field to conduct experimental research on the model of UAV being interfered by the electromagnetic field. According to the operation requirements, the UAV is fixed on the side of the transmission wire, and self-inspection is completed. Then, we gradually increase the voltage and current strength according to the experimental requirements, after which we observe the stability and controllability of the UAV. During the test, the model DJI-NAZA quadrotor experimental drone was used. The drone used for the experiment was 0.68 m long, 0.68 m wide, and 0.4 m high. It was made of carbon fiber and had good hover stability. The experimental steps are shown in Figure 9.

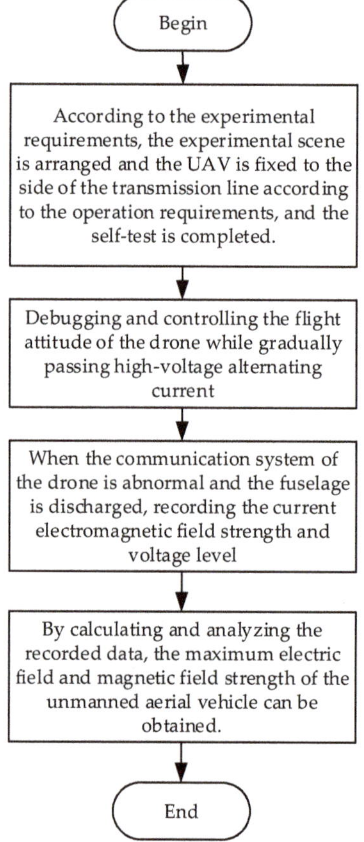

Figure 9. Experimental steps for a safe distance of UAVs against electromagnetic interference.

The experiment is completed according to the experimental steps shown in Figure 9, where the experimental layout equivalent diagram is shown in Figure 10, and the experimental scene diagram is shown in Figure 11. The UAV is fixed at a distance of 1 m from the conducting wire. During the experiment, the voltage control console is used to control the transformer to generate a high voltage of 0–80 kV. During the voltage regulation process, we observe the drone controllability and the transmission system stability while recording the voltage level and electric and magnetic field strength at each moment.

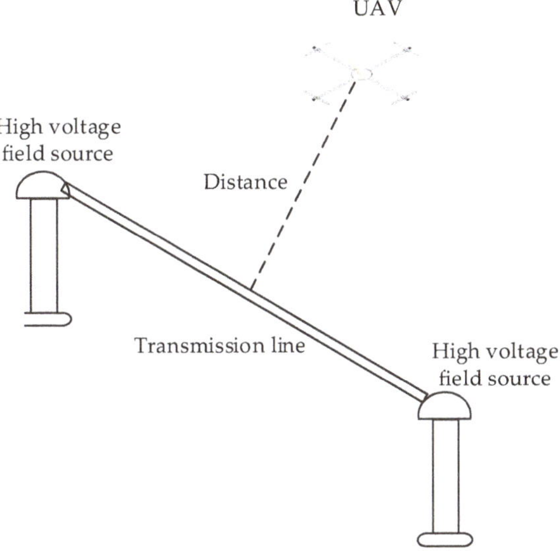

Figure 10. Schematic diagram of the electromagnetic field experiment.

Figure 11. Experimental scene diagram.

During the experiment, when the field strength around the UAV is 70–80 kV/m, the UAV platform starts to shake, and the phenomenon of delay occurs when the unmanned aerial vehicle is controlled by the ground control terminal. The image transmission has streaks, snowflake spots, and occasionally the black screen appears; the picture transmission is interrupted, and the UAV communication system and the control system are disturbed, which affects the stability of the data transmission system, and the drone cannot perform normal inspection operations. When the magnetic field strength is 200–210 μT, the UAV platform starts to shake, the magnetometer starts to be disturbed, and the drone slowly drifts toward the wire side; subsequently, abnormal handling performance and abnormal data transmission occur, and the inspection operation cannot be performed normally. In the actual implementation of the inspection operation, in order to ensure the safety inspection of the UAV, a certain safety margin should be left. Experimental research and simulation analysis show that in order to prevent the electromagnetic pulses generated between the UAV and the overhead wires from causing interference to the drone's communication and control system, and to avoid the electromagnetic field around the power line from affecting the normal inspection operations of the UAV. It is reasonable to set the safety threshold of the field strength to 50 kV/m and the magnetic field strength to 180 μT.

3. Adaptive Security Threshold Data Fusion Algorithm

3.1. Analysis of the Safe Distance of the UAV Patrol Power Line

In order to ensure the stability of the inspection of the unmanned aerial vehicle, in addition to the electronic components to meet certain anti-electromagnetic interference performance, the flight speed, wind speed, navigation positioning error, and the size of the drone structure and these factors have a great influence on the determination of the safe distance of the UAV line [20].

Assume that the positioning error of the navigation system of the patrol drone is D_1. In order to indicate the influence of the structure of the unmanned aerial vehicle on the determination of the safety distance, the length of the maximum diagonal of the lined drone is D_2. At the maximum wind speed of the allowed flight, the deviation due to the wind speed V_1 is D_3, the patrol speed of the drone is V_2, the communication time is recorded as t, The maximum electric field strength experienced by a UAV equipped with an electronic device during normal operation is E, and the maximum magnetic field strength experienced by the electronic device is B. The analysis in this paper is based on the D-S multi-source data fusion algorithm; the sensor data is collected and processed to give decision results in real time, and the adaptive adjustment of the safety distance threshold is realized. The block diagram of the decision system is shown in Figure 12.

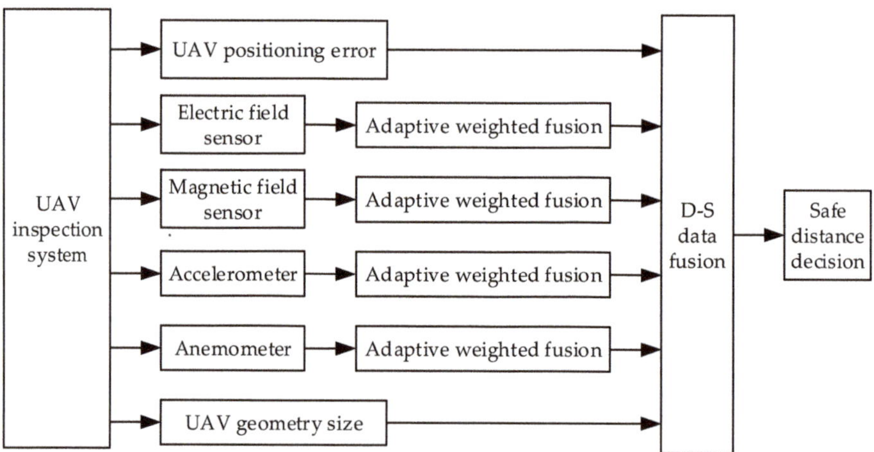

Figure 12. Block diagram of multi-sensor data fusion decision system.

3.2. Homogeneous Multi-Source Data Adaptive Weighted Fusion Algorithm

Each sensor can extract a part of the obstacle-avoiding navigation information, and the extracted part of the information cannot accurately reflect the accurate information of the target or can only extract a certain part of the target information due to interference of other factors. The state of the target object cannot be accurately determined. Multi-sensors are used to detect and fuse various navigation information during the inspection process, and different sensors are used to collect and detect changes of the external environment and extract meaningful information from the collected information. When different sensors distributed in different spatial positions of the unmanned aerial vehicle are used to measure obstacles such as transmission lines, towers, and electromagnetic field strength, the sensors will be different at different times and in different spaces to form a set of observations [26,27]. It is assumed that there are m sensors in the process of patrolling to measure the same variable. In a certain period of time, a single sensor obtains n measurement data, and the set is represented by Z as:

$$
Z = \begin{bmatrix}
Z_1(1) & Z_1(2) & \cdots & Z_1(n) \\
Z_2(1) & Z_2(2) & \cdots & Z_2(n) \\
\vdots & \vdots & \ddots & \vdots \\
Z_m(1) & Z_m(2) & \cdots & Z_m(n)
\end{bmatrix}
\tag{5}
$$

where $Z_m(n)$ represents the measured value of the *n-th* sensor at the *m-th* moment.

Suppose there is an estimated true value X. At the first moment, the estimated mean and variance of the X acquired by the same sensor are:

$$
\begin{cases}
Z_i^+ = Z_i(1) \\
P_i^+ = \sigma_i^2(1)
\end{cases}.
\tag{6}
$$

Based on the time recursive estimation theory, after the sensor collects the information, the estimated mean Z_i^+ and variance P_i^+ are used as the statistical characteristics of the next measurement, and the new measurement data is used to correct the one-time recursive calculation. The time fusion estimation value $Z_i^+(k)$ and variance $P_i^+(k)$ after k measurement of the same sensor can be obtained as follows:

$$
\begin{aligned}
Z_i^+(k) = {} & \frac{P_i^+(k-2) + \sigma_i^2(k)}{2[P_i^+(k-2) + P_i^+(k-1) + \sigma_i^2(k)]} Z_i^+(k-1) \\
& + \frac{P_i^+(k-1) + P_i^+(k-2)}{2[P_i^+(k-2) + P_i^+(k-1) + \sigma_i^2(k)]} Z_i^+(k)
\end{aligned}
\tag{7}
$$

$$
P_i^+(k) = \frac{P_i^+(k-1)\sigma_i^2(k)}{P_i^+(k-1) + \sigma_i^2(k)}
\tag{8}
$$

where the measurement variance can be obtained by calculating the auto-covariance factor and cross-covariance factor of a single sensor. For the *k-th* measurement, the estimation value of *m* sensors is $Z_1^+(k), Z_2^+(k), \ldots, Z_m^+(k)$; the mean square error is $P_1^+(k), P_2^+(k), \ldots, P_m^+(k)$, the weighting coefficient is $W_1(k), W_2(k), \ldots, W_m(k)$, and the optimal result of sensor space fusion is $\bar{Z}(k)$:

$$
\bar{Z}(k) = \sum_{i=0}^{m} W_i(k) Z_i^+(k), \quad \sum_{i=0}^{m} W_i(k) = 1.
\tag{9}
$$

Since the measured estimates $Z_1^+(k), Z_2^+(k), \ldots, Z_m^+(k)$ are independent of each other, the total mean square error of the *k-th* multi-sensor measurement spatiotemporal fusion is:

$$
\sigma^2(k) = E[(X - \bar{Z}(k))^2] = \sum_{i=0}^{m} W_i^2 P_i^+(k).
\tag{10}
$$

From the above formula, $\sigma^2(k)$ is a quadratic function of $W_i (i = 1, 2, \ldots, m)$; then, the minimum value of $\sigma^2(k)$ can be obtained by using the weighting coefficients $W_1(k), W_2(k), \ldots, W_m(k)$ to satisfy the constraint multivariate function extremum. The weighting factor of the total mean square error is shown in Equation (11):

$$W_i(k) = 1/(P_i^+(k)\sum_{j=1}^{m}\frac{1}{P_j^+(k)}), \ i = 1, \ 2, \ \ldots, \ m. \tag{11}$$

The corresponding total mean square error at this time is as shown in Formula (12):

$$\sigma^2_{min}(k) = 1/(\sum_{j=1}^{m}\frac{1}{P_j^+(k)}). \tag{12}$$

3.3. Evidence-Based Multi-Source Information Fusion Algorithm

Multi-source information fusion is based on the comprehensive processing of multiple kinds of information to accurately determine the state of the target object. D-S evidence theory is a reasoning method that was first proposed by Dempster and further developed by his student Shafer; it is also known as Dempster–Shafer evidence theory (D-S evidence theory). As a reasoning method, D-S (Dempster–Shafer) evidence theory can combine heterogeneous data information based on Bayes theory. It has outstanding features related to solving the problem of uncertainty, satisfies the conditions weaker than Bayes probability theory, and has the ability to directly express "uncertainty". The core idea is the combination rule of evidence theory. D-S (Dempster–Shafer) evidence theory is widely used in many fields such as medical diagnosis, target recognition, military command, etc., which need to comprehensively consider uncertain information from multiple sources [28]. The three basic elements of D-S (Dempster–Shafer) evidence theory are as follows: the basic probability assignment function m, the reliability function Bel(A), and the likelihood function pl(A). Firstly, the pre-processing of various sensor data is performed to solve the basic probability distribution function m, the belief function Bel(A), and the plausibility function Pl(A). Then, according to the combination rule, each value under the comprehensive evidence is solved. Finally, the reasonable judgment principle is adopted to take the maximum assumption of the belief function and the plausibility function as the final result of the data processing [29–31].

Let the set of all possible propositions in the universe be a recognition frame Θ. The basic probability distribution on the recognition frame Θ is a function m of $2^{\Theta} \rightarrow [0, 1]$, which satisfies:

$$m(\phi) = 0 \tag{13}$$

$$\sum_{A \subseteq \Theta} m(A) = 1 \tag{14}$$

where Φ represents an empty set; m (A) is a basic probability distribution function, which represents the degree of trust of the focus element A. The belief function Bel (A) and the plausibility function Pl (A) respectively represent the degree of support and possibility for the proposition A, and the expressions are expressed as follows.

Assuming $\forall A \subseteq 2^{\Theta}$, then:

$$Bel(A) = \sum_{B \subseteq A} m(B) \tag{15}$$

$$Pl(A) = \sum_{A \cap B \neq \phi} m(B). \tag{16}$$

The synthetic rules of D-S (Dempster–Shafer) evidence theory provide a synthetic formula that can fuse multiple evidence elements to provide evidence:

$$K = \sum_{A_i \cap B_j = \phi} m_1(A_i)m_2(B_j) < 1. \tag{17}$$

Then:

$$m(A) = \begin{cases} \dfrac{\sum\limits_{B_i \cap C_j} m_1(B_i)m_2(C_j)}{1-K} & A \neq \phi \\ U & A - \phi \end{cases}. \tag{18}$$

K is a collision factor, reflecting the degree of conflict of evidence, and $1/(K - 1)$ is called a normalization factor. This combination rule is equivalent to assigning an empty set (conflict) to each set in the combination.

In view of the actual working conditions of the UAV while inspecting an overhead transmission line, the various influencing factors of the safety distance are mutually influential, and one of the factors will cause the other factors to change. For example, the inspection speed of the drone will affect the measurement of the relative wind speed and affect the detection of the electric field and magnetic field strength. When the evidence conflict is more serious, the D-S (Dempster–Shafer) algorithm cannot focus according to different weights. Some identification frameworks may have a probability of 0. The combined result will often have a large error with the actual situation, which makes the system unable to make a more accurate judgment on the safe distance of the inspection. Therefore, this paper improves the D-S (Dempster–Shafer) evidence theory synthesis Formula (18), and uses the improved evidence theory synthesis formula as the global fusion algorithm of each environmental factor information data. Assuming the degree of conflict between evidences in the formula is represented by the parameter K, the evidence trust degree is represented by the parameter ε, and the evidence average trust degree function is represented by $q(A)$, then:

$$m(\phi) = 0 \tag{19}$$

$$m(A) = p(A) + K\varepsilon q(A) \tag{20}$$

$$p(A) = \sum_{B_i \cap C_j} m_1(B_i)m_2(C_j) \tag{21}$$

$$q(A) = \frac{1}{n}\sum_{i=1}^{n} m_i(A) \tag{22}$$

$$\varepsilon = e^{-s} \tag{23}$$

$$s = \frac{K}{n(n-1)/2} \tag{24}$$

where s is the sum of each pair of evidence sets in n evidence sets, which reflects the degree of conflict between the two sides of the evidence. Meanwhile, ε is a decreasing function of s, reflecting the credibility of the evidence; when the conflict between the evidences increases, the degree of trust of the evidence will decrease; s is different from K in the D-S (Dempster–Shafer) theory. K can reflect the overall degree of conflict of evidence. When K increases, s does not necessarily increase.

In the D-S (Dempster–Shafer) decision theory to solve the decision process based on multi-objective fusion, the evidence acquisition is very difficult. It embodies the scientific rationality of constructing the basic probability distribution function, and whether the allocation of basic probability numbers is reasonable directly determines whether the conclusions after the combination of evidence are credible [32,33]. The traditional solution is to ask for help in the field, but the credibility of the information given by the authoritativeness of the experts has a certain deviation. In order to avoid unnecessary deviation caused by human subjectivity, this paper uses the typical sample-based method to automatically perform basic probability distribution. This paper also uses the distance between the measured value and the typical value to construct the density function of the normal distribution curve, and then constructs the basic probability distribution function. In this paper, the BPA (Basic Probability Assignment) is obtained by constructing a normal curve from the Hamming distance between each evidence and a typical sample. Based on the above analysis, the specific steps of implementing the adaptive safety threshold algorithm for the overhead transmission line of the UAV are as follows.

It is assumed that there are m sensors in the process of patrolling to measure the same variable. In a certain period of time, a single sensor obtains n measurement data, and the set is represented by Z as:

$$Z = \{Z_i(k)\}(i = 1, 2, \ldots, m; k = 1, 2, \ldots, n). \tag{25}$$

Then, we perform the first-level fusion of the data measured by the same variable according to the homogenous multi-source data adaptive weighting fusion algorithm described in the paper, and obtain the estimation result after the fusion of each parameter variable.

Assuming that the number of target patterns is p, the number of evidence is l, and the construction recognition framework is $\Theta = \{X_1, X_2, \cdots, X_p\}$, in this framework, the typical value of each target pattern X_j on each evidence is $\{Z_{1j}, Z_{2j}, \cdots, Z_{lj}\}$ ($j = 1, 2, \cdots, p$), and the results of the adaptive weighted fusion of the measured values of the evidences are $\{Z_1, Z_2, \ldots, Z_l\}$.

The average Hamming distance between them is solved based on the measured fusion results and typical sample values:

$$H_{ij} = \left|Z_i - Z_{ij}\right|, \ H_i = \frac{1}{p}\sum_{j=1}^{p} H_{ij}. \tag{26}$$

Using the normal curve to find the similarity between the evidence Z_i and the corresponding typical value Z_{ij} in the target pattern X_j:

$$P_{ij} = e^{-Ak^2} \tag{27}$$

where

$$A = 0.5, \ k = H_{ij}/H_i. \tag{28}$$

The basic probability of evidence Z_i and the corresponding target mode X_j is:

$$M_i(X_j) = P_{ij} \Big/ \sum_{j=1}^{p} P_{ij}. \tag{29}$$

According to the D-S (Dempster–Shafer) evidence fusion algorithm, the target mode A basic probability distribution function $m(A)$ can be calculated and solved, and the decision is made based on this.

4. Results

Data Fusion Algorithm Verification and Analysis

In order to detect the safety distance of the inspection UAV according to the real-time data of each sensor under the current unmanned aerial vehicle inspection condition, judging the security status of the current location according to the matching of the data collected by various heterogeneous sensors and the evaluation rules, the patrol UAV can conduct an autonomous obstacle avoidance navigation control according to the decision result, or transmit the decision result to the ground terminal. Based on the decision result, the operator can make accurate judgment on the real working condition of the patrol inspection of the UAV and carry out a remote control of it. When performing data fusion of heterogeneous sensors, the identification framework should be constructed reasonably. The identification framework constructed in this paper is assumed to be:

$$\Theta = \{A, B, C\} \tag{30}$$

where A = {Farther away}, B = {Suitable distance}, and C = {Dangerous distance}.

According to a large number of experiments and principle analysis, and using the expert knowledge to give the basic probability assignment, the attribute interval of each detection condition is set as shown in Table 1 below.

Table 1. Conditional attribute discrete interval setting.

Conditional Attribute	Electric Field Strength (kV/m)	Magnetic Field Strength (μT)	Wind Speed (m/s)	Navigation Error (m)	Inspection Speed (m/s)	UAV Size (m)
Farther away	(0, 50)	(0, 180)	(0, 3)	(0, 1.5)	(3, 5)	(~, 1.5)
Suitable distance	(50, 150)	(180, 225)	(3, 8)	(1.5, 3)	(5, 10)	(1.5, 3)
Dangerous distance	(150, ~)	(225, ~)	(8, ~)	(3, ~)	(10, ~)	(3, ~)

In the experiment, four power frequency electric field sensors are used to collect the electric field intensity signals around the transmission lines. Four magnetic field sensors collect the magnetic field strength signals around the transmission lines, and the accelerometer is used to measure the speed of the UAV. The measurement of wind speed and direction is realized by the anemometer, and data fusion of various sensors such as the structural size parameters of the drone and the positioning error is used to realize the decision of the safety distance.

First, we select a set of data collected during the inspection process of a drone as the verification of the data fusion algorithm. Then, we select the data collected by each sensor within a certain period of time, according to the adaptive weighted fusion method described in the paper, by firstly merging the same sensor in time, and then adapting the fused data result between homogenous sensors. The result of the weighted fusion data is used as the detection data of this type of sensor. In a certain period of time, the data of each type of sensor after adaptive weighted fusion is selected as the D-S evidence theory algorithm to verify the UAV patrol overhead transmission line conditions.

In this paper, based on the typical sample method, the density function of normal distribution is constructed by elastic Hamming distance [26], and then the basic probability distribution function is constructed. The basic probability assignment of each influencing factor obtained through the processing of the data is shown in Table 2.

Table 2. Basic probability assignment assignment table.

	A	B	C
Electric field strength	0.752	0.203	0.045
Magnetic field strength	0.876	0.113	0.011
Inspection speed	0.658	0.241	0.101
Wind speed	0.381	0.528	0.091
Positioning error	0.274	0.456	0.27
UAV size	0.525	0.373	0.102

The evidences of electric field strength, magnetic field strength, wind speed, inspection speed, and UAV positioning error are m_1, m_2, m_3, m_4, and m_5, and the value of the obtainable probability assignment function *m(A)* is calculated according to the DS fusion algorithm. The *m(A)* obtained by fusing each proposition case is shown in the following table.

From the analysis of the fusion results and propositional hypotheses in Table 3, we can see that the UAV inspection status is proposition A; that is, the UAV is far away from the transmission line, and the inspection sensor carried by the UAV is not optimal. The inspection effect should control the drone to approach the field source to achieve a proper inspection distance, which can ensure the safety of the inspection of the drone, and also enable the detection sensor to reach the optimal detection state.

Table 3. Fusion results under various propositional conditions.

Proposition Hypothesis	A	B	C
Fusion result	0.97	0.025	0.005

At the same time, for this theory, we take the 220 kV and 500 kV high-voltage AC transmission lines of the patrol line unmanned aerial vehicle inspection as an example to analyze the safety distance of the inspection. In the case where all error parameters take the maximum value, the positioning error of the navigation system of the UAV inspection system is set to 2.5 m; the maximum wind speed of the patrol UAV that allows the drone to operate during wind-resistant flight under the normal inspection and the standard deviation allowed is 1.5 m. Therefore, the deviation due to the wind speed is 1.5 m, and the distance between the unmanned aerial vehicle and the wire is 3 m under the condition of the AC 220 KV voltage level and the maximum electric field strength required for the normal operation of the electronic equipment of the UAV. When the electronic device is working normally under the maximum magnetic field strength required, the distance between the drone and the wire is 1.2 m; under the condition of an AC 500 KV voltage level and the maximum electric field strength required for the normal operation of the electronic equipment of the drone, the distance between the UAV and the wire is 5 m.

Under the condition of the maximum magnetic field strength required for normal operation of electronic equipment, the distance between the UAV and the conductor is 3 m. In this model, the maximum diagonal length of the unmanned aerial vehicle (taking into account the length of the wing) is 1 m. Here, we assume that the patrol unmanned aerial vehicle speed of the aircraft is v = 5 m/s, and the communication time between the UAV and the ground is t = 0.2 s. Therefore, the safety inspection distance of patrol unmanned aerial vehicle inspection at the 220 kV voltage level is about 8 m; at the 500 kV voltage level, the safety inspection distance of the inspection line drone is about 11 m. However, it is less likely that the error parameters will take the maximum value at the same time during the actual inspection. It is unreasonable to judge the safety status of the high-voltage overhead transmission line of the UAV with a fixed safety threshold as a standard. This method adapts to a single working condition and does not achieve the optimal inspection effect.

The safety distance threshold is different under different environmental factors. If the decision is made with a fixed threshold, in some cases, the safety distance threshold is too close to cause the inspection accident, while in other cases, the fixed safety threshold is larger. The inspection equipment carried by the drone can not exert the best test performance, so that the situation of missed detection and misjudgment will not be achieved, and the expected inspection effect will not be achieved. The multi-sensor data fusion algorithm is used to perform multi-source data fusion on each sensor data and intelligently determine the current safety distance threshold to avoid problems caused by a single safety distance threshold. According to the analysis of the fusion results, the probability assignment of the corresponding propositions is more concentrated after D-S (Dempster–Shafer) evidence theory fusion, and the credibility distinction is more obvious. The adaptive security threshold algorithm has good rationality and scientificity.

5. Discussion and Conclusions

(1) In this paper, we establish a physical model of the UAV in the complex electromagnetic field environment, combined with the anti-electromagnetic interference capability and the influence of the electric field distortion of the electronic components that constitute the UAV inspection system. Finally, the electric field strength of the UAV safety inspection is 50 kV/m, and the magnetic field strength is 180 μT. The comprehensive analysis of the common voltage level of the UAV inspection shows that the relationship between the UAV safety inspection distance d (m) and the through current I (kA) in the wire is approximately:

$$d = I. \tag{31}$$

(2) Analysis of the simulation results shows that the electric field distribution on the surface of the UAV is not uniform. The field strength distribution at the tip of the rotating motor and the wing is concentrated, and the electric field distortion is large. Therefore, in the design of the UAV, the installation of the electric field measuring device should be as far as possible. Here, we avoid burrs or tips to prevent electric field distortion from affecting the measurement and determination of safety distance.

(3) This paper adopts a multi-sensor data fusion decision-making method to the data fusion of the main factors affecting UAV flight speed, wind speed, navigation positioning error, and structural size. According to the fusion result and the set proposition, the UAV inspection status is determined, and the UAV is adaptively adjusted according to the inspection operation environment. Then, the algorithm is verified. According to the analysis of the fusion results, the reliability of the multi-sensor adaptive fusion is more obvious, which better solves the single safety threshold adaptation condition and brings about a large cruise error due to the field source positioning error.

By studying the safety distance of overhead transmission lines for unmanned aerial vehicles, it can provide reference for improving the safety and reliability of UAV power inspection operations, and ensure the safe and stable operation of UAVs, power grids, and inspection equipment. The important significance provides a theoretical basis for the next step of studying with UAV.

Author Contributions: Conceptualization, W.Z., Y.N., and C.S.; Methodology, W.Z. and Y.N.; Formal Analysis, C.S.; Data Curation, Y.N.; Writing—Review and Editing, Y.N.; Project Administration, C.S.

Funding: This research was funded by the China Southern Power Grid Corporation Science and Technology Project, grant number (YNKJXM20170180).

Acknowledgments: Authors would also like to thank China Southern Power Grid for project support.

Conflicts of Interest: The authors declare no conflict of interest.

References

1. Luque-Vega, L.F.; Castillo-Toledo, B.; Loukianov, A. Power line inspection via an unmanned aerial system based on the quadrotor helicopter. In Proceedings of the MELECON 2014–2014 17th IEEE Mediterranean Electrotechnical Conference, Beirut, Lebanon, 13–16 April 2014.
2. Lu, J.; Huang, Q.; Mao, X.; Tan, Y.; Zhu, S.; Zhu, Y. Optimized Design of Modular Multilevel DC De-Icer for High Voltage Transmission Lines. *Electronics* **2018**, *7*, 204. [CrossRef]
3. Qayyum, A.; Malik, A.S.; Saad, N.M.; Abdullah, M.F.B.; Jaafar, M.Y.H.; Abdullah, A.R.B.A.; Rasheed, W.; Iqbal, M. Measuring height of high-voltage transmission poles using unmanned aerial vehicle (uav) imagery. *Imaging Sci. J.* **2017**, *65*, 1–14. [CrossRef]
4. Zhou, Y.X.; Liu, R.; Zhang, Y.X.; Zhang, X.; Guo, D.W. Corresponding Relation Between Aging Micrographs and Partial Discharge Properties of Electrical Trees in Silicone Rubber. *High Volt. Eng.* **2015**, *41*, 159–166.
5. Zhang, X.Q.; Su, J.J. *UAV Inspection Technology of Overhead Transmission Line*; China Electric Power Press: Beijing, China, 2016.
6. Azevedo, F.; Dias, A.; Almeida, J.; Oliveira, A.; Ferreira, A.; Santos, T.; Martins, A.; Silva, E. LiDAR-Based Real-Time Detection and Modeling of Power Lines for Unmanned Aerial Vehicles. *Sensors* **2019**, *19*, 1812. [CrossRef] [PubMed]

7. Nuriev, M.G.; Gizatullin, Z.M.; Gizatullin, R.M. Physical modeling of electromagnetic interferences in the unmanned aerial vehicle in the case of high-voltage transmission line impact. *Russ. Aeronaut.* **2017**, *60*, 292–298. [CrossRef]

8. Bhola, R.; Krishna, N.H.; Ramesh, K.N. Detection of the power lines in UAV remote sensed images using spectral-spatial methods. *J. Environ. Manag.* **2017**, *206*, 1233–1242. [CrossRef]

9. Qin, X.Y.; Wu, G.P.; Lei, J.; Fan, F.; Ye, X.H.; Mei, Q.J. A Novel Method of Autonomous Inspection for Transmission Line based on Cable Inspection Robot LiDAR Data. *Sensors* **2018**, *18*, 596. [CrossRef]

10. Song, B.; Li, X. Power line detection from optical images. *Neurocomputing* **2014**, *129*, 350–361. [CrossRef]

11. Chen, C.; Peng, X.Y.; Song, S.; Wang, K.; Qian, J.J.; Yang, B.S. Safety Distance Diagnosis of Large Scale Transmission Line Corridor Inspection Based on LiDAR Point Cloud Collected With UAV. *Power Syst. Technol.* **2017**, *41*, 2746–2753.

12. Tian, F.; Wang, Y.; Zhu, L. Power line recognition and tracking method for UAVs inspection. In Proceedings of the 2015 IEEE International Conference on Information and Automation (ICIA), Lijiang, China, 8–10 August 2015.

13. Matikainen, L.; Lehtomäki, M.; Ahokas, E.; Hyyppä, J.; Karjalainen, M.; Jaakkola, A.; Heinonen, T. Remote sensing methods for power line corridor surveys. *ISPRS J. Photogramm. Remote Sens.* **2016**, *119*, 10–31. [CrossRef]

14. Larrauri, J.I.; Sorrosal, G.; González, M. Automatic system for overhead power line inspection using an Unmanned Aerial Vehicle. In Proceedings of the International Conference on Unmanned Aircraft Systems, Atlanta, GA, USA, 28–31 May 2013.

15. Zheng, T.R.; Sun, L.M.; Lou, T.T.; Guo, X.; Liu, Q.H. Determination Method of Safe Flight Area for UAV Inspection for Transmission Line Based on the Electromagnetic Field Calculation. *Shandong Electr. Power* **2018**, *45*, 27–30.

16. Zhang, J.; Wang, S.Y.; Chen, X.; Xu, H.D.; Yu, D.K.; Yang, Z. Obstacle Avoidance for UAV Power Line Inspection Based on Cylindrical Space and Support Vector Machine. *Electr. Power* **2015**, *66*, 56–60.

17. Liu, Z.; Du, Y.; Chen, Y.; Ma, J.G.; Wu, X.D.; Yao, J.S. Simulation and Experiment on the Safety Distance of Typical µ500 kV DC Transmission Lines and Towers for UAV Inspection Gaodianya Jishu. *High Volt. Eng.* **2019**, *45*, 426–432.

18. Shen, J.; Liu, W.D.; Liu, H.B.; Wang, J.; Wu, Y. Safety distances of inspections on AC UHV lines of 1000 kV with a helicopter. *Electr. Power* **2011**, *44*, 41–45.

19. Wu, X.; Wan, B.Q. *Electromagnetic Environment of Transmission and Transformation Engineering*; China Electric Power Press: Beijing, China, 2009.

20. Dong, X.C.; Wu, Z.K.; Chen, Z.M.; Qu, F.R.; Li, Y.F.; Liu, G.; Zhu, N.X. Analysis of induction discharge distance between uav and overhead lines. *Adv. Technol. Electr. Eng. Energy* **2018**, *37*, P75–P81.

21. Zhang, X.D.; Hao, R.R.; Liu, S.S. *Protection Circuit against Strong Electromagnetic Pulse Interference*; Southeast University: Nanjing, China, 2010.

22. Qiu, Z.B.; Yan, J.J.; Xu, W.J.; Hang, C.P. Hybrid Prediction of the Breakdown Voltages of Short Air Gaps with Typical Electrodes. *High Volt. Eng.* **2018**, *44*, 2012–2018.

23. Peng, X.T.; Qian, J.J.; Wang, K.; Mai, X.M.; Yi, L. Multi-sensor Full-automatic Inspection System for Large Unmanned Helicopter and Its Application in 500 kV Lines. *Guangdong Electr. Power* **2016**, *29*, 8–15.

24. Silva, J.M.; Whitney, B. Evaluation of the Potential for Power Line Carrier (PLC) to Interfere with Use of the Nationwide Differential GPS Network. *IEEE Power Eng. Rev.* **2002**, *22*, 62.

25. Liu, X.F.; Yin, H.; Wu, X. Test and Analysis on Effect of High Voltage Transmission Lines Corona Radio Interference and Scattering to GPS Signal. *High Volt. Eng.* **2011**, *37*, 2937–2944.

26. Guo, L.; Ma, Y.H.; Zhang, X.E. Spatial-temporal estimation algorithm for multisensor data fusion. *Syst. Eng. Electron.* **2005**, *27*, 2016–2018.

27. Zhai, Y.L.; Dai, Y.S. Study of Adaptive Weighted Fusion Estimated Algorithm of Multi-sensor Data. *Acta Metrol. Sin.* **1998**, *19*, 69–75.

28. Yuan, M.; Wan, C.L.; Wei, L.S. Superiority of empirical Bayes estimator of the mean vector in multivariate normal distribution. *Sci. China Math.* **2016**, *59*, 1175–1186. [CrossRef]

29. Shafer, G. A mathematical theory of evidence. *Technometrics* **1978**, *20*, 242.

30. Xinyang, D.; Wen, J.; Zhen, W. Zero-sum polymatrix games with link uncertainty: A Dempster-Shafer theory solution. *Appl. Math. Comput.* **2019**, *340*, 101–112.

31. Shao, L. The Multi-Sensor Data Fusion Technology Research in Animal Building Environmental Monitoring System. Master's Thesis, Hebei Agriculture University, Baoding, China, 2013.
32. Wang, J.L.; Zhang, J.Y. Multisensor Target Identification Based on Mass Function of Statistical Evidence and D-S Evidence Theory. *Chin. J. Sens. Actuators* **2006**, *19*, 862–864.
33. Luan, C.J.; Wang, X.F.; Zhang, H.Z.; Jin, S.S.; Li, F. Approach for constructing basic probability assignment for evidence theory in problems of ordered propositions. *Comput. Eng. Appl.* **2012**, *48*, 217–221.

Article

Cascade Second Order Sliding Mode Control for Permanent Magnet Synchronous Motor Drive

Adel Merabet

Division of Engineering, Saint Mary's University, Halifax, NS B3H 3C3, Canada; adel.merabet@smu.ca;
Tel.: +1-902-420-5712

Received: 1 November 2019; Accepted: 5 December 2019; Published: 9 December 2019

Abstract: This paper presents a cascade second-order sliding mode control scheme applied to a permanent magnet synchronous motor for speed tracking applications. The control system is comprised of two control loops for the speed and the armature current control, where the command of the speed controller (outer loop) is the reference of the q-current controller (inner loop) that forms the cascade structure. The sliding mode control algorithm is based on a single input-output state space model and a second order control structure. The proposed cascade second order sliding mode control approach is validated on an experimental permanent magnet synchronous motor drive. Experimental results are provided to validate the effectiveness of the proposed control strategy with respect to speed and current control. Moreover, the robustness of the second-order sliding mode controller is guaranteed in terms of unknown disturbances and parametric and modeling uncertainties.

Keywords: permanent magnet synchronous motor; second-order sliding mode control; cascade control; robustness

---

1. Introduction

Permanent magnet synchronous motors (PMSM) have been broadly deployed in variable speed drives due to their efficiency, high power density and fast dynamics. However, the control efficiency of PMSM drives is affected by parametric inaccuracies, load disturbances and uncertainties in real applications. Furthermore, they are nonlinear multivariable systems with modeling uncertainties and parameters changing during operation.

High performance control of PMSMs is difficult to attain, using model-based methods, due to the above drawbacks. Therefore, various robust and advanced control methods have been conducted to improve the PMSM efficiency. In [1], an adaptive speed regulator was designed with no information about the PMSM parameters, the load and the torque values. Another adaptive control method was investigated in [2] through adapting the feedforward compensation gain with respect to the identified inertia and an extended state observer was used to estimate both the states and the disturbances simultaneously. A disturbance torque estimation and nonlinear control was developed in [3] and predictive control strategy in [4]. Similar to other works, tracking performance and robustness were performed through disturbance estimation, which alleviate the computation burden of such controllers [5,6].

High order sliding mode control (SMC) has emerged as an efficient control method due to its insensitivity with respect to system inaccuracies and unknown disturbances. Furthermore, it has advantages over the traditional SMC to overcome the chattering problems [7,8]. Different second-order SMC strategies have been developed for PMSMs. In [9] and [10], SMC-based disturbance observer, for uncertainties compensation, was proposed to reduce chattering and designed for the speed control loop. In [11], a second order SMC was presented as a direct torque and flux controller for PMSM without chattering. In [12], a second order SMC algorithm, using proportional-plus-integral sliding plane, was developed to improve the tracking and robustness performance compared to conventional first order SMC. In [13], a second order SMC was designed using an integral manifold for the speed control from the mechanical equation of the permanent magnet synchronous generator-based wind energy system. Other strategies, related to SMC applied to deferent drives and systems, are available in the literature [14–18]. In [19], a feedback linearization was combined with optimization to deal with parametric changes and applied to a pendulum like a robot. In [20], a standard SMC scheme was derived from the SISO case and used as a decoupled input-to-flat-output model for a robot manipulator with good performance. Despite the good performance of such control strategies, it is difficult to assess the behavior of SMC strategies when dealing with both mechanical and electrical models of the PMSM system. In general, the SMC method is applied to the speed equation of the machine and the conventional vector control, PI-based method, which is used for the electrical equation for the current control loop due to the computational burden and implementation limitation when applying advanced controllers [13]. This matter will be explored in this work by considering both mechanical and electrical systems to develop a cascade structure of an advanced control strategy and using the powerful Opal-RT real time experimental system for control implementation. Furthermore, this work deals with a cascade structure and interaction between different SMC controllers for multivariable control, which will help in understanding the implementation issues of such controller in the area of multivariable control. The second-order SMC scheme provides higher capabilities for dealing with nonlinearities and uncertainties compared with conventional control methods, such as PI control and linearization control, as changes in the system do not require tuning of its parameters or adding mechanisms to deal with parametric uncertainties.

In this paper, a second-order SMC is designed for multivariable control, in a cascade scheme, of PMSM drive. The output tracking error dynamic, for rotational speed and d-q components of the current, is modelled under a single input-output state space model and decoupled by considering the nonlinear terms in new variable inputs. Furthermore, the unknown load torque effect is rejected by the proposed second-order SMC through integral action in the controller without the need for observing the load torque or identifying the inertia seen by the rotor shaft. In general, any other uncertainties in the mechanical and electrical systems of the PMSM, such as unmodeled quantities, parameters' variations and external disturbance, will be compensated by the proposed control scheme, which constitutes the contribution of this work compared to [9–13]. The proposed controller for speed and current tracking is robust against parametric variations and uncertainties in the system.

2. PMSM Model

The PMSM model, in the d-q rotating reference frame, is described by

$$\frac{d}{dt}\begin{bmatrix} i_{sd} \\ i_{sq} \end{bmatrix} = \begin{bmatrix} -\frac{R}{L_d} & \frac{L_q}{L_d}p\omega_r \\ -\frac{L_d}{L_q}p\omega_r & -\frac{R}{L_q} \end{bmatrix}\begin{bmatrix} i_{sd} \\ i_{sq} \end{bmatrix} + \begin{bmatrix} \frac{1}{L_d} & 0 \\ 0 & \frac{1}{L_q} \end{bmatrix}\begin{bmatrix} v_{sd} \\ v_{sq} - \varphi_v p\omega_r \end{bmatrix} \tag{1a}$$

$$\frac{d\omega_r}{dt} = \frac{1}{J}\left(p\left(\varphi_v i_{sq} + \left(L_d - L_q\right)i_{sd}i_{sq}\right) - f\omega_r - T_L\right) \tag{1b}$$

where, i_{sd} is the d-axis of the stator current; i_{sq} is the q-axis of the stator current; v_{sd} is the d-axis control voltage; v_{sq} is the q-axis control voltage; φ_v is the permanent magnet magnetic flux linkage; p is the poles pairs number; R is the stator resistance; L_d is the d-axis of the inductance of the stator winding; L_q is the q-axis of the inductance of the stator winding; ω_r is the rotational speed; J is the moment of inertia; f is the viscous friction coefficient; and T_L is the load torque.

The variables to be controlled are the rotor speed ω_r and the d-axis of the stator current i_{sd} to track external references and the q-axis of the stator current i_{sq} to track a reference produced by the outer loop of speed control. Furthermore, the load torque is an unknown disturbance and will be compensated by the proposed controller.

3. Cascade Second-Order SMC

3.1. Control Developement

The state space model, based on single input single output (SISO), has the following expression

$$\dot{x}(t) = -Ax(t) + Bv(t) + d(t) \tag{2}$$

where, x is the state variable, v is the input variable and d is an unknown bounded disturbance.

The control objective is to ensure a zero steady state error such as

$$e(t) = x_{\text{ref}}(t) - x(t) = 0 \tag{3}$$

where, x_{ref} is the reference, and e is the tracking error.

The tracking error dynamic is carried out using (2) and (3) and expressed by

$$\dot{e} = \dot{x}_{\text{ref}} - \dot{x} = -Ae + u - d \tag{4}$$

where, u is the new control input.

The new control input is expressed by

$$u = -Bv + Ax_{\text{ref}} + \dot{x}_{\text{ref}} \tag{5}$$

Then, let us consider the second-order SMC for perturbation elimination [8,18]

$$\begin{cases} u = -k_1|e|^{\frac{1}{2}}\text{sgn}(e) + w \\ \dot{w} = -k_2\text{sgn}(e) \end{cases} \tag{6}$$

where, k_1 and k_2 are positive constants, and sgn is the switching function

$$\text{sgn}(x) = \begin{cases} 1 & x > 0 \\ 0 & x = 0 \\ -1 & x < 0 \end{cases} \tag{7}$$

Finally, from (5) and (6), the command v in (2) for regulating x to track the reference x_{ref} is provided by

$$\begin{cases} v = \frac{1}{B}\left[Ax_{\text{ref}} + \dot{x}_{\text{ref}} + k_1|e|^{\frac{1}{2}}\text{sgn}(e) - w\right] \\ \dot{w} = -k_2\text{sgn}(e) \end{cases} \tag{8}$$

The stability of the second order SMC (8) applied to (4) to ensure $e(t)\rightarrow 0$ can be demonstrated in a similar manner as in [8].

The advantage of the second-order SMC compared to the traditional SMC is that it takes into consideration the nonlinearities and uncertainties with better performance as it includes an integral action. It is well known that an integral action enhances the control efficiency in dealing with uncertainties.

3.2. Cascade Second Order SMC Structure

The cascade control scheme, depicted in Figure 1, includes two loops for the control systems such as: (1) the outer loop control for speed regulation provides the q-axis of the current reference applied to the inner control loop. (2) The inner loop control for current regulation that provides the voltage commands to generate the six pulses to control the DC-AC converter.

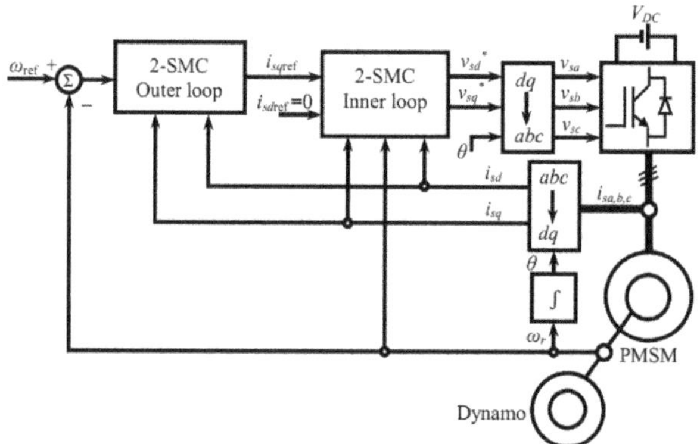

Figure 1. Block diagram of the proposed cascade second order SMC for a PMSM.

3.2.1. Outer Loop Control

The second-order SMC is developed using the motion equation. The control system provides the q-axis current as a reference for the inner loop in order to track the desired speed reference.

The motion Equation (1b) is expressed under the form (2) as

$$\dot{\omega}_r(t) = -A_\omega \omega_r(t) + B_\omega i_{sq}(t) + d_\omega(t) \tag{9}$$

where, $A_\omega = \frac{f}{J}$, $B_\omega = \frac{p}{J}\left(\varphi_v + \left(L_d - L_q\right)i_{sd}\right)$, $d_\omega = -\frac{T_L}{J}$.

The speed tracking error dynamic is expressed by

$$\dot{e}_\omega = -A_\omega e_\omega + u_\omega - d_\omega \tag{10}$$

where, $e_\omega = \omega_{\text{ref}} - \omega_r$ is the speed tracking error, and the new control input u_ω is provided by

$$u_\omega = -B_\omega i_{sq} + A_\omega \omega_{\text{ref}} + \dot{\omega}_{\text{ref}} \tag{11}$$

Therefore, the q-axis current reference based on the 2-SMC law (8) is expressed as

$$\begin{cases} i_{sqref} = \frac{1}{B_\omega}\left[A_\omega \omega_{\text{ref}} + \dot{\omega}_{\text{ref}} + k_1 |e|^{\frac{1}{2}} \text{sgn}(e_\omega) - w_\omega\right] \\ \dot{w}_\omega = -k_2 \text{sgn}(e_\omega) \end{cases} \tag{12}$$

3.2.2. Inner Loop Control

The sliding mode control is applied to the electrical model in order to provide the components of the stator voltage which minimizes the difference between the stator current components (i_{sd}, i_{sq}) and the current references (i_{sdref}, i_{sqref}), respectively.

The electrical equation (1.a) is expressed, under the form (2), as

$$\begin{cases} di_{sd}(t)/dt = -A_d i_{sd}(t) + B_v v_d(t) + d_d(t) \\ di_{sq}(t)/dt = -A_q i_{sq}(t) + B_q v_q(t) + d_q(t) \end{cases} \tag{13}$$

where, $A_d = \frac{R}{L_d}$, $B_d = \frac{1}{L_d}$, $A_q = \frac{R}{L_q}$, $B_q = \frac{1}{L_q}$.

The input variables (v_d, v_q) represent the decoupling components to linearize the system (1a) and are expressed as

$$\begin{cases} v_d = L_q p \omega_r i_{sq} + v_{sd} \\ v_q = -L_d p \omega_r i_{sd} - \varphi_v p \omega_r + v_{sq} \end{cases} \tag{14}$$

The current tracking error dynamics are expressed by

$$\begin{cases} \dot{e}_d = -A_d e_d + u_d - d_d \\ \dot{e}_q = -A_q e_q + u_q - d_q \end{cases} \tag{15}$$

where, $e_d = i_{sdref} - i_{sd}$ and $e_q = i_{sqref} - i_{sq}$ are the (d, q) current tracking errors.

The new control inputs are given by

$$\begin{cases} u_d = -B_d v_d + A_d i_{sdref} + di_{sdref}/dt \\ u_q = -B_q v_q + A_q i_{sqref} + di_{sqref}/dt \end{cases} \tag{16}$$

The second-order SMC for d-q current control is carried out from (14)–(16), using the control design (5)–(8), and the voltage command components are given by

$$\begin{cases} v_{sd} = L_q p \omega_r i_{sq} + L_d \left[A_d i_{sdref} + di_{sdref}/dt + k_1 |e_d|^{\frac{1}{2}} \text{sgn}(e_d) - w_d \right] \\ \dot{w}_d = -k_2 \text{sgn}(e_d) \end{cases} \tag{17a}$$

$$\begin{cases} v_{sq} = L_d p \omega_r i_{sd} + L_q \left[A_q i_{sqref} + di_{sqref}/dt + k_1 |e_q|^{\frac{1}{2}} \text{sgn}(e_q) - w_q \right] \\ \dot{w}_q = -k_2 \text{sgn}(e_q) \end{cases} \tag{17b}$$

3.3. Chattering Reduction

The reduction of the chattering phenomenon in the control laws (12) and (17), the sgn function is modified such as

$$\text{sgn}(e) \rightarrow \text{sat}\left(\frac{e}{\alpha}\right) = \begin{cases} \text{sgn}(e) & \text{if } |e| \geq \alpha \\ \frac{e}{\alpha} & \text{if } |e| < \alpha \end{cases} \tag{18}$$

where, $\alpha > 0$ represents the boundary layer thickness.

4. Robustness to Uncertainties

The PMSM dynamics are uncertain and include uncertainties such as unmodeled quantities, parametric variations, and unknown and external disturbances. Therefore, the control design must be robust to compensate all these uncertainties and enhance the tracking performance.

The PMSM limitations, related to the boundedness of the rotor acceleration and speed, the electrical-mechanical parameters in (1), and the load torque, during its operation are described by [13]

$$|\dot{\omega}_r| \leq \dot{\omega}_r^M \text{ and } |\omega_r| \leq \omega_r^M \tag{19}$$

$$\begin{cases} R = \bar{R} + \Delta R; \ L_{d,q} = \bar{L}_{d,q} + \Delta L_{d,q} \\ |\Delta R| \leq \delta_R; \ |\Delta L| \leq \delta_L \end{cases} \begin{cases} J = \bar{J} + \Delta J; \ f = \bar{f} + \Delta f \\ |\Delta J| \leq \delta_J; \ |\Delta f| \leq \delta_f \end{cases} \tag{20}$$

$$|T_L| \leq \delta_T \omega_r^M \tag{21}$$

where, ω_r^M is the upper value of the rotor speed, and $\dot{\omega}_r^M$ is the upper value for the rotor acceleration.

Note: the symbol "−" is used for the nominal value, Δ represents the uncertainty, and $\delta_{(.)}$ are bounded positive constants.

Stability analysis for perturbed systems, controlled by a second-order SMC, has been detailed in [8]. The asymptotic stability of the closed-loop system is guaranteed by selecting control gains that satisfy the following conditions

$$\begin{cases} k_1 > 2\delta \\ k_2 > k_1 \frac{5k_1\delta + 4\delta}{2(k_1 - 2\delta)} \end{cases} \tag{22}$$

Let us define the system (1) with uncertainties

$$\frac{d}{dt}\mathbf{i}_s = (\mathbf{A} + \Delta\mathbf{A})\mathbf{i}_s + (\mathbf{B} + \Delta\mathbf{B})\mathbf{v}_s + \mathbf{d}_s \tag{23a}$$

$$\frac{d}{dt}\omega_r = -(A_\omega + \Delta A_\omega)\omega_r + (B_\omega + \Delta B_\omega)i_{sq} + d_\omega \tag{23b}$$

where, $\mathbf{i}_s = \begin{bmatrix} i_{sd} \\ i_{sq} \end{bmatrix}$, $\mathbf{A} = \begin{bmatrix} -\frac{R}{L_d} & \frac{L_q}{L_d}p\omega_r \\ -\frac{L_d}{L_q}p\omega_r & -\frac{R}{L_q} \end{bmatrix}$, $\mathbf{B} = \begin{bmatrix} \frac{1}{L_d} & 0 \\ 0 & \frac{1}{L_q} \end{bmatrix}$, $\mathbf{v}_s = \begin{bmatrix} v_{sd} \\ v_{sq} - \varphi_v p\omega_r \end{bmatrix}$, $\mathbf{d}_s = \begin{bmatrix} d_d \\ d_q \end{bmatrix}$ and $\Delta(\cdot)$ are parametric uncertainties.

The equations in (23) are rearranged with all uncertainties included in single terms such as

$$\frac{d}{dt}\mathbf{i}_s = (\mathbf{A} + \Delta\mathbf{A})\mathbf{i}_s + (\mathbf{B} + \Delta\mathbf{B})\mathbf{v}_s + \mathbf{d}_{snew} \tag{24a}$$

$$\frac{d}{dt}\omega_r = -(A_\omega + \Delta A_\omega)\omega_r + (B_\omega + \Delta B_\omega)i_{sq} + d_{\omega new} \tag{24b}$$

where, $\mathbf{d}_{snew} = \Delta\mathbf{A}\mathbf{i}_s + \Delta\mathbf{B}\mathbf{v}_s + \mathbf{d}_{snew}$ and $d_{\omega new} = -\Delta A_\omega \omega_r + \Delta B_\omega i_{sq} + d_{\omega new}$.

The new perturbation terms, in (24), include parametric uncertainties, unmodeled quantities and unknown disturbances. The SMC laws (12) and (17) will reject the bounded perturbations, which is one of the major characteristics of the SMC [7]. The advantage of the proposed cascade second-order SMC, compared to model based methods, such as optimal, feedback and predictive control, is that the knowledge about the load torque is not required, by measurement or estimation, and will be rejected to achieve zero steady-state tracking error.

5. Opal-RT HIL and Experimental System

The experimental system is shown in Figure 2, and the hardware connection is depicted in Figure 3. It includes a PMSM drive; speed encoder; power supply; insulated-gate bipolar transistor (IGBT) inverter; inductor line; dynamometer connected to the rotor shaft acting as a load; and data acquisition board (OP8660) with inputs for current, voltage and speed measurements and outputs for inverter pulses. The system is driven by the Opal-RT real-time simulator (OP5600). The schematic of the hardware connections between all modules is shown in Figure 4. The proposed cascade second-order SMC scheme is implemented in the Simulink/MATLAB and RT-LAB real-time simulation software environment, executed in the OP5600 and deployed to the system drive through the OP5600. The hardware in the loop (HIL) system, that includes the PMSM drive, the OP8660 and the OP5600, is illustrated in Figure 4. Details about the communication and deployment are available in [21].

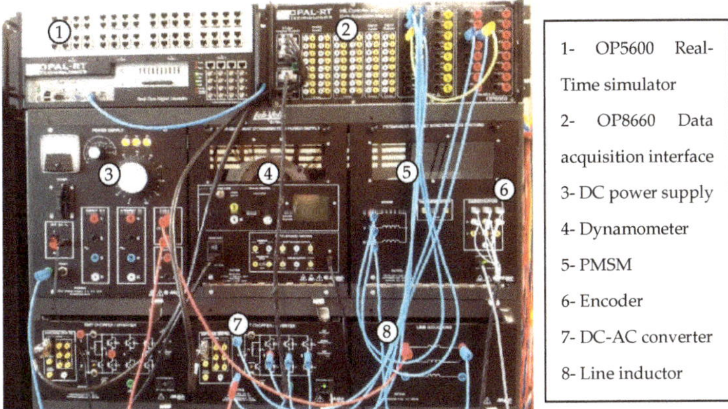

Figure 2. Experimental PMSM drive.

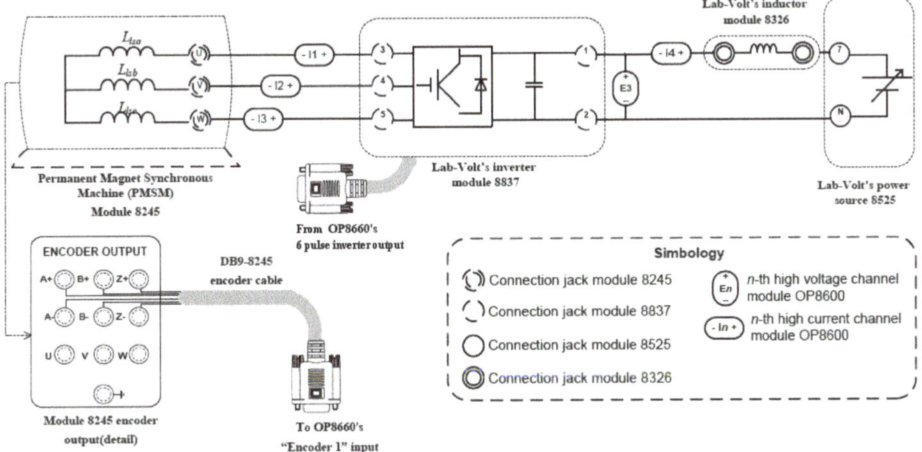

Figure 3. Schematic of the experimental PMSM drive.

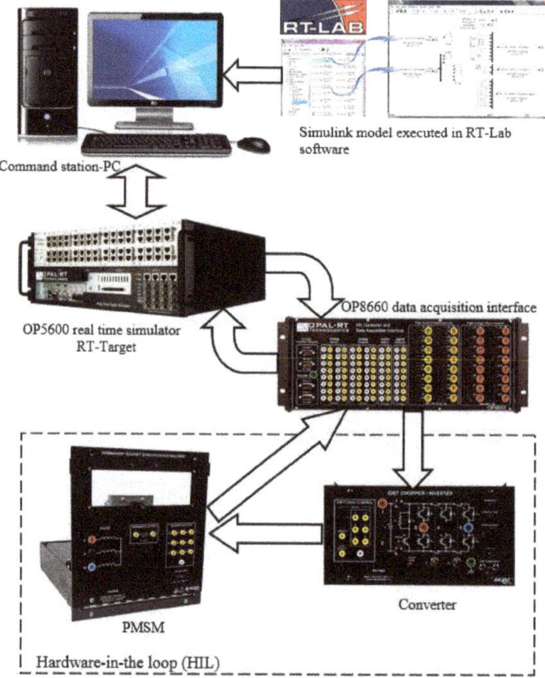

Figure 4. HIL PMSM drive system.

6. Experimental Results and Discussions

Experiments were conducted to validate the proposed cascade second-order SMC scheme under multiple operating conditions that include tracking performance under variable speed, robustness to parametric uncertainties (electrical and mechanical), rejection of external disturbance (sudden change in the power supply voltage), and unknown load torque. The PMSM parameters are provided in Table 1. In all scenarios, the control gains, provided in Table 2, were selected by trial and error method and remain constant for all experiments. The experimental results are illustrated in Figures 5–10.

Table 1. PMSM parameters.

Parameter	Unit	Value
Rated power	W	260
Rated current	A	3
Stator resistance	Ω	1.3
Stator d-axis inductance	mH	1.5
Stator q-axis inductance	mH	1.5
Flux linkage	Wb	0.027
Number of pole pairs	–	3
Moment of inertia	kg·m^2	1.7×10^{-6}
Friction coefficient	Nm·s/rad	0.3141×10^{-6}

Table 2. Control gains.

Gain	Value
Speed control (outer loop)	
k_1, k_2	$10^3, 10^4$
δ	0.2
α	0.01
Current controller (inner loop)	
k_1, k_2 (d-axis)	$10^2, 10^3$
k_1, k_2 (q-axis)	$10^2, 10^3$

6.1. Speed Tracking Performance for Variable Reference

A variable step reference was generated to check the speed tracking performance at sudden changes for the motor operating at no load. First, the second order-SMC control law was tested using the *sgn* function. It can be observed from the results of the speed tracking and *d-q* axes currents control, shown in Figure 5a,b, respectively, that the steady state error was eliminated by the controller without any overshoot and a fast response time at the reference signal transitions. However, the chattering occurs in the speed tracking error, Figure 5a, the speed control law (i_q command), and *d-q* currents, Figure 5b.

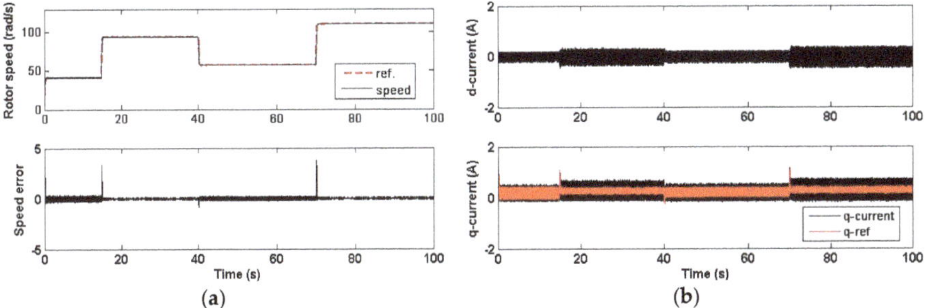

Figure 5. Tracking response for a variable speed reference (SMC with *sgn* function). (**a**) Speed tacking and error; (**b**) *d-q* currents tracking and *q*-command.

Then, the second-order SMC law was tested using the modified *sgn* function (18). It can be observed from the results of the speed tracking, Figure 6a, and the speed control law (i_q command), Figure 6b, that the chattering has been significantly reduced; however, the speed error at the transitions has slightly increased but without diminishing the overall performance of the control law.

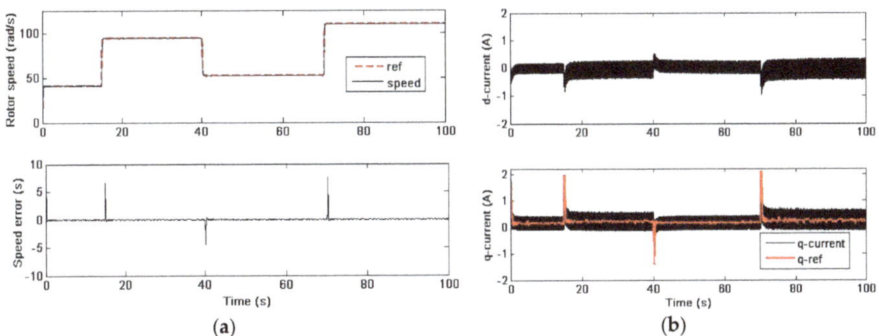

Figure 6. Tracking response for a variable speed reference (SMC with modified *sgn* function). (**a**) Speed tacking and error; (**b**) *d-q* currents tracking and *q*-command.

The only exception, in both cases, is the chattering in the d-q currents response, which was not eliminated by the second version of the control law even with control gains' change in multiple trials. The reason that the phenomenon is not only software from the control law, but hardware as the power supply, in the experimental set-up, includes fluctuations in its voltage signal as shown in Figure 8. Also, noises in data acquisition and IGBT pulses contribute to the chattering. Compared to similar works in this field on PMSM, as in [4] and [21], the current response is acceptable.

Overall performance of the cascade control strategy is satisfactory as the main objective is to track a speed reference while maintaining the current within its limit.

6.2. Parameter Uncertainties

The robustness of the proposed control scheme against parametric uncertainties was verified in this experiment. The uncertainties include the mechanical parameters (moment of inertia and viscous coefficient) and the electrical parameters (magnetic-flux, resistance and inductance). The parameters were set in the control laws with different values than the real ones of the machine. Random and extreme variation were chosen, where the values of the mechanical parameters were increased by 50% in the controller and the values of the electrical parameters' variations were changed separately (φ_v increased by 20%, R decreased by 20% and $L_{d,q}$ increased by 40%). Practically, the two important parameters to be tested are the resistance, which changes during the motor's operation due to the heat, and the rotor inertia, which varies when coupled with an unknown load. In this work, variations of different parameters were tested in order to prove the control strategy effectiveness. Figure 7a shows the speed response under parametric variations, where a good speed tracking performance is reached with a zero steady-state error. Figure 7b represents the q-current command to ensure robustness against parametric variations.

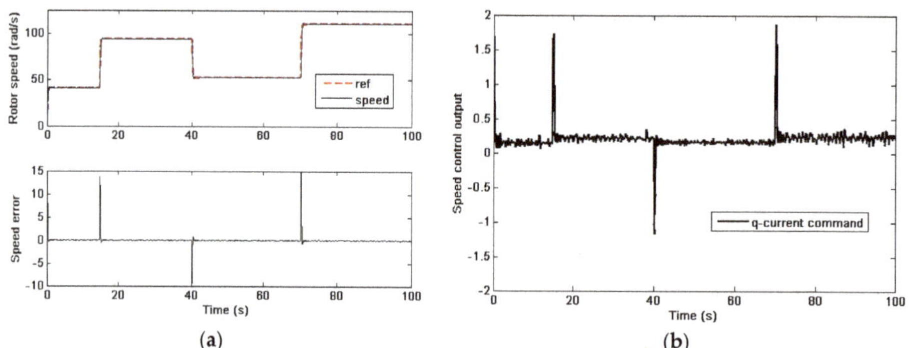

Figure 7. Speed tracking response under mismatched model. (**a**) Speed tacking and error; (**b**) Speed control output (q-command).

6.3. External Disturbance

In order to check the disturbance rejection of the proposed second-order SMC, the voltage V_{DC} of the power supply, considered constant, is varied along its nominal value with ± 10% change as shown in Figure 8. This change is considered extreme as in general the power supply is well stabilised. The DC voltage variation can be presented as an external disturbance because the controller has no information about this change. The system is controlled to track a constant reference and it can be observed from the result, shown in Figure 9a, that the speed response was affected by the variation of the power supply, however, the zero steady-state error was established after each variation due to the 2-SMC speed controller output, shown in Figure 9b, which proves the capability of the proposed controller to reject external disturbance with a good performance.

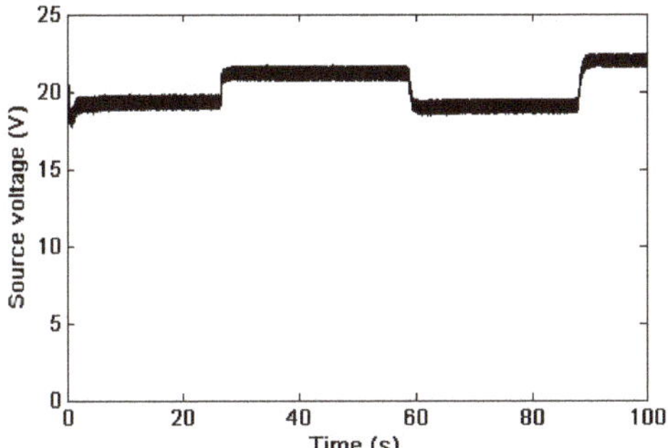

Figure 8. DC power supply variation (External disturbance).

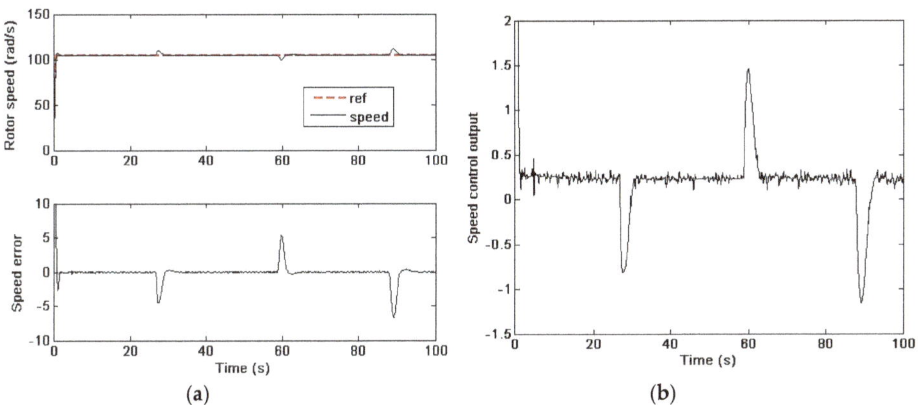

(a)   (b)

Figure 9. Speed tracking response under external disturbance. (**a**) Speed tacking and error; (**b**) Speed control output (q-command).

6.4. Unknown Load

Finally, to verify the effects of an unknown load, the parameters of a dynamometer, coupled with the motor through a belt with a gear ratio 1:2, were not included in the controller design. Figure 10a,b represent the speed tracking response and speed controller output, where it can be observed that the speed tracking performance is successfully achieved even with the unknown perturbation.

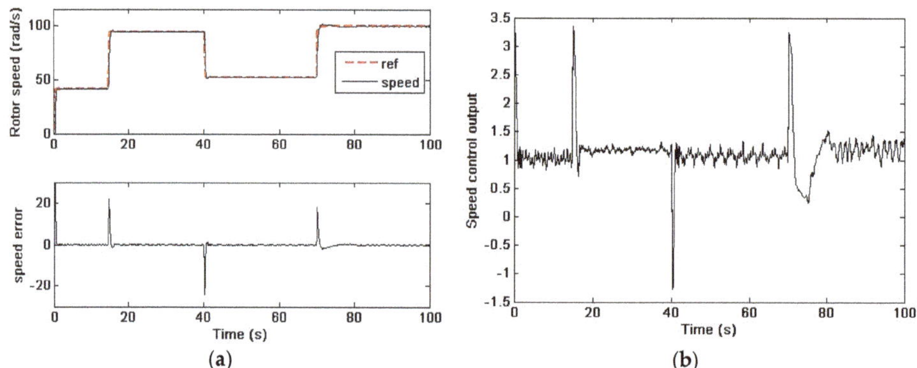

Figure 10. Speed tracking response under unknown load disturbance. (**a**) Speed tacking and error; (**b**) Speed control output (*q*-command).

7. Conclusions

A cascade second-order sliding mode control scheme for a permanent magnet synchronous motor drive has been presented. The outer loop control is a speed controller to track the rotor speed trajectory, while the inner loop control, a current control, regulates the *d*-axis current. The aim of the proposed second-order SMC is to achieve accurate speed tracking despite the system inaccuracies such as parametric uncertainties and unknown perturbations. The second-order SMC law is developed for a state space model and applied to the PMSM drive by decoupling and linearizing its model. Opal-RT real time technology-based HIL is used to implement the control strategy. An experimental setup was used to experimentally validate of the proposed second-order SMC under different conditions. Experimental results have proved the effectiveness and the robustness of the proposed control scheme with respect to reference variations and system inaccuracies. The second-order SMC provides more features concerning system uncertainties compensation as it includes an integral action.

Funding: This research was funded in part by the CANADA FOUNDATION FOR INNOVATION, grant number 30527.

Conflicts of Interest: The author declare no conflict of interest.

References

1. Choi, H.H.; Vu, N.T.T.; Jung, J.W. Digital implementation of an adaptive speed regulator for a PMSM. *IEEE Trans. Ind. Electron.* **2011**, *26*, 3–8. [CrossRef]
2. Li, S.; Liu, Z. Adaptive speed control for permanent-magnet synchronous motor system with variations of load inertia. *IEEE Trans. Ind. Electron.* **2009**, *56*, 3050–3059.
3. Solsona, J.; Valla, M.I.; Muravchik, C. Nonlinear control of a permanent magnet synchronous motor with disturbance torque estimation. *IEEE Trans. Energy Convers.* **2000**, *18*, 163–168. [CrossRef]
4. Errouissi, R.; Ouhrouche, M.; Chen, W.H.; Trzynadlowski, A.M. Robust nonlinear predictive controller for permanent-magnet synchronous motors with an optimized cost function. *IEEE Trans. Ind. Electron.* **2012**, *59*, 2849–2858. [CrossRef]
5. Abdel-Rady, Y.; Mohamed, I. Design and implementation of a robust digital current control scheme for a PMSM vector drive with a simple adaptive disturbance observer. *IEEE Trans. Ind. Electron.* **2007**, *54*, 1981–1988.
6. Abdel-Rady, Y.; Mohamed, I. A newly designed instantaneous-torque control of a direct-drive PMSM servo actuator with improved torque estimation and control characteristics. *IEEE Trans. Ind. Electron.* **2007**, *54*, 2864–2873.
7. Rivera, J.; Garcia, L.; Mora, C.; Raygoza, J.J.; Ortega, S. Super-twisting sliding mode in motion control systems. In *Sliding Mode Control*; Bartoszewicz, A., Ed.; InTech: Rijeka, Croatia, 2011; pp. 237–254.

8. Moreno, J.A.; Osorio, M. A Lyapunov approach to a second-order sliding mode controllers and observers. In Proceedings of the 47th IEEE Conference on Decision and Control, Cancun, Mexico, 9–11 December 2008; pp. 2856–2861.

9. Li, S.; Zhou, M.; Yu, X. Design and implementation of terminal sliding mode control method for PMSM speed regulation system. *IEEE Trans. Ind. Inform.* **2012**, *9*, 1879–1891. [CrossRef]

10. Zhang, X.; Sun, L.; Zhao, K.; Sun, L. Nonlinear speed control for PMSM System using sliding-mode control and disturbance compensation techniques. *IEEE Trans. Power Electron.* **2013**, *28*, 1358–1365. [CrossRef]

11. Lascu, C.; Boldea, I.; Blaabjerg, F. Super-twisting sliding mode control of torque and flux in permanent magnet synchronous machine drives. In Proceedings of the 39th Annual Conference of the IEEE Industrial Electronics Society, Vienna, Austria, 10–13 November 2013; pp. 3171–3176.

12. Valenciaga, F.; Puleston, P.F. High-order siding control for a wind energy conversion system based on a permanent magnet synchronous generator. *IEEE Trans. Energy Convers.* **2008**, *23*, 860–867. [CrossRef]

13. Merabet, A.; Ahmed, K.; Ibrahim, H.; Beguenane, R. Implementation of sliding mode control system for generator and grid sides control of wind energy conversion system. *IEEE Trans. Sustain. Energy* **2016**, *7*, 1327–1335. [CrossRef]

14. Wang, Q.; Yu, H.; Wang, M.; Qi, X. An improved sliding mode control using disturbance torque observer for permanent magnet synchronous motor. *IEEE Access* **2019**, *7*, 36691–36701. [CrossRef]

15. Merabet, A. Adaptive sliding mode speed control for wind energy experimental system. *Energies* **2018**, *11*, 2238. [CrossRef]

16. Pisano, A.; Dávila, A.; Fridman, L.; Usai, E. Cascade control of PM DC drives via second-order sliding-mode technique. *IEEE Trans. Ind. Electron.* **2008**, *55*, 3846–3854. [CrossRef]

17. Merabet, A.; Islam, M.A.; Beguenane, R.; Ibrahim, H. Second-order sliding mode control for variable speed wind turbine experiment system. In Proceedings of the International Conference on Renewable Energies and Power Quality (ICREPQ'14), Cordoba, Spain, 8–10 April 2014.

18. Merabet, A.; Labib, L.; Ghias, A.M.Y.M.; Aldurra, A.; Debbouza, M. Dual-mode operation based second-order sliding mode control for grid-connected solar photovoltaic energy system. *Int. J. Electr. Power* **2019**, *111*, 459–474. [CrossRef]

19. Scalera, L.; Gasparetto, A.; Zanotto, D. Design and experimental validation of a 3-DOF underactuated pendulum-like robot. *IEEE/ASME Trans. Mechatron.* **2019**, 1. [CrossRef]

20. Sira-Ramírez, H.; Aguilar-Orduña, M.A.; Zurita-Bustamante, E.W. On the sliding mode control of MIMO nonlinear systems: An input-output approach. *Int. J. Robust Nonlinear Control* **2019**, *29*, 715–735. [CrossRef]

21. Merabet, A.; Ahmed, K.T.; Ibrahim, H.; Beguenane, R.; Ghias, A.M.Y.M. Energy management and control system for laboratory scale microgrid based wind-PV-battery. *IEEE Trans. Sustain. Energy* **2017**, *8*, 145–154. [CrossRef]

Article

Unified Predictive Current Control of PMSMs with Parameter Uncertainty

Peng Tang [1], Yuehong Dai [1,2,3,*] and Zhaoyang Li [1]

[1] School of Aeronautics and Astronautics, University of Electronic Science and Technology of China (UESTC), Chengdu 611731, China; tp_study@hotmail.com (P.T.); lzy_uestc@hotmail.com (Z.L.)
[2] Research Center of Intelligent Aircraft System Technology and Application of UESTC, Chengdu 611731, China
[3] Aircraft Swarm Intelligent Sensing and Cooperative Control key laboratory of Sichuan Province, Chengdu 611731, China
* Correspondence: daiyh@uestc.edu.cn; Tel.: +86-139-8185-3598

Received: 12 November 2019; Accepted: 9 December 2019; Published: 12 December 2019

Abstract: Predictive current control (PCC) applied on permanent magnet synchronous motors (PMSMs) has been developed into mainly three methods: the conventional finite-control-set PCC, the double voltage vectors PCC, and deadbeat PCC. However, each approach has its particular calculation way for voltage vectors selection and respective execution duration. This paper, based on the deadbeat idea, presents a unified predictive current control scheme of PMSMs. Under this scheme, the prior three classes are able to be clearly unified into one frame with lower calculation effort. Furthermore, to cope with problem of parameter mismatch in dq-axis current predictive model, a integrated identification method is proposed. Firstly, data selectors are designed to reject abnormal data of sampling signals, and then the interval-varying multi-innovation least squares algorithm is combined with forgetting factor (V-FF-MILS) to approximate the error terms caused by electromagnetic parameters error. The estimated results are online fed to the model of PMSM to enhance its accuracy. Finally, the processor in loop (PIL) simulation results verify that the proposed integrated scheme has advantages in current control of PMSMs with large-scale parameter uncertainty.

Keywords: PMSM; current control; model predictive control; parameter identification

1. Introduction

Owing to its advantages on efficiency, power density, and compact structure, PMSMs have been widely employed in various electrical drive appliances in vehicle, aerospace, and other industrial fields. There are a variety of control strategies designed to improve PMSMs' output performance. Field-oriented control (FOC), also called vector control, and direct torque control (DTC) are the two most common control frames together with many other modified methods. It is well known that FOC and DTC have limitations on necessity of independent d-q current controllers with a pulse modulator and considerable torque ripple, respectively [1]. Recently, thanks to the continuous advancement of digital signal processing technology over the last decades [2], the model predictive control (MPC), with intuitive operation principle and good transient performance, has gained attention on regulating the output current or torque of PMSMs, yielding the predictive current control (PCC). Overall, the PCC can be divided into three categories: the conventional finite-control-set PCC, the Double VVs, and deadbeat PCC. They have been successfully implemented on controllers of PMSMs [3,4].

The conventional finite-control-set PCC, also known as direct PCC, takes advantage of the inherent nature of inverter that has a finite number of switching states $\mathbf{S}_n$, where $n = 8$ for the commonly used two-level voltage source inverter; thus, there are eight voltage vectors (VVs) including six active

VVs and two null VVs. All the VVs are used to predict possible stator currents $\mathbf{i}^p(k+1)$ based on the discrete model of motor. The one that minimizes a cost function evaluating the error between the reference current $\mathbf{i}^*(k+1)$ and the predictions $\mathbf{i}^p(k+1)$ and other constraints is selected as the action of the next control period. Although the conventional PCC has feature of intuitive principle, there are several drawbacks, e.g. high current ripple as amplitude, phase of the action VV cannot be adjusted, and computational burden result from the enumerated way to search the optimal VV. The inverter is operated in a variable switching frequency because of only one VV per period with no need for a modulator, causing large harmonics [2]. To overcome this, the Double VVs PCC is introduced by taking account of exerting two voltage vectors using a PWM modulator in a control period. Combination forms of two VVs have been explored. The authors of [3–5] considered one active vector and a successively null vector as a set, thus amplitude of the resultant vector can be adjusted. However, Zhang and Yang [6] showed it has an inferior steady-state performance at high speed. Therefore, Liu *et al.* [7] considered all the feasible combinations of two VVs so that the phase of the composed vector can be adjusted, but the situation that two non-adjacent vectors are combined would increase the switching loss of inverter. The duty cycle of each vector is calculated based on the current dead-bead principle in the above methods, which considers the current will reach the reference value at the end of this operation.

The Deadbeat PCC is similar to the FOC scheme, which usually has a space vector PWM modulator, and its optimal control set composed by two adjacent active VVs and a null VV is determined by the reference voltage, which is directly obtained from the predictive model of PMSMs based on the deadbeat idea [3], while that of FOC is from a specific PI or other complex current controller.

The performance differences about the three PCC schemes can be found in [3]. Each scheme has its unique advantage, such as less nonlinearity of inverter introduced for direct PCC in each control period. In this paper, the above three methods are simply denoted as 1-PCC, 2-PCC and 3-PCC, respectively. The numeric symbols represent the number of voltage vectors applied per control period. Currently, there are numerous variations of those schemes implemented in different computation ways. For example, the 1-PCC is implemented on $\alpha\beta$-frame based on an enumerated way, while the 2-PCC usually takes d-q coordinate, which is not convenient for engineers. One study on unifying the 1-PCC and 2-PCC method in one frame [8] reveals that an alternative cost function about voltage vector is equivalent to the conventional one, but its calculation process is still not intuitive. This paper proposes a generalized predictive current control scheme based on the principle of deadbeat control and vector composition in an explicit way. It is shown that 3-PCC can easily be reduced to 2-PCC and 1-PCC by judging simple linear equations. The durations of VVs for 2-PCC can be calculated simultaneously. Consideration for the system uncertainty can also be included.

An inevitable issue is that the PCC heavily relies on the model parameters of motor, resulting in current tracking performance deterioration when occurring in parameter mismatch. The nonlinear control is widely used to cope with this situation. For example, the sliding mode observers are frequently designed to estimate the disturbance caused by parameter uncertainty [9–11]. The uncertain term in the predictive model can be replaced by nonlinear function [12]. Advanced robust techniques are often introduced to design control law, such as the control-Lyapunov function [13], disturbance estimation [14], and differential flatness [15]. Another strategy lies in the parameter identification, such as a recursive inductance estimator [9], multi-parameter identification with a decoupling method [16], model reference adaptive technique [17], extended Kalman filter [18], modified particle swarm optimization [19], Adaline neural network [20], etc. All those methods aim to acquire accurate parameters, which is not necessary in the PCC schemes. In this paper, we estimate three model error terms caused by parameter perturbation. Thus, the identification model can be simplified into a linear regression model. Then, a integrated identification method is proposed to approximate the error terms.

This paper is organized as follows. The predictive model of a surface-mounted PMSM and the calculation of the inverter's output voltage are presented in Section 2. Section 3 proposes the generalized predictive current control scheme. The model identifying method based on a modified multi-innovation recursive least square to estimate the model error is presented in Section 4. Section 5 presents the PIL test results of proposed method. Finally, the conclusions are drawn in Section 6.

2. System Model

2.1. Discrete Predictive Model of PMSMs

The stator voltage equation of a surface-mounted PMSM described in the synchronous reference can be found in [3]. After discretizing it by Euler forward difference method, the current predictive model in *dq* frame can be expressed as follows,

$$\mathbf{i}^p(k+1) = \mathbf{A}(k)\mathbf{i}(k) + \mathbf{B}\bar{\mathbf{U}}(k) + \mathbf{H}\omega(k) \tag{1}$$

where $\mathbf{A}(k) = \mathbf{A}_0(k) + \mathbf{A}_\delta$, $\mathbf{B} = \mathbf{B}_0 + \mathbf{B}_\delta$, $\mathbf{H} = \mathbf{H}_0 + \mathbf{H}_\delta$, and

$$\mathbf{A}_0(k) = \begin{bmatrix} 1 - \frac{R_0}{L_0}T_s & \omega_e(k)T_s \\ -\omega_e(k)T_s & 1 - \frac{R_0}{L_0}T_s \end{bmatrix} \qquad \mathbf{B}_0 = \mathrm{diag}\,(T_s/L_0, T_s/L_0) \qquad \mathbf{H}_0 = [0, -T_s\psi_0/L_0]^T$$

$$\mathbf{A}_\delta = \mathrm{diag}\,(\delta_1, \delta_1) \qquad \mathbf{B}_\delta = \mathrm{diag}\,(\delta_2, \delta_2) \qquad \mathbf{H}_\delta = [0, \delta_3]^T$$

$$\mathbf{i}(k) = [i_d(k), i_q(k)]^T \qquad \bar{\mathbf{U}}(k) = [\bar{u}_d(k), \bar{u}_q(k)]^T \qquad \omega(k) = \omega_e(k)$$

$\mathbf{i}^p(k+1)$ is the predictive current at instant $k+1$ with respect to the control input $\bar{\mathbf{U}}(k)$, $\mathbf{i}(k)$ is the sampling current at instant k and the electrical angular velocity $\omega_e(k)$, and T_s represents the sample period of current. The matrices of nominal model $\mathbf{A}_0(k)$, $\mathbf{B}_0$, and $\mathbf{H}_0$ are determined by the values of R_0, L_0, and ψ_0, which represent the electromagnetic parameters, with stator resistance, inductance, and magnetic flux, respectively. $\mathbf{A}_\delta$, $\mathbf{B}_\delta$, and $\mathbf{H}_\delta$ are the perturbation parts composed of three error terms $\delta_1, \delta_2, \delta_3$ independently, which result from differences between nominal parameters and real values. For instance, the estimation errors of stator resistance and inductance jointly lead to the error term δ_1. The symbol diag(.) means it is a diagonal matrix. The control input $\bar{\mathbf{U}}(k)$ is produced by a switching sequence of inverter per control period, thus $\bar{u}_d(k)$ and $\bar{u}_q(k)$ are mean voltages at the d-axis and q-axis, respectively.

2.2. Inverter Model

As shown in Figure 1, the PMSM is fed by a two-level voltage source inverter, which is composed of three phase legs (a, b, c), and each leg has two opposite switching states. Since there are eight different switching sates, which can be defined as $\mathbf{S}_n = [S_a, S_b, S_c]^T$, n = 0, ..., 7, the index number satisfies $n = S_a \cdot 2^0 + S_b \cdot 2^1 + S_c \cdot 2^2$, and each one generates a voltage vector $\mathbf{V}_n$. The stator voltage in dq-axis can be expressed as a function of the switching states $\mathbf{S}_n$

$$\mathbf{U}_n(k) = \frac{2}{3}v_{dc}\mathbf{T}(k)\mathbf{D}\mathbf{S}_n \tag{2}$$

where

$$\mathbf{T}(k) = \begin{bmatrix} \cos\theta_e(k) & \sin\theta_e(k) \\ -\sin\theta_e(k) & \cos\theta_e(k) \end{bmatrix} \qquad \mathbf{D} = \begin{bmatrix} 1 & -\frac{1}{2} & -\frac{1}{2} \\ 0 & \frac{\sqrt{3}}{2} & -\frac{\sqrt{3}}{2} \end{bmatrix}$$

v_{dc} is the DC bus voltage of inverter and $\theta_e(k)$ represents electrical angular of rotor at instant k. The transformation method is based on the amplitude unchanged principle, thus the coefficient is $2/3$. It would be $\sqrt{2/3}$ if the power unchanged principle were adopted. Obviously, when n equals 0 or 7, $\mathbf{U}_n(k) = \mathbf{0}$.

After gaining stator voltage, the control input $\bar{\mathbf{U}}(k)$ can be obtained under the principle of volt-second balance

$$\bar{\mathbf{U}}(k) = \frac{t_i}{T_s}\mathbf{U}_i(k) + \frac{t_j}{T_s}\mathbf{U}_j(k) + \frac{t_0}{T_s}\mathbf{U}_z(k)$$
$$= \bar{d}_i\mathbf{U}_i(k) + \bar{d}_j\mathbf{U}_j(k) + d_0\mathbf{U}_z(k) \tag{3}$$

$$T_s = t_i + t_j + t_0 \tag{4}$$

where $i,j \in \{1,2,..,6\}, z \in \{0,7\}$. t_i, t_j, and t_0 are durations of the respective stator voltage vector during the kth control period. $\mathbf{U}_z(k)$ is zero vector, thus

$$\bar{\mathbf{U}}(k) = \mathbf{U}_{ij}(k)\bar{\mathbf{d}}(k) \tag{5}$$

where the $\bar{\mathbf{d}}(k) = [\bar{d}_i, \bar{d}_j]^T = [t_i/T_s, t_j/T_s]$ is duty cycle vector of active voltage vectors and $\mathbf{U}_{ij}(k) = [\mathbf{U}_i(k), \mathbf{U}_j(k)]$. We define $\mathbf{S}_{ij} = [\mathbf{S}_i, \mathbf{S}_j]$; combined with Equation (2), $\mathbf{U}_{ij}(k)$ can be expressed as

$$\mathbf{U}_{ij}(k) = \frac{2}{3}v_{dc}\mathbf{T}(k)\mathbf{DS}_{ij} \tag{6}$$

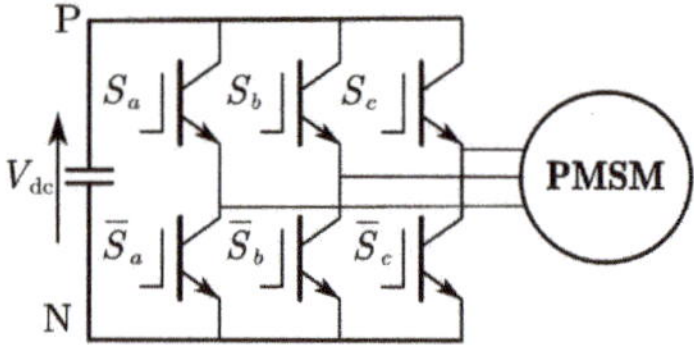

Figure 1. PMSM fed by a two-level voltage source inverter.

3. The Proposed Control Scheme

The conventional one-horizon PCC method has a time delay between the measurements and actuation, which causes current ripple [21]. A simple solution to compensate this delay is to consider a two-step-ahead prediction. The predictive current at instant $k + 2$ can be obtained by

$$\mathbf{i}^p(k+2) = \mathbf{A}(k+1)\mathbf{i}^p(k+1) + \mathbf{B}\bar{\mathbf{U}}(k+1) + \mathbf{H}\omega(k+1) \tag{7}$$

The angular velocity $\omega_e(k)$ is deemed as a constant during the two sampling periods. Because current dynamics are much faster than those of speed, $\mathbf{A}(k) = \mathbf{A}(k+1)$. The control aim is achieved by a cost function that is defined as follows:

$$g_i = \|\mathbf{i}^* - \mathbf{i}^p(k+2)\|_2 \tag{8}$$

where $\mathbf{i}^*$, as the output of speed controller, is the reference current that is also deemed as a constant from kth to $(k + 2)$th period. In this paper, we take the Euler 2-norm to define the cost function, thus the objective is to search the control action to minimize the distance of between reference current and predictive value.

3.1. Deadbeat PCC

The reference stator voltage in *dq*-frame is obtained based on deadbeat principle and the inverse model of PMSM, which is shown as follows:

$$\begin{aligned}
\mathbf{V}^*(k+1) &= \mathbf{B}^{-1}[\mathbf{i}^* - \mathbf{A}(k+1)\mathbf{i}^p(k+1) - \mathbf{H}\omega(k+1)] \\
&= \mathbf{B}^{-1}[\mathbf{i}^* - \mathbf{A}(k)\mathbf{i}^p(k+1) - \mathbf{H}\omega(k)]
\end{aligned} \tag{9}$$

The matrix $\mathbf{B}$ is obviously invertible. We can define an alternative cost function with respect to motor voltage

$$g_v = \|\mathbf{V}^*(k+1) - \bar{\mathbf{U}}(k+1)\|_2 = \|\mathbf{V}_E(k+1)\|_2 \tag{10}$$

It is easily seen that, if the control input $\bar{\mathbf{U}}(k+1)$ gradually approximates the reference voltage $\mathbf{V}^*(k+1) = [v_d^*(k+1), v_q^*(k+1)]^T$, the resultant current $\mathbf{i}(k+2)$ will approach the reference current $\mathbf{i}^*$. Therefore, the cost function g_v is equivalent to g_i. After obtaining the reference voltage, we need to make sure its position in the voltage vector plane that is divided into six sectors, as shown in Figure 2. The sector position of reference voltage can be calculated by solving an arc-tangent function

$$\theta_v = \arctan\left[v_d^*(k+1)/v_q^*(k+1)\right] + \theta_e \tag{11}$$

The reference voltage is composed of the two active voltage vector

$$\mathbf{V}^*(k+1) = \bar{\mathbf{U}}(k+1) = \mathbf{U}_{ij}(k+1)\bar{\mathbf{d}}(k+1) \tag{12}$$

where the $\mathbf{U}_i$ and $\mathbf{U}_j$ are two adjacent vectors, and $i < j$ to flexibility produce pulse sequence; for example, when the reference voltage is located in Sector II, then $i = 2, j = 6$. Clearly, $\mathbf{U}_{ij}(k+1)$ is invertible, thus the duty cycles of the vectors are calculated by

$$\begin{aligned}
\bar{\mathbf{d}}(k+1) &= \mathbf{U}_{ij}^{-1}(k+1)\mathbf{V}^*(k+1) \\
&= \frac{3}{2v_{dc}}\mathbf{F}_N^{-1}\mathbf{T}^{-1}(k+1)\mathbf{V}^*(k+1)
\end{aligned} \tag{13}$$

where $\mathbf{F}_N = \mathbf{DS}_{ij}$, which can be calculated offline to reduce the computational burden. N denotes the sector index. $\mathbf{T}^{-1}(k+1)$ can be directly obtained from the elements of the park transformation matrix $\mathbf{T}$. If $\bar{d}_i + \bar{d}_j > 1$, it means the reference voltage has gone over the scope of voltage vector plane, and a common overmodulation method can be used $\bar{d}_i = \bar{d}_i/(\bar{d}_i + \bar{d}_j), \bar{d}_j = \bar{d}_j/(\bar{d}_i + \bar{d}_j)$. The duty cycle of the null VV is calculated by $\bar{d}_0 = 1 - \bar{d}_i - \bar{d}_j$, and then a space vector pulse width modulation (SVPWM) is applied to produce the switching sequence of inverter.

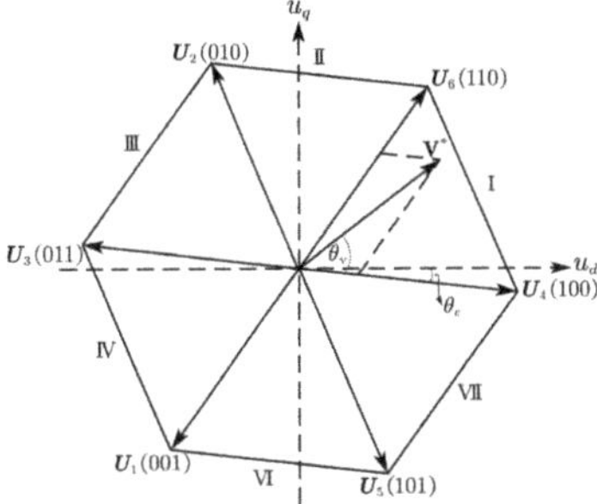

Figure 2. Position of reference voltage in space voltage vector plane.

3.2. Direct PCC

The direct PCC selects only one optimal voltage vector that minimizes the cost function in Equation (10), which aims to search error voltage vector $\mathbf{V}_E$ with shortest length. After locating position of reference voltage, the two adjacent active VV and null VV in the sector are feasible optimal VV and necessary to be checked by the cost function. For example, we assume reference voltage is located in Sector I, as shown in Figure 3a. The triangular sectors are divided into three equal parts by their three center lines. The reference voltage is located in the third part. The shortest $\mathbf{V}_E$ is the result of the difference between $\mathbf{V}^*$ and active voltage vector $\mathbf{U}_6$, which is depicted by the red dashed line. As a result, $\mathbf{U}_6$ is selected as the optimal VV. Similarly, if reference VV lies in the first part or second part of Sector I, then the optimal VV is $\mathbf{U}_z$ or $\mathbf{U}_4$, respectively. Therefore, once we confirm to which part the reference voltage belongs, the optimal actuation can be directly obtained. The three parts in a sector are mapped in the $\bar{d}_i - \bar{d}_j$ coordinate axis because of the linear feature of vector synthesis. As shown in Figure 3b, each part can be judged by a set of inequalities as follows.

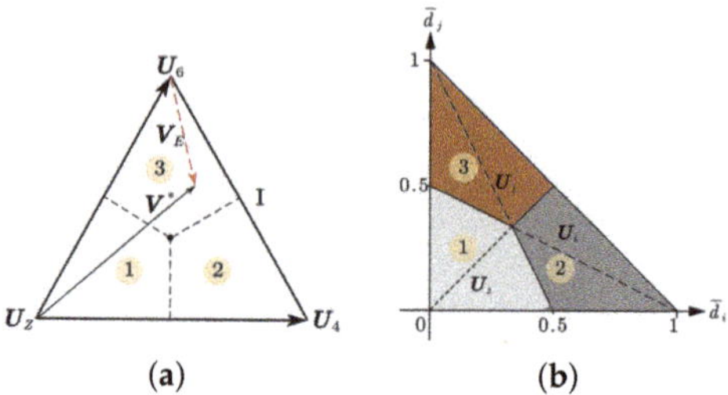

Figure 3. Selection of optimal VV: (a)Distribution of reference voltage, (b) Confirm the optimal VV.

$$\textcircled{1}\;\begin{cases} \bar{d}_i + 2\bar{d}_j - 1 \leq 0 \\ 2\bar{d}_i + \bar{d}_j - 1 \leq 0 \end{cases} \Rightarrow \mathbf{U}_z \quad \textcircled{2}\;\begin{cases} \bar{d}_i - \bar{d}_j \geq 0 \\ 2\bar{d}_i + \bar{d}_j - 1 > 0 \end{cases} \Rightarrow \mathbf{U}_i \quad \textcircled{3}\;\begin{cases} \bar{d}_i - \bar{d}_j < 0 \\ \bar{d}_i + 2\bar{d}_j - 1 > 0 \end{cases} \Rightarrow \mathbf{U}_j \quad (14)$$

Consequently, after getting $\bar{d}_i$ and $\bar{d}_j$ with Equation (13), we can directly select the optimal VV from in the above inequalities of duty cycles, which greatly reduce the computational burden compared with the enumeration-based method, and makes it convenient to switch control method from Deadbeat PCC to Direct PCC.

3.3. Double VVs PCC

This method needs to apply two VVs in every control period, which can either be one active VV and one null VV or two active VVs, and the resultant voltage vector is located on one side of a triangular sector. Thus, the exploration to find the minimum cost function is to identify the shortest distance between the reference VV and the near sides. Obviously, $\mathbf{V}_E$ needs to be perpendicular to the nearest side. For instance, as depicted in Figure 4a, the reference voltage $\mathbf{V}_1^*$ is located in the second part, thus it is close to $\mathbf{U}_4$. $\mathbf{V}_E$ is normal to $\mathbf{U}_4$ and obtained by

$$\mathbf{V}_E = \mathbf{V}_1^* - d_4\mathbf{U}_4 = \bar{d}_6\mathbf{U}_6 + (\bar{d}_4 - d_4)\mathbf{U}_4 \tag{15}$$

It is noted that $\mathbf{V}_E$ is in the same direction as vector $2\mathbf{U}_6 - \mathbf{U}_4$, thus

$$d_4 = \frac{2\bar{d}_4 + \bar{d}_6}{2} \tag{16}$$

d_4 is the actual duty cycle, the residual duration $1 - d_4$ belongs to the null voltage vector U_Z, and the principle to choose a null VV is subject to minimum switching number. Similarly, when reference voltage $\mathbf{V}_2^*$ lies in the first part in Figure 4a, the optimal vector combination is $(\mathbf{U}_4, \mathbf{U}_6)$. $\mathbf{V}_E$ equals $(\bar{d}_6 - d_6)\mathbf{U} + (\bar{d}_4 - d_4)\mathbf{U}_4$, which is in the same direction as vector $-(\mathbf{U}_4 + \mathbf{U}_6)$, and $d_6 = 1 - d_4$. Thus, the duration of $\mathbf{U}_4$ can be calculated as follows:

$$d_4 = \frac{1 + \bar{d}_4 - \bar{d}_6}{2} \tag{17}$$

In summary, we firstly need to identify in which part the reference voltage is located, and then confirm the optimal vector combination. Finally, we calculate the respective duty cycle. The process is summarized as below:

$$\textcircled{1} \quad \begin{cases} \bar{d}_i + 2\bar{d}_j - 1 > 0 \\ 2\bar{d}_i + \bar{d}_j - 1 > 0 \end{cases} \Rightarrow (\mathbf{U}_i, \mathbf{U}_j) \Rightarrow \begin{cases} d_i = \dfrac{1 + \bar{d}_i - \bar{d}_j}{2} \\ d_j = \dfrac{1 - \bar{d}_i + \bar{d}_j}{2} \end{cases} \tag{18}$$

$$\textcircled{2} \quad \begin{cases} \bar{d}_i - \bar{d}_j \geq 0 \\ \bar{d}_i + 2\bar{d}_j - 1 \leq 0 \end{cases} \Rightarrow (\mathbf{U}_i, \mathbf{U}_z) \Rightarrow \begin{cases} d_i = \dfrac{2\bar{d}_i + \bar{d}_j}{2} \\ d_0 = 1 - d_i \end{cases} \tag{19}$$

$$\textcircled{3} \quad \begin{cases} \bar{d}_i - \bar{d}_j < 0 \\ 2\bar{d}_i + \bar{d}_j - 1 \leq 0 \end{cases} \Rightarrow (\mathbf{U}_j, \mathbf{U}_z) \Rightarrow \begin{cases} d_j = \dfrac{\bar{d}_i + 2\bar{d}_j}{2} \\ d_0 = 1 - d_j \end{cases} \tag{20}$$

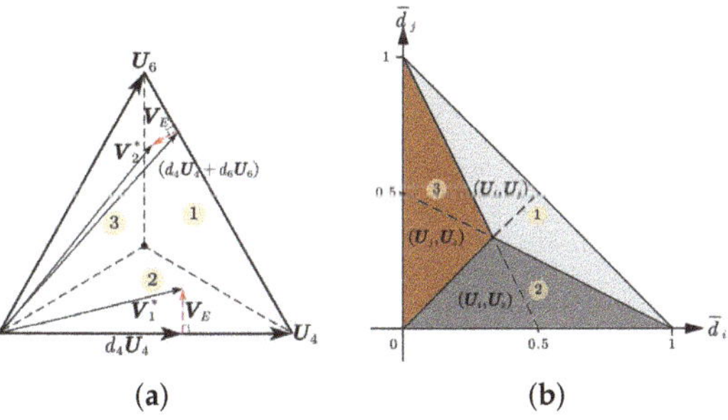

Figure 4. Selection of optimal VVs: (**a**) The shortest error VV, (**b**) Confirm the optimal VVs.

Supplementing other sectors in the voltage vector plane, we can acquire the maps to directly select optimal actuation according to the position of reference voltage. Figure 5a shows the distribution of optimal voltage vector of direct PCC method, and its duration is the whole control period with no need for a pulse modulator. The distribution of optimal two VVs combination is shown in Figure 5b. The execution time of two VVs is $\mathbf{t} = T_s\mathbf{d}$; thus, a pulse modulator is needed to produce switching signals. It can be seen that the above three PCCs are unified into a frame with little calculation effort, and we can easily adjust the current control method in practical application.

Inevitably, predictive current control heavily depends on the accuracy of motor model, and an exact predictive model is the precondition of this control scheme. Therefore, an identification algorithm aiming to approximate the uncertain terms in model is incorporated into the scheme. A block diagram of the unified predictive current control (UPCC) scheme is depicted in Figure 6. Before acquiring accurate model, 1-PCC is chosen to be the operation method because of its better robustness performance [3], and the fewer switches also reduce the influence of dead band of inverter per period. The results of identification are fed into the PMSM's model online, thus enhancing the model accuracy. The detailed estimation method is introduced in the next section.

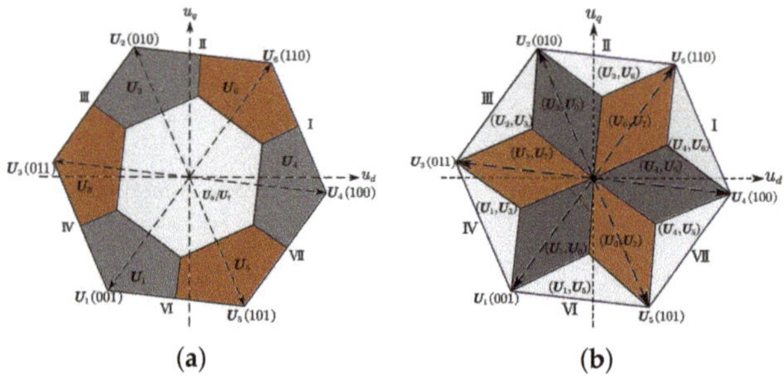

Figure 5. Distribution of optimal VVs: (**a**) Optimal VV of Direct PCC, (**b**) Optimal VVs of Double VVs PCC.

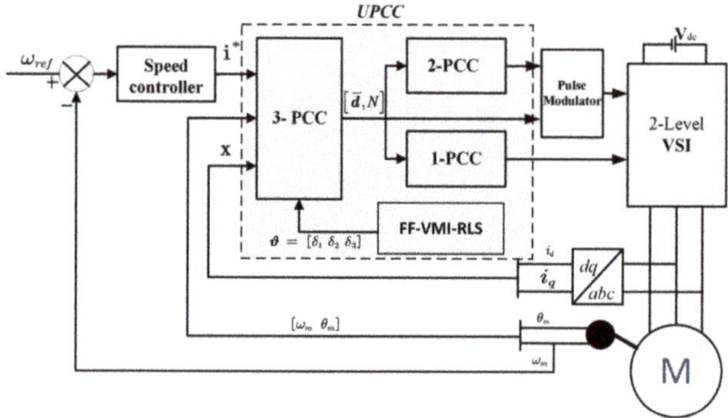

Figure 6. The proposed control scheme.

4. Identification for Model Uncertain Term

In this section, we estimate the three error terms caused by inaccurate resistance, inductance, and flux linkage in the current error predictive model, rather than identify the electromagnetic parameter in the nonlinear model. Based on Equation (1), we can predict the $(k+1)$th current $\mathbf{i}^p(k+1)$ at the kth instant, and acquire the real value $\mathbf{i}(k+1)$ at the next instant of the predictive current error model $\Delta \mathbf{i}(k+1) = \mathbf{A}_0 \mathbf{i}(k) + \mathbf{B}_0 \bar{\mathbf{U}}(k) + \mathbf{H}_0 \omega(k)$, namely

$$\begin{bmatrix} \Delta i_d^p(k+1) \\ \Delta i_q^p(k+1) \end{bmatrix} = \begin{bmatrix} i_d(k) & u_d(k) & 0 \\ i_q(k) & u_q(k) & w_e(k) \end{bmatrix} \begin{bmatrix} \delta_1 \\ \delta_2 \\ \delta_3 \end{bmatrix} \tag{21}$$

It can be seen that we can estimate the whole error vector $\boldsymbol{\vartheta} = [\delta_1, \delta_2, \delta_3]^T$ just by the q-axis error equation. Since the sample time of angular velocity is much slower than electrical dynamic, $w_e(k)$ staying constant during several sampling intervals of current is not accurate if only the q-axis dynamic information is used. Therefore, we firstly utilize the d-axis electrical information to identify the first two error term δ_1 and δ_2, and then independently analyze the third term δ_3.

Most of identification methods make full use of the dynamic information of motor system, but the obtained current and rotor position of the system is commonly accompanied by large noise signal, which causes failure of some ideal methods despite the presence of the filter. An effective method of interval-varying multi-innovation least squares (V-MILS) was proposed by Ding *et al.* [22]

to overcome missing measurement data. Thus, we combine a data selector and interval-varying multi-innovation least squares with forgetting factor (V-FF-MILS) algorithm to identify the error vector in d-axis prediction error equation.

Firstly, we define an integer sequence t_s, s=0, 1, 2... to label the signals passed from selector S_1 as

$$S_1[i_d(t_s), \Delta i_d^p(t_s + 1)] = \begin{cases} i_d(k) \in [a \quad b] \\ \Delta i_d^p(k + 1) \in [c \quad d] \end{cases} \tag{22}$$

t_s satisfies $0 = t_0 < t_1 < t_2 < \ldots < t_{s-1} < t_s$, and the interval $t_s - t_{s-1} \geq 1$. $[a \quad b]$ and $[c \quad d]$ are the reasonable ranges of d-axis current and its predictive error, respectively. Based on the principle of V-MILS, the predictive error model of d-axis current can be represented in linear regressive form,

$$y(p, t_s) = \Phi^T(p, t_s) \vartheta + v(p, t_s) \tag{23}$$

where

$$y(p, t_s) = [y(t_s), y(t_s - 1), \ldots, y(t_s - p + 1)]^T$$
$$= \left[\Delta i_d^p(t_s + 1), \Delta i_d^p(t_s), \ldots, \Delta i_d^p(t_s - p + 2)\right]^T \in \mathbb{R}^p$$
$$\Phi(p, t_s) = [\varphi(t_s), \varphi(t_s - 1), \ldots, \varphi_s(t_s - p + 1)] \in \mathbb{R}^{2 \times p}$$
$$\varphi(t_s) = [i_d(t_s), u_d(t_s)]^T, \vartheta = [\delta_1, \delta_2]^T, v(t_s) = [v_1(t_s), v_2(t_s)]^T$$

The interval-varying multi-innovation least squares with forgetting factor algorithm can be written as follows,

$$\hat{\vartheta}(t_s) = \hat{\vartheta}(t_{s-1}) + P(t_s) \Phi(p, t_s) \left[y(p, t_s) - \Phi^T(p, t_s) \hat{\vartheta}(t_{s-1}) \right] \tag{24}$$

$$P(t_s) = \frac{1}{\eta} P(t_{s-1}) - \frac{1}{\eta} L(t_s) \Phi^T(p, t_s) P(t_{s-1}) \tag{25}$$

$$L(t_s) = P(t_{s-1}) \Phi(p, t_s) \left[\eta I_p + \Phi^T(p, t_s) P(t_{s-1}) \Phi(p, t_s) \right]^{-1} \tag{26}$$

The forgetting factor satisfies $0 < \eta \leq 1$. The initial condition is set as $\hat{\vartheta}(0) = [1/p_0, 1/p_0]^T$, $P(0) = p_0 I_3$, $p_0 = 10^6$. The two error terms are approximated after finite recursive estimation. A termination principle of identification is designed as

$$\frac{t_{max} - t_{min}}{t_{max} + t_{min}} \leq \epsilon \tag{27}$$

t_{max} and t_{min} are the maximum and minimum value of identification during an interval with enough length. ϵ can reflect the fluctuation amplitude of online identifying results. After acquiring estimated δ_1 and δ_2, the third error term can be derived by real-time information of q-axis current equation. Our strategy is to command the angular velocity of rotor in a constant values; choose the effective q-axis data by a similar selector $S_2[i_q(k), \Delta i_q(k + 1), \omega_e(k)];$, and then online analyze the proportional relation between the error term result from δ_3 and ω_e in a linear regression technique. The ratio is the estimated value of δ_3.

5. PIL Test Results

The proposed methods were verified using processor-in-loop (PIL) simulation, which means the algorithm was conducted in a real microprocessor. The mathematical model of the plant was run in the PC host and exchanged data by serial communication. The real parameters of this PMSM were $R = 0.33 \ \Omega$, $L = 1.8$ mH, $\psi = 0.0145$ Wb, and $p = 4$. The DC bus voltage was $V_{dc} = 36$V. The speed controller utilized PI regulator with $k_p = 0.01$ and $k_i = 0.18$. Its sample time was set equal to 1 ms, and for current controller was 100 µs. The TI TMS32F28335 was selected as control processor. The identification for three uncertain term in motor model was tested in two designed cases. In Case 1, the nominal resistance R_0 was enlarged two times, with three times flux linkage, and the inductance was a quarter of real values. In Case 2, we assumed that inductance was known but a small deviation $\delta_2 = -0.2$ was artificially created to generate enough current prediction error. The nominal resistance R_0 and flux linkage ψ_0 were enlarged as $R_0 = 10R$ and $\psi_0 = 5\psi$, independently. Both cases considered the disturbance of noise signal. All the sampling signals were accompanied with a noise signal of which expectation equaled zero, and variance was $\sigma^2 = 0.1^2$. Results are presented and compared in the following figures. The labels 1-PCC, 2-PCC, and 3-PCC represent the proposed predictive current control methods. C-PCC, FOC, and DTC-SVM denote the conventional PCC, vector control, and direct torque control with a space vector modulator, respectively.

Figures 7a and 8 show the transient performances at the start-up phase after the speed command of 1000 r/min. It is illustrated that all methods have the same current transient response in Figure 7a,b, in which both approximate current reference value at the time of 0.005 s. In Figure 8, performance of speed response at start-up phase of all methods are compared. They all show a similar capability that reaches the speed command at about 0.2 s, which verifies that the proposed PCCs keep the superior transient performance of classical current regulation strategies.

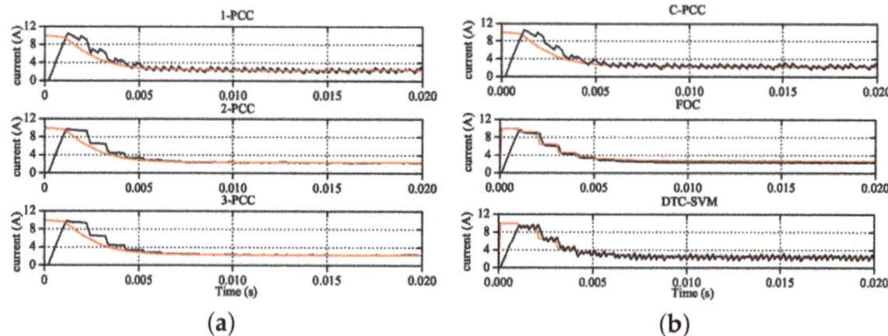

Figure 7. q-axis current response at start-up: (**a**) Performance of proposed PCCs, (**b**) Performance of comparison group.

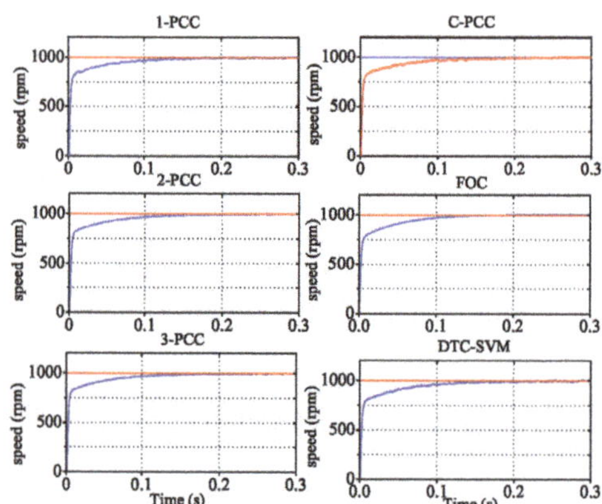

Figure 8. Speed response comparison at start-up.

In Figures 9 and 10, the transient performance of the proposed controller is further verified under the change of load and speed. Figure 9 compares the q-axis current tracking capability, where the red line denotes the reference values i_q^* from speed controller, it can be seen that C-PCC and 1-PCC have almost the same current response when the load is suddenly changed from 0.2 Nm to 0.4 Nm at the instant of 0.5 s. Performance of 3-PCC is close to that of FOC, which are both slightly better than 2-PCC with less ripple during tracking the reference current. DTC-SVM has the largest ripple during the transient process. In Figure 10, we gradually reduce the rotational speed from 1000 r/min to −1000 r/min at a constant rate within 1 s. The u-phase current regulation capabilities of 1-PCC and C-PCC show same tracking accuracy, but they both encounter large current perturbation when the motor is in counter rotation at the time of 4.0 s. 2-PCC, 3-PCC, and FOC show higher accuracy whether tracking phase current or speed. Similarly, DTC-SVM presents large ripple of u-phase current with the speed adjusted.

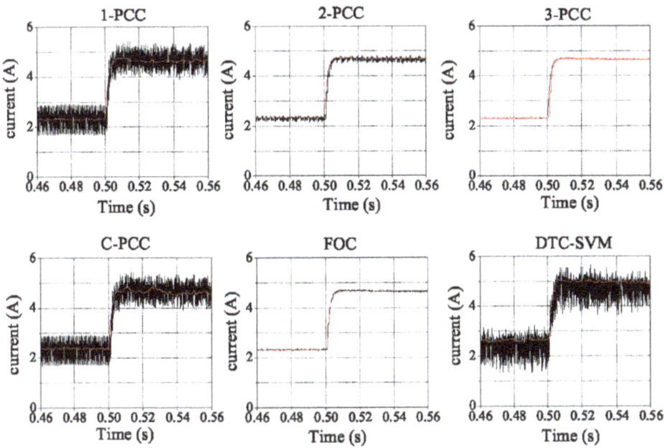

Figure 9. *q*-axis current tracking when occurring load perturbation.

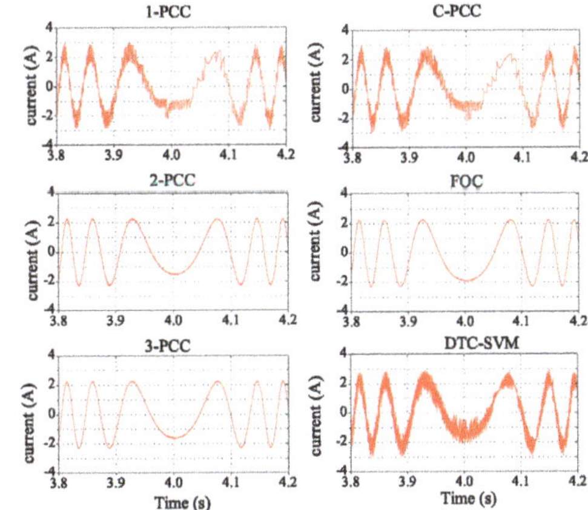

Figure 10. u-phase current response comparison when speed adjusted.

Figure 11a,b compares the steady-state performance of output torque within 0.1 s, and the standard deviations of *q*-axis current that represent the degree of torque ripple are 0.3689 and 0.3687 for C-PCC and 1-PCC, respectively, both inevitably leading to large output torque ripple. Relatively, the steady-state errors in 2-PCC and 3-PCC are much smaller, with 0.0576 and 0.0181 standard differences, respectively. These results are in accordance with other researcher's work [3]. FOC shows a high accuracy for torque, while DTC-SVM is the worst case. It is noted that the control parameters of FOC in this test were searched by optimization algorithm in Matlab software to make sure of the best output performance.

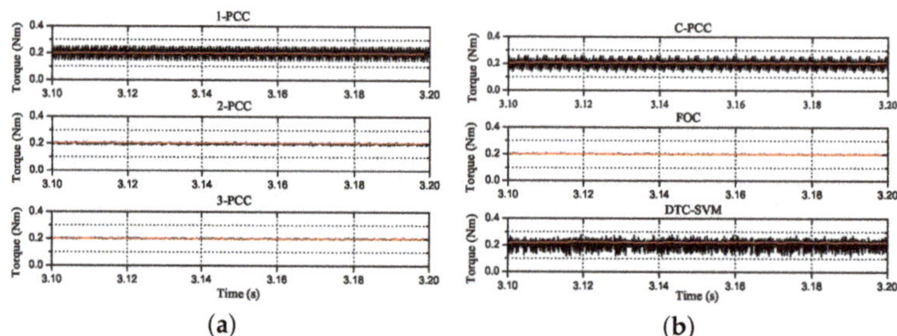

Figure 11. Output torque in steady-state: (**a**) Performance of proposed PCCs, (**b**) Performance of comparison group.

In Figure 12, the left column depicts total harmonic distortion (THD) of u-phase current of the proposed PCCs in steady state, with 20.05%, 5.84%, and 1.28% for 1-PCC, 2-PCC, and 3-PCC, respectively. The right column presents the corresponding THD of comparison group. Obviously, the optimal FOC has minimum THD, which is roughly equal to that of 3-PCC. The largest THD belongd to DTC-SVM. The harmonic quality of C-PCC and 1-PCC are similar based on the same current response and THD values. Therefore, the results in Figures 11a,b and 12 show that the proposed approaches have advantage in steady-state output accuracy, especially 2-PCC and 3-PCC.

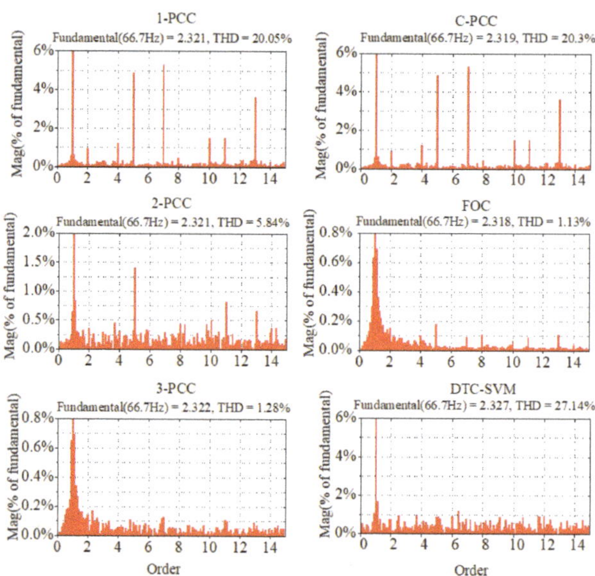

Figure 12. Harmonic quality comparison.

The quantifiable metrics in speed response time in start-up phase and the steady performance verified by harmonic component and current standard deviation of each approach are listed in Table 1. It can be seen that the transient response has no obvious difference. 2-PCC and 3-PCC present superior steady output accuracy. Although FOC has slightly better performance, it is difficult to obtain the optimal control parameters in practical application, while the proposed PCCs have no need of parameter tuning.

Table 1. Quantifiable metrics comparison.

	Speed Response Time	Harmonic Component	Current Standard Deviation
1-PCC	0.16 s	20.05%	0.3687
2-PCC	0.16 s	5.84%	0.0576
3-PCC	0.16 s	1.28%	0.0181
C-PCC	0.17 s	20.3%	0.3689
FOC	0.16 s	1.13%	0.0119
DTC-SVM	0.16 s	27.14%	0.4356

The execution duration of PCC has important influence on its output capability. The conventional predictive current controller is based on an enumerated search strategy causing a high computation time. As shown in Figure 13, the execution time of C-PCC is 88.7 µs, followed by 62.8 µs for 2-PCC and 60.8 µs for 3-PCC. 1-PCC takes the minimum computing time as it does not include the pulse modulator comparing with 2-PCC and 3-PCC. Therefore, the proposed calculation frame has advantage in alleviating computational burden.

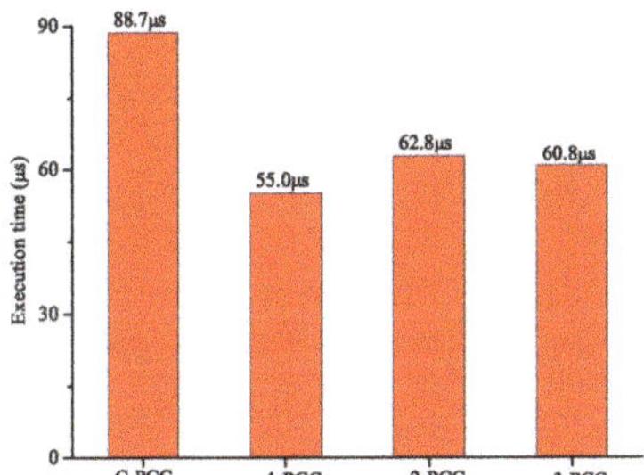

Figure 13. Execution time for each algorithm.

Figures 14a–18b show the performance of the proposed integrated identification algorithm. Estimated values of δ_1 and δ_2 are shown in Figure 14a,b, respectively. It can be seen that both estimated values in ideal environment have high steady-state accuracy of less than 5% error rates under the wave rate $\epsilon = 5\%$. However, when they are faced with noise condition, the results show small fluctuation and slight deviation from the actual values. When the termination parameter was set as $\epsilon = 15\%$, the final results are $\hat{\vartheta} = [0.1215, -0.1618]^T$. Compared with the actual values $\delta_1 = 0.1283$ and $\delta_2 = -0.1667$, the error rates are 5.3% and 2.9%, respectively.

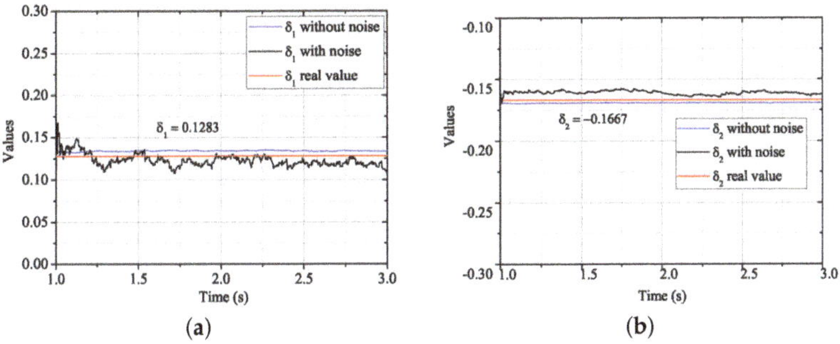

Figure 14. Error terms estimation in Case 1: (**a**) Estimation of δ_1, (**b**) Estimation of δ_2.

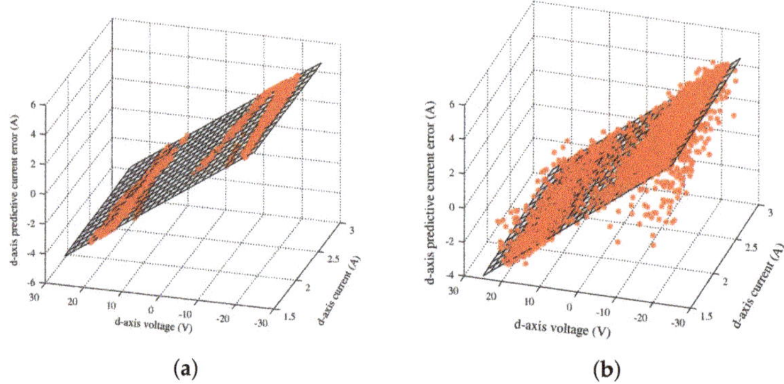

(a)   (b)

Figure 15. Original data of d-axis current error equation in Case 1: (**a**) Data in ideal environment, (**b**) Data in noise environment.

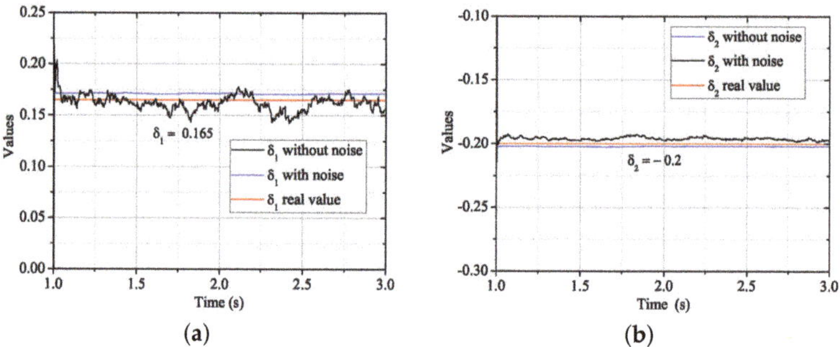

(a)   (b)

Figure 16. Error terms estimation in Case 2: (**a**) Estimation of δ_1, (**b**) Estimation of δ_2.

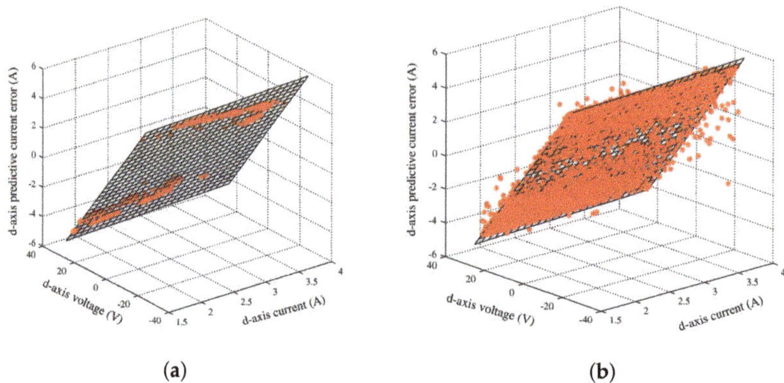

(a)   (b)

Figure 17. Original data of d-axis current error equation in Case 2: (**a**) Data in ideal environment, (**b**) Data in noise environment.

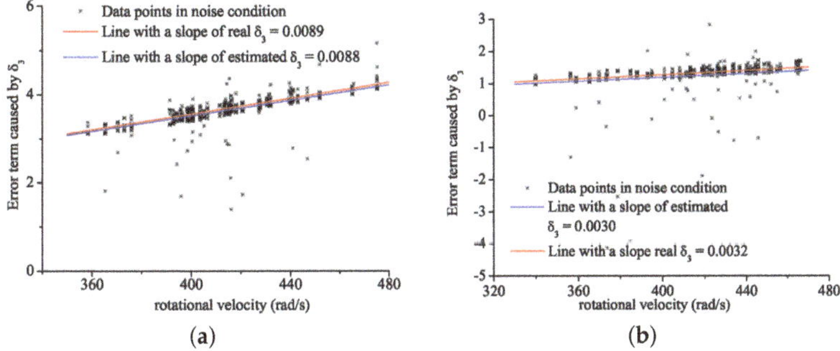

Figure 18. δ_3 estimation in noise condition: (**a**) Estimation results in Case 1, (**b**) Estimation results in Case 2.

The estimated results are the coefficients of planes in Figure 15a,b, which show the original data in different condition of *d*-axis current, voltage, and lumped predictive current error caused by δ_1 and δ_2. The left figure indicates that signals were sampled without the disturbance of noise, thus the data are almost located in the estimated plane, while some points deviate from estimated plane in the noisy environment.

Similarly, Figure 16a,b presents the identification process of Case 2. The ideal results also show an opposite steady-state error because of the deviation of *d*-axis current by the out-of-step rotor position signal. Estimated values in noise condition fluctuate in an allowable range $\epsilon = 15\%$, with $\hat{\delta}_1 = 0.1605$ and $\hat{\delta}_2 = -0.1955$, independently. The real values are $\vartheta = [0.165, -0.2]^T$, thus the identification error rates are 2.7% and 2.3%, respectively. Figure 17a,b shows the original data of *d*-axis current error equation in Case 2, which have the same results in Figure 15a,b.

After identifying the first two error factors, the motor is running at near constant velocity, and the *q*-axis signals are recorded online. The third error factor can be directly derived based on enough data during a short time in a simple linear regressive method. The slopes of the blue lines in Figure 18a,b represent estimated results of the third error factor $\hat{\delta}_3$ in noise condition of Cases 1 and 2, respectively. The estimated value of $\hat{\delta}_3$ equals 0.0088 in Case 1, which achieves a satisfactory accuracy comparing with the real value of 0.0089. In Case 2, the result also has a high precision with estimation of 0.0030 compare to the real value of 0.0032.

The results of identification are fed to the PMSM's model online to enhance the accuracy of predictive model. For instance, when the PMSM system was operated in the direct predictive current control (1-PCC) scheme, the nominal parameters took that of Case 1. In Figure 19a,b, before identification is finished, each predictive current in *dq*-axis presents a biggish error. However, errors of predictive *dq*-axis current are both substantially decreased after the feedback of the previous three estimated values at instant 1.7 s. As shown in Figure 20a, the magnitude of *d*-axis current error in whole frequency range is highly decreased, which is also shown for the error of *q*-axis current in Figure 20b. On the other hand, the fall of power spectral density (PSD) of errors of predictive current with noise shown in Figure 21a,b further verifies that identification and compensation are able to enhance the efficiency of electric drive system.

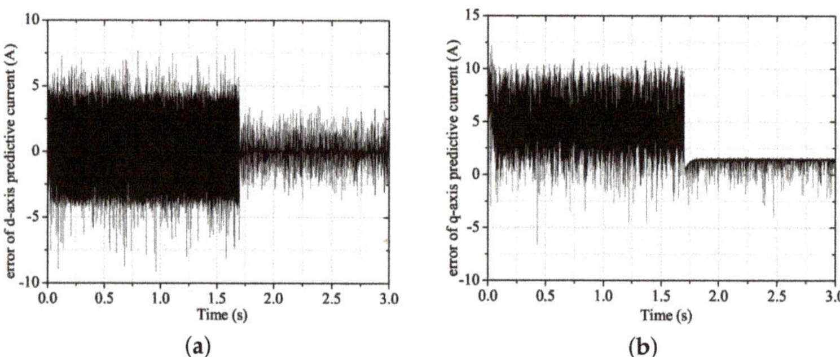

Figure 19. Errors of predictive current: (**a**) Error of *d*-axis predictive current, (**b**) Error of *q*-axis predictive current.

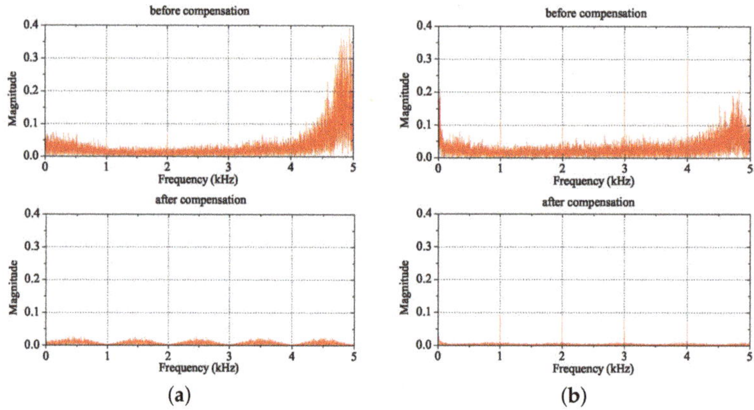

Figure 20. Analysis in frequency domain of current errors: (**a**) Changes of *d*-axis current error after compensation, (**b**) Changes of *d*-axis current error after compensation.

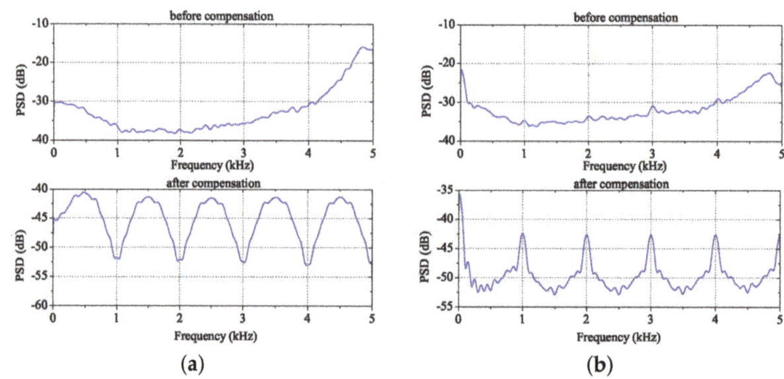

Figure 21. Analysis of power spectral density (PSD) of current errors with noise: (**a**) PSD of *d*-axis current error, (**b**) PSD of *q*-axis current error.

6. Conclusions

In this paper, a unified predictive current control scheme based on deadbeat principle for PMSM drive system is proposed. We define an alternative cost function that simplifies the calculation of optimal vectors. The results show that the performance of proposed 1-PCC is not inferior to the conventional PCC. Moreover, its computation time has an advantage over that of C-PCC. The merits of the proposed 2-PCC and 3-PCC are reflected in their much better steady-state accuracy with a little increase computing burden comparing with the 1-PCC. Detailed comparisons of transient and steady capabilities with optimal FOC and DTC-SVM were also conducted, and the results verify that the proposed PCCs have superior output performance. The proposed integrated identification method aiming to approximate the error term in predictive model shows allowable accuracy even in noisy environments. After compensation by the estimated values, the precision of predictive current was significantly improved. The analysis results in frequency domain and PSD further show the increase of system efficiency. In summary, the proposed unified control scheme with compensation is able to ease the application of model predictive control in electric drive system with large-scale parameter uncertainty.

Author Contributions: Conceptualization, P.T. and Y.D.; methodology, P.T.; software, P.T.; validation, P.T. and Z.L.; formal analysis, P.T.; investigation, P.T. and Z.L.; resources, Y.D.; data curation, Y.D.; writing—original draft preparation, P.T.; writing—review and editing, P.T. and Y.D.; visualization, P.T. and Z.L.; supervision, Y.D.; project administration, Y.D.; and funding acquisition, Y.D.

Funding: This research was funded by the Fundamental Research Funds for the Central Universities under Grant A03013023001015.

Conflicts of Interest: The authors declare no conflict of interest.

References

1. Bozorgi, A.M.; Farasat, M.; Jafarishiadeh, S. Model predictive current control of surface-mounted permanent magnet synchronous motor with low torque and current ripple. *IET Power Electron.* **2017**, *10*, 1120–1128. [CrossRef]

2. Kouro, S.; Cortes, P.; Vargas, R.; Ammann, U.; Rodriguez, J. Model Predictive Control-A Simple and Powerful Method to Control Power Converters. *IEEE Trans. Ind. Electron.* **2009**, *56*, 1826–1838. [CrossRef]

3. Morel, F.; Lin-Shi, X.; Retif, J.; Allard, B.; Buttay, C. A Comparative Study of Predictive Current Control Schemes for a Permanent-Magnet Synchronous Machine Drive. *IEEE Trans. Ind. Electron.* **2009**, *56*, 2715–2728. [CrossRef]

4. Zhang, Y.; Xu, D.; Liu, J.; Gao, S.; Xu, W. Performance Improvement of Model-Predictive Current Control of Permanent Magnet Synchronous Motor Drives. *IEEE Trans. Ind. Appl.* **2017**, *53*, 3683–3695. [CrossRef]

5. Davari, S.A.; Khaburi, D.A.; Kennel, R. An Improved FCS-MPC Algorithm for an Induction Motor With an Imposed Optimized Weighting Factor. *IEEE Trans. Power Electron.* **2012**, *27*, 1540–1551. [CrossRef]

6. Zhang, Y.; Yang, H. Model predictive torque control of induction motor drives with optimal duty cycle control. *IEEE Trans. Power Electron.* **2014**. [CrossRef]

7. Liu, Y.; Cheng, S.M.; Zhao, Y.Z.; Liu, J.; Li, Y.S. Optimal two-vector combination-based model predictive current control with compensation for PMSM drives. *Int. J. Electron.* **2019**, *106*, 880–894. [CrossRef]

8. Zhang, Y.; Xu, D.; Huang, L. Generalized Multiple-Vector-Based Mode Predictive Control for PMSM Drives. *IEEE Trans. Ind. Electron.* **2018**, *65*, 9356–9366. [CrossRef]

9. Chen, Z.; Qiu, J.; Jin, M. Adaptive finite-control-set model predictive current control for IPMSM drives with inductance variation. *IET Electr. Power Appl.* **2017**, *11*, 874–884. [CrossRef]

10. Lyu, M.; Wu, G.; Luo, D.; Rong, F.; Huang, S. Robust Nonlinear Predictive Current Control Techniques for PMSM. *Energies* **2019**, *12*. [CrossRef]

11. Zhang, X.; Zhang, L.; Zhang, Y. Model Predictive Current Control for PMSM Drives With Parameter Robustness Improvement. *IEEE Trans. Power Electron.* **2019**, *34*, 1645–1657. [CrossRef]

12. Türker, T.; Buyukkeles, U.; Bakan, A.F. A Robust Predictive Current Controller for PMSM Drives. *IEEE Trans. Ind. Electron.* **2016**, *63*, 3906–3914. [CrossRef]

13. Hoach The, N.; Jung, J.W. Finite Control Set Model Predictive Control to Guarantee Stability and Robustness for Surface-Mounted PM Synchronous Motors. *IEEE Trans. Ind. Electron.* **2018**, *65*, 8510–8519. [CrossRef]

14. Liu, X.; Zhang, Q. Robust Current Predictive Control-Based Equivalent Input Disturbance Approach for PMSM Drive. *Electronics* **2019**, *8*. [CrossRef]

15. Stumper, J.F.; Hagenmeyer, V.; Kuehl, S.; Kennel, R. Deadbeat Control for Electrical Drives: A Robust and Performant Design Based on Differential Flatness. *IEEE Trans. Power Electron.* **2015**, *30*, 4585–4596. [CrossRef]

16. Nalakath, S.; Preindl, M.; Emadi, A. Online multi-parameter estimation of interior permanent magnet motor drives with finite control set model predictive control. *IET Electr. Power Appl.* **2017**, *11*, 944–951. [CrossRef]

17. Gatto, G.; Marongiu, I.; Serpi, A. Discrete-Time Parameter Identification of a Surface-Mounted Permanent Magnet Synchronous Machine. *IEEE Trans. Ind. Electron.* **2013**, *60*, 4869–4880. [CrossRef]

18. Sim, H.W.; Lee, J.S.; Lee, K.B. On-line Parameter Estimation of Interior Permanent Magnet Synchronous Motor using an Extended Kalman Filter. *J. Electr. Eng. Technol.* **2014**, *9*, 600–608. [CrossRef]

19. Liu, Z.H.; Wei, H.L.; Zhong, Q.C.; Liu, K.; Xiao, X.S.; Wu, L.H. Parameter Estimation for VSI-Fed PMSM Based on a Dynamic PSO With Learning Strategies. *IEEE Trans. Power Electron.* **2017**, *32*, 3154–3165. [CrossRef]

20. Liu, K.; Zhu, Z.Q. Position-Offset-Based Parameter Estimation Using the Adaline NN for Condition Monitoring of Permanent-Magnet Synchronous Machines. *IEEE Trans. Ind. Electron.* **2015**, *62*, 2372–2383. [CrossRef]

21. Cortes, P.; Rodriguez, J.; Silva, C.; Flores, A. Delay Compensation in Model Predictive Current Control of a Three-Phase Inverter. *IEEE Trans. Ind. Electron.* **2012**, *59*, 1323–1325. [CrossRef]

22. Ding, F.; Liu, P.X.; Liu, G. Multiinnovation Least-Squares Identification for System Modeling. *IEEE Trans. Syst. Man Cybern. Part B (Cybern.)* **2010**, *40*, 767–778. [CrossRef] [PubMed]

 *electronics*

Article

A Novel Control Method of Clutch During Mode Transition of Single-Shaft Parallel Hybrid Electric Vehicles

Jingang Ding [1,2] and Xiaohong Jiao [1,*]

[1] School of Electrical Engineering, Yanshan University, Qinhuangdao 066004, China; dingjingang@126.com
[2] Power Center, Baic Motors CO.LTD Automotive Research Institute, Beijing 101106, China
* Correspondence: jiaoxh@ysu.edu.cn

Received: 23 November 2019; Accepted: 27 December 2019; Published: 30 December 2019

Abstract: The mode transition of single-shaft parallel hybrid electric vehicles (HEVs) between engine and motor has an important impact on power and drivability. Especially, in the process of mode transition from the pure motor-drive operating mode to the only engine-drive operating mode, the motor starting engine and the clutch control problem have an important influence on driving quality, and solutions have a bit of room for improving dynamic performance. In this paper, a novel mode transition control method is proposed to guarantee a fast and smooth mode transition process in this regard. First, an adaptive sliding mode control (A-SMC) strategy is presented to obtain the desired torque trajectory of the clutch transmission. Second, a proportional-integral (PI) observer is designed to estimate the actual transmission torque of the clutch. Meanwhile, a fractional order proportional-integral-differential (FOPID) controller with the optimized control parameters by particle swarm optimization (PSO) is employed to realize the accurate position tracking of the direct current (DC) motor clutch so as to ensure clutch transmission torque tracking. Finally, the effectiveness and adaptability to system parameter perturbation of the proposed control approach are verified by comparison with the traditional control strategy in a MATLAB environment. The simulation results show that the driving quality of the closed-loop system using the proposed control approach is obviously improved due to fast and smooth mode transition process and better adaptability.

Keywords: hybrid electric vehicles (HEVs); mode transition; adaptive sliding mode control (A-SMC); clutch actuator; PI observer; fractional order proportional-integral-differential (FOPID)

---

1. Introduction

Hybrid electric vehicles (HEVs) have received extensive attention from the automotive industry and academia, and are widely regarded as one of the most effective solutions to the growing use of petroleum fuels for transportation and environmental problems [1,2]. HEVs possessing multiple power sources have different working modes, the control strategy for the vehicle determines the working mode according to the driver's intention and the driving state of the car [3]. The hybrid system equipped with an automatic clutch and an electric motor (EM) has to perform the mode transition process from the pure electric driving mode to the engine-on driving mode in most conditions, in which the engine needs to be started and engaged into the driveline via the clutch [4–6]. In this process, the differences in response characteristics of each power source and the change of the state of the transmission or the clutch may cause a sudden change in engine or motor torque, which may lead to a non-negligible impact on the powertrain system [7,8]. Therefore, control approaches with high performance need to be developed to guarantee the fast and smooth process of mode transition.

During the mode transition of the single-shaft hybrid powertrain, the engine can be started and engaged into the driveline by the clutch. The performance of mode transition mainly relies on the quality of the clutch control [9,10], hence, the control of clutch engagement remains an important and challenging issue during mode transition. Some researchers have dedicated much valuable work to achieve fast and smooth clutch engagement control of mode transition for HEVs. For example, a nonlinear feedforward-feedback controller and PID algorithm are developed in [10] to effectively improve both riding comfort and engine-start time of a P2 hybrid vehicle. A robust control-based hierarchical mode transition control approach is proposed in [11] to reduce the mode transition time and obtain acceptable vehicle jerk. The double closed-loop control strategy is designed in [12] to control the clutch transmitting required torque of engine–start process, which contains a fuzzy method as an outer loop and a modified predictive functional method as an inner loop. Although the dynamic process and the control objectives of the clutch engagement process of mode transition of the single-shaft parallel hybrid powertrain are significantly different from that of the conventional vehicle, the dry clutch engagement control design of the conventional vehicle could be a reference to the clutch control of mode transition process. For the clutch control of the conventional vehicle, many effective approaches have been put forward. A model predictive control approach is developed in [13] to comply with constrains to ensure a comfortable lockup and to avoid the stall of the engine as well as to reduce the clutch engagement time. For example, a model predictive control method with the correction of clutch wear is proposed in [14] based on the estimation of resistance torque to achieve precise position control of the clutch. A modified predictive functional control approach with a sliding mode observer is proposed in [15] to obtain a satisfying performance for the automated clutch control of vehicle. The optimal control method is designed in [16] to generate the reference trajectories of the clutch slip speed and motor torque to provide satisfactory performance even under large variation of vehicle mass and road grade. Moreover, another approach has been extensively studied and utilized, such as triple-step control method [17,18] and backstepping control method [19] are adopted to solve the optimal trajectory tracking problem to achieve accurate clutch control.

Motivated by the aforementioned analysis, in this paper, aiming at further improving the control performance of clutch engagement during the mode transition from the pure motor-drive operating mode to the only engine-drive operating mode, a novel control method is proposed to solve the vehicle jerk and clutch slipping energy loss issues of mode transition process. The main feature of the proposed method has three aspects. One is that adaptive sliding mode control (A-SMC) is introduced into the clutch transmission torque command to generate the desired clutch transmission torque trajectory based on the rotating speed difference between the clutch driving and driven discs. The second is that the unmeasurable actual transmission torque of the clutch is estimated by a PI observer employing the EM speed. The third is that the realization of the clutch transmission torque tracking is accomplished through a fractional order proportional-integral-differential (FOPID) controller with the optimized control parameters by particle swarm optimization (PSO), which results in performing the accurate position tracking of the direct current (DC) motor clutch.

The rest of the paper is organized as follows. In Section 2, the dynamic models of the driveline of a single-shaft parallel powertrain and the automatic dry clutch actuating mechanism are built. In Section 3, the mode transition control strategy is designed. Simulation results and discussions are given in Section 4. Finally, the paper is concluded in Section 5.

2. System Modeling

In this paper, a single-shaft parallel powertrain with the automated mechanical transmission (AMT) is investigated, as shown in Figure 1. It consists of a conventional engine, automatic dry clutch, electric motor (EM), and a five-speed AMT. The clutch is controlled to perform the mode transition. In this system, the clutch is placed inside the drive motor. This HEV can operate in a variety of driving modes: pure electric driving mode, pure engine driving mode, hybrid driving mode, engine active charging mode, and regenerative braking mode. When the vehicle speed reaches the switching threshold, the hybrid control unit (HCU) will send the mode transition signal to the transmission control unit (TCU). When the TCU receives the mode transition signal, the clutch will quickly start to engage. Furthermore, the HCU will send the torque command to the EM control unit (MCU) and the engine control unit (ECU) to satisfy the driving torque of the vehicle. The main parameters of the single-shaft parallel vehicle are given in Table 1.

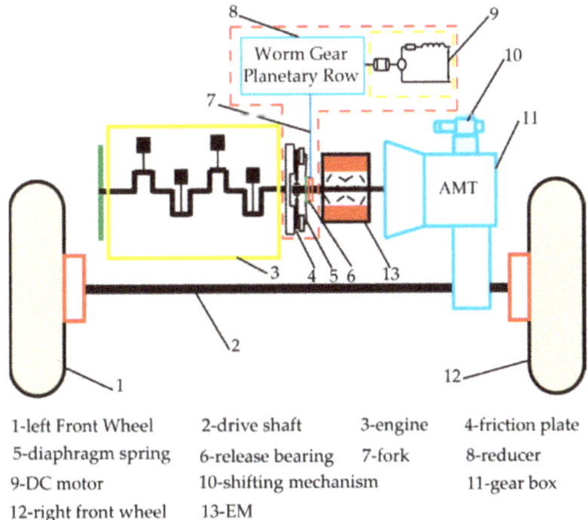

1-left Front Wheel   2-drive shaft   3-engine   4-friction plate
5-diaphragm spring   6-release bearing   7-fork   8-reducer
9-DC motor   10-shifting mechanism   11-gear box
12-right front wheel   13-EM

Figure 1. Structure of single-shaft parallel hybrid powertrain.

Table 1. Main parameters of single-shaft parallel hybrid vehicle.

Components	Description
Engine	Max torque: 150 Nm, peak power: 78 kW.
Electric motor (EM)	Permanent magnet synchronous motor (PMSM), max torque: 80 Nm, peak power: 30 kW.
Clutch	Max torque: 180 Nm.
Automated mechanical transmission (AMT)	Gear ratios: 3.583, 1.956, 1.327, 0.968, 0.702. Final gear ratio: 4.05.
Battery	Lithium iron phosphate battery, capacity: 10 Ah, nominal voltage: 288 V.

2.1. Powertrain System Dynamic Modeling

The models of main components of the single-shaft parallel hybrid powertrain related to the mode transition process are given in this part. Based on the configuration characteristics of the single-shaft parallel hybrid powertrain, the simplified schematic of the powertrain model in a mode transition process is presented for the design of the control strategy in Figure 2. Where T_e, T_c, T_m and T_r are the engine torque, the clutch torque, the EM torque and the vehicle resistant torque, respectively. ω_e and ω_m represent the engine speed and the EM speed, respectively. I_e, I_m and I_r represent the engine inertia moment, the EM inertia moment and the vehicle inertia moment, respectively. i_a and i_g are the gear ratio and the final gear ratio of AMT, respectively.

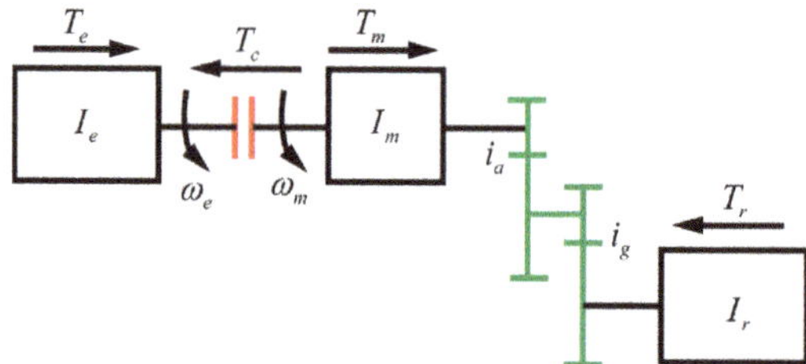

Figure 2. Simplified model sketch of mode transition.

Assuming that the tires slip is ignored, and according to the vehicle longitudinal dynamics, the vehicle resistant torque can be calculated as follows [20]:

$$T_r = (mgf \cos \alpha + \frac{C_D A v^2}{21.15} + mg \sin \alpha) r_w \tag{1}$$

where m, g, f, α, C_D, A, v and r_w are vehicle mass, gravity acceleration, rolling resistance coefficient, road slope angle, air resistance coefficient, frontal area of the vehicle, vehicle speed, and wheel radius, respectively.

In this paper, according to the clutch operating characteristics [12], the mode transition process from EM driving mode to engine driving mode is divided into three stages:

(i) Clutch disengaged stage

During this stage, the engine is still, the clutch is in the disengaged state, and the vehicle is solely driven by the EM. The dynamical equation of this stage can be expressed as follows:

$$T_m = T_v + (I_v + I_m)\dot{\omega}_m \tag{2}$$

where $I_v = I_r/(i_a^2 i_g^2)$ is the equivalent inertia moment of the vehicle at the EM output shaft. $T_v = T_r/(i_a i_g)$ is the equivalent load torque of hybrid powertrain on the EM output shaft.

(ii) Clutch slipping stage

In this stage, the clutch starts to engage after receiving the command from the TCU, and the EM is used to start the engine and drive the vehicle simultaneously. The dynamical equation during this stage can be expressed as follows:

$$T_m = T_v + T_c + (I_v + I_m)\dot{\omega}_m \tag{3}$$

$$T_c = I_e \dot{\omega}_e - T_e \tag{4}$$

$$\begin{cases} T_e = -b_e \omega_e, & \omega_e \leq \omega_{fire} \\ T_e = f(\beta, \omega_e), & \omega_e > \omega_{fire} \end{cases} \tag{5}$$

where b_e, ω_{fire} and β are engine frictional coefficient, engine idle speed and engine throttle opening angle, respectively. $f(\beta, \omega_e)$ represents that the engine torque is a nonlinear function of the engine speed and the throttle opening angle.

The clutch torque can be expressed as follows [21]:

$$T_c = \mu_k r_c n F_n sign(\omega_m - \omega_e) \tag{6}$$

where μ_k is the frictional coefficient, r_c is the effective radius of the clutch disk, n is the number of clutch plate, F_n is the pressing force on the plates. $sign(\cdot)$ is the sign function described as follows:

$$sign(\omega_m - \omega_e) = \begin{cases} 1, & \omega_m - \omega_e > 0 \\ 0, & \omega_m - \omega_e = 0 \\ -1, & \omega_m - \omega_e < 0 \end{cases} \tag{7}$$

At this stage, the engagement speed of the clutch has an important influence on the quality of the mode transition. In order to meet the longitudinal jerk of the vehicle, the engagement time should be minimized to avoid unnecessary energy loss of slipping friction.

(iii) Clutch engaged stage

In this stage, the clutch is engaged fast, the vehicle is driven by both engine torque and EM torque. The dynamic equation can be described as follows:

$$T_e + T_m = T_v + (I_e + I_m + I_v)\dot{\omega}_m \tag{8}$$

2.2. Clutch System Dynamic Modeling

The schematic diagram of the automatic dry clutch system is shown in the red dotted frame of Figure 1. As shown in Figure 1, the clutch system consists of dry clutch and actuating mechanism. The actuating mechanism includes a direct current (DC) motor used as impetus, worm gear and planetary row employed as transmission mechanism to reduce speed. The rotational motion of the motor is transformed into linear motion via the transmission mechanism, and the release bearing can be released by the release fork.

According to Newton's second law and Kirchhoff's law, the DC motor's voltage balance equation and torque balance equation [17] can be described as follows:

$$L_d \dot{i}_d = u_d - R_d i_d - K_b \omega_d \tag{9}$$

$$I_d \dot{\omega}_d = K_a i_d - K_c \omega_d - T_{CL} \tag{10}$$

where L_d, ω_d, I_d, i_d, R_d and u_d are the armature inductance, the motor rotational angle velocity, the motor rotational inertia moment, the armature current, the armature resistance and the battery voltage, respectively. K_a, K_b and K_c are the torque constant, the Back electromotive force (EMF) constant and the frictional coefficient, respectively. T_{CL} is the load torque of the motor.

Assuming that the actuating mechanism has no mechanical deformation, and according to mechanical transmission principle, the dynamics of the actuating mechanism is described as follows:

$$T_{CL} = I_{eq} \dot{\omega}_d + \frac{F_{xth} L}{\eta_t i_t} \tag{11}$$

$$\theta_d = \frac{y_c}{L} i_t \tag{12}$$

where I_{eq} is the equivalent inertia moment at the DC motor output shaft. L, i_t and η_t is the release fork arm length, the transmission ratio and the transmission efficiency of the actuating mechanism, respectively. θ_d is the motor rotational angle. F_{xth} is the return force of diaphragm spring. y_c is the release bearing position. Moreover, it is difficult to build an accurate model to calculate the return force of diaphragm spring, so the test dada from the clutch manufacturer is used in this paper to regard as a lookup table in simulation, the relationship between F_{xth} and y_c is shown in Figure 3.

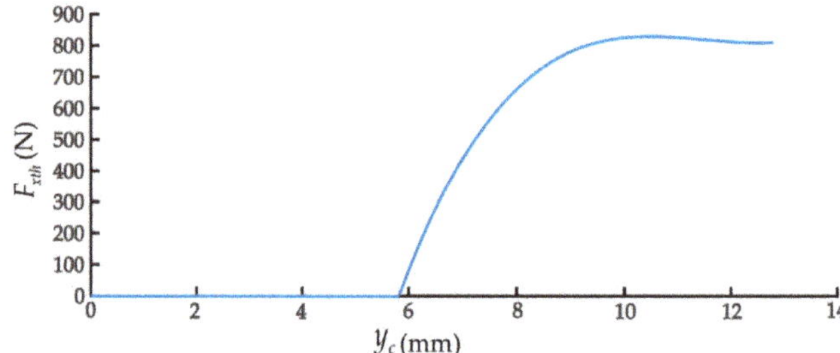

Figure 3. Relationship between spring pressure and release bearing position.

3. Mode Transition Controller Design

According to the stage division described in Section 2.1, the flow chart of the mode transition process from EM driving mode to engine driving mode is shown in Figure 4.

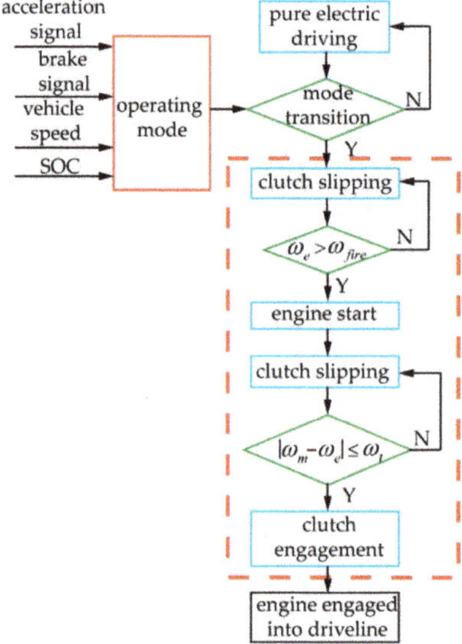

Figure 4. Flow chart of the mode transition from EM driving to engine driving.

Firstly, the clutch disengaged begins to engage. Due to the driven disk not touching the active disk, the clutch doesn't transmit torque and the engine speed is zero. Secondly, as the clutch position continuously increases, the driven disk of clutch moves past the free displacement, the clutch begins to slip and transmit torque to the engine for speed increase. When the engine speed reaches and exceeds the idle speed, the engine begins to fire and then will be synchronized with EM via the ECU. Finally, the clutch engages rapidly and the engine is engaged into the powertrain. However, the fixed clutch engagement speed will cause excessive slipping energy loss and vehicle jerk. Therefore, the mode transition control problem could be converted to clutch control problem to achieve the fast and smooth mode transition. The vehicle jerk and the slipping energy loss are crucial indexes of mode transition performance, which can be described as follows:

$$\begin{cases} j = \frac{da}{dt} \\ w = \int_{t1}^{t2} T_c |\omega_m - \omega_e| dt \end{cases} \tag{13}$$

where j is the vehicle jerk. w is the slipping energy loss. t_1 and t_2 are the starting time and stopping time of the clutch sliding process, respectively. a is the vehicle acceleration.

The effect of clutch control is characterized by the following requirement:

(1) The vehicle jerk is controlled within 10 m/s^3 during the clutch engagement process;
(2) The mode transition time should be within 1 s;
(3) The engine doesn't turn off during the mode transition process;
(4) The clutch slipping energy loss should be as little as possible during the mode transition process.

The schematic diagram of the proposed control approach to mode transition from EM driving mode to engine driving mode is shown in Figure 5.

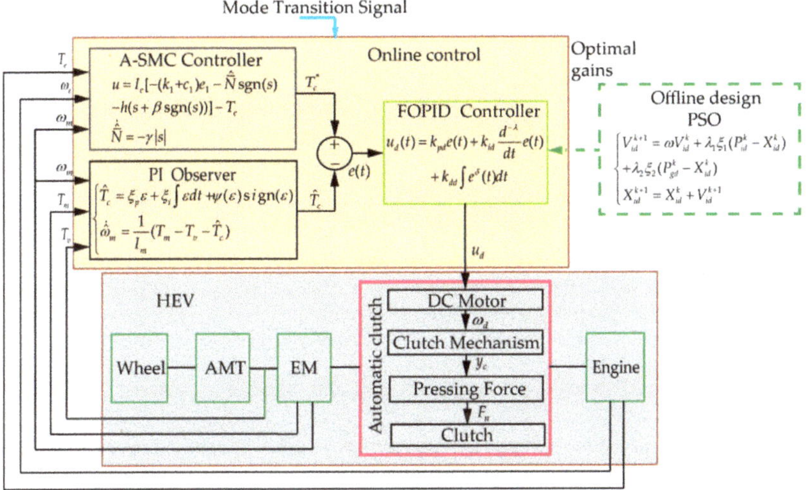

Figure 5. Schematic diagram of the proposed control method.

From Figure 5, it can be seen that the proposed control strategy consists of two parts: the design of clutch torque command and the torque tracking control. In the consideration of the nonlinear dynamics of the clutch engagement and some uncertainties exiting in this process, the A-SMC technique is adopted to generate the clutch torque command. Moreover, considering that the transmitted torque of the dry clutch is usually not measurable, a PI observer is designed to estimate the actual clutch torque. Meanwhile, a FOPID controller is adopted to control the torque tracking. Because FOPID is recognized to guarantee better performance besides the simplicity realization and fine stability of the classical PID controller due to the more flexibility and more robust in presence of parameters' uncertainties both in the controlled system and the controller itself [22,23]. Similar to classical PID, the controller parameters of FOPID should be defined in advance by the designer. To overcome time-consuming of the manual tuning and fractional calculus, an optimal tuning method is required to determine the control parameters of FOPID. In this paper, PSO is adopted to optimize the control parameters of FOPID offline.

3.1. Design of Clutch Torque Command based on A-SMC

Engine and EM should be considered together to design the clutch torque command. In light of nonlinear dynamics of the clutch engagement and uncertainty during the mode transition process, an A-SMC-based generator of the clutch torque command is designed.

According to the powertrain dynamic Equations (3) and (4), the state variables and the control variable of the powertrain system are selected as:

$$x = \omega_e, \ x_d = \omega_m, \ e_1 = x - x_d, \ e_{1\alpha} = \int (x - x_d)\, dt, \ u = T_c^*$$

the system model can be expressed as follows:

$$\begin{cases} \dot{e}_{1\alpha} = e_1 \\ \dot{e}_1 = \frac{1}{I_e}(u + T_e) + N \end{cases} \tag{14}$$

where N represents the whole disturbance resulting from the clutch engagement process including the effect from the EM, which is not measureable but bounded with $|N| \leq \overline{N}$, $\overline{N}$ is regarded as an unknown constant although it can be estimated by the limit value of the worst case, which can be handled by adaptive technique instead of robust control.

Therefore, the control law $u = T_c^*$ will be designed based on the sliding mode control and the adaptive backstepping technique. Furthermore, there is the following proposition.

Proposition 1. *For the powertrain system (3)–(4) operating the mode transition from the motor driving to engine driving, the reference trajectory of the clutch torque during the clutch engagement process can be generated by the output of an adaptive sliding mode controller designed as follows:*

$$T_c^* = u = I_e[-(k_1 + c_1)e_1 - \hat{\overline{N}}\mathrm{sgn}(s) - h(s + \beta\mathrm{sgn}(s))] - T_e \tag{15}$$

with adaptive update law:

$$\dot{\hat{\overline{N}}} = -\gamma|s| \tag{16}$$

and a sliding mode surface:

$$s = k_1 e_{1\alpha} + e_2 \tag{17}$$

where $e_2 = e_1 + c_1 e_{1\alpha}$. $\hat{\overline{N}}$ is the estimate of the unknown constant $\overline{N}$. c_1, k_1, h, β and γ are adjustable positive real values, and c_1, k_1, h should be chosen satisfying the following condition:

$$c_1 + hk_1^2 > 0, \ h(c_1 + k_1) > \frac{1}{4} \tag{18}$$

Proof. Firstly, a Lyapunov function is constructed as follows:

$$V_1 = \frac{1}{2}e_{1\alpha}^2 \tag{19}$$

considering e_2 as the virtual control, the time derivative of V_1 is calculated as:

$$\dot{V}_1 = e_{1\alpha}\dot{e}_{1\alpha} = e_{1\alpha}e_2 - c_1 e_{1\alpha}^2 \tag{20}$$

And then, a Lyapunov function for the system (14) is constructed as follows:

$$V_2 = V_1 + \frac{1}{2}s^2 \tag{21}$$

considering (20), it follows:

$$\dot{V}_2 = e_{1\alpha}e_2 - c_1 e_{1\alpha}^2 + s\left[k_1 e_1 + \frac{1}{I_e}(u + T_e) + N - \dot{x}_d + c_1 e_1\right] \tag{22}$$

Therefore, when the control law u is designed as (15) with (17), the following inequality holds.

$$\dot{V}_2 \le e_{1a}e_2 - c_1 e_{1a}^2 - hs^2 - h\beta|s| + \overline{N}|s| - \hat{\overline{N}}|s| = -e^T Q e - h\beta|s| + \widetilde{N}|s| \tag{23}$$

where

$$Q = \begin{bmatrix} c_1 + hk_1^2 & hk_1 - \frac{1}{2} \\ hk_1 - \frac{1}{2} & h \end{bmatrix}, e^T = [e_{1a}, e_2], \widetilde{N} = \overline{N} - \hat{\overline{N}}$$

Finally, a Lyapunov function for the whole closed-loop system is constructed as follows:

$$V_3 = V_2 + \frac{1}{2\gamma}\widetilde{N}^2 \tag{24}$$

considering (23), the time derivative of V_3 satisfies the following inequality.

$$\dot{V}_3 \le -e^T Q e - h\beta|s| + \widetilde{N}|s| - \frac{1}{\gamma}\widetilde{N}\dot{\hat{\overline{N}}} \tag{25}$$

then, considering the adaptive update law (16), it follows:

$$\dot{V}_3 \le -e^T Q e - h\beta|s| \tag{26}$$

Due to the condition (18), Q is a positive definite matrix so that $\dot{V}_3 \le 0$, $\forall e, s, \widetilde{N}$, thus, the whole system is Lyapunov stable at the equilibrium $(e_{1a} = 0, s = 0, \widetilde{N} = 0)$. Furthermore, based on the principle of LaSalle's invariant set, it follows that $e_1 \to 0$, namely, $\omega_e \to \omega_m$ as $t \to \infty$.

It means that if the clutch torque output is rendered as the trajectory T_c^* (15), the whole driveline dynamics during the clutch engagement process under the mode transition from motor driving to engine driving is stable and the engine speed can be synchronized with motor speed. Meanwhile, it should be noted from (26) that the convergence rate of $\omega_e \to \omega_m$ is dependent on Q, β, namely, is closely related to the choice of the adjustable parameters c_1, h, k_1, β. Thus, for better convergence performance, these parameters should not be too small, but they should not be too large to avoid oscillations in the trajectory under the premise of satisfying the condition (18). $\square$

3.2. Design of PI Observer

The dry clutch system has a strong nonlinearity and many factors have an effect on the clutch engagement performance, such as the plate temperature and the slipping speed. Therefore, it is difficult to accurately calculate and measure the transmitted torque of the dry clutch in engineering. For this regard, according to [24,25], a PI observer is designed to estimate the transmitted torque of the dry clutch.

The actual transmitted torque of the dry clutch can be expressed as follows:

$$\begin{cases} T_c = T_{c0} + \Delta(\varepsilon) \\ |\Delta(\varepsilon)| \le \psi(\varepsilon) \\ \varepsilon = \hat{\omega}_m - \omega_m \end{cases} \tag{27}$$

where T_{c0} is the initial transmitted torque, $\psi(\varepsilon)$ is the bound of the unknown function $\Delta(\varepsilon)$, and $\hat{\omega}_m$ is the estimate of the EM speed. Thus, a PI observer of the transmitted torque is designed as follows:

$$\begin{cases} \hat{T}_c = \xi_p\varepsilon + \xi_i\int \varepsilon dt + \psi(\varepsilon)\text{sign}(\varepsilon) \\ \dot{\hat{\omega}}_m = \frac{1}{I_m}(T_m - T_v - \hat{T}_c) \end{cases} \tag{28}$$

where ξ_p and ξ_i are the proportional gain and the integral gain of the PI observer, respectively.

The selection of the two parameters follows the principle that they should not be too small for the better convergence accuracy of the observer, but they should not be too large to avoid overshoot, that is to say, the compromise is required between better transient and steady-state performance. The detail can refer to the adjusting rules of PI gains presented in [25].

3.3. Design of Clutch Torque Tracking Controller

For guaranteeing the clutch transmission torque fast tracking the desired torque trajectory obtained by the SMC-based adaptive controller (21), a FOPID controller in the following form [26] is adopted to generate the voltage u_d of the DC motor to actuate the clutch.

$$u_d(t) = k_{pd}e(t) + k_{id}\frac{d^{-\lambda}}{dt}e(t) + k_{dd}\int e^\delta(t)dt \tag{29}$$

where $e(t) = T_c^*(t) - \hat{T}_c(t)$. k_{pd}, k_{id} and k_{dd} are the proportional, integral and differential gains, respectively, and λ and δ are positive coefficients of the fractional-order. The detail on the adjusting rules of FOPID parameters can refer to the description in [27], and to avoid the fractional calculus, an optimal tuning method is usually required to determine the control parameters of FOPID.

In this paper, the PSO algorithm is adopted to seek optimal gains of FOPID controller [28]. It should be noted that the PSO algorithm is executed offline to obtain the optimal parameters of FOPID. The schematic diagram of the parameter optimization offline and the flowchart of the PSO algorithm are shown in Figure 6.

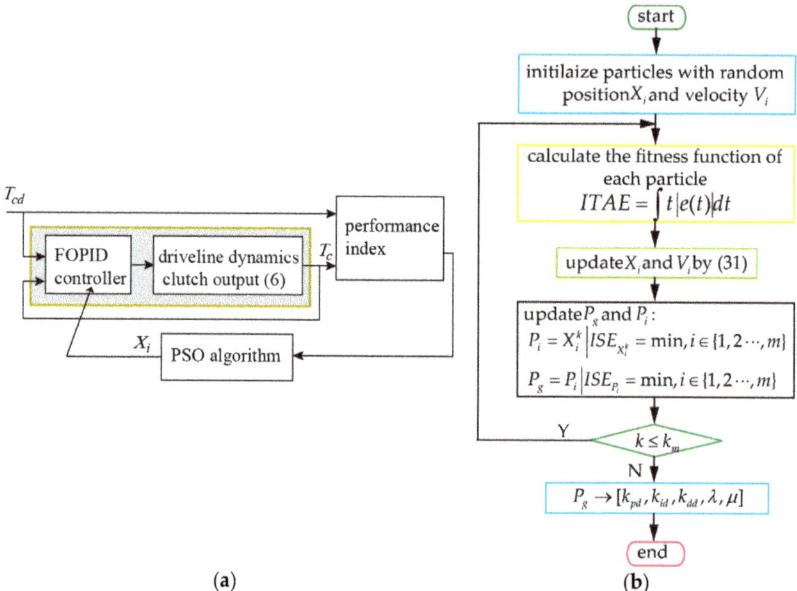

Figure 6. Schematic diagram of the particle swarm optimization (PSO) algorithm. (**a**) Schematic diagram of the parameter optimization offline, (**b**) flowchart of PSO.

In the algorithm, m is chosen as the number of particles, and each particle $X_i = [k_{pd}, k_{id}, k_{dd}, \lambda, \mu]$. The integrated time absolute error (ITAE) criterion is utilized as the fitness function, which can be described as follows:

$$ITAE = \int t|e(t)|dt \tag{30}$$

where $e(t) = T_{cd}(t) - T_c(t)$, $T_{cd}(t)$ is given as the idea trajectory of the clutch torque during the clutch engaging process, and $T_c(t)$ is the clutch torque output of the model (6) with the driveline dynamics established in Simulink. The update principles of the velocity V_i and position X_i of each particle during the iterations can be described as follows:

$$\begin{cases} V_{id}^{k+1} = \omega V_{id}^k + \lambda_1 \xi_1 (P_{id}^k - X_{id}^k) + \lambda_2 \xi_2 (P_{gd}^k - X_{id}^k) \\ X_{id}^{k+1} = X_{id}^k + V_{id}^{k+1} \end{cases} \tag{31}$$

where $i = 1, 2, \ldots, m$, $d = 1, 2, \ldots, 5$. k is iteration steps, $k = 0, 1, 2, 3, \ldots, k_m$. P_i, P_g are the individual optimal position and global optimal position, respectively. λ_1 and λ_2 are acceleration factors that are non-negative constants. ξ_1 and ξ_2 are random numbers between 0 and 1.

In this paper, considering the calculation accuracy and the large computing burden of PSO, choose $m = 20$, $k_m = 100$, select $X_{imax} = [1,1,1,2,2]$ and $X_{imin} = [0.00001,0.00001,0.00001,0.00001,0.00001]$. The parameters are offline optimized as: $k_{pd} = 0.5$, $k_{id} = 0.5$, $k_{dd} = 0.0001$, $\lambda = 0.97$, $\delta = 0.83$.

4. Simulation Results

In this section, the simulation model of single-shaft parallel hybrid powertrain is built based on MATLAB/Simulink. It is worth mentioning that the system modelling presented in Section 2 is a control-oriented model, but the model with high fidelity in the simulation is further detailed in order to simulate the real vehicle powertrain more realistically, such as more realistic nonlinearity and complexity are modeled for the dynamical description of the whole driveline system, and the delays of the control transmission and measurement are also taken into account. The simulations of launching and accelerating process are carried out to verify the effectiveness of the proposed method. The driver's demand torque in simulation is shown in Figure 7. The main system parameters of the studied hybrid powertrain are given in Table 2.

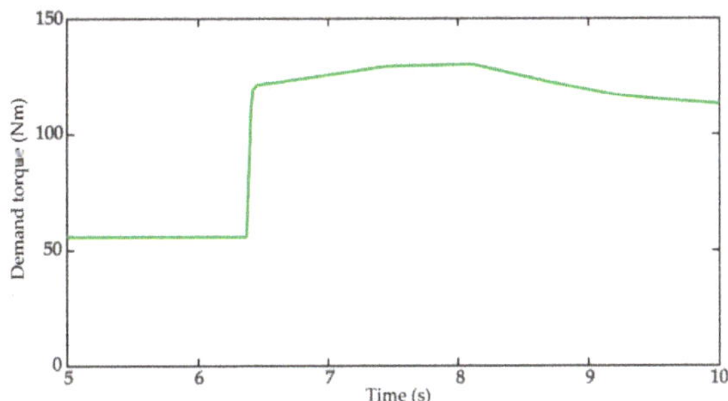

Figure 7. Driver's demand torque.

Table 2. Main parameters of the simulation model.

Parameter	Value	Parameter	Value
m	1250 kg	ω_{fire}	800 r/min
g	9.8 m/s^2	L_d	0.0003 H
f	0.022	I_d	0.0001 kg m^2
α	0	R_d	0.2 Ω
C_D	0.9	K_a	0.041 V/(rad/s)
A	2.4 m^2	K_b	0.041 Nm/A
r_w	0.293 m	K_c	0.001 Nm/(rad/s)
I_e	0.2 kg m^2	J_e	0.003 kg m^2
I_m	0.1 kg m^2	L	0.045 m
I_r	1.9 kg m^2	i_t	135.8
b_e	0.025 Nm/(rad/s)	η_t	0.61

Moreover, in the simulation, according to the selecting principle presented as in the previous section, the control parameters of the A-SMC, PI observer and FOPID controller are chosen as:

$$c_1 = 0.8, \ k_1 = 0.3, \ h = 0.9, \ \beta = 2.5, \ \gamma = 0.1, \ \xi_p = 0.005, \ \xi_i = 1.1,$$
$$k_{pd} = 0.5, \ k_{id} = 0.5, \ k_{dd} = 0.0001, \ \lambda = 0.97, \ \delta = 0.83 \tag{32}$$

4.1. Comparative Analysis

To evaluate the control performance of the proposed method, the comparison with the conventional engineering control method is presented. Firstly, Figure 8 shows the effect of the observer of the clutch torque.

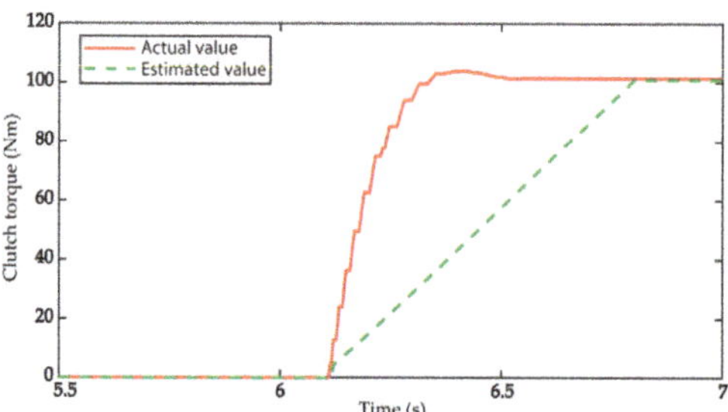

Figure 8. Result of PI observer.

As seen in the Figure 8, the estimated value of clutch torque converges to the actual value quickly in approximate 0.7 s, and the estimate error is within an acceptable scale.

Further, the comparison result during the mode transition from EM driving mode to engine driving mode is shown in Figure 9. Meanwhile, for clarifying the comparison on the control performance index, these comparison results are summarized in Table 3.

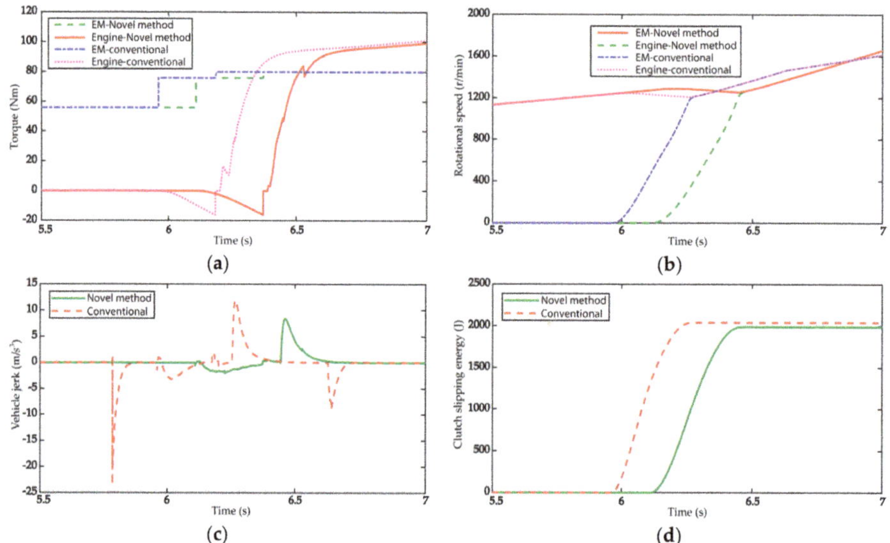

Figure 9. Comparison result in simulation. (**a**) Engine and EM torque, (**b**) engine and EM speed, (**c**) vehicle jerk, (**d**) clutch slipping energy loss.

Table 3. Summary of comparison results for two methods.

Performance Index	Proposed Novel Control	Conventional Control
Mode transition time [s]	0.75	0.6
Vehicle jerk [m/s^2]	8.1	23.9
Clutch slipping energy loss [J]	1976	2051

As shown in Figure 9a,b, and Table 3, the mode transition time of the proposed control method is 0.75 s, longer than that 0.6s of the conventional engineering control method, which is helpful to reduce the vehicle jerk. The whole mode transition process takes no more than 0.8 s and vehicle keeps accelerating in the novel method, which meets the control target. It can be seen from Figure 9c and Table 3, the maximum value of vehicle jerk 8.1 m/s^3 is less than 10 m/s^3 in the novel method during the mode transition process, which is far less than that 23.9 m/s^3 of conventional method. Moreover, the novel method generates less slipping energy loss 1976 J than that of conventional method 2015 J, shown in Figure 9d and Table 3. Comparison of vehicle jerk and slipping energy loss indicates the fast and smooth advantages of the novel control method.

4.2. Adaptability Analysis

Some system parameters may change as the system and road conditions change during the actual mode transition process of single-shaft parallel hybrid electric vehicles, such as the frictional coefficient b_e, vehicle mass m, rolling resistance coefficient f_r, road slope angle α, clutch mechanical deformation and so on. In this subsection, the vehicle jerk and clutch slipping energy loss of mode transition with different b_e, m and f_r are studied to verify the adaptability of the proposed novel control method. The simulation result is shown in Figure 10 and summarized in Table 4.

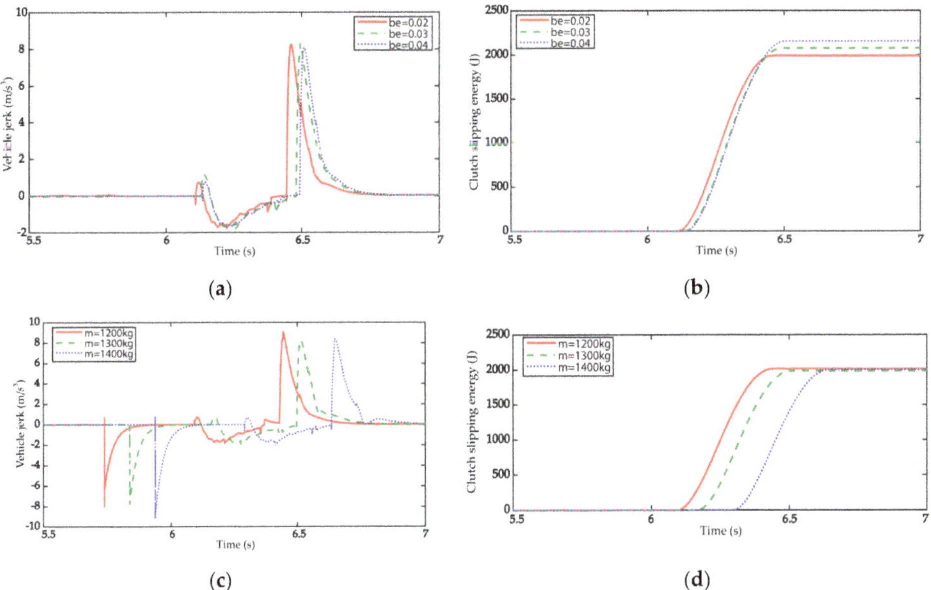

(a)

(b)

(c)

(d)

Figure 10. *Cont.*

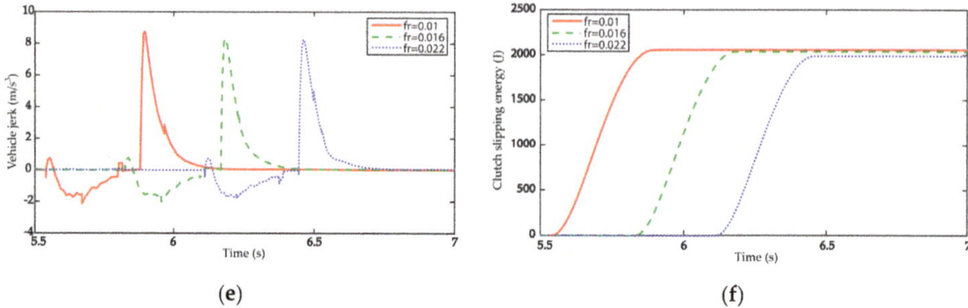

Figure 10. Comparison result for adaptability analysis. (**a**) Vehicle jerk under different engine frictional coefficients, (**b**) clutch slipping energy loss under different engine frictional coefficients, (**c**) vehicle jerk under different vehicle mass, (**d**) clutch slipping energy loss under different vehicle mass, (**e**) vehicle jerk under different rolling resistance coefficient, (**f**) clutch slipping energy loss under different rolling resistance coefficient.

Table 4. Summary of comparison results for different parameter variations.

Physical Name	Parameter Variation	Vehicle Jerk	Clutch Slipping Energy Loss
Frictional coefficient	$b_e = 0.02$	8.34	1990
	$b_e = 0.03$	8.31	2071
	$b_e = 0.04$	8.08	2152
Vehicle mass	$m = 1200$	9.01	2013
	$m = 1300$	8.21	1998
	$m = 1400$	8.45	2003
Rolling resistance coefficient	$f_r = 0.01$	8.95	2046
	$f_r = 0.016$	8.26	2031
	$f_r = 0.022$	8.23	1997

From Table 4, Figure 10a,b, it can be seen that both vehicle jerk and clutch slipping energy loss are very small in the different engine frictional coefficients 0.02, 0.03, 0.04. It indicates that the proposed novel control method is adaptive to different engine frictional coefficients. Similarly, as seen from Table 4, Figure 10c,d, and Figure 10e,f, the vehicle jerk and the clutch slipping energy loss are all very small in different vehicle masses 1200 kg, 1300 kg, 1400 kg, and different rolling resistance coefficients 0.01, 0.015, 0.022. These demonstrate that the proposed novel control method is adaptive to different vehicle masses and rolling resistance coefficients.

5. Conclusions

A novel control strategy of mode transition from EM driving to engine driving has been developed to reduce vehicle jerk and slipping friction energy loss for a single-shaft parallel hybrid powertrain in this paper. The proposed control strategy includes a SMC-based adaptive controller obtaining the desired clutch torque, a FOPID controller actuating the clutch mechanism to track the desired clutch torque, and a PI observer estimating the unmeasurable actual clutch torque. These three parts cooperate well to accomplish the mode transition from the EM driving mode to the engine driving mode with satisfactory control performance. For better tracking control performance, a PSO algorithm is used to optimize the control parameters of the FOPID offline. Both theoretical analysis and the simulation comparison with the traditional control strategy verify this point. Moreover, the simulation results under various physical parameter variations also show that the proposed novel control method is adaptive to system parameter perturbation.

In this paper, the effects of the clutch wear isn't taken into account in the controller design, so mode transition control considering more performance index including clutch wear will be studied in future work.

Author Contributions: J.D. wrote the original draft, designed the control method, drew the figures and performed the simulations. X.J. was responsible for supervising this research and involved in exchanging ideas, reviewing and revising the article draft. All authors have read and agreed to the published version of the manuscript.

Funding: This research was funded by the National Natural Science Foundation of China (Grant No. 61573304 and No. 61973265) and the Natural Science Foundation of Hebei Province (Grant No. F2017203210).

Conflicts of Interest: The authors declare no conflict of interest.

References

1. Fu, X.; Ma, J.; Zhang, Q.; Wang, C.; Tang, J. Torque Coordination Control of Hybrid Electric Vehicles Based on Hybrid Dynamical System Theory. *Electronics* **2019**, *8*, 712. [CrossRef]
2. Zhou, Y.; Ravey, A.; Péra, M. A Survey on Driving Prediction Techniques for Predictive Energy Management of Plug-in Hybrid Electric Vehicles. *J. Power Sources* **2019**, *412*, 480–495. [CrossRef]
3. Lei, Z.; Sun, D.; Liu, Y.; Qin, D. Analysis and Coordinated Control of Mode Transition and Shifting for A Full Hybrid Electric Vehicle Based on Dual Clutch Transmissions. *Mech. Mach. Theory* **2017**, *114*, 125–140. [CrossRef]
4. Berkel, K.; Veldpaus, F.; Hofman, T.; Vroemen, B. Fast and Smooth Clutch Engagement Control for A Mechanical Hybrid Powertrain. *IEEE Trans. Control Syst. Technol.* **2014**, *22*, 1241–1254. [CrossRef]
5. Yang, C.; Song, J.; Li, L.; Li, S. Economical Vehicle Launching and Accelerating Control for Plug-in Hybrid Electric Bus with Single-shaft Parallel Hybrid Powertrain. *Mech. Syst. Signal Process.* **2016**, *76–77*, 649–664. [CrossRef]
6. Yang, C.; Jiao, X.; Li, L.; Zhang, Y. Robust Coordinated Control for Hybrid Electric Bus with Single-Shaft Parallel Hybrid Powertrain. *IET Control Theory Appl.* **2015**, *9*, 270–282. [CrossRef]
7. Jauch, C.; Tamilarasan, S.; Bovee, K.; Güvenc, L. Modeling for Drivability and Drivability improving Control of HEV. *Control Eng. Pract.* **2018**, *70*, 50–62. [CrossRef]
8. Zhao, Z.; Lei, D.; Chen, J.; Li, H. Optimal Control of Mode Transition for Four-Wheel-Drive Hybrid Electric Vehicle with Dry Dual-Clutch Transmission. *Mech. Syst. Signal Process.* **2018**, *105*, 68–89. [CrossRef]
9. Smith, A.; Bucknor, N.; Yang, H.; He, Y. Controls Development for Clutch-Assisted Engine Starts in A Parallel Hybrid Electric Vehicle. *SAE Tech. Pap.* **2011**. [CrossRef]
10. Xu, X.; Liang, Y.; Jordan, M.; Tenberge, P. Optimized Control of Engine Start Assisted by The Disconnect Clutch in A P2 Hybrid Automatic Transmission. *Mech. Syst. Signal Process.* **2019**, *124*, 313–329. [CrossRef]
11. Yang, C.; Jiao, X.; Li, L.; Zhang, Y. A Robust H_∞ Control-Based Hierarchical Mode Transition Control System for Plug-in Hybrid Electric Vehicle. *Mech. Syst. Signal Process.* **2018**, *99*, 326–344. [CrossRef]
12. Wang, X.; Li, L.; Yang, C. Hierarchical Control of Dry Clutch for Engine-Start Process in A Parallel Hybrid Electric Vehicle. *IEEE Trans. Transp. Electrif.* **2016**, *2*, 231–243. [CrossRef]
13. Pisaturo, M.; Cirrincione, M.; Senatore, A. Multiple Constrained MPC Design for Automotive Dry Clutch Engagement. *IEEE/ASME Trans. Mechatron.* **2015**, *20*, 469–480. [CrossRef]
14. Li, L.; Wang, X.; Qi, X.; Li, X. Automatic Clutch Control Based on Estimation of Resistance Torque for AMT. *IEEE/ASME Trans. Mechatron.* **2015**, *21*, 1–13. [CrossRef]
15. Li, L.; Wang, X.; Hu, X.; Chen, Z. A Modified Predictive Functional Control with Sliding Mode Observer for Automated Dry Clutch Control of Vehicle. *J. Dyn. Syst. Meas. Control* **2016**, *138*. [CrossRef]
16. Gao, B.; Xiang, Y.; Chen, H.; Liang, Q. Optimal Trajectory Planning of Motor Torque and Clutch Slip Speed for Gear Shift of a Two-Speed Electric Vehicle. *J. Dyn. Syst. Meas. Control* **2015**, *137*. [CrossRef]
17. Gao, B.; Chen, H.; Liu, Q.; Chu, H. Position Control of Electric Clutch Actuator Using a Triple-Step Nonlinear Method. *IEEE Trans. Ind. Electron.* **2014**, *61*, 6995–7003. [CrossRef]
18. Gao, B.; Li, X.; Zeng, X.; Chen, H. Nonlinear Control of Direct-drive Pump-controlled Clutch Actuator in Consideration of Pump Efficiency Map. *Control Eng. Pract.* **2019**, *91*, 1–11. [CrossRef]
19. Gao, B.; Chen, H.; Sanada, K.; Hu, Y. Design of Clutch-Slip Controller for Automatic Transmission Using Backstepping. *IEEE/ASME Trans. Mechatron.* **2011**, *16*, 498–508. [CrossRef]
20. Park, J.; Choi, S.; Oh, J.; Eo, J. Adaptive torque tracking control during slip engagement of a dry clutch in vehicle powertrain. *Mech. Mach. Theory* **2019**, *134*, 249–266. [CrossRef]
21. Gatta, A.; Iannelli, L.; Pisaturo, M.; Senatore, A. A survey on modeling and engagement control for automotive dry clutch. *Mechatronics* **2018**, *55*, 63–75. [CrossRef]

22. Biswas, A.; Das, S.; Abraham, A.; Dasgupta, S. Design of fractional-order PID controllers with an improved differential evolution. *Eng. Appl. Artif. Intell.* **2008**, *11*, 1–8.

23. Monje, C.A.; Vinagre, B.A.; Feliu, V.; Chen, Y. Tuning and auto-tuning of fractional order controllers for industry applications. *Control Eng. Pract.* **2008**, *16*, 798–812.

24. He, L.; Shen, T.; Yu, L.; Feng, N. A Mode-Predictive-Control-Based Torque Demand Control Approach for Parallel Hybrid Powertrains. *IEEE Trans. Veh. Technol.* **2013**, *62*, 1041–1052. [CrossRef]

25. Li, Y.; Ang, K.; Chong, C. PID Control system analysis and design. *IEEE Control Syst. Mag.* **2006**, *26*, 32–41.

26. Igor, P. Fractional-Order Systems and $PI^{\lambda}D^{\mu}$-Controllers. *IEEE Trans. Autom. Control* **1999**, *44*, 208–214.

27. Ranganayakulu, R.; Uday, G.; Seshagiri, A. A comparative study of fractional order $PI^{\lambda}/PI^{\lambda}D^{\mu}$ tuning rules for stable first order plus time delay processes. *Resour. Eff. Technol.* **2016**, *2*, 136–152. [CrossRef]

28. Zafer, B.; Oguzhan, K. Comparison of PID and FOPID controllers tuned by PSO and ABC algorithms for unstable and integrating systems with time delay. *Optim. Control Appl. Methods* **2018**, *39*, 1431–1450.

Article

Effective Position Control for a Three-Phase Motor

Patxi Alkorta [1,*], Oscar Barambones [2], José Antonio Cortajarena [1], Itziar Martija [3] and Fco. Javier Maseda [3]

[1]   Engineering School of Gipuzkoa, University of the Basque Country; Otaola hirib. 29, 20600 Eibar, Spain; josean.cortajarena@ehu.eus
[2]   Engineering School of Vitoria, University of the Basque Country; Nieves Cano 12, 01006 Vitoria, Spain; oscar.barambones@ehu.eus
[3]   Engineering School of Bilbao, University of the Basque Country; Rafael Moreno 3, 48013 Bilbao, Spain; itziar.martija@ehu.eus (I.M.); fcjavier.maseda@ehu.eus (F.J.M.)
*   Correspondence: patxi.alkorta@ehu.eus; Tel.: +34-943-033-027

Received: 31 December 2019; Accepted: 27 January 2020; Published: 1 February 2020

Abstract: This document presents an efficient proportional derivative (PD) position controller for three-phase motor drives. The regulator has been designed in frequency domain, employing the direct–quadrature (d–q) synchronous rotating reference frame and the indirect vector control. The presented position regulator is easy to tune and incorporates a feed forward (FF) term to compensate effectively the effect of the load disturbance. This position controller has been validated experimentally by using two industrial three-phase motors: an induction motor (IM) of 7.5 kW and a permanent magnet synchronous motor (PMSM) of 3.83 kW. The inner proportional integral (PI) current loops of both machines have also been designed in the frequency domain. Each machine has connected in its shaft an incremental encoder of 4096 pulses per revolution, to measure the position. Several simulations and experimental tests have been carried out with both motors, in favorable conditions and also with various types of adversities (parametric uncertainties, unknown load disturbance and measurement noise in the position and current loops), getting very good results and suggesting that this controller could be used in the research area and also in the industry.

Keywords: indirect vector control; position control of motor; induction motor; permanent magnet synchronous motor

---

1. Introduction

Three-phase motors are widely used in civil engineering and in industry applications, where the most employed are the induction motor (IM) and permanent magnet synchronous motor (PMSM). IM are the most used machines in general (cranes, heavy industrial processes, motor pumps, trains, etc.), but the use of permanent magnet synchronous motors are increasing considerably last years (tool machines, electrical vehicles, etc.). IM offers good characteristics like a low maintenance, low moment of inertia, robust architecture, low torque ripple and low price. On the other hand, PMSM provides very good torque–weight ratio, high efficiency and accuracy. Depending on the application, both kinds of motors are controlled by using the Direct Torque Control (DTC) or Field Oriented Control (FOC) techniques. The FOC or vector control strategy is well settled such in civil engineering as in the industry, due to the fact that this technique gets excellent dynamics performance of the three-phase machine. That is due to using FOC techniques the three-phase machine is controlled like an independent excitation Direct Current (DC) machine, in which the electromagnetic torque and field flux are controlled separately [1]. This way, in the three-phase machine are also controlled independently the electromagnetic torque and rotor flux. The position control of the three-phase motor is an application that takes advantage the vector control benefits. One of the first research works related is [2], in which the experimental implementation by using FOC for speed and position control of IM and PMSM is presented. Very interesting position control schemes for IM are presented during the last decades, by using the classical proportional derivative (PD)/PID [3,4], and by employing advanced position regulators as a fuzzy [5], passivity-based [6], model predictive control [7], sliding-mode control [8,9], optimization-based sensorless [10] and poles placement [11]. Regarding PMSM, it is typical to find sensorless-based research works due to that position sensorless techniques are considerably easier to implement in these machines than in induction ones, where the classical PD/PID algorithms are employed in the position regulation loops [12,13]. However, to get high accuracy is necessary to have installed the position sensor in the system. Thus, several advanced position controllers as a sliding-mode control [14,15], fuzzy [16–18] and model predictive control [19,20] are presented. All these position controllers provide good dynamics and high accuracy, in spite of much of them not detailing the last one explicitly. Taking similar conditions in experiments, for IM [8] offers best steady-steady state accuracy, getting between 0 and 0.00038 rad of the position error, [7] and [11] provide an accuracy of 0.005 rad, while the steady-steady state accuracy of the presented proposal for IM is very close to it, 0.007 rad. On the other hand, for PMSM, [14] offers a position accuracy between 0.002 and 0.005 rad, [17] provides a position accuracy around 0.017 rad, while the presented proposal provides an accuracy of 0.002 rad. This way, advanced controllers provide excellent performance to these machines, but their control laws and tuning can be too complex to use them experimentally, i.e., in the industry, much of them do not offer a clear way or rules to tune. The real applications area needs efficient controllers but at the same time be easy to tune. The presented position regulator provides good dynamics and high accuracy, it is easy to tune with a simple control law and it is valid for different three-phase motors. In [4], a PD position controller tuned in the frequency domain for IM was presented and experimentally validated. The presented proposal in this paper takes this PD regulator and extends to the PMSM the experimental validation. Moreover, practical tuning recommendations for both motors are given and experimental comparatives between them are done.

The goal of this paper is to offer an effective and easy to tune PD position controller for the most employed three-phase machines, that is IM and PMSM. The paper is organized as follows. Section 2 describes the dynamics of the motors and the position and current regulators design in the frequency domain with the study of stability. Section 3 presents the employed two tests benches, and the tests (simulations and real experiments) done with both motors. Finally, Section 4 discusses and gets the conclusions.

2. Three-Phase Motor Position Proportional Derivative Controller Design

2.1. Induction Motor Dynamics

Induction motor dynamics can be expressed by employing the next five differential equations, expressed in the direct–quadrature (d–q) synchronous rotating reference frame and taking that the q (quadrature) component of the stator current (electromagnetic torque current component, i_{sq}) and the d (direct) component of the stator current (rotor flux current component, i_{sd}) are decoupled, due to that the q component of rotor flux is null, $\psi_{rq} = 0$ and this way the rotor flux is the same as its d (direct) component, $\psi_r = \psi_{rd}$, [21],

$$T_e - T_L = J\frac{d^2\theta_m}{dt} + B_v\frac{d\theta_m}{dt} \tag{1}$$

$$T_e = \frac{3}{4}p\frac{L_m}{L_r}\psi_{rd}i_{sq} \tag{2}$$

$$\frac{d\psi_{rd}}{dt} = \frac{R_r}{L_r}L_m i_{sd} - \frac{R_r}{L_r}\psi_{rd} \tag{3}$$

$$v_{sd} = R_s i_{sd} + \sigma L_s\frac{di_{sd}}{dt} + \frac{L_m}{L_r}\frac{d\psi_{rd}}{dt} - \omega_s\sigma L_s i_{sq} \tag{4}$$

$$v_{sq} = R_s i_{sq} + \sigma L_s \frac{di_{sq}}{dt} + \omega_s \frac{L_m}{L_r} \psi_{rd} + \omega_s \sigma L_s i_{sd} \tag{5}$$

where T_L is the load disturbance, θ_m is the mechanical position, ω_s is the synchronous speed, v_{sd} is the d component of the stator voltage and v_{sq} is the q component of the stator voltage. The rest of the parameters are described in Section 3.1.1. As it is well known, the i_{sd} current is employed to regulate the ψ_r rotor flux of the machine. This flux keeps constant its value, the typically rated value. Then, considering the steady state and taking Equation (3), it is possible to get the relation between both variables,

$$\psi_{rd} = L_m i_{sd} \tag{6}$$

and taking Equation (2) of the electromagnetic torque, it can be expressed as

$$T_e = K_T i_{sq} \tag{7}$$

where the K_T torque constant is defined as

$$K_T = \frac{3p}{4} \frac{L_m}{L_r} \psi_{rd} = \frac{3p}{4} \frac{L_m^2}{L_r} i_{sd} \tag{8}$$

Now Equation (7) can be replaced in Equation (1), obtaining the following Expression (9), which contains the first three equations.

$$K_T i_{sq} = J \frac{d^2 \theta_m}{dt} + B_v \frac{d\theta_m}{dt} + T_L \tag{9}$$

This way, Equation (9) contains the electromagnetic and mechanical equations, while Equations (4) and (5) contain the electrical equations.

2.2. Permanent Magnet Synchronous Motor Dynamics

PMSM motor dynamics are shown by using the next four differential equations, where like in the previous case (IM), all of them are expressed in a d–q synchronous rotating reference frame and supposing that the i_{sq} electromagnetic torque current component and the i_{sd} rotor flux current component are decoupled, that is $\psi_{rq} = 0$ and $\psi_r = \psi_{rd}$. Moreover, as the PMSM has its own rotor flux generated by its rotor magnets, that is ψ_M (constant), then rotor flux generated electrically is usually fixed to zero, and consequently $i_{sd} = 0$, [22],

$$T_e - T_L = J \frac{d^2 \theta_m}{dt} + B_v \frac{d\theta_m}{dt} \tag{10}$$

$$T_e = \frac{3p}{4} \psi_M i_{sq} \tag{11}$$

$$v_{sd} = R_s i_{sd} + L_{sd} \frac{di_{sd}}{dt} - \omega_r L_{sq} i_{sq} \tag{12}$$

$$v_{sq} = R_s i_{sq} + L_{sq} \frac{di_{sq}}{dt} + \omega_r (L_{sd} i_{sd} + \psi_M) \tag{13}$$

where ω_r is the electrical rotor speed, and the rest of the parameters are described in Section 3.1.2. As the ψ_M is constant, then Equation (11) of the electromagnetic torque can be rewritten as follows,

$$T_e = K_T i_{sq} \tag{14}$$

where the K_T torque constant is,

$$K_T = \frac{3p}{4} \psi_M \tag{15}$$

As in the IM case, Equation (14) can be replaced in Equation (10), obtaining the following Expression (17), which contains the first two equations of the dynamics of PMSM,

$$K_T i_{sq} = J\frac{d^2\theta_m}{dt} + B_v\frac{d\theta_m}{dt} + T_L \tag{16}$$

In this sense, Equation (16) contains the electromagnetic and mechanical equations, while Equations (12) and (13) contain the electrical equations.

2.3. Position Proportional Derivative with a Feed-Forward Controller Design

Figure 1 shows the complete vector position control scheme for the three-phase motor scheme, employing the d–q synchronous rotating reference frame. The objective of this scheme is to regulate the θ_m mechanical position of the motor's shaft, which is the main variable (outer loop), by using the $PD_{\theta m}$ regulator. This regulator provides the best electromagnetic torque current reference, $i_{sq}{}^*$, according to (7) and (15). On the other hand, the rotor flux current reference, $i_{sd}{}^*$, imposes the value of the rotor flux that the motor needs. This way, the inner loops (current loops) are controlled by two PI regulators, where their objective is to get the best voltage references, $v_{sd}{}^*$ and $v_{sq}{}^*$, corresponding to their current references, $i_{sd}{}^*$ and $i_{sq}{}^*$, respectively. The $ABC \to dq$ block implements the Clarke–Park transformations, and the $dq \to ABC$ block the inverse Park–Clarke transformations. The voltage sourced inverter (*VSI*) block represents the three-phase Voltage Sourced Inverter, which receives the control signals from the *SVPWM* (Space Vector Pulse Width Modulation) block. This modulator generates the control signals according to the voltage references obtained from the inner loops, asking to the inverter to generate these voltages in its three-phase output. *Calc* θ_s calculates the d–q synchronous rotating reference frame's angle with respect the α–β stationary reference frame. For that it is used the indirect vector control, by integrating the ω_s synchronous speed. Thus for the PMSM motor, this speed is the same as the ω_r electrical speed

$$\omega_s = \omega_r \tag{17}$$

which is obtained from the ω_m mechanical speed (measured) by using this expression,

$$\omega_r = \omega_m\frac{p}{2} \tag{18}$$

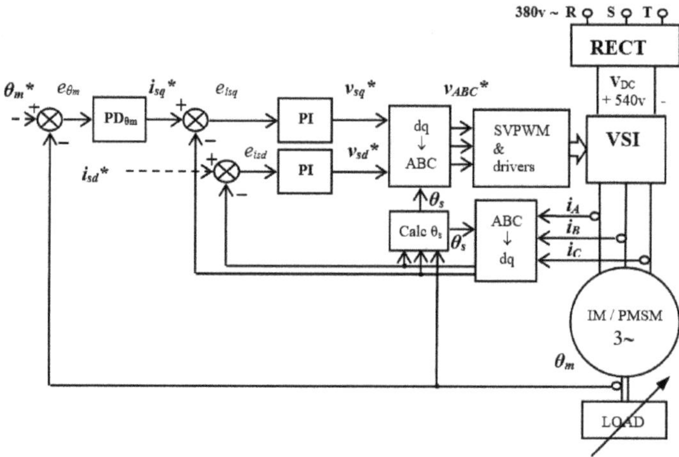

Figure 1. The proportional derivative position indirect vector control scheme for the three-phase motor.

On the other hand, for the IM motor, this speed is given by Equation (19),

$$\omega_s = \omega_r + \omega_{slip} \tag{19}$$

where the ω_{slip} slip speed is calculated easily with Equation (20)

$$\omega_{slip} = \frac{L_m}{\psi_r} \frac{R_r}{L_r} i_{sq} \tag{20}$$

As we have seen in the previous subsections, both three-phase machines have the same electromechanical Equations (9) and (16). This way, the design of the position regulator for both is the same. Taking into account that the dynamics of the two PI current regulators, the SVPWM modulator and the VSI inverter are faster than the electromechanical and PD dynamics, then the following simplified scheme can be obtained,

Thus, taking the electromechanical equation and considering null the load disturbance ($T_L = 0$) and applying Laplace's transformation, the following transfer function can be obtained

$$G(s) = \frac{\theta_m(s)}{I_{sq}(s)} = \frac{K_T/B_v}{1 + s\tau_m} \frac{1}{s} = \frac{K_T}{Js + B_v} \frac{1}{s} \tag{21}$$

where $\tau_m = J/B_v$ is the mechanical time constant. Now taking the causal PD regulator

$$PD_{\theta m}(s) = \left(Kp_{\theta m} + \frac{sKd_{\theta m}}{s + val} \right) \tag{22}$$

where $Kp_{\theta m}$ and $Kd_{\theta m}$ are the proportional and derivative coefficients, respectively, of the PD regulator, and *val* is the absolute value of the pole to convert this regulator in the causal transfer function. This value has to be large enough to simplify its dynamics compared with the zero's dynamics.

The open loop transfer function (OLTF) of the simplified diagram presented in Figure 2, can be obtained as:

$$OLTF_{\theta m}(s) = \left(Kp_\theta + \frac{sKd_{\theta m}}{s + val} \right) \frac{K_T}{Js + B_v} \frac{1}{s} \tag{23}$$

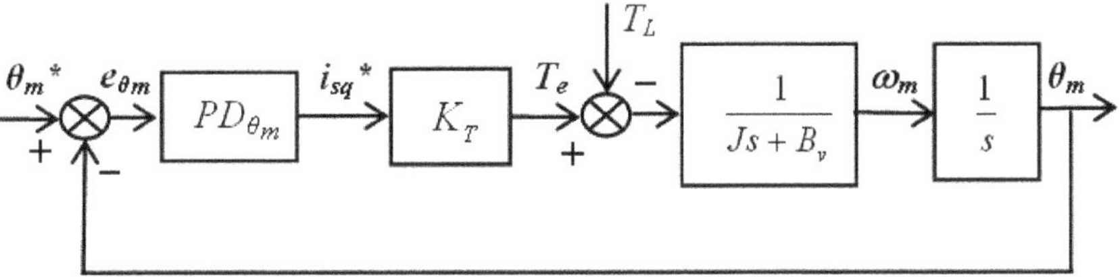

Figure 2. Simplified scheme for proportion derivative (PD) position indirect vector control for the three-phase motor.

It is well known from engineering control that at the $\omega_{GC\theta m}$ gain cross-over frequency the module is the unity

$$\left| OLTF_{\theta m}(s) \right|_{s=j\omega_{GC\theta m}} = 1 \tag{24}$$

and the phase Equation is (26), where $PM_{\theta m}$ is the phase margin of the position loop

$$\arg\left(OLTF_{\theta m}(s) \right)\big|_{s=j\omega_{GC\theta m}} = -180^\circ + PM_{\theta m} \tag{25}$$

Now, taking the system of these two Equations, (25) and (26), where the $\omega_{GC\theta m}$ and $PM_{\theta m}$ are the design data, and by solving them, the two coefficients of the PD regulator ($Kp_{\theta m}$ and $Kd_{\theta m}$) necessary for that will be obtained. For this resolution, the following group of intermediate variables are necessary (den, a, b)

$$den = (B_v + J \cdot val)^2 \omega_{GC\theta m}{}^4 + (B_v \cdot val \cdot \omega_{GC\theta m} - J \cdot \omega_{GC\theta m}{}^3)^2 \tag{26}$$

$$K_T{}^2 \cdot [(Kd_{\theta m} + Kp_{\theta m})^2 \omega_{GC\theta m}{}^2 + (Kp_{\theta m} \cdot val)^2] = den \tag{27}$$

$$a = \omega_{GC\theta m}{}^3 (B_v + J \cdot val) + \omega_{GC\theta m} tg(PM_{\theta m})(B_v \cdot val \cdot \omega_{GC\theta m} - J\omega_{GC\theta m}{}^3) \tag{28}$$

$$b = \left(val \cdot \omega_{GC\theta m}{}^2 \cdot tg(PM_{\theta m}) - \omega_{GC\theta m}{}^3\right)(B_v + J \cdot val) \\ - \left(\omega_{GC\theta m} tg(PM_{\theta m}) + val\right)(B_v \cdot val \cdot \omega_{GC\theta m} - J\omega_{GC\theta m}{}^3) \tag{29}$$

$$Kd_{\theta m} = \frac{1}{K_T} \sqrt{\frac{den}{\omega_{GC\theta m}{}^2(1 + 2 \cdot a/b) + (a/b)^2(\omega_{GC\theta m}{}^2 + val^2)}} \tag{30}$$

$$Kp_{\theta m} = Kd_{\theta m} \frac{a}{b} \tag{31}$$

The effect of the T_L load disturbance can produce important problems when the objective is to get an efficient position tracking. To avoid this problem, it can be added as a Feed-Forward (FF) action in the controller, where for that it is necessary to know the disturbance's value, Figure 3. This way, the control law contains both actions, PD and FF,

$$i_{sq}^*(t) = i_{sq\_pd}^*(t) + i_{sq\_ff}^*(t) = i_{sq\_pd}^*(t) + \hat{T}_L(t)/K_T \tag{32}$$

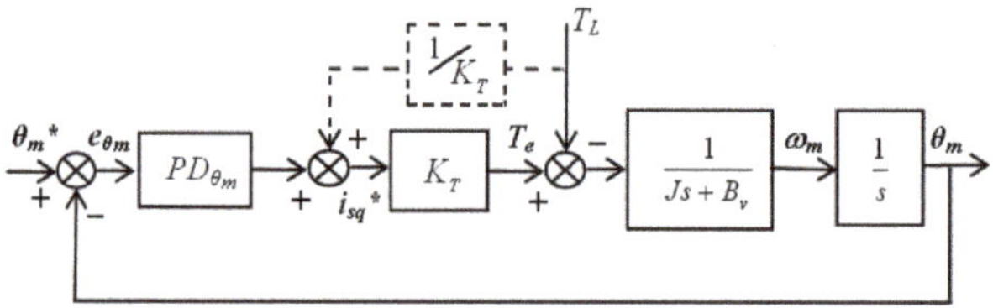

Figure 3. Simplified scheme for PD position control with Feed-Forward (FF) action for the three-phase motor.

As in many cases it is impossible to know the load disturbance, and therefore an easy way to solve this issue is to estimate it by using a load estimator or observer. A simple and effective load observer is obtained directly from the electromechanical Equation (9)/(16),

$$\hat{T}_L = T_e - J\frac{d^2\theta_m}{dt} - B_v\frac{d\theta_m}{dt} \tag{33}$$

2.4. Current Proportional Integral Controllers Design

The design of the current regulators or inner loops for both motors is very similar. In the IM motor, the design of the two loops is done only for one, that is for i_{sq} current, considering that for i_{sd} current the regulator is the same, [21]. This is due to that in both loops the σL_s factor is the same, but in the function of the expected results, it could be necessary for an independent tuning of both. Then for the IM machine, taking the Equations (4) and (5), and considering that the last two terms of each one are disturbances and neglecting them, where taking into account that the dynamics of the SVPWM modulator and VSI inverter are faster than the electrical and PI dynamics (Figure 1), then the following transfer function and simplified scheme can be obtained for the i_{sq} current loop,

$$\frac{I_{sq}(s)}{V_{sq}^*(s)} = \frac{1}{s\sigma L_s + R_s} \tag{34}$$

From Figure 4, the open loop transfer function (OLTF) of the simplified diagram is obtained,

$$OLTF_i(s) = \left(Kp_i + \frac{Ki_i}{s}\right)\left(\frac{1}{R_s + s\sigma L_s}\right) \tag{35}$$

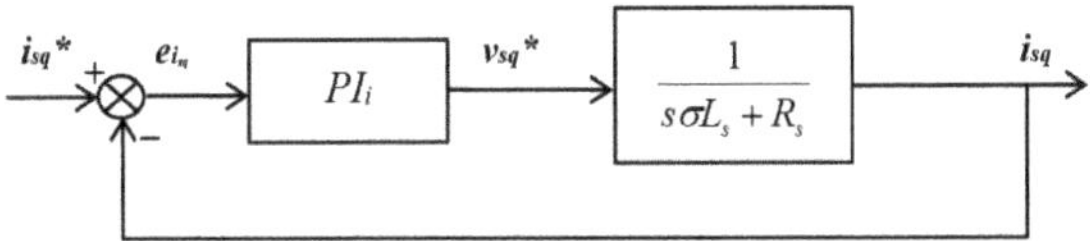

Figure 4. Simplified scheme of the PI regulator for i_{sq} and i_{sd} currents loops (induction motor (IM) machine case).

As in the position loop has been done, we know that at the ω_{GCi} gain cross-over frequency the module is the unity

$$\left|OLTF_i(s)\right|_{s=j\omega_{GCi}} = 1 \tag{36}$$

where the phase equation is (37), in which PM_i is the phase margin of the currents loop

$$\arg\left(OLTF_i(s)\right)\Big|_{s=j\omega_{GCi}} = -180° + PM_i \tag{37}$$

Next, taking these two equations, (36) and (37), being the ω_{GCi} and PM_i design data, and solving them, the two coefficients of PI regulator (Kp_i and Ki_i) necessaries to comply the imposed design are obtained, where for this resolution the δ intermediate variables is necessary

$$\delta = tg\left(PM_i - \frac{\pi}{2} + arctg\frac{\omega_{GC_i}L_s\sigma}{R_s}\right) \tag{38}$$

$$Ki_i = \frac{\omega_{ci}\sqrt{R_s^2 + (\omega_{GC_i}L_s\sigma)^2}}{\sqrt{1+\delta^2}} \tag{39}$$

$$Kp_i = \frac{Ki_i\delta}{\omega_{GC_i}} \tag{40}$$

In the PMSM machine, as L_{sd} and L_{sq} are different, it is correct to think that initially the tuning for each loop would be different. However, a way to start with a tuning can be using the same tuning for both, and go adjusting each one according to expected results. In this sense, an arithmetic mean value of both stator inductances is necessary to calculate, that is L_s,

$$L_s = (L_{sd} + L_{sq})/2 \tag{41}$$

Then for the PMSM machine, taking Equations (12) and (13), and considering the last term of each one as a disturbance and neglecting it, where as in the IM case, also taking into account that the dynamics of the SVPWM modulator and VSI inverter are faster than the electrical and PI dynamics

(Figure 1), then the following transfer function and simplified scheme can be obtained for the i_{sq} current loop,

$$\frac{I_{sq}(s)}{V_{sq}^*(s)} = \frac{1}{sL_s + R_s} \tag{42}$$

Like it was done for the IM machine, now the open loop transfer function (OLTF) of the simplified diagram of Figure 5 is obtained for the PMSM,

$$OLTF_i(s) = \left(Kp_i + \frac{Ki_i}{s}\right)\left(\frac{1}{R_s + sL_s}\right) \tag{43}$$

where for the ω_{GCi} gain cross-over frequency the module and phase equations are obtained, (44) and (45) respectively

$$\left| OLTF_i(s) \right|_{s=j\omega_{GCi}} = 1 \tag{44}$$

$$\arg\left(OLTF_i(s) \right)\big|_{s=j\omega_{GCi}} = -180° + PM_i \tag{45}$$

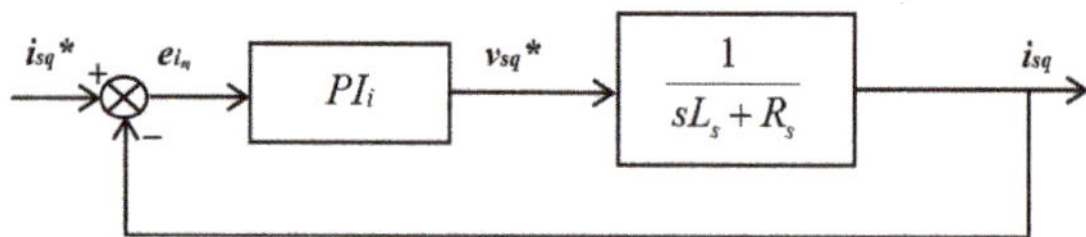

Figure 5. Simplified scheme of the PI regulator for i_{sq} and i_{sd} currents loops (permanent magnet synchronous motor (PMSM) machine case).

Finally, taking this two equations system, (43) and (44), and solving them, where the ω_{GCi} and PM_i are the design data, the two coefficients of the PI regulator (Kp_i and Ki_i) are obtained, where for this resolution the λ intermediate variables is necessary.

$$\lambda = tg\left(PM_i - \frac{\pi}{2} + arctg\frac{\omega_{GC_i}L_s}{R_s}\right) \tag{46}$$

$$Ki_i = \frac{\omega_{GC_i}\sqrt{R_s^2 + (\omega_{GC_i}L_s)^2}}{\sqrt{1+\lambda^2}} \tag{47}$$

$$Kp_i = \frac{Ki_i\lambda}{\omega_{GC_i}} \tag{48}$$

As it can be seen, these last three equations are very similar to (38), (39) and (40), where the unique difference is that σL_s term for the IM machine has been replaced by L_s for the PMSM case.

2.5. Estability Demonstration of PD Position and PI Current Controlled Systems

The stability of the position closed loop is guaranteed choosing a $PM_{\theta m}$ phase margin positive, and the currents closed loops are stable while their PM_i phase margin is positive.

3. Simulation and Experimental Results

3.1. Tests Benchees

The following two tests benches have been employed to do the experiments, Figure 6:

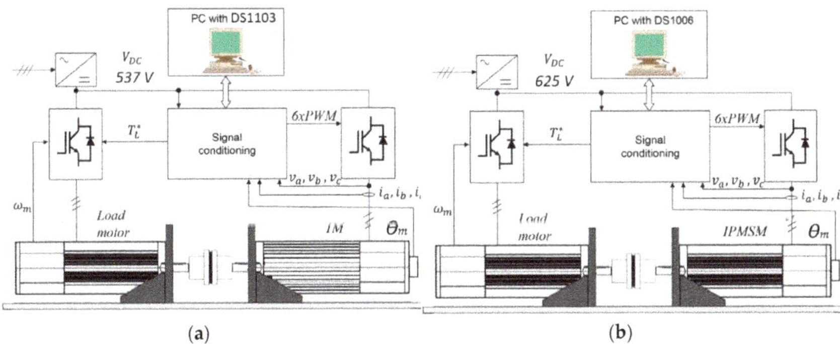

Figure 6. (a) IM tests bench and (b) PMSM tests bench.

3.1.1. IM Tests Bench

The IM tests platform is based on a M2AA 132M4 ABB industrial induction motor, of the die cast aluminum squirrel-cage type, Figure 6a, with these parameters:

- P, rated active power, 7.5 kW;
- V, rated stator voltage, 380 V;
- T_e, rated electromagnetic torque, 50 N·m;
- n, rated mechanical rotor speed, 1445 rpm (151.32 rad/s);
- R_s, stator resistance, 0.729 Ω;
- R_r, rotor resistance, 0.40 Ω;
- L_m, magnetizing inductance, 0.1125 H;
- L_s, stator inductance, 0.1138 H;
- L_r, rotor inductance, 0.1152 H;
- σ, coefficient of magnetic dispersion, 0.0346 (dimensionless);
- p, number of poles, 4;
- J, moment of inertia, 0.0503 kg·m^2;
- B_v, viscous friction coefficient, 0.0105 N·m/(rad/s);
- I_s, rated stator current, 15.24 A;
- I_{sq}, rated torque current, 20 A;
- I_{sd}, rated rotor flux current, 8.026 A;
- Ψ_r, rated rotor flux, 0.9030 Wb.

The load disturbance is generated by employing a commercial 190U2 Unimotor synchronous AC servo motor (controlled in torque) of 10.6 kW, which is mechanically connected in the shaft of the IM machine. Both machines are connected to a DC bus of 537 V by using three-phase VSI inverters working at 10 kHz of switching frequency. The control and monitoring tasks are done from a PC, in which are installed MatLab/Simulink, dsControl and a DS1103 real time interface of dSpace. For measuring the mechanical position, an incremental encoder of 4096 square pulses/revolution is used, where by quadruplicating them it is getting a resolution of 0.000385 rad (0.022°).

3.1.2. PMSM Tests Bench

The PMSM experiments platform is built on a 142U2C300BACAA165240 Unimotor synchronous AC servo motor (PMSM motor), Figure 6b, with the following parameters:

- P, rated active power, 3.83 kW;
- V, rated stator voltage, 400 V;

- T_e, rated electromagnetic torque, 12.2 N·m;
- n, rated mechanical rotor speed, 3000 rpm (314.16 rad/s);
- R_s, stator resistance, 0.49 Ω;
- L_{sd}, stator d inductance, 0.0039 H;
- L_{sq}, stator q inductance, 0.0069 H;
- p, number of poles, 6;
- J, moment of inertia, 0.0055 kg·m^2;
- B_v, viscous friction coefficient, 0.014 N·m/(rad/s);
- I_s, rated stator current, 7.62 A;
- Ψ_M, rotor magnets flux, 0.3556 Wb.

The load disturbance is implemented by using another 142U2C300BACAA165240 Unimotor commercial synchronous AC servo motor (controlled in torque) and is mechanically connected in the shaft of the PMSM motor. The DC bus employed for both machines is one of 625 V and a switching frequency of 10 kHz for each VSI. The control and monitoring system is also done from a PC, where are installed MatLab/Simulink, dsControl and a DS1006 real time interface of dSpace. The mechanical position is measured by employing the same incremental encoder and quadruplicated system that are in the IM tests bench, and consequently getting the same resolution of 0.000385 rad (0.022°).

3.2. D1 and D2 Position and Currents Regulators Design

The bandwidth of these designs are for closed loop systems, by using the cross-over frequency (ω_C), but as this frequency matches approximately with the open loop gain cross-over frequency (ω_{GC}), then by choosing ω_{GC} the closed loop bandwidth is chosen. The following two tables (Tables 1 and 2) show two designs for each type of motor. For both, D1 is used to work with low-medium load disturbances, and it takes a bandwidth ($\omega_{GC\theta m}$) between 40 and 50 rad/s, and a $PM_{\theta m}$ around 70°. For both kind of machines, D2 is employed for working with medium-height load disturbances, using a higher bandwidth than in D1, between 75 and 85 rad/s, and $PM_{\theta m}$ also higher than D1, around 80°, to compensate better than D1 the effect of a higher load torque. If there is any interest to get more derivative action (to get a faster response and to reduce oscillations in the response), keeping the same bandwidth, it is necessary to increase the $PM_{\theta m}$. Regarding the PI current regulators, the employed tuning is valid for both designs in both machines and it is enough faster than the PD position regulator and electromechanical dynamics (in order to neglect the fastest blocks, done in Section 2.3). The variable *val* takes a value of 1000, (22). All regulators have been designed in continuous time but to do the simulations and experiments, they have been discretized at 100 μs to take advantage of the switching frequency (10 kHz).

Table 1. D1 and D2 position controller designs for the IM machine.

Design	$\omega_{GC\theta m}$	$PM_{\theta m}$	ω_{GCi}	PM_i
D1	50 rad/s	74°	3000 rad/s	70°
	$Kp_{\theta m} = 11.1$,	$Kd_{\theta m} = 914.63$,	$Kp_i = 10.82$,	$Ki_i = 14401$
D2	85 rad/s	79°	3000 rad/s	70°
	$Kp_{\theta m} = 15.23$,	$Kd_{\theta m} = 1597$,	$Kp_i = 10.82$,	$Ki_i = 14401$

Table 2. D1 and D2 position controller designs for the PMSM machine.

Design	$\omega_{GC\theta m}$	$PM_{\theta m}$	ω_{GCi}	PM_i
D1	45 rad/s	70°	3000 rad/s	70°
	$Kp_{\theta m} = 2.8$,	$Kd_{\theta m} = 139.25$,	$Kp_i = 15$,	$Ki_i = 18004$
D2	75 rad/s	75°	3000 rad/s	70°
	$Kp_{\theta m} = 4.25$,	$Kd_{\theta m} = 248.15$,	$Kp_i = 15$,	$Ki_i = 18004$

3.3. Simulation and Experimental Results

Figure 7 shows both tests, simulation and experimental, using the D1 position regulator for IM. The tests were using a position square waveform of 2 rad of amplitude and 0.25 Hz of frequency during 6 s, and also, a constant load torque of 25 N·m (50% of rated torque) starting from 3rd second to the end. As it can be seen, the response of the motor was fast and without overshoot (a), (b) and (c). Moreover, the position error in the steady state was practically 0 rad (0°) for the simulation case and less than 0.002 rad (0.1146°) for the experimental one, without load disturbance. With disturbance the position error increased a bit, the simulation case error was less than 0.002 rad (0.1146°) and the experimental case error was around 0.007 rad (0.4011°), but the position tracking was still very good. Regarding the electromagnetic torque, (e), this was as smooth and effective as its current, (f), and the estimated load torque was very similar to the load disturbance, demonstrating that the estimator worked fine. Finally, the real rotor flux current, (g), tracked very properly its reference, consequently the machine generated the imposed rotor flux. Comparing both tests, it could be observed that they were very similar.

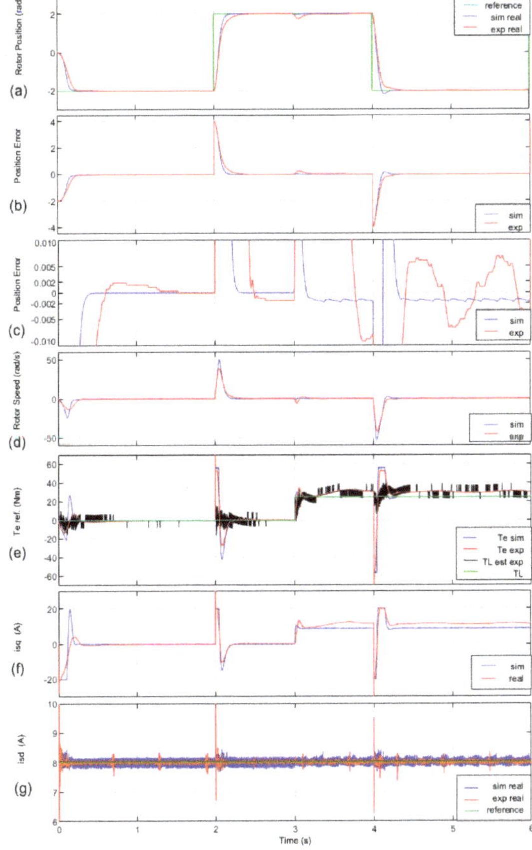

Figure 7. Simulation and experimental tests' performances employing the D1 design for the IM machine: (**a**) rotor position, reference and real (rad); (**b**) rotor position error (rad); (**c**) rotor position error zoom (rad); (**d**) rotor speed (rad/s); (**e**) electromagnetic torque (Te), estimated load torque (TL est) and load torque (TL), (N·m); (**f**) torque current, isq (A) and (**g**) rotor flux current, reference and real, isd (A).

Figure 8 shows the performance of the experimental test using the D2 position regulator for IM. The test employs a position square waveform of 2 rad of amplitude and 0.25 Hz of frequency during 10 s, and also, a square load torque of 37.5 N·m (75% of rated torque) of amplitude and 0.25 Hz of frequency. As it could be observed, the response of the motor was fast and with small overshoot a) and (b). The position error in the steady state was small or equal to 0.015 rad (0.8594°), getting very good tracking. In spite of high load disturbance, the electromagnetic torque, (c), was not aggressive and neither was its current, (d). The estimated load torque also worked properly in the hard condition, getting a very similar waveform to the load disturbance. The real rotor flux current, (e), also tracked very properly its reference in hard conditions, this way the machine produced the imposed rotor flux without any problem.

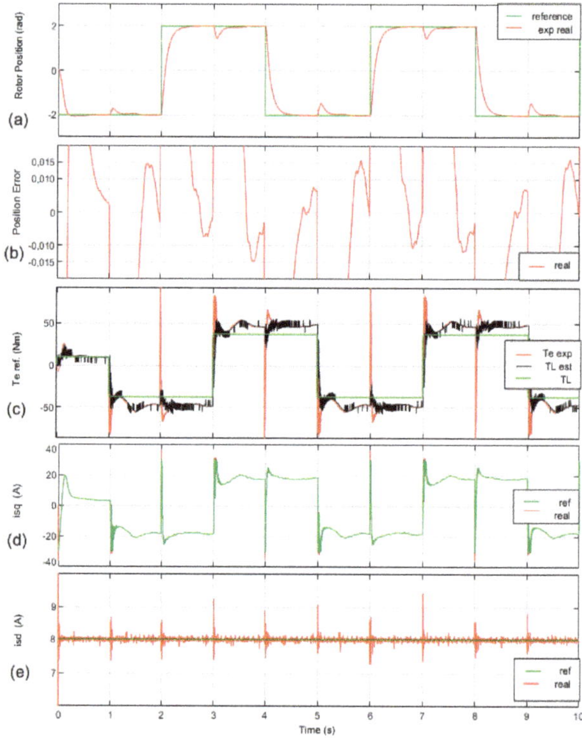

Figure 8. Experimental test employing the D2 design for the IM machine: (**a**) rotor position, reference and real (rad); (**b**) rotor position error (rad); (**c**) rotor position error zoom (rad); (**d**) electromagnetic torque (Te), estimated load torque (TL est) and load torque (TL), (N·m); (**e**) torque current, isq (A) and (**f**) rotor flux current, reference and real, isd (A).

Figure 9 shows the simulation and experimental tests, using the D1 position regulator for PMSM. The tests were using the same conditions employed in Figure 4, but now using a constant load torque of 6.1 N·m (50% of rated torque). Seeing (a), (b) and (c) graphs, it could be observed the fast and without overshoot response of the motor. Besides, the position error in the steady state tended to be 0 rad (0°) for the simulation and also for the experimental case, without load disturbance. In the presence of the disturbance, the position error increased a bit in the simulation case, producing an error of 0.002 rad (0.1146°) and in the experimental case with an error around 0 rad (0°), getting excellent position tracking. The electromagnetic torque, (e), was as smooth and effective as its current, (f), and the estimator of load torque provided a value similar to the real load torque, demonstrating that the estimator worked properly. The real rotor flux current, (g), presents good tracking to 0 A (as it has been designed), and as a consequence the machine generated only the magnets rotor flux.

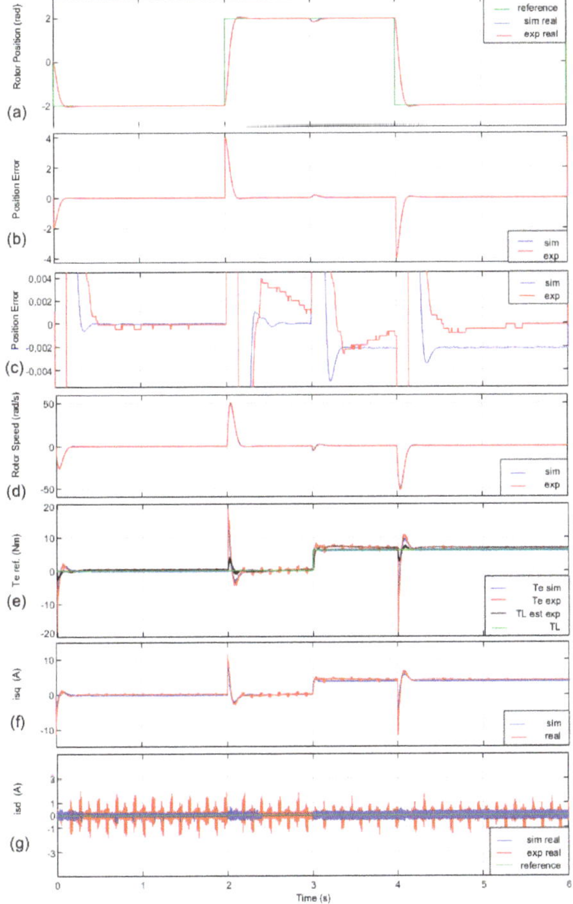

Figure 9. Simulation and experimental tests' performances employing the D1 design for the PMSM machine: (**a**) rotor position, reference and real (rad); (**b**) rotor position error (rad); (**c**) rotor position error zoom (rad); (**d**) rotor speed (rad/s); (**e**) electromagnetic torque (Te), estimated load torque (TL est) and load torque (TL), (N·m); (**f**) torque current, isq (A); (**g**) rotor flux current, reference and real, isd (A).

Observing the simulation test and the experiment, we could realize that both had a large similarity.

Figure 10 shows the experimental performance of the D2 position regulator for the PMSM machine. The test was the same as that employed and shown in Figure 5, but in this case using a square load torque of 9.15 N·m (75% of rated torque). Observing the response of the motor, it was fast and with small overshoot (a) and (b). The position error in the steady state was small or equal to 0.004 rad (0.2292°), obtaining excellent position tracking taking into account the hard load disturbance conditions. In spite of this, the electromagnetic torque, (c), was smooth, like its current, (d). The estimator of the load torque also worked very fine, obtaining a very similar disturbance's waveform. Finally, it could be seen that the real rotor flux current, (e), tracked very properly its reference in these hard situations, obtaining that the machine produced only the magnet rotor flux successfully.

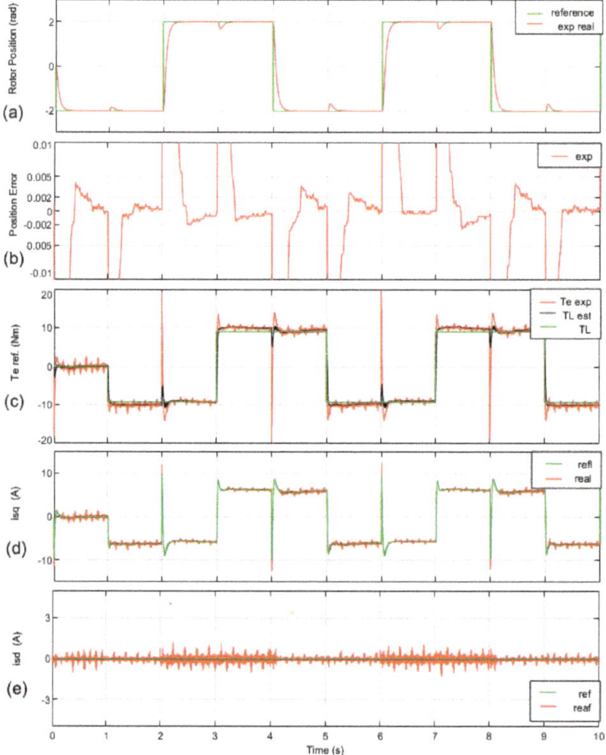

Figure 10. Experimental test employing the D2 design for the IM machine: (**a**) rotor position, reference and real (rad); (**b**) rotor position error (rad); (**c**) rotor position error zoom (rad); (**d**) electromagnetic torque (Te), estimated load torque (TL est) and load torque (TL), (N·m); (**e**) torque current, isq (A) and (**f**) rotor flux current, reference and real, isd (A)

4. Conclusions

Taking into account the results obtained in the simulation and in the experimental tests, the next conclusions could be remarked. For three-phase machines of several units of kW, a very good position tracking was obtained employing a bandwidth ($\omega_{GC\theta m}$) between 40 and 50 rad/s, and a phase margin ($PM_{\theta m}$) around of 70° for the PD regulator, when a step type position reference and the load torque was unknown and its value took up to 50% or less than the rated value of the electromagnetic torque. Really excellent results were obtained, with a position error of 0.007 rad (0.4011°) for IM and 0.002 rad (0.1146°) for PMSM. For load torque that was 75% or higher of the rated value of the electromagnetic torque, a bandwidth ($\omega_{GC\theta m}$) between 75 and 85 rad/s, and a phase margin ($PM_{\theta m}$) around of 80° were necessary, obtaining also excellent results: a position error of 0.015 rad (0.8594°) for IM and between 0.004 rad (0.2292°) and 0, for PMSM. The PMSM motor provided better accuracy than the IM machine by employing the same PD position controller. The implementation of the load torque estimator was essential to obtain these results. Regarding the two PI inner loops, the tuning of a bandwidth (ω_{GCi}) of 3000 rad/s and a phase margin (PM_i) of 70°, was adequate for both loops (i_{sd} and i_{sq}), in both machines, due to that the current tracking was also very good in all cases. Moreover, despite PMSM having different stator inductances (L_{sd} and L_{sq}), the solution was to get their arithmetic mean value (L_s) and employ it in tuning for both loops, obtaining good results.

The methodology presented in this paper consists of a simple way to tune the position PD regulator for the most employed kind of commercial/industrial three-phase motors, that is IM and PMSM, obtaining excellent results, as well as much other advanced position controllers. This methodology could be used for other three-phase machines like Switched Reluctance Motor (SRM), Brushless DC (BLDC), Double Feed Induction Machine (DFIM) and including DC motors, for machines of different power.

Author Contributions: Conceptualization, P.A., O.B. and J.A.C.; Data curation, F.J.M., Formal analysis, I.M. and F.J.M.; Investigation, P.A. and F.J.M.; Methodology, P.A., O.B., J.A.C. and I.M.; Software, O.B. and I.M.; Validation, P.A. and J.A.C.; Writing—original draft, P.A.; Writing—review & editing, O.B., J.A.C. and P.A. All authors have read and agreed to the published version of the manuscript.

Funding: This research was partially funded by the Basque Government through the project SMAR3NAK (ELKARTEK KK-2019/00051.

Acknowledgments: The authors wish to express their gratitude to the Basque Government, and to the UPV/EHU for supporting this work.

Conflicts of Interest: The authors confirm that there is not any conflict of interest with any institution or company.

References

1. Vas, P. The Space-Vector method. In *Electrical Machines and Drives: a Space-Vector theory approach*, 1st ed.; Oxford University Press: Oxford, UK, 1992; Volume 1, pp. 32–125.
2. Kume, T.; Iwakane, T. High-Performance Vector-Controlled AC Motor Drives: Applications and New Technologies. *IEEE Trans. Ind. Appl.* **1987**, *23*, 872–880. [CrossRef]
3. Stojic, M.; Vukosavic, S. Design of microprocessor-based system for positioning servomechanism with induction motor. *IEEE Trans. Ind. Electron.* **1991**, *38*, 369–378. [CrossRef]
4. Alkorta, P.; Barambones, O.; Cortajarena, J.A.; Vicandi, F.J.; Martija, I. Effective Proportional Derivative Position control of Induction Motor Drives. In Proceedings of the 2016 IEEE International Conference on Industrial Technology (ICIT), Taipei, Taiwan, 14–17 March 2016.
5. Huang, T.; El-Sharkawi, M. High performance speed and position tracking of induction motors using multi-layer fuzzy control. *IEEE Trans. Energy Convers.* **1996**, *11*, 353–358. [CrossRef]
6. Cecati, C. Position Control of the Induction Motor using a Passivity-based Controller. *IEEE Trans. Ind. Appl.* **2000**, *36*, 1277–1284. [CrossRef]
7. Rodriguez, P.; Dumur, D. Generalized predictive control robustification under frequency and time-domain constraints. *IEEE Trans. Control. Syst. Technol.* **2005**, *13*, 577–587. [CrossRef]
8. Veselić, B.; Peruničić-Draženović, B.; Milosavljevic, Č. High-Performance Position Control of Induction Motor Using Discrete-Time Sliding-Mode Control. *IEEE Trans. Ind. Electron.* **2008**, *55*, 3809–3817.
9. Barambones, O.; Alkorta, P.; De Durana, J.M.G. A real-time estimation and control scheme for induction motors based on sliding mode theory. *J. Frankl. Inst.* **2014**, *351*, 4251–4270. [CrossRef]
10. Callegaro, A.D.; Nalakath, S.; Srivatchan, L.N.; Luedtke, D.; Preindl, M. Optimization-Based Position Sensorless Control for Induction Machines. In Proceedings of the 2018 IEEE Transportation Electrification Conference and Expo (ITEC), Long Beach, CA, USA, 13–15 June 2018; pp. 460–466.
11. Egiguren, P.A.; Echeverria, J.A.C.; Caramazana, O.B.; Lopez, I.M.; Alcorta, J.C. Poles Placement Position Control of Induction Motor Drives. In Proceedings of the IECON 2019—45th Annual Conference of the IEEE Industrial Electronics Society, Lisbon, Portugal, 14–17 October 2019.
12. Ogasawara, S.; Akagi, H. Implementation and position control performance of a position-sensorless IPM motor drive system based on magnetic saliency. *IEEE Trans. Ind. Appl.* **1998**, *34*, 806–812. [CrossRef]
13. Hua, Y.; Sumner, M.; Asher, G.; Gao, Q.; Saleh, K. Improved sensorless control of a Permanent Magnet Synchronous Machine using fundamental Pulse Width Modulation excitation. *IET Electr. Power Appl.* **2011**, *5*, 259–370. [CrossRef]
14. Ghafari-Kashani, A.; Faiz, J.; Yazdanpanah, M. Integration of non-linear H∞ and sliding mode control techniques for motion control of a permanent magnet synchronous motor. *IET Electr. Power Appl.* **2010**, *4*, 267. [CrossRef]

15. Yin, Z.; Gong, L.; Du, C.; Liu, J.; Zhong, Y. Integrated Position and Speed Loops Under Sliding-Mode Control Optimized by Differential Evolution Algorithm for PMSM Drives. *IEEE Trans. Power Electron.* **2019**, *34*, 8994–9005. [CrossRef]

16. Lin, F.-J.; Lin, C.-H. A Permanent-Magnet Synchronous Motor Servo Drive Using Self-Constructing Fuzzy Neural Network Controller. *IEEE Trans. Energy Convers.* **2004**, *19*, 66–72. [CrossRef]

17. Lin, F.-J.; Sun, I.-F.; Chang, J.-K.; Yang, K.-J. Intelligent position control of permanent magnet synchronous motor using recurrent fuzzy neural cerebellar model articulation network. *IET Electr. Power Appl.* **2015**, *9*, 248–264. [CrossRef]

18. Yu, J.; Shi, P.; Yu, H.; Chen, B.; Lin, C. Approximation–based Discrete-time Adaptive Position Tracking Control for Interior Permanent Magnet Synchronous Motors. *IEEE Trans. Cybernetics.* **2015**, *45*, 1363–1371. [CrossRef] [PubMed]

19. Belda, K.; Vosmik, D. Explicit Generalized Predictive Control of Speed and Position of PMSM Drives. *IEEE Trans. Ind. Electron.* **2016**, *63*, 3889–3896. [CrossRef]

20. Mubarok, M.S.; Liu, T.-H. Implementation of Predictive Controllers for Matrix-Converter-Based Interior Permanent Magnet Synchronous Motor Position Control Systems. *IEEE Trans. Emerg. Sel. Topics Power Electron.* **2019**, *7*, 261–273. [CrossRef]

21. Mohan, N. *Mathematical Description of Vector Control in Induction Machines*; Wiley: Hoboken, NJ, USA, 2014; Volume 1, pp. 79–96.

22. Bose, B.K. AC Machines for Drives. In *Modern Power Electronics and AC Drives*, 1st ed.; Prentice Hall PTR: Upper Saddle River, NJ, USA, 2002; Volume 1, pp. 86–92.

 electronics

Article

Realization of the Sensorless Permanent Magnet Synchronous Motor Drive Control System with an Intelligent Controller

Hung-Khong Hoai [1,2], Seng-Chi Chen [1,*] and Hoang Than [3]

[1] Department of Electrical Engineering, Southern Taiwan University of Science and Technology, Tainan 71005, Taiwan; da62b201@stust.edu.tw

[2] Faculty of Electrical and Electronics Engineering, Ho Chi Minh City University of Transport, Ho Chi Minh City 70000, Vietnam

[3] Department of Electrical and Electronics Engineering, Hue Industrial College, Hue 49000, Vietnam; hthan@hueic.edu.vn

* Correspondence: amtfcsg123@stust.edu.tw; Tel.: +886-6-253-3131 (ext. 3324)

Received: 21 January 2020; Accepted: 18 February 2020; Published: 21 February 2020

Abstract: This paper presents the sensorless control algorithm for a permanent magnet synchronous motor (PMSM) drive system with the estimator and the intelligent controller. The estimator is constructed on the novel sliding mode observer (SMO) in combination with a phase-locked loop (PLL) to estimate the position and speed of the rotor. The intelligent controller is a radial basis function neural network (RBFNN)-based self-tuning PID (Proportional-Integral-Derivative) controller, applied to the velocity control loop of the PMSM drive control system to adapt strongly to dynamic characteristics during the operation with an external load. The *I-f* startup strategy is adopted to accelerate the motor from standstill, then switches to the sensorless mode smoothly. The control algorithm program is based on MATLAB and can be executed in simulations and experiments. The control system performance is verified on an experimental platform with various speeds and the dynamic load, in which the specified *I-f* startup mode and sensorless mode, inspected by tracking response and speed regulation. The simulation and experimental results demonstrate that the proposed method has worked successfully. The motor control system has smooth switching, good tracking response, and robustness against disturbance.

Keywords: permanent magnet synchronous motor; sensorless control; sliding mode observer; RBFNN-based self-tuning PID controller; *I-f* startup strategy

---

1. Introduction

Applying the advanced developments in power electronics, microprocessors and digital signal processors (DSP), permanent magnet synchronous motors (PMSMs) have been extensively used in industrial automatic control applications, from washing machines to electric vehicles, due to their efficient performance characteristics, high power transmission efficiency, large torque-to-weight ratio, and long service life. Among various PMSM drive control techniques, field-oriented control (FOC) has been the most essential and efficient scheme, in which the rotor's position and speed data are required. As a sensor-based solution, these data are managed by a typical sensor, such as a resolver, encoder, or Hall sensor, installed on the motor's shaft. However, this makes the PMSM drive control system more expensive and larger in size. In some cases, the sensors are environment-sensitive, reducing control reliability and adaptiveness. To solve these shortcomings, various back-EMF-based (back electromotive force based) sensorless control techniques are designed and applied, such as the extended Kalman filter (EKF) approaches [1–3] or the sliding mode observer (SMO) approaches [4–7]. The EKF involves a lot of recursive computations because it consists of prediction and innovation. Meanwhile, the SMO is robust against disturbance, has a high accuracy estimation potential, and is easy to be implemented. Generally, the conventional SMO solutions are impacted by chattering problems and noise effects because of using the on-off function and traditional arctangent calculation, so that the novel SMO and PLL combinations are widely applied to enhance the estimator's robustness and accuracy [6,8–10]. Due to the back EMF value being small at the standstill or the low-speed region, the startup strategies are implemented to speed up the motor to the specific speed threshold at which the back-EMF is large enough to be estimated precisely. A simple *I-f* startup strategy was presented [11] and applied [1,5]. It involves the closed-loop current control to ensure the motor starts successfully under different external load situations without initial rotor position estimation and machine parameters estimation.

A proper motor drive control system requires a wide adjustable speed range, adaption to load disturbances and parameter variations, high instantaneous torque response, and lower torque ripple during the operating condition. It is essential to improve the motor control algorithm to obtain optimal performance. Therefore, various controllers have been proposed, such as the ANFIS controller [12], the neural network controller [13], the hybrid fuzzy-PID (Proportional-Integral-Derivative) controller [14], and the backstepping controller [4]. Among these algorithms, the radial basis function neural network (RBFNN)-based self-tuning PID controller is considered to enhance the speed control quality of the PMSMs drive control system. This intelligent controller not only inherits the typical PID controller's structural simplicity but also is optimized by online adjusting the operating parameters based on the advantages of a neural network such as the ability to identify nonlinear system dynamics, the ability to learn, generalize, and adapt.

In the realization of the motor control system, the DSP of Texas Instruments provides a flexible solution, improving system reliability and efficiency. The DSP integrates highly optimized peripheral circuits, memory, and a single-chip CPU structure. It exhibits powerful processing and high performance for complex real-time control systems. In particular, with MATLAB's Embedded Coder Support Package utilities, it is resourceful to build up the motor control algorithm in Simulink. The DSP application development time is significantly shortened. Accordingly, in this paper, an RBFNN-based self-tuning PID speed controller is implemented to improve the performance of the DSP-based sensorless PMSM drive control system with an estimator based on the novel SMO in combination with a PLL (the novel SMO-PLL). The MATLAB-based implemented control algorithm is not only executed in simulation but also applied to the real-time experiment system. The *I-f* startup mode, tracking response, and speed regulation are investigated to evaluate the control system performance. The proposed control algorithm is verified on an experimental platform with a PMSM, inverter, control circuit, a Texas Instruments DSP F28379D, and the dynamic load.

This paper is organized into the following sections. Section 2 describes the sensorless PMSM drive control system with the mathematical model, the estimator based on the novel SMO-PLL, and the *I-f* startup strategy. Section 3 introduces the self-tuning PID controller based on an RBFNN. Section 4 presents the implementation of the control algorithm in MATLAB Simulink. In Section 5, the control system performance is inspected in the digital simulation to evaluate the correctness and the effectiveness of the control algorithm. In Section 6, the proposed algorithm is verified on the experimental platform. Finally, the conclusion is given in Section 7.

2. Description of the Sensorless PMSM Drive System

The architecture of the proposed DSP-based sensorless PMSM drive control system based on the novel SMO-PLL estimator and the RBFNN-based self-tuning PID controller is shown in Figure 1. As mentioned above, two control modes are investigated from low speed to high speed, including a startup mode and a sensorless mode. The first mode is the *I-f* startup strategy mode in the low-speed range. The switches are activated in position 1. The second mode is the sensorless control mode in the high-speed range. The switches are activated in position 2. In the sensorless control mode, there are two closed control loops—the inner current control loop and the outer velocity control loop. The current control loops are executed on the PI (Proportional-Integral) controller and the FOC algorithm. The velocity control loop is implemented by an RBFNN-based self-tuning PID controller. The detailed formulation and control algorithm are described as follows.

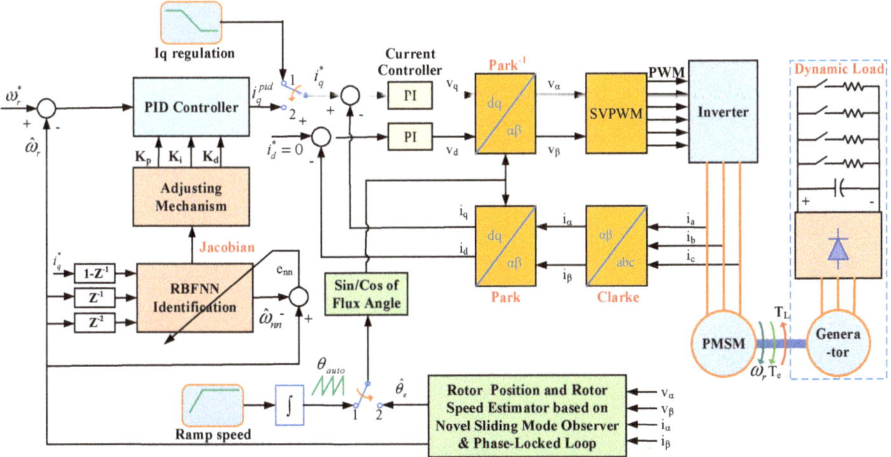

Figure 1. The architecture of a self-tuning PID (Proportional-Integral-Derivative) speed controller based on the radial basis function neural network (RBFNN) for the sensorless permanent magnet synchronous motor (PMSM) drive control system with the dynamic load.

2.1. Modeling of the PMSM

In general, the typical mathematical model of a PMSM is formulated in the *d-q* synchronous rotating coordinate as:

$$\begin{cases} v_d = r_s i_d + L_s \frac{d}{dt} i_d - \omega_e L_s i_q \\ v_q = r_s i_q + L_s \frac{d}{dt} i_q + \omega_e L_s i_d + \omega_e \lambda_f \end{cases} \tag{1}$$

where v_d, v_q are the voltages of the d and q axis, respectively; r_s is the stator winding resistance per phase; $L_s = L_d = L_q$ (for SPMSM) and L_d, L_q are the inductances of the d and q, axis respectively; i_d, i_q are the currents of the d and q axis respectively; ω_e is the rotating speed of magnet flux; λ_f is the permanent magnet flux linkage.

In Figure 1, the current control of PMSM is decoupled to torque control (i_q) and flux control (i_d) based on the Clarke and Park transformation. Therefore, the PMSM is controlled in a similar way to controlling a DC motor. When the current i_d is controlled to zero, the torque of PMSM can be expressed by the following equation:

$$T_e = \frac{3N_p}{4} \lambda_f i_q \triangleq K_t i_q \tag{2}$$

with:

$$K_t = \frac{3N_p}{4} \lambda_f \tag{3}$$

where T_e is the electromagnetic torque, K_t is the torque constant, N_p is the pole pairs.

The dynamic equation of a PMSM drive system with the mechanical load can be presented as:

$$\frac{d\omega_r}{dt} = \frac{1}{J}(T_e - T_L - B\omega_r) \tag{4}$$

where ω_r is the rotor speed, T_L is the load torque; J and B are the inertia and friction constants of the PMSM, respectively.

2.2. Novel Sliding Mode Observer

Applying the inverse Park transformation for (1), the circuit equation of PMSM on the $\alpha - \beta$ stationary coordinate can be represented by the following equation:

$$\begin{cases} v_\alpha = r_s i_\alpha + L_s \frac{d}{dt} i_\alpha + e_\alpha \\ v_\beta = r_s i_\beta + L_s \frac{d}{dt} i_\beta + e_\beta \end{cases} \tag{5}$$

where i_α, i_β are the current, v_α, v_β are the voltage of the α and β axis, and e_α, e_β are the back EMF, which are expressed by

$$\begin{cases} e_\alpha = -\omega_e \lambda_f \sin \theta_e \\ e_\beta = \omega_e \lambda_f \cos \theta_e \end{cases} \tag{6}$$

where θ_e is the electrical rotor position and $\dot{\theta}_e = \omega_e$.

Combining (5) and (6), the current equation can be derived as:

$$\begin{cases} \frac{d}{dt} i_\alpha = \frac{1}{L_s}(-r_s i_\alpha + v_\alpha - e_\alpha) \\ \frac{d}{dt} i_\beta = \frac{1}{L_s}(-r_s i_\beta + v_\beta - e_\beta) \end{cases} \tag{7}$$

According to the novel sliding mode observer theory, the current observer formulation is written as follows:

$$\begin{cases} \frac{d}{dt} \hat{i}_\alpha = \frac{1}{L_s}(-r_s i_\alpha + v_\alpha - kH(\hat{i}_\alpha - i_\alpha)) \\ \frac{d}{dt} \hat{i}_\beta = \frac{1}{L_s}(-r_s i_\beta + v_\beta - kH(\hat{i}_\beta - i_\beta)) \end{cases} \tag{8}$$

where $\hat{i}_\alpha, \hat{i}_\beta$ are the estimated current in the α and β axis. The k is the gain and $H(x)$ is the Sigmoid function, which is defined as:

$$H(x) = \frac{2}{1 + e^{-2\mu x}} - 1 = \frac{1 - e^{-2\mu x}}{1 + e^{-2\mu x}} \tag{9}$$

where μ is the shape parameter. Moreover, when (8) subtracts (7), the estimated current error can be represented in the dynamic equation as the following equation:

$$\begin{cases} \frac{d}{dt} \widetilde{i}_\alpha = \frac{1}{L_s}(-r_s \widetilde{i}_\alpha + e_\alpha - kH(\widetilde{i}_\alpha)) \\ \frac{d}{dt} \widetilde{i}_\beta = \frac{1}{L_s}(-r_s \widetilde{i}_\beta + e_\beta - kH(\widetilde{i}_\beta)) \end{cases} \tag{10}$$

where $\widetilde{i}_\alpha, \widetilde{i}_\beta$ are the estimated current error in the α and β axis, and are defined as: $\widetilde{i}_\alpha = \hat{i}_\alpha - i_\alpha$, and $\widetilde{i}_\beta = \hat{i}_\beta - i_\beta$.

To analyze the stability of the novel sliding mode observer, the Lyapunov function is chosen as:

$$V = \frac{1}{2}(\widetilde{i}_\alpha^2 + \widetilde{i}_\beta^2) \tag{11}$$

Taking the derivative of the Lyapunov function by the time, we obtain

$$\dot{V} = \widetilde{i}_\alpha \times \frac{d}{dt} \widetilde{i}_\alpha + \widetilde{i}_\beta \times \frac{d}{dt} \widetilde{i}_\beta \tag{12}$$

Furthermore, if we design $k > max(|e_\alpha|, |e_\beta|)$ and use (10), we have

$$\begin{cases} \widetilde{i}_\alpha \times \frac{d}{dt} \widetilde{i}_\alpha = \frac{\widetilde{i}_\alpha}{L_s}(-r_s \widetilde{i}_\alpha + e_\alpha - kH(\widetilde{i}_\alpha)) \le -\frac{r_s}{L_s} \widetilde{i}_\alpha^2 \le 0 \\ \widetilde{i}_\beta \times \frac{d}{dt} \widetilde{i}_\beta = \frac{\widetilde{i}_\beta}{L_s}(-r_s \widetilde{i}_\beta + e_\beta - kH(\widetilde{i}_\beta)) \le -\frac{r_s}{L_s} \widetilde{i}_\beta^2 \le 0 \end{cases} \tag{13}$$

Therefore, $\dot{V} \leq 0$, the estimated current will converge to the actual current. The back EMF can be obtained as:

$$\begin{cases} e_\alpha \approx \hat{e}_\alpha = kH(\hat{i}_\alpha - i_\alpha) = kH(\widetilde{i_\alpha}) \\ e_\beta \approx \hat{e}_\beta = kH(\hat{i}_\beta - i_\beta) = kH(\widetilde{i_\beta}) \end{cases} \tag{14}$$

where $\hat{e}_\alpha, \hat{e}_\beta$ are the estimated back EMF.

Moreover, a low pass filter is implemented to reduce the noise effect in estimated back EMF.

$$\begin{cases} \hat{e}_{\alpha F} = \frac{\omega_c}{s+\omega_c}\hat{e}_\alpha \\ \hat{e}_{\beta F} = \frac{\omega_c}{s+\omega_c}\hat{e}_\beta \end{cases} \tag{15}$$

where $\omega_c = 2\pi f_c$ and f_c is the cut-off frequency of the low pass filter.

Once the back EMF can be estimated, the phase-locked loop is executed to estimate the rotor position. The architecture of the novel SMO with a PLL is shown in Figure 2.

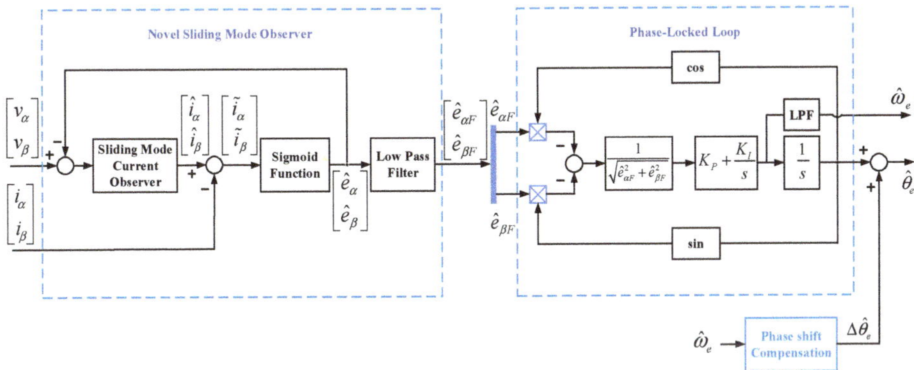

Figure 2. Novel sliding mode observer with a phase-locked loop.

From (6), an estimated error is defined by

$$\varepsilon = -\hat{e}_{\alpha F}cos(\hat{\theta}_e) - \hat{e}_{\beta F}sin(\hat{\theta}_e) \tag{16}$$

If this error is controlled to zero by a PI controller, the rotor speed is estimated by

$$\begin{cases} \hat{\omega}_e = K_P f_I(\varepsilon) + K_i \int f_I(\varepsilon)dt \\ f_I(\varepsilon) = \frac{\varepsilon}{\sqrt{\hat{e}_{\alpha F}^2 + \hat{e}_{\beta F}^2}} \end{cases} \tag{17}$$

and, the estimated rotor position can be obtained by

$$\hat{\theta}_e = \int \hat{\omega}_e dt \tag{18}$$

Finally, a phase shift compensation is calculated to correct the estimated rotor position, rejecting the phase delay effect in the back EMF's low pass filter.

$$\Delta\hat{\theta}_e = arctan\left(\frac{\hat{\omega}_e}{\omega_c}\right) \tag{19}$$

2.3. The I-f Startup Strategy

During the inter-mode transition from the startup stage to the sensorless FOC algorithm stage, the *I-f* startup strategy smooths torque and speed transition. It is based on the closed-loop current controller architecture. According to the study of [11], the *I-f* startup method is implemented in a three-stepped sequence, shown in Figure 3, and described as follows:

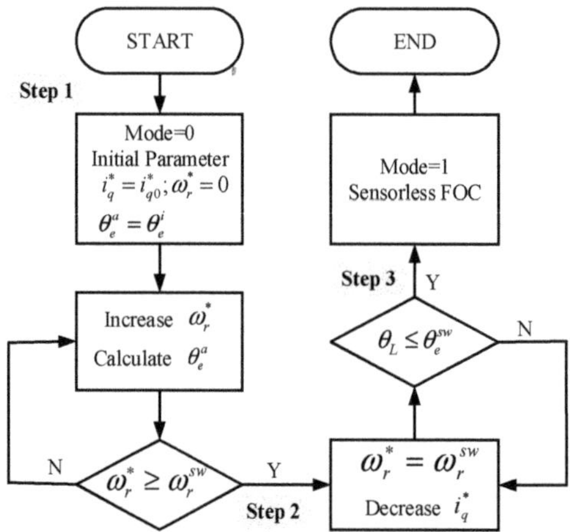

Figure 3. The *I-f* startup strategy flowchart.

Step 1: Ramping up reference speed with the constant i_q^* regulation at the initial start. This step aims to accelerate the motor from standstill to the switching reference speed where the back-EMF signal is large enough for the novel SMO-PLL estimator, which can obtain the rotor position and speed information accurately. The reference current i_q^* and reference speed ω_r^* are set to:

$$i_q^*(k) = i_{q0}^* \tag{20}$$

$$\omega_r^*(k+1) = \begin{cases} \omega_r^*(k) + K_{ramp}T_s & if \ \omega_r^* < \omega_r^{sw} \\ \omega_r^{sw} & if \ \omega_r^* \geq \omega_r^{sw} \end{cases} \tag{21}$$

$$\theta_e^a(k+1) = \theta_e^a(k) + \omega_r^*T_s \ \ with \ \theta_e^a(0) = \theta_e^i \tag{22}$$

where i_{q0}^* is the initial current in q-axis; θ_e^a, θ_e^i are the auto-generated rotor position and its initial value; ω_r^{sw} is the specific reference speed for switching to the sensorless mode, and K_{ramp} is the proportional gain for ramping up the reference rotor speed.

Step 2: Keeping reference speed constant with decreasing i_q^* axis current at the end of step 1. The purpose of this step is to reduce the current i_q^*, while the same torque is generated as before and balanced the load torque. In addition, the load torque angular θ_L reduces to a critical threshold to guarantee the smooth switching condition.

$$i_q^*(k+1) = i_q^*(k) - K_aT_s \tag{23}$$

$$\omega_r^*(k+1) = \omega_r^*(k) = \omega_r^{sw} \tag{24}$$

while:

$$\theta_L = \hat{\theta}_e - \theta_e^a > \theta_e^{sw} \tag{25}$$

where θ_e^{sw} is the angular threshold value of switching condition and K_a is the proportional gain for ramping down the current command i_q^*.

Step 3: Switching to the sensorless FOC mode operation based on the novel SMO-PLL estimator at the end of step 2.

3. The Self-Tuning PID Controller Based on an RBFNN

3.1. PID Controller

In Figure 1, the velocity loop is executed with a PID controller to regulate the reference current i_q^* in the q-axis. The algorithm of an incremental PID controller can be written as the following discrete-time equation:

$$i_q^*(k) = i_q^*(k-1) + k_p \widetilde{e}_p(k) + k_i \widetilde{e}_i(k) + k_d \widetilde{e}_d(k) \tag{26}$$

with:

$$\begin{cases} \widetilde{e}_p(k) = e(k) - e(k-1) \\ \widetilde{e}_i(k) = e(k) \\ \widetilde{e}_d(k) = e(k) - 2e(k-1) + e(k-2) \end{cases} \tag{27}$$

where the rotor speed error is defined as:

$$e(k) = \omega_r^*(k) - \hat{\omega}_r(k) \tag{28}$$

and k_p, k_i, k_d are the proportional, integral, and differential gains of the controller, respectively. Meanwhile $\widetilde{e}_p(k), \widetilde{e}_i(k), \widetilde{e}_d(k)$ are the proportional, integral, and differential variations in one sampling time, successively.

3.2. Radial Basis Function Neural Network

To get a self-tuning PID controller, the PMSM drive control system is identified by an RBFNN. Then, the controller's parameters should be adjusted suitably. Therefore, the RBFNN structure is shown in Figure 4. It comprises the feedforward neural network architecture, consisting of three layers—an input layer with three inputs, a hidden layer with five neurons, and a single output layer.

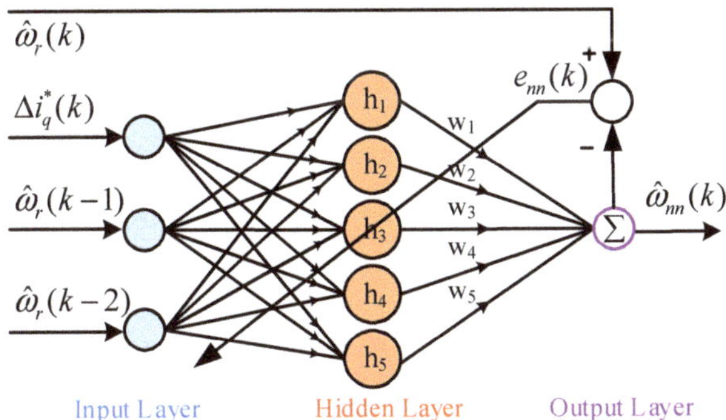

Figure 4. The RBFNN for tuning the PID parameters.

The input vector is defined as:

$$x(k) = \left[\Delta i_q^*(k) \ \ \hat{\omega}_r(k-1) \ \ \hat{\omega}_r(k-2) \right]^T \tag{29}$$

with:

$$\Delta i_q^*(k) = i_q^*(k) - i_q^*(k-1) \tag{30}$$

In the hidden layer, the activation function is implemented by a Gaussian function

$$h_j(k) = exp\left(-\frac{x(k) - C_j(k)^2}{2b_j^2(k)}\right) \tag{31}$$

where $C_j(k) = \left[c_{1j}(k) \; c_{2j}(k) \; c_{3j}(k)\right]^T$ and b_j, c_{ij} are the width and center of the Gaussian function. In the output layer, the output is a linear sum of hidden nodes:

$$\hat{\omega}_{nn}(k) = \sum_{j=1}^{5} w_j(k)h_j(k) \tag{32}$$

Furthermore, to simplify the online adaptive updating law for the parameter of RBFNN, the cost function is defined as:

$$J(k) = \frac{1}{2}(\hat{\omega}_r(k) - \hat{\omega}_{nn}(k))^2 = \frac{1}{2}e_{nn}^2(k) \tag{33}$$

According to the stochastic gradient descent (SGD) method, the learning algorithm can be calculated and updated by the below equations [15]:

$$\begin{cases} \Delta\omega_j(k) = -\eta\frac{\partial J(k)}{\partial\omega_j(k)} = \eta e_{nn}(k)\omega_j(k) \\ \omega_j(k) = \omega_j(k-1) + \Delta\omega_j(k) + \alpha(\omega_j(k-1) - \omega_j(k-2)) \end{cases} \tag{34}$$

$$\begin{cases} \Delta b_j(k) = -\eta\frac{\partial J(k)}{\partial b_j(k)} = \eta e_{nn}(k)\omega_j(k)h_j(k)\frac{x(k)-C_j(k)^2}{2b_j^3(k)} \\ b_j(k) = b_j(k-1) + \Delta b_j(k) + \alpha(b_j(k-1) - b_j(k-2)) \end{cases} \tag{35}$$

$$\begin{cases} \Delta c_{ij}(k) = -\eta\frac{\partial J(k)}{\partial c_{ij}(k)} = \eta e_{nn}(k)\omega_j(k)\frac{x_i(k)-c_{ij}(k)}{b_j^2(k)} \\ c_{ij}(k) = c_{ij}(k-1) + \Delta c_{ij}(k) + \alpha(c_{ij}(k-1) - c_{ij}(k-2)) \end{cases} \tag{36}$$

where $i = 1,2,3$; $j = 1,2...5$; η is the learning rate; α is the momentum factor.

Furthermore, the Jacobian transformation can be obtained as:

$$\frac{\partial\hat{\omega}_r}{\partial\Delta i_q^*} \approx \frac{\partial\hat{\omega}_{nn}}{\partial\Delta i_q^*} = \sum_{j=1}^{5} w_j(k)h_j(k)\frac{c_{1j}(k) - \partial\Delta i_q^*(k)}{b_j^2(k)} \tag{37}$$

3.3. Adjusting Mechanism Of PID Controller

In the closed-loop control, the parameters of the PID controller are adjusted online to minimize the square rotor speed error between the reference speed and the estimated rotor speed. Therefore, the cost function can be expressed as:

$$E(k) = \frac{1}{2}(\omega_r^*(k) - \hat{\omega}_r(k))^2 = \frac{1}{2}e^2(k) \tag{38}$$

Then, using the SGD method, the parameters of the PID controller can be updated and optimally tuned as the following equations [16]:

$$\begin{cases} \Delta k_p = -\eta\frac{\partial E}{\partial k_p} = \eta\frac{\partial E}{\partial\hat{\omega}_r}\frac{\partial\hat{\omega}_r}{\partial\Delta i_q^*}\frac{\partial\Delta i_q^*}{\partial k_p} \approx \eta e(k)\frac{\partial\hat{\omega}_{nn}}{\partial\Delta i_q^*}\tilde{e}_p \\ k_p(k) = k_p(k-1) + \Delta k_p \end{cases} \tag{39}$$

$$\begin{cases} \Delta k_i = -\eta\frac{\partial E}{\partial k_i} = \eta\frac{\partial E}{\partial\hat{\omega}_r}\frac{\partial\hat{\omega}_r}{\partial\Delta i_q^*}\frac{\partial\Delta i_q^*}{\partial k_i} \approx \eta e(k)\frac{\partial\hat{\omega}_{nn}}{\partial\Delta i_q^*}\tilde{e}_i \\ k_i(k) = k_i(k-1) + \Delta k_i \end{cases} \tag{40}$$

$$\begin{cases} \Delta k_d = -\eta \frac{\partial E}{\partial k_d} = \eta \frac{\partial E}{\partial \hat{\omega}_r} \frac{\partial \hat{\omega}_r}{\partial \Delta i_q^*} \frac{\partial \Delta i_q^*}{\partial k_d} \approx \eta e(k) \frac{\partial \hat{\omega}_{nn}}{\partial \Delta i_q^*} \widetilde{e}_d \\ k_d(k) = k_d(k-1) + \Delta k_d \end{cases} \tag{41}$$

where η is the learning rate.

4. Implementation of a Sensorless Speed Control Algorithm in MATLAB Simulink

Among the motor control system realization approaches, the DSP-executed hardware solution requires transforming the system description equation from the continuous time domain to the discrete-time domain. The sliding mode observer is represented by the discrete-time equation as:

$$\begin{cases} \hat{i}_\alpha(k+1) = e^{-\frac{r_s}{L_s}T_s}\hat{i}_\alpha(k) + \frac{1}{r_s}(1-e^{-\frac{r_s}{L_s}T_s})(v_\alpha(k)-\hat{e}_\alpha(k)) \\ \hat{i}_\beta(k+1) = e^{-\frac{r_s}{L_s}T_s}\hat{i}_\beta(k) + \frac{1}{r_s}(1-e^{-\frac{r_s}{L_s}T_s})(v_\beta(k)-\hat{e}_\beta(k)) \end{cases} \tag{42}$$

where T_s is the sampling time.

The motor control system expression is standardized in the per-unit system, so the current observer is rewritten as:

$$\begin{cases} \hat{i}_{\alpha,pu}(k+1) = F\hat{i}_{\alpha,pu}(k) + G(v_{\alpha,pu}(k)-\hat{e}_{\alpha,pu}(k)) \\ \hat{i}_{\beta,pu}(k+1) = F\hat{i}_{\beta,pu}(k) + G(v_{\beta,pu}(k)-\hat{e}_{\beta,pu}(k)) \end{cases} \tag{43}$$

with: $F = e^{-\frac{r_s}{L_s}T_s}$ and $G = \frac{U_m}{I_m}\frac{1}{r_s}(1-e^{-\frac{r_s}{L_s}T_s})$

where U_m and I_m are the base phase voltage and current, F and G are the feedback and gain factors of the SMO block system, respectively.

Additionally, the back EMF's low pass filter is also represented by the discrete equation, as:

$$\begin{cases} \hat{e}_{\alpha F}(k+1) = \hat{e}_{\alpha F}(k) + K_{LPF}[\hat{e}_\alpha(k)-\hat{e}_{\alpha F}(k)] \\ \hat{e}_{\beta F}(k+1) = \hat{e}_{\beta F}(k) + K_{LPF}[\hat{e}_\beta(k)-\hat{e}_{\beta F}(k)] \end{cases} \tag{44}$$

where $K_{LPF} = \omega_c T_s$.

Generally, a DSP-based control system is programmed in C/C++ languages, which requires some coding and debugging skills. However, in the case of using the Texas Instruments' DSP controller, the algorithm is not only being implemented directly by Code Composer Studio (CCS) software but is also enhanced by utilizing MATLAB and CCS software. The algorithm is firstly designed in MATLAB Simulink, then compiled the block system into C/C++ languages by Embedded Coder. Successively, the code is imported into CCS to execute, debug, and monitor the real-time system. Moreover, with enhanced functions of the Embedded Coder Support Package for Texas Instruments in MATLAB, it is very convenient to develop a motor control algorithm with Simulink. This shortens the DSP application development time significantly.

For developing the FOC algorithm, the Digital Control Motor (DCM) blocks are provided with the pre-built functions of not only the Clarke, Park, and Inverse Park transformations, but also for SVPWM, and current PI controllers. The FOC block is not shown in this paper with these functions, because it is already described in MATLAB manuals. Moreover, the TI IQmath library is supported for highly optimized and high precision mathematical functions of cosine, sine, magnitude, etc. Therefore, the sensorless PMSM control algorithm is easily developed in MATLAB Simulink for both simulation and experiment. Figure 5 presents the detailed block diagram of the novel SMO, the sigmoid function, the low pass filter, and the PLL in Simulink.

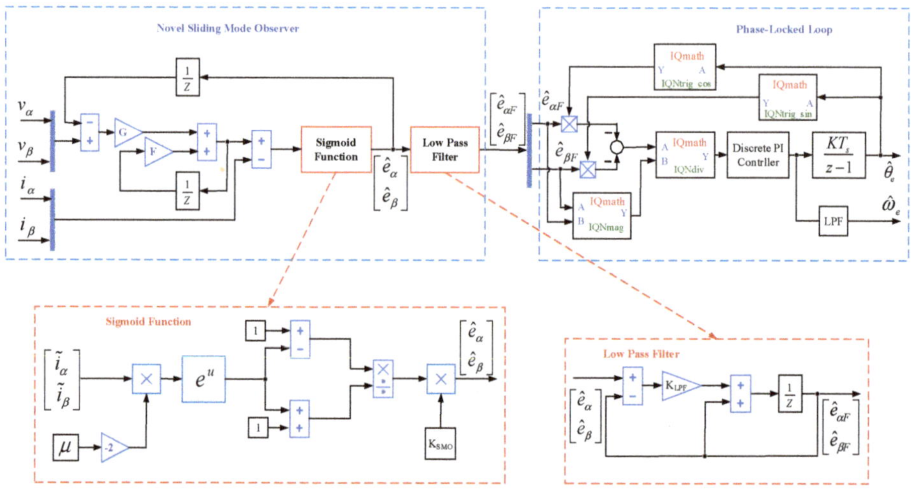

Figure 5. The block diagram of the novel sliding mode observer (SMO) with a phase-locked loop (PLL) in Simulink.

According to the design technique in the literature [15,17], the discrete PID controller algorithm in (26)–(28), (39)–(41) and an RBFNN identification algorithm in (29)–(37) are implemented as two function blocks, integrated in the closed velocity control loop (Figure 6).

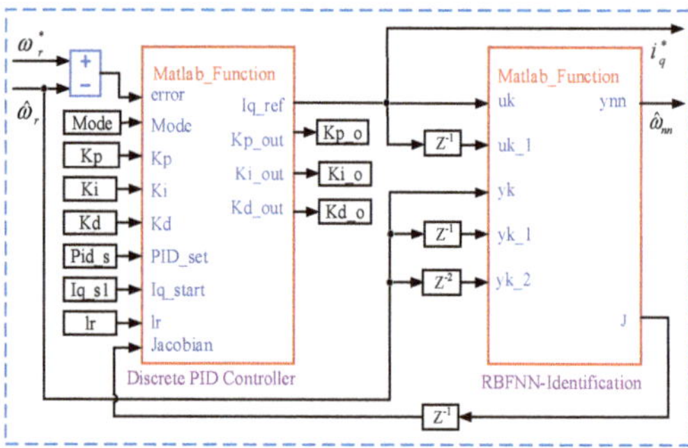

Figure 6. The block diagram of PID and RBFNN based on MATLAB function.

5. The Simulation Results

The platform of a motor speed control system was built on the PMSM's parameters, shown in Table 1. The control algorithm is verified in both the simulation and experiment under three cases. The first case is the *I-f* startup mode and switching to the sensorless FOC algorithm. The second one is the tracking response for checking the transient specification and the steady-state error. The last one is the speed regulation for evaluating the system stability regarding external disturbances. All the instances are operated with the dynamic load system, which includes an electrical circuit and a generator, coupled to the PMSM. The electrical circuit consists of a rectifier, a capacitor, and resistors, which are connected in parallel.

Table 1. Parameters.

Parameters	Values
Rated power	750 W
Input voltage	220 VAC
Rated current	4.24 A
Rated speed	2000 rpm
Rated torque	3.655 Nm
Phase resistance	1.326 Ω
Phase inductance	2.952 mH
BEMF constant	56.5 ($V_{L\text{-}L}$/krpm)
Torque constant	0.86 Nm/A
Inertia	3.63 Kgcm^2
Slot/poles	12S/8P

The complete design of the sensorless PMSM drive control system is shown in Figure 7, which includes three blocks. The first block (A) is built for real-time platform modeling. The discrete times of the PMSM model, the inverter, and the generator are 50 µs. The second block (B) is the motor control algorithm, which is executed in simulation and the real-time experiment. The third block (C) is used to monitor and acquire the simulation data. Within the control algorithm block, five subblocks are executed as the reference speed regulator, the discrete RBFNN-based self-tuning PID controller, the i_q^* regulator of *I-f* mode, the FOC algorithm, and the novel sliding mode observer with a phase-locked loop, respectively. The sampling time of the velocity loop is 1ms, while the sampling times of the FOC algorithm and the novel SMO-PLL estimator are 50 µs. In the current control loop, two PI controllers are adopted for the currents in *d-q* axis. Those PI controllers' parameters are set as $K_{Pd} = 0.25$, $K_{Id} = 0.025$, $K_{Pq} = 0.25$, $K_{Iq} = 0.025$. In the speed control loop, it is implemented by a self-tuning PID controller based on an RBFNN. The neuron parameters of those compositional five hidden nodes are initially set, as node centers ($c_1 = -0.01$, $c_2 = -0.005$, $c_3 = 0.0$, $c_4 = 0.005$, $c_5 = 0.01$), node widths ($b_1 = b_2 = b_3 = b_4 = b_5 = 0.005$), connective weights ($w_1 = w_2 = w_3 = w_4 = w_5 = 0.000625$). The learning rate is 0.475, and the momentum factor is 0.95. Additionally, the discrete PID's initial parameters are designed as $K_{Ps} = 0.725$, $K_{Is} = 0.0105$, $K_{Ds} = 0.483$. These initial values are tuned during the operating time. The dynamic load system is set up with an initial resistance load of 100 Ω and a capacitor of 470 µF.

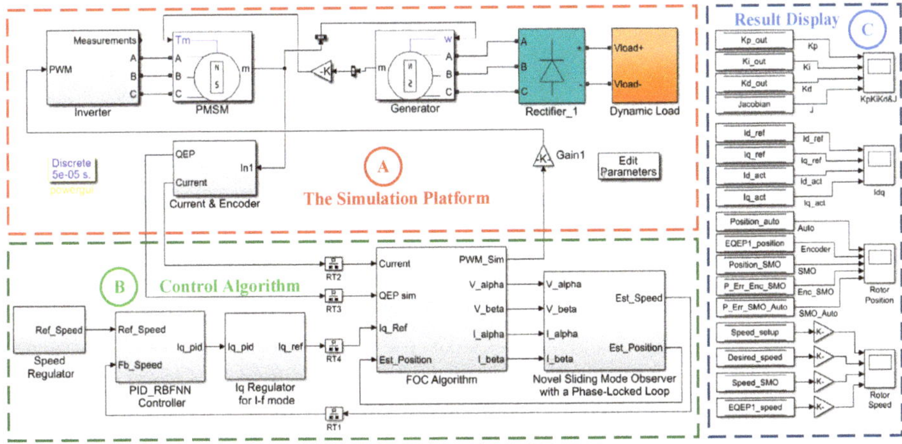

Figure 7. The simulation structure of the sensorless speed control algorithm for the PMSM drive system with a dynamic load system in MATLAB Simulink.

Figure 8 shows the motor's speed responses while speeding up the rotor speed from 0 to 2000 rpm. The reference speed is changed in a period of 1s with a sequence of 200 → 400 → 700 → 1000 → 700 → 1000 → 1400 → 1700 → 2000 rpm. Figure 8a refers that the novel SMO-PLL estimated rotor speed is closely tracked to the reference speed, and overlaps the actual rotor speed, calculated by the encoder. Figure 8b presents the current response in the d-q axis. The current i_q is varied and regarded to the reference speed. When the rotor speed is increased, the external load becomes higher. Therefore, the more torque is required in the motor, and the current i_q is regulated to be larger. The current i_d almost is equal to zero and has some small pulses when changing the speed. Figure 8c–f illustrates the electrical positions of the rotor, separately calculated by the novel SMO-PLL estimator and the encoder, with the estimated errors at the speeds of 700, 1000, 1400, and 2000 rpm. There are 7, 10, 14 and 20 position cycles in a period of 0.15s, respectively. Correspondingly, the rotation frequencies are 46.67, 66.67, 93.33, and 133.33 Hz. These values are suitable for the mentioned rotor speed. The estimated and actual positions are approximated to each other; thus, the estimated error is equal to zero. The PID gains are updated during the operating time. At = 3.5 s (as ω_r = 700 rpm), the PID gains are tuned to K_{Ps} = 0.7424, K_{Is} = 0.0107, and K_{Ds} = 0.4806. At = 4.5 s (as ω_r = 1000 rpm), the PID gains are tuned to K_{Ps} = 0.7694, K_{Is} = 0.0110, and K_{Ds} = 0.4768. At t = 7.5 s (as ω_r = 1400 rpm), the PID gains are tuned to K_{Ps} = 0.8549, K_{Is} = 0.0117, and K_{Ds} = 0.4648. Lastly, the rotor speed reaches 2000 rpm and the PID gains are also tuned to K_{Ps} = 1.0288, K_{Is} = 0.0131, and K_{Ds} = 0.4400 at t = 9.5 s.

Figure 9 demonstrates the detailed speed response for the startup motor mode and switching to the sensorless control mode at the speed of 200 rpm. The ramp-up speed ratio is set at 500 rpm/s. As mentioned in step 1 of the I-f strategy, the rotor speed is increased, following the ramp function, with the initial reference current i_q^* of 0.63 A. The actual rotor speed reaches 200 rpm in 0.415 s. At t = 0.441 s, the actual rotor speed reaches the maximum of 207.6 rpm, so the maximum error is +7.6 rpm. In step 2 of the I-f strategy, from t = 0.5 s, the reference current i_q^* decreases down to 0.183 A, with a current down ratio of 0.42 A/s. The actual rotor speed is kept close to 200 rpm. At t = 1.585 s, the actual rotor speed is reduced to a minimum of 191.3 rpm, so the minimum error is −9.7 rpm. At t = 1.682 s, the switching operation occurs, when the deviation of the estimated position and the auto-generated position is 3.6 degrees. Three rotor position curves are almost overlapped together. The current i_q has a small ripple, while the small pulse (about 0.05 A) appears in the current i_d. Therefore, the switching transient is smooth. Then, the motor begins to operate in the sensorless control mode.

Figure 10 shows the motor's speed responses while slowing down the rotor speed from 2000 to 400 rpm. The reference speed is sequentially varied as 2000 → 1700 → 1400 → 1200 → 800 → 1000 → 1200 → 1400 → 1000→ 700 → 400 rpm. Figure 10a shows that the estimated rotor speed is closely tracked by the reference speed and overlaps the actual rotor speed. Figure 10b presents the current response in the d-q axis. The current i_q is also varied to the rotor speed. When the rotor speed is decreased, the external load becomes lower. Therefore, the less torque is required to the motor, and the current i_q is regulated to be smaller. Moreover, the self-tuning PID controller's parameters are also tuned successfully. At t = 12.5 s (as ω_r = 1200 rpm), the PID gains are tuned to K_{Ps} = 0.8490, K_{Is} = 0.0115, K_{Ds} = 0.4666. At t = 17.5 s (as ω_r = 1000 rpm), the PID gains are tuned to K_{Ps} = 0.6567, K_{Is} = 0.0094, K_{Ds} = 0.4947. Lastly, the rotor speed decreases to 400 rpm and the PID gains are also tuned to K_{Ps} = 0.6020, K_{Is} = 0.0087, K_{Ds} = 0.5026 at t = 19.5 s.

Figure 11 presents the system performance when changing the external load. In the beginning, the dynamic load system is operated with a capacitor of 470 μF, and the total resistor of 100 Ω, connected in the electrical circuit. The motor starts up and increases the rotor speed following the sequence of 200 → 500 → 1000 → 1500 → 2000 rpm. The rotor speed reaches 2000 rpm after 5 s. The current i_q is approximately 1.79 A. At t = 6 s, the resistance load is changed from 100 Ω to 50 Ω by turning on some resistors in parallel connection. The external load is enhanced. The rotor speed reduces to the minimum of 1838 rpm, at t = 6.037 s, and stabilizes at 2000 rpm again at t = 6.468 s. The speed reduction is 162 rpm, and the recovery time is 0.431 s. The current i_q is increased to 2.51 A. Additionally, at t = 8 s, the resistance load is returned to the initial value of 100 Ω by turning off the same resistors.

The motor speed is increased to the maximum of 2170 rpm at t = 8.041 s and stabilized at 2000 rpm again at t = 8.492 s. The speed increment is 170 rpm, and the recovery time is 0.451 s. The current i_q is decreased to the previous value of 1.79 A because the additional external load is removed. It can be inferred that the motor still operates stably with the sensorless control algorithm by the novel SMO-PLL estimator in the dynamic load condition.

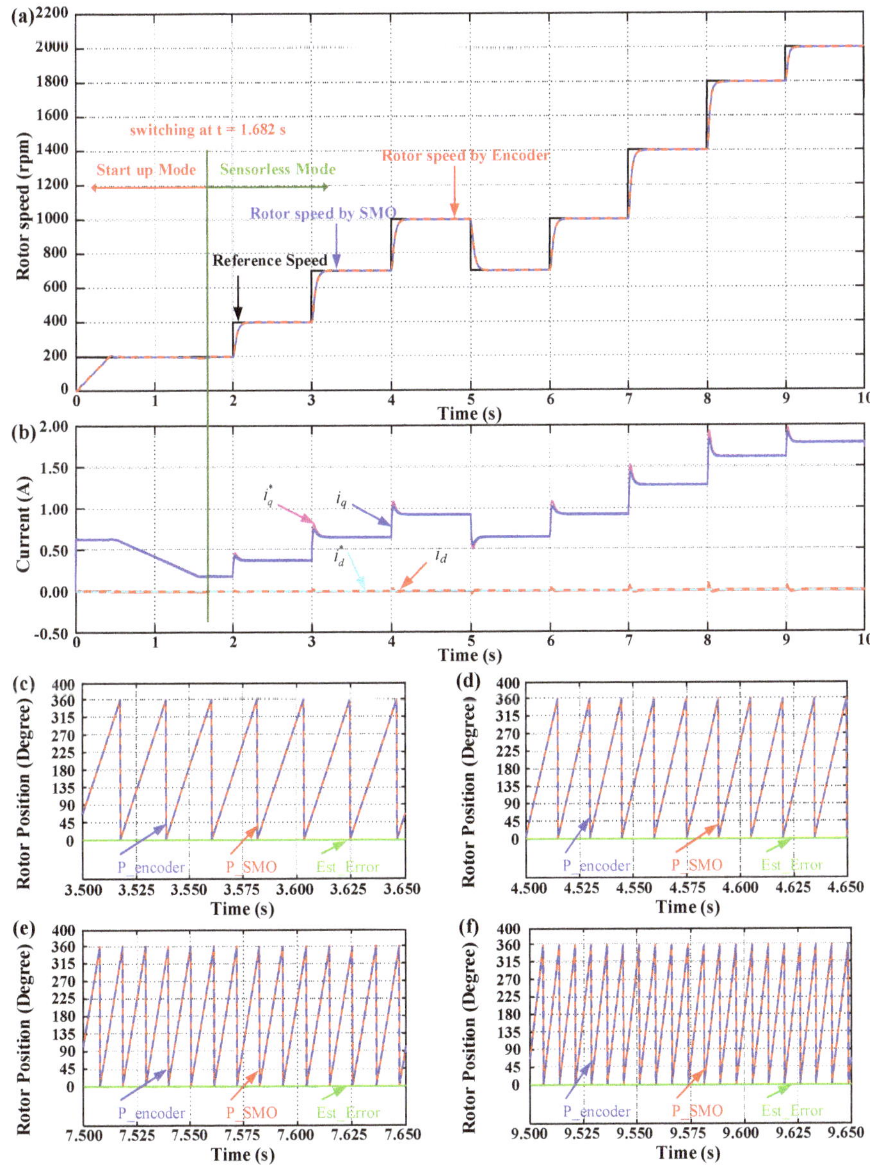

Figure 8. Simulation results in the speed increasing strategy, from 0 to 2000 rpm, for (**a**) speed response, (**b**) current response, and rotor position response at (**c**) 700, (**d**) 1000, (**e**) 1400, (**f**) 2000 rpm.

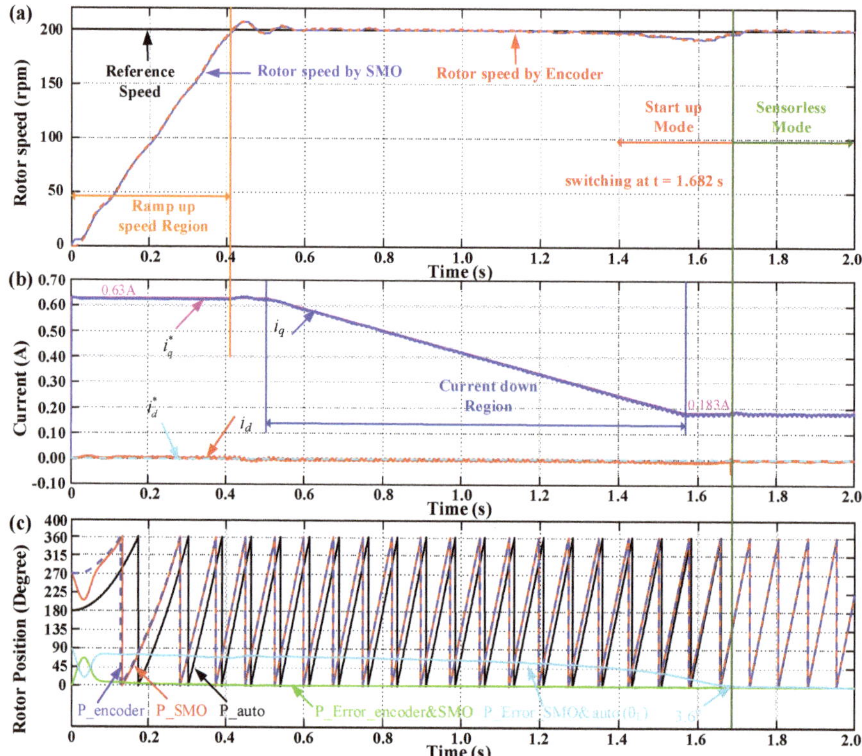

Figure 9. Simulation results in the startup mode and sensorless switching, from 0 rpm to 200 rpm for (**a**) speed response, (**b**) current response, (**c**) rotor position response.

As a summary, the various reference speed changes and the dynamic load condition are investigated in Figures 8–11 to analyze the controller performance. Referring to those results, it is easy to find that the rotor speed almost tracks to the commands very well, in which all the steady-state errors approach zero, and the overshoot or undershoot is also too small. The estimator works successfully because the estimated position approaches the actual position. The estimated error is almost zero. Moreover, the system has a good mode transition and robust performance against disturbance. The simulation results confirm that the proposed estimation and control algorithm for the sensorless PMSM drive system are correct and effective.

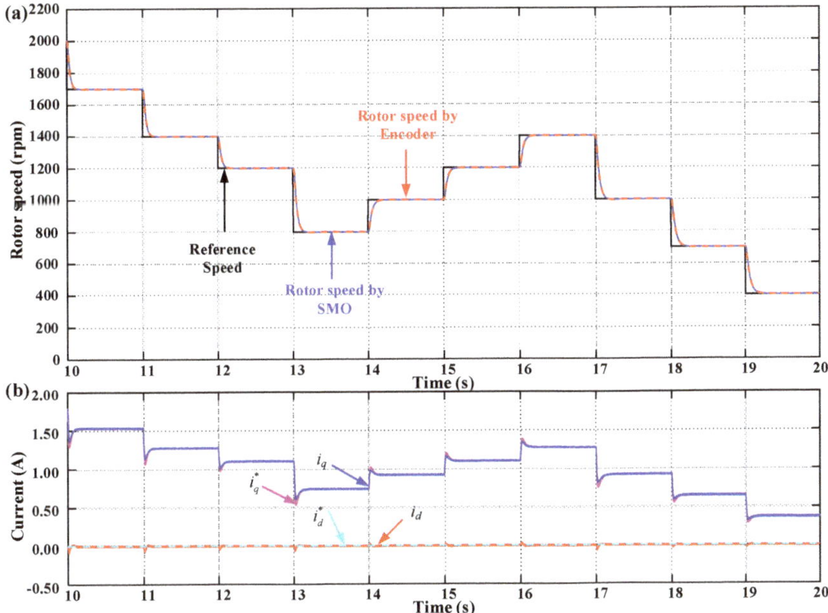

Figure 10. Simulation results in the speed decreasing strategy, from 2000 rpm to 400 rpm for (**a**) speed response, (**b**) current response.

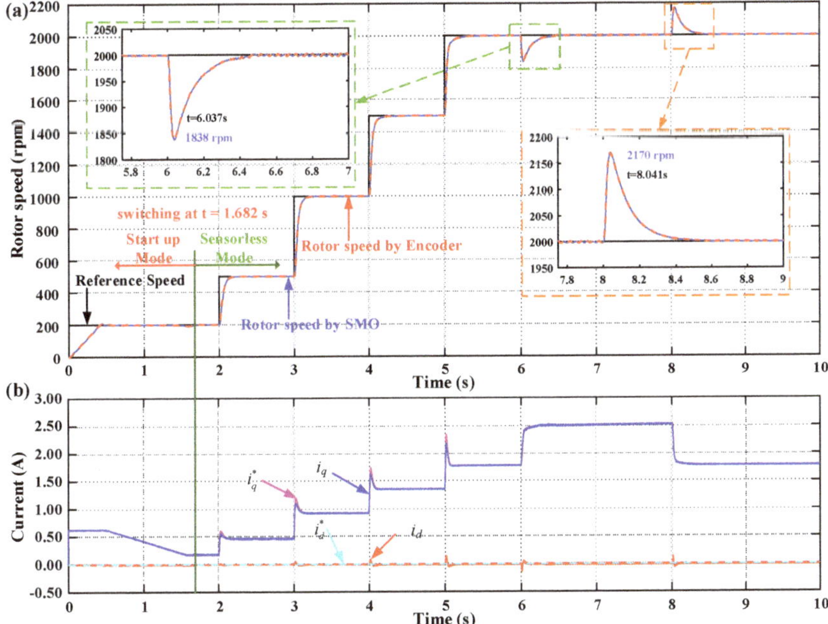

Figure 11. Simulation results in the case of the varied external load ($R_L = 100 \leftrightarrow 50\ \Omega$) at 2000 rpm for (**a**) speed response, (**b**) current response.

6. The Experimental Verification and Results

After finishing the simulation, to verify the algorithm and analyze the system performance, the motor control algorithm is implemented by the platform in Figure 12. The hardware platform consists of a PMSM coupled to a generator, with an electrical load system, an inverter, a DSP F28379D (as TI-DSP), and a control circuit. The parameters of PMSM are listed in Table 1. The DSP F28379D is equipped with 200 MHz dual C28xCPUs and dual CLAs, 1 MB Flash, 16-bit/12-bit ADCs, comparators, 12-bit DACs, HRPWMs, eCAPs, eQEPs, CANs, etc. The control circuit is designed to isolate the PWM signal between the inverter and TI-DSP, process the phase current measurement, and protect the overcurrent status. When the overcurrent happens, the PWM control signal generated by TI-DSP will be locked. The electrical load system includes a rectifier, a capacitor of 470 µF-450 V, and power resistors of 100 Ω-100 W.

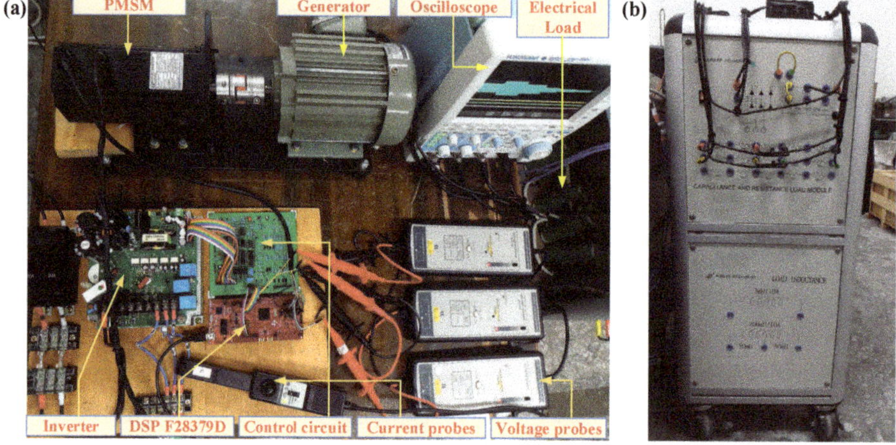

Figure 12. Experimental system with (**a**) real platform, (**b**) electrical load.

In the simulation structure in Figure 7, only the second block (B) is compiled to generate the algorithm code in C/C++ language, which is imported into the CCS software. The CCS software was responsible for connecting, downloading, debugging, and monitoring online the variables during the operation of the system. Moreover, some data in the block (C) are acquired online by MATLAB, based on the SCI function.

The sampling times of the velocity loop, the current loop and the novel SMO-PLL estimator, are still set as the same as the simulation values. The inverter's switching frequency is set at 15 Khz. The parameters of PI current controller are set as $K_{Pd} = 0.25$, $K_{Id} = 0.025$, $K_{Pq} = 0.25$, and $K_{Iq} = 0.025$. The RBFNN's initial parameters are setup as the node centers ($c_1 = -0.0025$, $c_2 = -0.00125$, $c_3 = 0.0$, $c_4 = 0.00125$, $c_5 = 0.0025$), the node widths ($b_1 = b_2 = b_3 = b_4 = b_5 = 0.5$), the connective weights ($w_1 = w_2 = w_3 = w_4 = w_5 = 0.0000625$). The learning rate is 0.435 and the momentum factor is 0.75. Additionally, the discrete PID's initial parameters are designed as $K_{Ps} = 0.735$, $K_{Is} = 0.00435$, and $K_{Ds} = 0.478$. All experimental results are implemented with an initial resistance load of 100Ω-400 W, a capacitor of 470µF-450V in the dynamic load system.

Figure 13 shows the motor's speed responses while speeding up the rotor speed from 0 to 2000 rpm. The reference speed is changed in a period of 5s with the sequences as $300 \rightarrow 500 \rightarrow 700 \rightarrow 1000 \rightarrow 800 \rightarrow 1200 \rightarrow 1600 \rightarrow 1800 \rightarrow 2000$ rpm. Figure 13a refers that the novel SMO-PLL estimated rotor speed is closely tracked to the reference speed, and overlapped the actual rotor speed, measured by the encoder. Although there is the oscillation of the current responses in Figure 13b, their average value still follows the current command. Figure 13c–f illustrates the electrical positions of the rotor, separately calculated by the novel SMO-PLL estimator and the encoder, with the estimated errors at the speeds of 500, 1000, 1600, 2000 rpm. There are approximately 5, 10, 16 and 20 position cycles in a period of 0.15s, respectively. Correspondingly, the rotation frequencies are 33.33, 66.67, 106.67, and 133.33 Hz. These values are suitable for the rotor speed. The estimated and actual positions are approximated to each other, so the estimated error approaches zero. The self-tuning PID controller's parameters are also updated during the operating time. The PID gains are tuned to $K_{Ps} = 0.7378$, $K_{Is} = 0.0039$, $K_{Ds} = 0.4751$ at t = 12 s (as $\omega_r = 500$ rpm); $K_{Ps} = 0.7583$, $K_{Is} = 0.0039$, $K_{Ds} = 0.4692$ at t = 22 s (as $\omega_r = 1000$ rpm); $K_{Ps} = 0.8408$, $K_{Is} = 0.0034$, $K_{Ds} = 0.4619$ at t = 37.5 s (as $\omega_r = 1600$ rpm). Lastly, rotor speed reaches 2000 rpm, and $K_{Ps} = 0.8672$, $K_{Is} = 0.0034$, $K_{Ds} = 0.4487$ at t = 46 s.

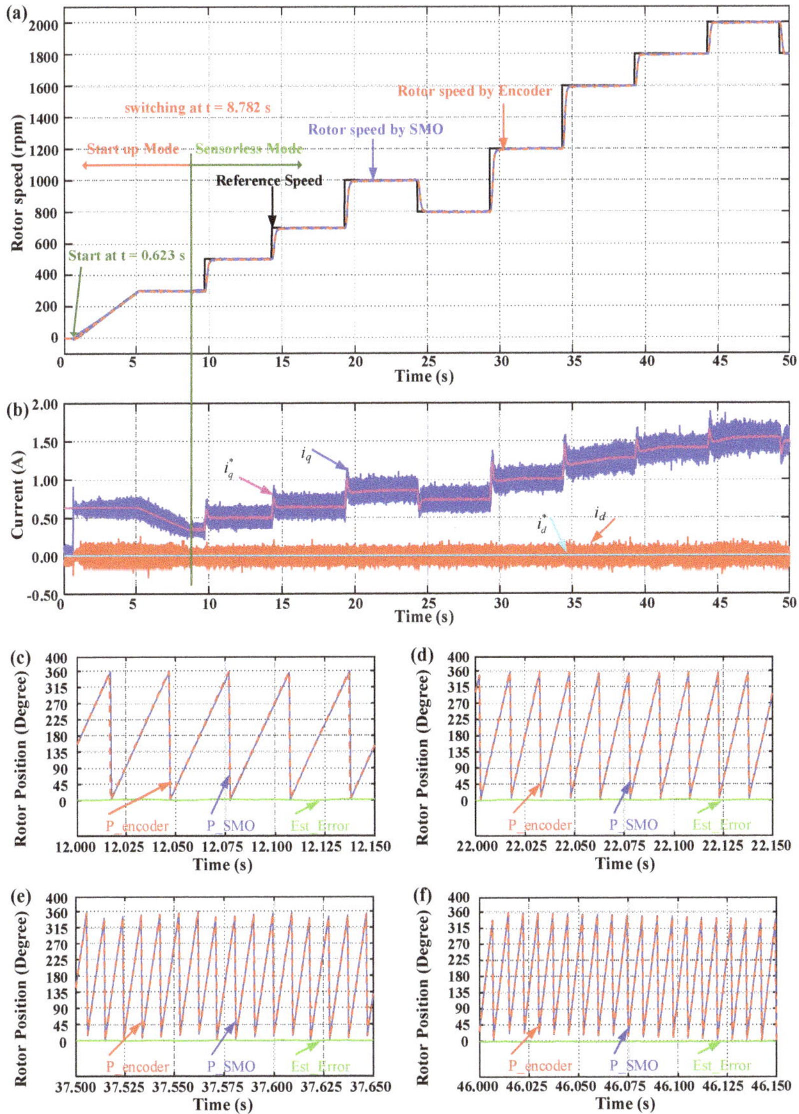

Figure 13. Experimental results in the speed increasing strategy, from 0 to 2000 rpm for (**a**) speed response, (**b**) current response, and rotor position response at (**c**) 500, (**d**) 1000, (**e**) 1600, (**f**) 2000 rpm.

Figure 14 demonstrates the detailed speed response for the startup motor mode and switching to the sensorless control mode at the speed of 300 rpm. The ramp-up speed ratio is set up at 66.67 rpm/s. At t = 0.623 s, the motor starts to speed up. The rotor speed is increased, following the ramp function with the initial reference current i_q^* of 0.635 A. The actual rotor speed reaches 295 rpm at t = 5.2 s. Then, the reference current i_q^* decreases down to 0.336 A with a current down ratio of 0.085 A/s. The reference speed is kept at 300 rpm, and the actual rotor speed is still close to 300 rpm. At t = 6.403 s, the actual rotor speed reaches the maximum of 302.7 rpm. The maximum speed error is +2.7 rpm. At t = 8.832 s, the actual rotor speed drops to a minimum of 286.1 rpm. The minimum speed error is −13.9 rpm. The estimated position leads the auto position by 27.95 degrees at t = 5 s, while the estimated position error is 39.02 degrees. These values are reduced in the current down region. The switching operation occurs at t = 8.782 s when the deviation of estimated position and auto position (θ_L) is also equal to 3.6 degrees. The estimated position error is only 0.175 degrees. The ripple of the current i_q is roughly 0.06 A, while the current i_d fluctuates around zero within a range of 0.1 A. The motor switches the control mode smoothly. After this, the motor begins to operate in the sensorless control mode.

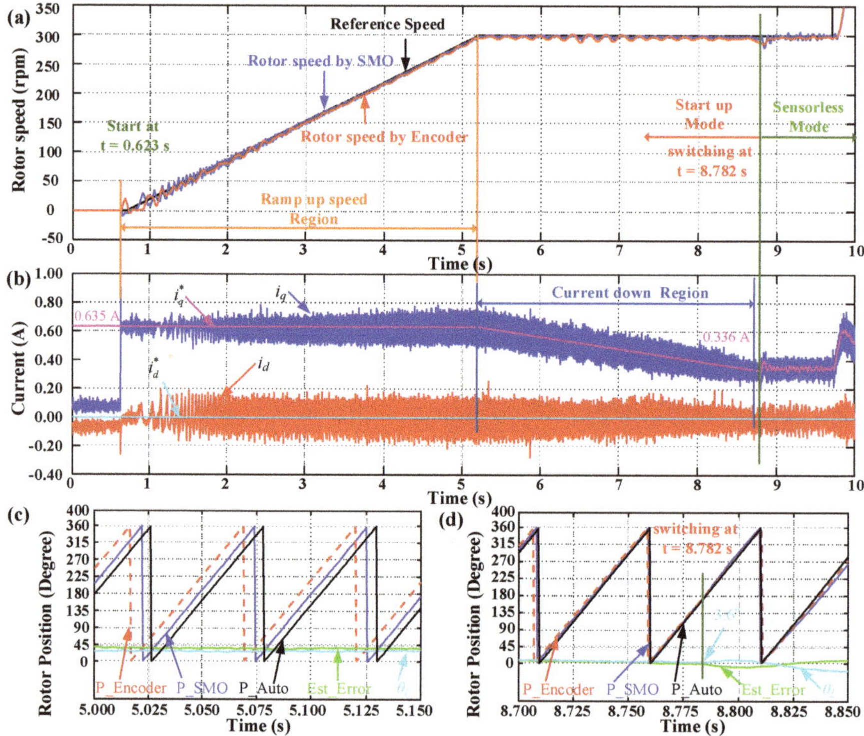

Figure 14. Experimental results in the startup mode and sensorless switching, from 0 rpm to 300 rpm, for (**a**) speed response, (**b**) current response, (**c**) rotor position response.

Figure 15 shows the motor's speed responses while slowing down the rotor speed from 1800 to 300 rpm. The reference speed is sequentially varied as 1800 → 1600 → 1400 → 1000 → 1200 → 800 → 500 → 300→ 500 → 700 → 1000 rpm. Figure 15a shows that the estimated rotor speed is closely tracked by the reference speed and overlaps the actual rotor speed. Figure 15b presents the current response in the *d–q* axis. The current i_q is also varied according to the rotor speed. When the rotor speed is decreased, the external load becomes lower. Therefore, the less torque is required by the motor, and the current i_q is regulated to be smaller. Moreover, the self-tuning PID controller's parameters are also tuned effectively. The PID gains are tuned to $K_{Ps} = 0.832$, $K_{Is} = 0.0039$, $K_{Ds} = 0.4531$ at t = 62 s (as $\omega_r = 1400$ rpm); $K_{Ps} = 0.7280$, $K_{Is} = 0.0044$, $K_{Ds} = 0.4580$ at t = 77 s (as $\omega_r = 800$ rpm); and $K_{Ps} = 0.7046$, $K_{Is} = 0.0044$, $K_{Ds} = 0.4639$ at t = 87 s (as $\omega_r = 300$ rpm).

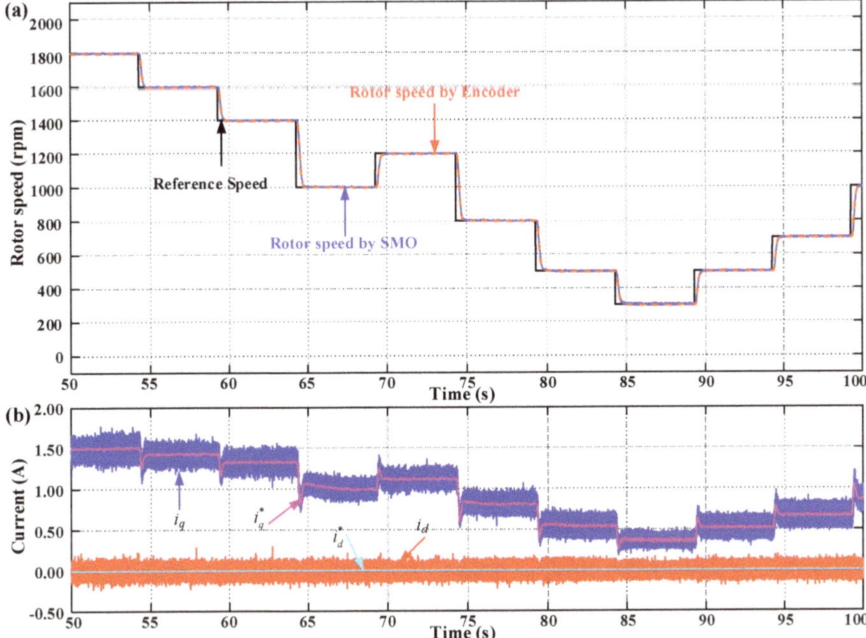

Figure 15. Experimental results in the speed decreasing strategy, from 1800 rpm to 300 rpm for (**a**) speed response, (**b**) current response.

Figure 16 presents the system performance when the external load is varied. In the beginning, the dynamic load is set up with a capacitor of 470 µF-450 V, the total resistor of 100 Ω-400 W. The motor starts up and increases the rotor speed, reaching 2000 rpm. The current i_q fluctuates around 1.56 A. At t = 37.99 s, the total resistance load is changed to 50 Ω-800 W by turning on more resistors. At t = 38.11 s, the actual rotor speed (red line) drops to a minimum of 1797 rpm while the minimum estimated value (blue line) is 1817 rpm. The recovery time is 2.99 s when the motor reaches 1995 rpm again at t = 40.98 s. The average current i_q reaches 2.42 A due to the external load increment. In addition, at t = 45.60 s, the total resistance load is returned to the initial value of 100 Ω-400 W by turning off the same resistor. The actual rotor speed is increased to the maximum of 2171 rpm at t = 45.71 s, while the maximum estimated value is 2157 rpm. The recovery time is 1.06 s when the motor stabilizes at 2000 rpm again at t = 46.66 s. The average current i_q is decreased to 1.57 A because of the external load reduction. Within the speed regulation at 2000 rpm, the estimate speed error is about +5 rpm, while this error is larger at the up peak (20 rpm) or the low peak (14 rpm). The maximum control speed steady-state error is ±5 rpm at 2000 rpm. Although the system is affected by the disturbance of the external load, the motor still operates stably and successfully with the sensorless control algorithm based on the novel SMO-PLL estimator.

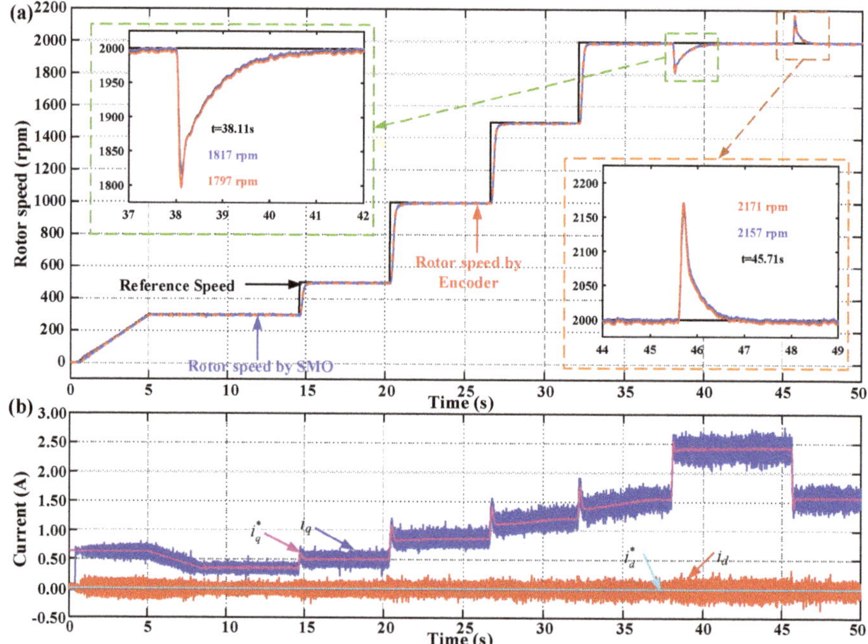

Figure 16. Experimental results in the case of the varied external load (RL = 100 ↔ 50 Ω) at 2000 rpm for (**a**) speed response, (**b**) current response.

Finally, similar to the simulation, the experimental results in Figures 13–16 show that the rotor speed almost tracks to the command very well, all the steady-state errors approach zero (within ±5 rpm in tolerance), and the overshoot or undershoot is also too small. The novel SMO-PLL estimator works successfully. The estimated position approaches the actual position. The estimated error is close to zero. Additionally, the real-time system has a good mode transition and robust performance against disturbance. The experimental results again confirm that the proposed estimation and control algorithm of the PMSM system are correct and effective in the real-time system. Furthermore, the DSP application for the PMSM drive control system is built in MATLAB Simulink properly, and is deployed to CCS software to realize the real-time system successfully. This deployment method shortens the application development time.

7. Conclusions

In this paper, a self-tuning PID controller based on a radial basis function neural network and a rotor position estimator based on the novel SMO in combination with a PLL for the sensorless PMSM drive control system have been described and developed successfully. The control algorithm, consisting of the *I-f* startup strategy, novel SMO-PLL estimator, RBFNN-based self-tuning PID controller, and FOC algorithm, was deployed properly in both the simulation and the real-time hardware, established on a DSP F28379D. The system performance has been verified in three terms: startup mode, tracking response, and speed regulation with the dynamic load system. The simulation and experimental results indicate that the motor control system has a smooth transition from the startup mode to the sensorless control mode with the low ripple in the current and the rotor speed. The novel SMO-PLL estimator is stable, and the estimated position approximates to the actual position, so the estimated error is almost minimal and negligible. The PID gains are tuned effectively. The rotor speed tracks properly to the reference speed, and the overshoot or undershoot is very small. The steady-state errors approach zero. Additionally, the system provides a robust performance against disturbance. Accordingly, the system performance confirms that the proposed intelligent control algorithm and the position estimator for the sensorless PMSM drive control system are correct and effective. Furthermore, the DSP application for the PMSM drive control system is implemented based on the combination of MATLAB and CCS software. Therefore, we can easily develop additional intelligent controllers for the motor control system. It also has the advantage of faster testing, monitoring and acquisition of data online, and troubleshooting of systems. This motivates the improvement of the proposed algorithm for the wide speed range control in further works.

Author Contributions: H-K.H. wrote this article, designed the control method, implemented the hardware platform, drew the figures and performed the simulations as well as the experimental data. S-C.C. supervised, coordinated the investigations, checked up the manuscript's logical structure. H.T. was responsible for exchanging ideas, reviewing the article draft. All authors have read and agreed to the published version of the manuscript.

Funding: This research received no external funding.

Conflicts of Interest: The authors declare no conflict of interest.

References

1. Risfendra, K.Y.-S.; Wang, M.-S.; Huang, L.-C. Realization of a sensorless speed controller for permanent magnet synchronous motor drives based on field programmable gate array technology. *Adv. Mech. Eng.* **2017**, *9*. [CrossRef]
2. Than, H.; Kung, Y.-S. FPGA-realization of an RBF-NN tuning PI controller for sensorless PMSM drives. *Microsyst. Technol.* **2019**. [CrossRef]
3. Zhang, Y.; Cheng, X.-F. Sensorless Control of Permanent Magnet Synchronous Motors and EKF Parameter Tuning Research. *Math. Probl. Eng.* **2016**, *2016*, 1–12. [CrossRef]
4. Chen, C.-X.; Xie, Y.-X.; Lan, Y.-H. Backstepping control of speed sensorless permanent magnet synchronous motor based on slide model observer. *Int. J. Autom. Comput.* **2015**, *12*, 149–155. [CrossRef]
5. Kung, Y.-S.; Risfendra, R.; Lin, Y.-D.; Huang, L.-C. FPGA-realization of a sensorless speed controller for PMSM drives using novel sliding mode observer. *Microsyst. Technol.* **2018**, *24*, 79–93. [CrossRef]
6. Lu, Q.; Quan, L.; Zhu, X.; Zuo, Y.; Wu, W. Improved Sliding Mode Observer for Position Sensorless Open-Winding Permanent Magnet Brushless Motor Drives. *Prog. Electromagn. Res.* **2019**, *77*, 147–156. [CrossRef]
7. Wang, Y.; Xu, Y.; Zou, J. Sliding-Mode Sensorless Control of PMSM With Inverter Nonlinearity Compensation. *IEEE Trans. Power Electron.* **2019**, *34*, 10206–10220. [CrossRef]
8. An, Q.; An, Q.; Liu, X.; Zhang, J.; Bi, K. Improved Sliding Mode Observer for Position Sensorless Control of Permanent Magnet Synchronous Motor. In Proceedings of the 2018 IEEE Transportation Electrification Conference and Expo, Asia-Pacific (ITEC Asia-Pacific), Bangkok, Thailand, 6–9 June 2018.
9. Chen, Y.; Li, M.; Gao, Y.-W.; Chen, Z.-Y. A sliding mode speed and position observer for a surface-mounted PMSM. *ISA Trans.* **2019**, *87*, 17–27. [CrossRef] [PubMed]
10. Kung, Y.-S.; Lin, Y.-D.; Huang, L.-C. FPGA-based sensorless controller for PMSM drives using sliding mode observer and phase locked loop. In Proceedings of the 2016 International Conference on Applied System Innovation (ICASI), Okinawa, Japan, 26–30 May 2016.
11. Wang, Z.; Lu, K.; Blaabjerg, F. A Simple Startup Strategy Based on Current Regulation for Back-EMF-Based Sensorless Control of PMSM. *IEEE Trans. Power Electron.* **2012**, *27*, 3817–3825. [CrossRef]
12. Wang, M.-S.; Chen, S.-C.; Shih, C.-H. Speed control of brushless DC motor by adaptive network-based fuzzy inference. *Microsyst. Technol.* **2016**, *24*, 33–39. [CrossRef]
13. Wang, M.-S.; Syamsiana, I.N.; Lin, F.-C. Sensorless Speed Control of Permanent Magnet Synchronous Motors by Neural Network Algorithm. *Math. Probl. Eng.* **2014**, *2014*, 1–7. [CrossRef]
14. Ananthamoorthy, N.; Baskaran, K. High performance hybrid fuzzy PID controller for permanent magnet synchronous motor drive with minimum rule base. *J. Vib. Control* **2013**, *21*, 181–194.
15. Liu, J. *Radial Basis Function (RBF) Neural Network Control for Mechanical Systems: Design, Analysis and Matlab Simulation*; Springer Science+Business Media: Berlin/Heidelberg, Germany, 2013.

16. Zhang, Y.; Song, J.; Song, S.; Yan, M. Adaptive PID Speed Controller Based on RBF for Permanent Magnet Synchronous Motor System. In Proceedings of the 2010 International Conference on Intelligent Computation Technology and Automation, Changsha, China, 11–12 May 2010.

17. Liu, J. *Intelligent Control Design and MATLAB Simulation*; Springer Nature: Gateway East, Singapore, 2018.

 *electronics*

Article

Design of a Low-Order FIR Filter for a High-Frequency Square-Wave Voltage Injection Method of the PMLSM Used in Maglev Train

He Zhao [1], Liwei Zhang [1,*], Jie Liu [2], Chao Zhang [1], Jiao Cai [3] and Lu Shen [1]

[1] School of Electrical Engineering, Beijing Jiaotong University, Beijing 100044, China; 17121540@bjtu.edu.cn (H.Z.); 18117026@bjtu.edu.cn (C.Z.); 19117017@bjtu.edu.cn (L.S.)
[2] Anhui Huaying Automotive Technology Co., Ltd., Anhui 239400, China; liujiebjh@dingtalk.com
[3] Information Technology Center, Beijing Jiaotong University, Beijing 100044, China; caijiao@bjtu.edu.cn
* Correspondence: lwzhang@bjtu.edu.cn; +86-135-5268-3918

Received: 4 April 2020; Accepted: 26 April 2020; Published: 28 April 2020

Abstract: In position sensorless control based on a high-frequency pulsating voltage injection method, filters are used to complete the extraction of high-frequency response signals for position observation. A finite impulse response (FIR) filter has the advantages of good stability and linear phase. However, the FIR filter designed by using traditional methods has a high order which will cause a large time delay. This paper proposes a low-order FIR filter design method for a high-frequency signal injection method in the permanent magnet linear synchronous motor. Based on the frequency characteristics of the current signal, the requirement that the FIR filter needs to meet were analyzed. According to the amplitude–frequency characteristic of the FIR filter, these requirements were converted into constraint equations. By solving these equations, the coefficient of the FIR filter could be obtained. The simulation and experiment results showed the effectiveness of this low-order FIR filter.

Keywords: PMLSM; position sensorless control; high-frequency square-wave voltage injection; FIR filter; maglev train

---

1. Introduction

Due to the fact of its high efficiency and high-power density, the permanent magnet linear synchronous motor (PMLSM) is used in the drive system of the maglev train. For the PMLSM control system, position detection is crucial. There are many traditional position detection methods [1–5]. The inductive looped-cable method was used in References [1–3] which requires signal process units, crossed-looped cables, and antennas. In addition, the crossed-looped cables need to be placed along the rails which will make the system very expensive to construct and maintain [4]. In References [4,5], radio millimeter waves and Doppler radar are used to detect the position of the maglev train. Although they do not need to lay lines, they need additional equipment and design such as radar and mobile, base, and central stations. This will also increase the cost and the complexity of mechanical installation [6]. Therefore, the position sensorless control method has received increasing research attention.

Position sensorless control can be classified into two types. The first type is based on the back electromotive force (EMF) of the motor [7]. To improve the performance of the position sensorless control system, the extended back EMF method [8], sliding mode observer [9], and extended Kalman filter [10] are used in this method. However, when the motor is at zero or low speed, the back EMF is too small to detect which means this method is not applicable. The second type is based on motor saliency which can realize position estimation at zero or low speed, and the high-frequency signal injection method is representative of this type. Currently, research on the high-frequency signal injection method is mainly concentrated on the permanent magnet synchronous motor (PMSM). As the PMLSM and

PMSM have the same working principles and similar mathematical models, these high-frequency signal injection methods can be directly used for PMLSM. Injecting a high-frequency rotating voltage into a stationary reference frame and injecting a high-frequency pulsating voltage into an estimated synchronous reference frame are two traditional methods of high-frequency signal injection method [11]. In addition, after years of research, many improvements have appeared [12–16]. In Reference [12], the high-frequency pulsating signal was injected into the estimated fixed-frequency rotating reference frame instead of the estimated synchronous reference frame. In addition, the high-frequency pulsating signal was injected into the stationary reference frame in Reference [13] and was injected into the ABC reference frame in Reference [14]. In Reference [15], a bidirectional rotating signal was used, and a high-frequency square-wave voltage signal was used in Reference [16]. The differences among these methods lay in the form of the injection signal and the selection of the injection reference frame. However, all methods are required in order to extract the high-frequency response signal which carries magnetic pole position information from the motor current. This extraction process directly determines the accuracy of the position estimation.

In the PMLSM control system, the inverter will induce a large amount of harmonics. When the high-frequency square-wave voltage injection method is used to realize position estimation, these harmonics will affect the estimation accuracy [16]. Increasing the amplitude of injection voltage can increase the signal-to-noise ratio (SNR) which can reduce the effect of harmonics. However, the voltage utilization will be reduced [17]. Moreover, when PMLSM is used in the maglev train drive system and the high-frequency square-wave voltage is injected, the current will fluctuate and generate normal force fluctuation. This means that increasing the amplitude of injection voltage will increase the impact on the suspension system. In addition to increasing the amplitude of injection voltage, digital filters can also help to improve the SNR. The infinite impulse response (IIR) filters were designed in References [18,19]. However, the IIR filter is a nonlinear phase filter and requires output feedback which increases the complexity of the control system. Furthermore, according to the research in Reference [20], the FIR filter has a better performance than the IIR filter in the position estimation system. An FIR filter was designed in Reference [21], but its order is as high as 17 which will cause a large time delay. A moving average filter was designed in Reference [22] which can provide the system with a higher bandwidth and better harmonics suppression capability. But it is also an FIR filter with a high order of 21. In order to reduce the time delay, many studies have proposed different improvements. An algebraic operation was used in References [23,24] to separate carrier signals, but this calculation process is ideal without considering the influence of harmonics. A direct signal demodulation method was proposed in Reference [25] which can remove the low-pass filter, but it is mainly used in the initial position estimation. Reference [26] designed an all-pass filter to calculate the position which can eliminate the time delay caused by the band-pass filter and low-pass filter, but the nonlinearity of the inverter is not considered either.

After considering the influence of the inverter and the time delay caused by the filter, this paper proposes a low-order FIR filter design method. Different from the traditional design method, this new method does not need to determine the passband and stopband of the filter. Based on the analysis of the motor current components, the amplitude–frequency characteristics at some special frequencies were mainly considered. Compared with the traditional design method, this new method can reduce the order of the FIR filter which will reduce the time delay. Moreover, compared with the filterless estimation system, the amplitude of the injection voltage can be reduced which will increase the voltage utilization [17] and reduce the normal force fluctuation.

2. Basic Principles of High-Frequency Square-Wave Voltage Injection Method

The high-frequency square-wave voltage injection method is based on the saliency of the PMLSM. In this paper, a surface-mounted motor was used in the experiment, so the saliency was saturated saliency. The generation of saturated saliency comes from the characteristics of the magnetization curve of silicon steel. As the magnetic field strength H increases, the magnetic field in the silicon steel gradually saturates, and the rising rate of the magnetic induction B becomes slower. According to the definition of magnetic permeability, μ = B/H, μ also becomes smaller as H increases, thereby reducing the inductance. Since the permanent magnet is in the *d*-axis magnetic circuit, the *d*-axis inductance will be less than the *q*-axis inductance at saturation.

By injecting a high-frequency square-wave signal into the virtual *d*-axis, the current of PMLSM will carry the high-frequency signal that contains position information of the magnetic pole. By extracting this high-frequency current signal, the position estimation can be realized.

The voltage equation of PMLSM can be described in the *d*–*q* synchronization reference frame (SRF) as shown in Equation (1) [27].

$$\begin{bmatrix} u_d \\ u_q \end{bmatrix} = \begin{bmatrix} R + pL_d & -\frac{\pi v}{\tau}L_q \\ \frac{\pi v}{\tau}L_d & R + pL_q \end{bmatrix} \begin{bmatrix} i_d \\ i_q \end{bmatrix} + \frac{\pi v}{\tau} \begin{bmatrix} 0 \\ \psi_f \end{bmatrix} \tag{1}$$

where u_d, u_q, i_d, i_q, L_d, L_q are *d*-axis voltage, *q*-axis voltage, *d*-axis current, *q*-axis current, *d*-axis inductance, and *q*-axis inductance, respectively. And R, v, τ, ψ_f, and p are the resistance of motor winding, speed, pole pitch, flux linkage of the permanent magnet, and differential operator, respectively.

With high-frequency excitation, Equation (1) can be simplified. The frequency of the injected square-wave voltage is very high, which means the voltage drop in the resistor is much smaller than the voltage drop in the motor inductor. Therefore, the voltage drop in the resistor can be ignored. In addition, the high-frequency square-wave voltage injection method is used when PMLSM is at low speed, which means speed v is approximately 0. Therefore, the parts related to v can be also ignored in Equation (1). After simplification, Equation (1) can be expressed as:

$$\begin{bmatrix} u_{dh} \\ u_{qh} \end{bmatrix} = \begin{bmatrix} pL_d & \\ & pL_q \end{bmatrix} \begin{bmatrix} i_{dh} \\ i_{qh} \end{bmatrix} \Rightarrow \frac{d}{dt} \begin{bmatrix} i_{dh} \\ i_{qh} \end{bmatrix} = \begin{bmatrix} \frac{1}{L_d} & \\ & \frac{1}{L_q} \end{bmatrix} \begin{bmatrix} u_{dh} \\ u_{qh} \end{bmatrix} \tag{2}$$

where i_{dh}, i_{qh}, u_{dh}, u_{qh}, denote *d*, *q* components of high-frequency response current and *d*, *q* components of high-frequency injection voltage in SRF, respectively.

Because PMSM and PMLSM generate the three-phase synthetic magnetic field in the same way, the working principle of them are the same. In addition, vector control was used based on the rotating reference frame. Therefore, PMLSM can be analyzed using a rotating reference frame just like PMSM. In order to achieve this goal, the *d*–*q* axis in PMLSM was converted to a rotating reference frame in Figure 1. In addition, the $\hat{d} - \hat{q}$ reference frame is also defined in Figure 1 which represents the estimated magnetic pole position. $\Delta\theta = \Delta x/\tau * \pi$ indicates the estimation error which was converted to the rotating reference frame, and $\omega_r = v/\tau * \pi$ indicates the motor speed which was converted to angular velocity.

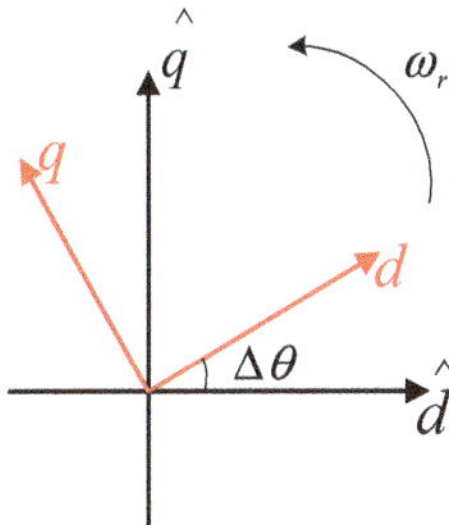

Figure 1. Relationship of the two reference frames.

The variables in the two coordinates shown in Figure 1 can be converted to each other via Equation (3).

$$\begin{bmatrix} x_d \\ x_q \end{bmatrix} = H_t \begin{bmatrix} x_{\hat{d}} \\ x_{\hat{q}} \end{bmatrix} = \begin{bmatrix} \cos \Delta\theta & \sin \Delta\theta \\ -\sin \Delta\theta & \cos \Delta\theta \end{bmatrix} \begin{bmatrix} x_{\hat{d}} \\ x_{\hat{q}} \end{bmatrix} \tag{3}$$

where x_d and x_q denote the physical quantities in the d–q reference frame and $x_{\hat{d}}$ and $x_{\hat{q}}$ denote the physical quantities in the $\hat{d} - \hat{q}$ reference frame. H_t is the transformation matrix.

By using the coordinate transformation in (3), Equation (2) can be derived as:

$$\frac{d}{dt}\begin{bmatrix} i_{\hat{d}h} \\ i_{\hat{q}h} \end{bmatrix} = H_t^{-1} \begin{bmatrix} \frac{1}{L_d} & \\ & \frac{1}{L_q} \end{bmatrix} H_t \begin{bmatrix} u_{\hat{d}h} \\ u_{\hat{q}h} \end{bmatrix} = \frac{1}{\sum L^2 - \Delta L^2} \times \begin{bmatrix} \sum L + \Delta L \cos 2\Delta\theta & \Delta L \sin 2\Delta\theta \\ \Delta L \sin 2\Delta\theta & \sum L - \Delta L \cos 2\Delta\theta \end{bmatrix} \begin{bmatrix} u_{\hat{d}h} \\ u_{\hat{q}h} \end{bmatrix} \tag{4}$$

where $\sum L = (L_d + L_q)/2$ and $\Delta L = (L_q - L_d)/2$.

A high-frequency square-wave voltage, which is shown in Figure 2, was injected into the virtual d-axis. Its magnitude and period were V_h and T_h, respectively. Due to the simple form of the square wave signal, its frequency could be increased to the inverter switching frequency level. From the above analysis, it can be seen that under high-frequency excitation, the influence of resistance is ignored, and the model of the motor is simplified. The effect of resistance decreases as the frequency of the injected signal increases. Therefore, when the high-frequency square-wave signal injection method is used, it is beneficial to simplify the high-frequency model of PMLSM. Moreover, as the signal frequency increases, the saliency ratio of the motor increases [28] which can enhance the performance of position sensorless control based on saliency.

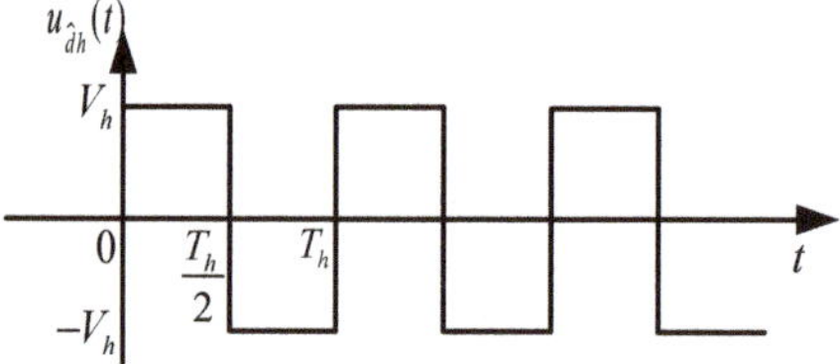

Figure 2. High-frequency square-wave voltage signal diagram.

This injected signal can be expressed as:

$$\begin{bmatrix} u_{\hat{d}h} \\ u_{\hat{q}h} \end{bmatrix} = \begin{bmatrix} V_h(-1)^k \\ 0 \end{bmatrix} (k = 0,1,2,3 \ldots) \tag{5}$$

From Equations (4) and (5), the high-frequency current response can be derived as:

$$\frac{d}{dt}\begin{bmatrix} i_{\hat{d}h} \\ i_{\hat{q}h} \end{bmatrix} = \frac{V_h(-1)^k}{\sum L^2 - \Delta L^2} \begin{bmatrix} \sum L + \Delta L \cos 2\Delta\theta \\ \Delta L \sin 2\Delta\theta \end{bmatrix} \tag{6}$$

It can be seen from Equation (6) that the high-frequency response current includes $\Delta\theta$ which is the difference between the estimated value and actual value of the magnetic pole position. In order to get the actual position, an observer is needed, as Figure 3 shows.

In Figure 3, $i_{\hat{q}fir}$ is the filtered current of $i_{\hat{q}}$ which is used for current loop feedback. $i_{\hat{q}h}$ is the high-frequency response current caused by injection voltage. It will be eliminated by the FIR filter at first, and then it will be obtained by signal processing. Subsequently, the differential of $i_{\hat{q}h}$ is multiplied by $(-1)^k$. The result is shown below.

$$f(\Delta\theta) = \Delta i_{\hat{q}h} \times (-1)^k = \frac{V_h \Delta L T_s}{\sum L^2 - \Delta L^2} \sin(2\Delta\theta) \tag{7}$$

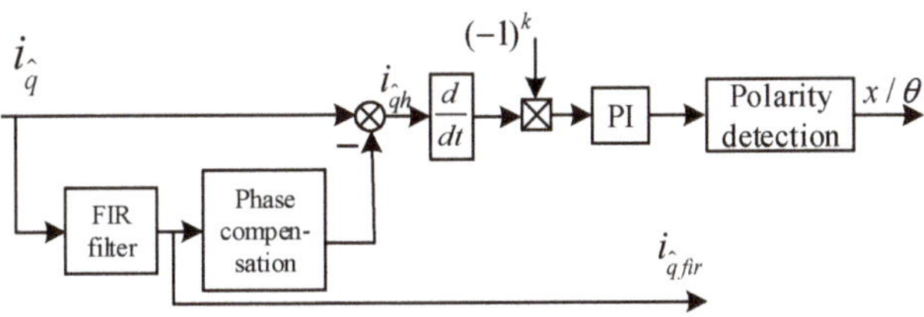

Figure 3. Structure of the position observer.

Finally, the processed signal passes through the PI regulator to change the estimated angle $\hat{\theta}$ so that $f(\Delta\theta)$ becomes 0. At this time, $\Delta\theta$ will converge to 0 or π which means the actual position of the magnetic pole coincides with the estimated position or differs by one pole distance. Therefore, in order to obtain an accurate magnetic pole position, polarity detection is necessary. This paper used the d-axis current peak value comparison method to achieve this goal. Figure 4 shows the basic principle of this method. When $\Delta\theta$ converges to 0, the positive d-axis current will increase the magnetic field. At this time, the saturation of the d-axis magnetic circuit is increased, so the d-axis inductance is reduced. Therefore, the positive current peak $|i_{dmax}|$ will be greater than the absolute value of negative current peak $|i_{dmin}|$. Conversely, when $\Delta\theta$ converges to π, the negative d-axis current will increase the magnetic field which means $|i_{dmax}|$ will be smaller than $|i_{dmin}|$. Therefore, by comparing $|i_{dmax}|$ and $|i_{dmin}|$, the magnetic pole polarity can be detected.

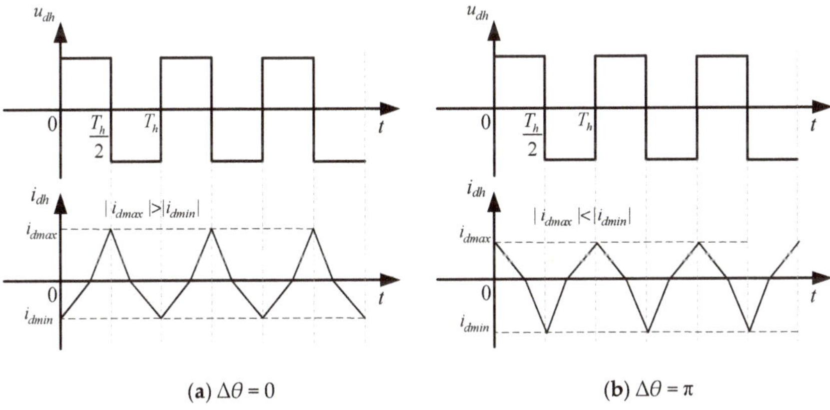

(**a**) $\Delta\theta = 0$ (**b**) $\Delta\theta = \pi$

Figure 4. Ideal d-axis current waveform at different $\Delta\theta$.

For the d-axis current, when the observed value of the position converges to the actual value, it can be expressed as in Equation (8).

$$\frac{di_{\hat{d}h}}{dt} = \frac{V_h(-1)^k}{\sum L - \Delta L} \tag{8}$$

As the injection voltage increases, the fluctuation of the d-axis current becomes larger. The relationship between linear motor normal force and d-axis current can be approximated as in Equation (9) [29]

$$\begin{cases} F_y \propto (F_d + 2F_f)^2 \\ F_d \propto i_d \end{cases} \tag{9}$$

where F_d denotes the magnetomotive force of the d-axis, and F_f denotes the magnetomotive force of the permanent magnet. Therefore, fluctuations in the d-axis current can cause fluctuations in the normal force of PMLSM which will affect the suspension system.

3. FIR Filter Analysis and Design

3.1. Current Frequency Spectrum Analysis

Since the filter design was based on the frequency characteristics and the position was estimated using $i_{\hat{q}}$, the frequency characteristics of $i_{\hat{q}}$ needed to be analyzed. When using the space vector pulse width modulation (SVPWM) strategy to control the motor, it was necessary to modulate the calculated reference wave with a high-frequency carrier to generate a pulse-width modulating (PWM) signal to control the inverter. In this process, harmonic components related to the carrier wave are generated. When considering the influence of SVPWM harmonics, $i_{\hat{q}}$ can be expressed as:

$$i_{\hat{q}} = i_{\hat{q}f} + i_{\hat{q}h} + i_{\hat{q}hm} \tag{10}$$

where $i_{\hat{q}f}$, $i_{\hat{q}h}$, and $i_{\hat{q}hm}$ are a fundamental current, a high-frequency response current generated by the injection voltage, and a harmonic current generated by SVPWM, respectively.

When the motor runs stably, $i_{\hat{q}f}$ remains basically unchanged. So, $i_{\hat{q}f}$ can be regarded as a DC component. Even if the motor's operating state changes, $i_{\hat{q}f}$ changes much slower than $i_{\hat{q}hm}$ and $i_{\hat{q}h}$. Therefore, the frequency of $i_{\hat{q}f}$ can be approximately considered as 0 Hz [30].

The harmonics of SVPWM are mainly the sideband harmonics near the first and second carrier frequencies [31]. Considering the motor speed is low, it can be approximated that the frequency of $i_{\hat{q}hm}$ is concentrated on the first and second carrier frequencies.

It can be seen from Equation (6) that the frequency of $i_{\hat{q}h}$ is the same as that of the high-frequency injection signal. By using the Fourier transform, the frequency domain expression of the signal shown in Figure 2 can be incorporated into Equation (11), which indicates the frequency of $i_{\hat{q}h}$ is mainly the odd-numbered multiples of the injection signal.

$$v(\omega) = b_n \sin\left(\frac{2\pi n}{T_h}t\right) \tag{11}$$

where $b_n = \begin{cases} 0 (n \text{ is even number}) \\ \frac{4}{n\pi} (n \text{ is odd number}) \end{cases}$.

In this paper, the carrier frequency of SVPWM was 10 kHz, the injection signal frequency was 5 kHz, and the current sampling frequency was 50 kHz. The components less than the Nyquist frequency were considered. The frequency distribution of the three components in $i_{\hat{q}}$ are expressed in Table 1.

Table 1. Distribution of the components of $i_{\hat{q}}$.

Component	Frequency (kHz)
$i_{\hat{q}f}$	Approximate 0
$i_{\hat{q}hm}$	10, 20
$i_{\hat{q}h}$	5, 15

3.2. Principle of FIR Filter

The essence of the FIR filter is to weigh and sum the input values of the past time. Compared with the IIR filter, there is no need to feedback the output, so the structure is simple and easy to implement in programming. The FIR filter can be described as:

$$y[n] = \sum_{k=0}^{M} b_k x[n - k] \tag{12}$$

where $y[n]$ is the filter output of the current moment, $x[n - k]$ is the system input of the k sampling moment before the current moment, M is filter order, and b_k are the filter coefficients.

As can be seen from Equation (12), the FIR filter uses the current input and the past M samples to calculate the system output. The filter coefficient b_k represents the "weight" of the system input at different times. The determination of b_k is the core of filter design.

When b_k is determined, the frequency response of the FIR filter can be described as:

$$H(e^{j\hat{\omega}}) = \sum_{k=0}^{M} b_k e^{-j\hat{\omega}k} \tag{13}$$

where $\hat{\omega} = \omega T_s$ denotes the normalized angular frequency, and ω and T_s are the actual angular frequency and sampling frequency, respectively. When the coefficients of the filter are even and symmetrical, approximately M/2 (i.e., $b_k = b_{M-k}$ (k = 0,1,2,$\cdots$ M/2)), the FIR filter will be a linear phase filter. Using the Euler formula, Equation (13) can be derived as:

$$H(e^{j\hat{\omega}}) = e^{-j\frac{M}{2}\hat{\omega}} A(\hat{\omega}) \tag{14}$$

where

$$A(\hat{\omega}) = \begin{cases} b_{\frac{M}{2}} + \sum_{k=0}^{\frac{M}{2}-1} 2b_k \cos\hat{\omega}(\frac{M}{2} - k) & M = 2,4,6,8\cdots \\ \sum_{k=0}^{\frac{M-1}{2}} 2b_k \cos\hat{\omega}(\frac{M}{2} - k) & M = 1,3,5,7\cdots \end{cases} \tag{15}$$

In Equation (15), $A(\hat{\omega})$ is a real number which has no influence on the phase of the output signal. The norm of $e^{-j\frac{M}{2}\hat{\omega}}$ is 1 which means it has no effect on the amplitude of the output signal. Therefore, the amplitude–frequency characteristic and phase–frequency characteristic of the frequency response function in Equation (14) can be expressed as in Equation (16).

$$\begin{cases} \left|H(e^{j\hat{\omega}})\right| = \left|A(\hat{\omega})\right| \\ \angle H(e^{j\hat{\omega}}) = -\frac{M}{2}\hat{\omega} \end{cases} \tag{16}$$

It can be seen from the phase–frequency characteristic in Equation (16) that the time delay caused by the FIR filter is proportional to the filter order M. As the filter order decreases, the time delay caused by the filter also decreases. Therefore, it is necessary to reduce the order of the filter when the requirements are met.

3.3. FIR Filter Design Method

According to the signal processing flow illustrated in Figure 3, $i_{\hat{q}h}$ needs to be removed at first. Therefore, at the frequency of $i_{\hat{q}h}$, the amplitude response of the designed filter should be as small as possible. So, it is better to change b_k to make the amplitude response zero at the frequency of $i_{\hat{q}h}$. In Table 1, since the frequencies of $i_{\hat{q}h}$ are mainly 5 kHz and 15 kHz, the amplitude–frequency characteristics of the FIR filter should satisfy Equation (17):

$$\begin{cases} \left|H(e^{j\hat{\omega}_{5k}})\right| = 0 \\ \left|H(e^{j\hat{\omega}_{15k}})\right| = 0 \end{cases} \tag{17}$$

where $\hat{\omega}_{5k} = 0.2\pi$ and $\hat{\omega}_{15k} = 0.6\pi$ denotes the normalized frequency of 5 kHz and 15 kHz, respectively.

In Figure 3, $i_{\hat{q}h}$ is regained by subtraction. To ensure the harmonic signals introduced by SVPWM are eliminated at the same time, the amplitude–frequency characteristics at the frequencies of 10 kHz and 20 kHz must be the same. Therefore, the amplitude–frequency response of the filter should also satisfy Equation (18).

$$\left|H(e^{j\hat{\omega}_{10k}})\right| = \left|H(e^{j\hat{\omega}_{20k}})\right| \tag{18}$$

where $\hat{\omega}_{10k} = 0.4\pi$ and $\hat{\omega}_{20k} = 0.8\pi$ denotes the normalized frequency of 10 kHz and 20 kHz, respectively.

Combining Equations (17) and (18) with Equation (15) and (16), a system of equations in matrix form is obtained. When M is an even number, it can be described as:

$$
\begin{bmatrix}
2\cos\frac{M}{2}\hat{\omega}_{5k} & \cdots & 2\cos\hat{\omega}_{5k} & 1 \\
2\cos\frac{M}{2}\hat{\omega}_{15k} & \cdots & 2\cos\hat{\omega}_{15k} & 1 \\
sub(\frac{M}{2}) & \cdots & sub(1) & 0
\end{bmatrix}
\begin{bmatrix}
b_0 \\
\vdots \\
b_{\frac{M}{2}}
\end{bmatrix}
= 0
\tag{19}
$$

When M is an odd number, it can be described as:

$$
\begin{bmatrix}
\cos\frac{M}{2}\hat{\omega}_{5k} & \cdots & \cos 1.5\hat{\omega}_{5k} & \cos 0.5\hat{\omega}_{5k} \\
\cos\frac{M}{2}\hat{\omega}_{15k} & \cdots & \cos 1.5\hat{\omega}_{15k} & \cos 0.5\hat{\omega}_{15k} \\
sub(\frac{M}{2}) & \cdots & sub(1.5) & sub(0.5)
\end{bmatrix}
\begin{bmatrix}
b_0 \\
\vdots \\
b_{\frac{M-1}{2}}
\end{bmatrix}
= 0
\tag{20}
$$

where $sub(n) = |\cos n\hat{\omega}_{10k} - \cos n\hat{\omega}_{20k}|$ $n = M/2, M/2\text{-}1, M/2\text{-}2, \cdots$ $(n > 0)$.

Equations (19) and (20) are homogeneous linear equations. When the filter order M is determined, unknown quantities are only b_k, the filter coefficients. When the rank of the coefficient matrix is smaller than the number of unknown quantities, the system of equations has a non-zero solution. When $M = 6$ or $M = 7$, the number of unknown quantities is four. Since the maximum rank of the coefficient matrix of Equations (17) and (18) is three, the equations must have a non-zero solution. Therefore, the maximum filter order is only seven in this design. The solution process of the filter coefficients can be expressed with the flowchart in Figure 5, and the obtained solution vector is b_k.

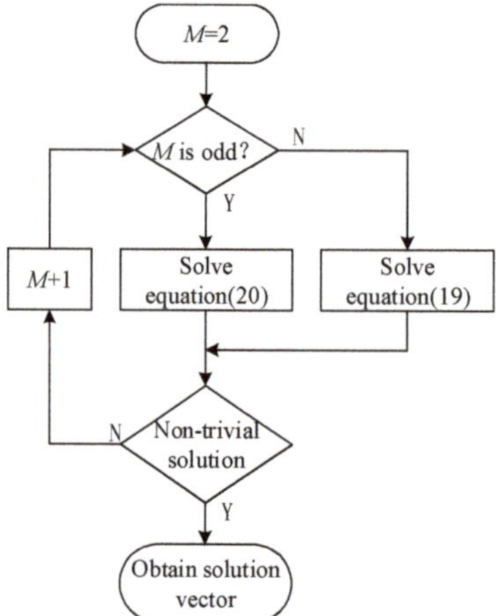

Figure 5. Solution process of filter coefficients.

According to this method, a non-zero solution appears when $M = 5$, i.e. $b_k = [1,0,0,0,0,1]^{\mathrm{T}}$ and M is less than seven, consistent with the theoretical analysis. This designed FIR filter can be described as in Equation (21).

$$
y[n] = x[n] - x[n-5]
\tag{21}
$$

Its amplitude–frequency response curve and phase–frequency response curve are shown in Figure 6.

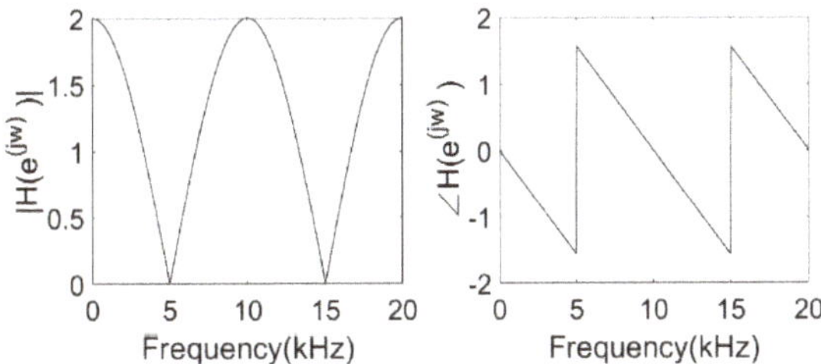

Figure 6. Amplitude–frequency response curve (**left**) and phase–frequency response curve (**right**).

From Figure 6, it can be seen that the amplitude–frequency response of the filter meets the expected requirements at 5 kHz, 10 kHz, 15 kHz, and 20 kHz. Since there is no phase shift at 10 kHz and 20 kHz, the phase compensation can be omitted. In other words, the phase compensation is 0 in this design. Meanwhile, the amplitude gain of this filter at 10 kHz and 20 kHz is two. Thus, the filter output should be multiplied by 0.5 before subtraction in the process of extracting $i_{\hat{q}h}$.

4. Simulation and Experiment Results

In order to verify the effectiveness of the FIR filter designed in this paper, simulation and experimental verification were carried out. The frequency of injection signal and the switching frequency of inverter were 5 kHz and 10 kHz, respectively. The topology of the inverter based on MOSFET was a three-phase full bridge, and the digital controller used for experiments was based on a TMS320F28335. The experiment platform is shown in Figure 7, and the parameters of PMLSM are in Tables 2 and A1.

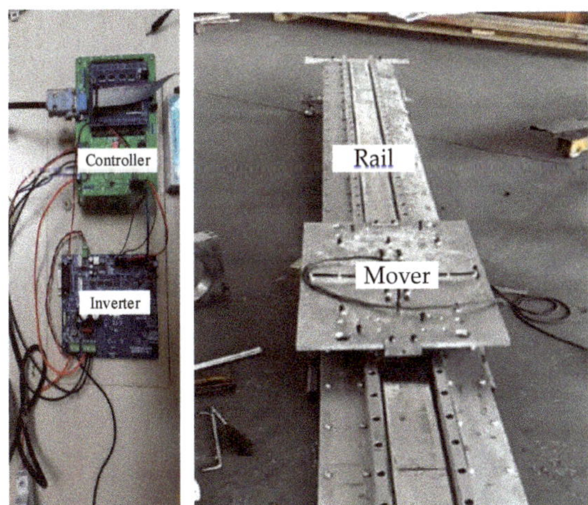

Figure 7. Experimental platform used for the test.

Table 2. Parameters of PMLSM.

Parameter	Value
Resistance (Ω)	1.3
Inductance (mH)	7.8
Continuous current (A)	7.5
Pole pitch (m)	0.018
Thrust constant (N/A)	65.5
Mover mass (kg)	5.8

Figure 8 is the overall block diagram of the position sensorless control system. The observer in this control system is shown in Figure 3.

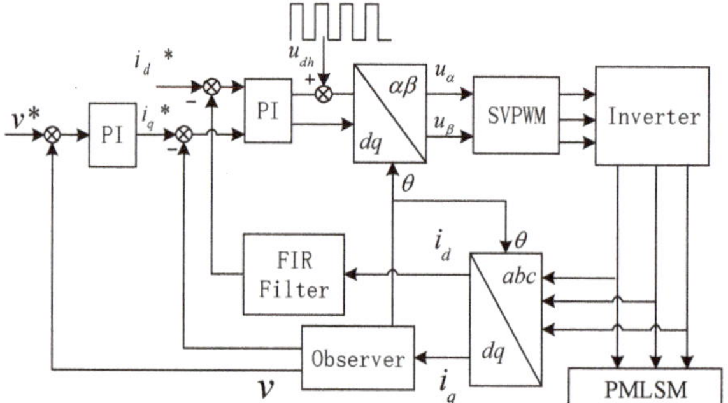

Figure 8. Block diagram of the position-sensorless control system based on high-frequency square-wave voltage injection.

With the help of the fast Fourier transform tool in MATLAB/Simulink, spectrum analysis of current is given. Figure 9 shows the frequency spectrum comparison of $i_{\hat{q}}$ before and after signal filtering in simulation which verifies the effect of FIR filter and signal separation. In PMLSM, 0 Hz, 10 kHz, and 20 kHz components were the main interference in q-axis current, which is shown in Figure 9a. After filtering, 5 kHz and 15 kHz components were eliminated, leaving only 0 Hz, 10 kHz, and 20 kHz components. The amplitude and phase of 0 Hz, 10 kHz, and 20 kHz components in original current and filtered current were the same. Therefore, by subtracting the filtered current from the original current, these components disappeared and only 5 kHz and 15 kHz components were left, which is shown in Figure 9b. Although there were other components besides 5 kHz and 15 kHz components in Figure 9b, the amplitude of these noise components was much smaller than that of 0 Hz, 10 kHz, and 20 kHz components which means the main interference was eliminated. Therefore, the main goal was achieved, and the simulation results were consistent with the theoretical analysis.

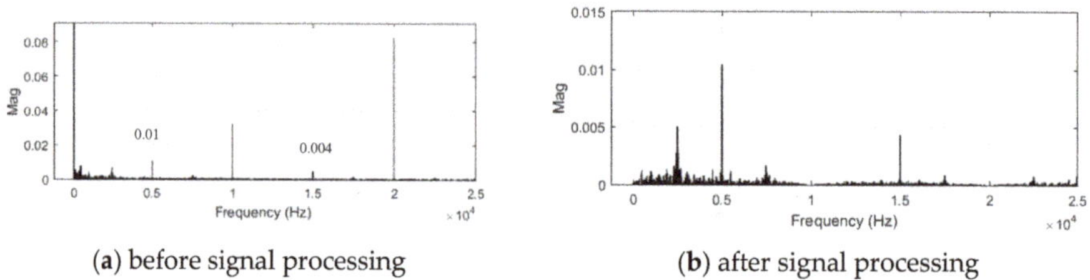

(**a**) before signal processing

(**b**) after signal processing

Figure 9. Frequency spectrum comparison of $i_{\hat{q}}$ before and after signal processing.

In order to further verify the effectiveness of the designed filter, position estimation experiments were conducted on the experimental platform shown in Figure 7. After the experiment was over, the experiment results were exported for plotting which are shown in Figures 10–14.

The initial position of the magnetic pole was set to 6 mm ahead of A-phase axis, and the position observation was divided into two stages. At 0~0.3 s, the initial position was estimated. After 0.3 s, the motor started to move at low speed. The experiment results of position estimation after using this low-order FIR filter is shown in Figure 10. It can be seen that the estimated position value converged near the actual value. Therefore, the position estimation of PMLSM can be achieved by using this low-order FIR filter.

Figures 11–13 shows the results of magnetic pole position estimation under different conditions. In order to describe the error better, this paper used the absolute value of the ratio of deviation to pole pitch as the standard for measuring accuracy. In Figure 11, the position observation error exceeded 7% of the pole pitch with no filter used when the amplitude of the injection voltage was 50 V. When the amplitude of the injection voltage increased to 100 V, the position observation result is shown in Figure 12. The maximum error of the observation did not exceed 4% of the pole pitch, improving the observation accuracy. Figure 13 shows the observation result after adding the low-order FIR filter designed in this paper with an injection voltage of 50 V. In this case, the maximum error was only slightly higher than 4% of the pole pitch. Therefore, with this low-order FIR filter, the position estimation accuracy can be improved to an approximate level compared with the method of increasing injection voltage amplitude.

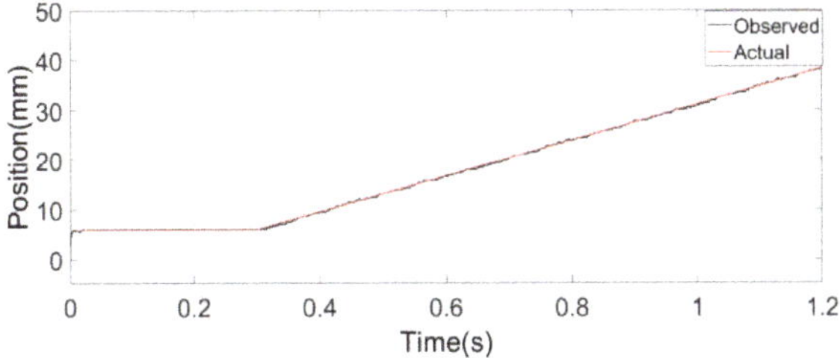

Figure 10. Position observation result at zero and low speed.

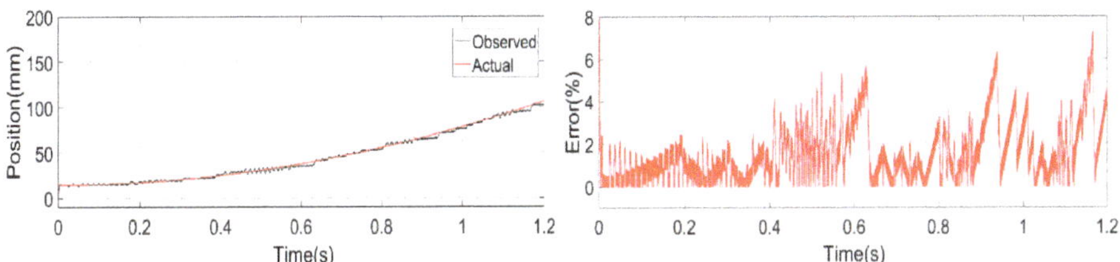

Figure 11. Position observation results at an injection voltage of 50 V without filter.

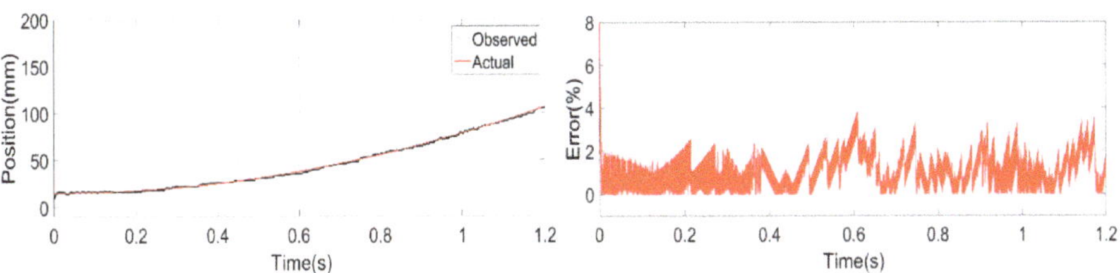

Figure 12. Position observation results at an injection voltage of 100 V without a filter.

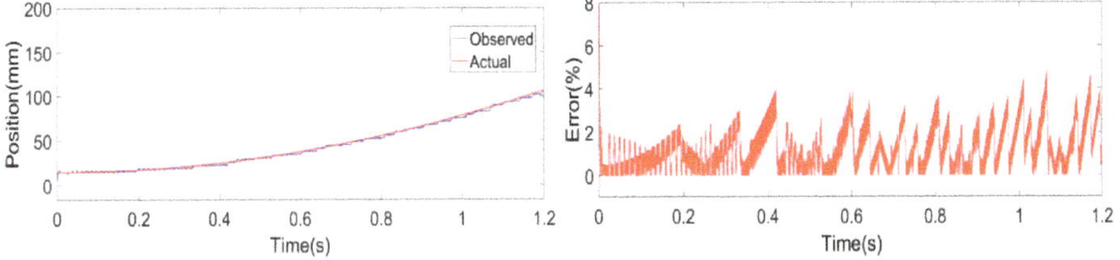

Figure 13. Position observation results at an injection voltage of 50 V with the FIR filter designed in this paper.

In order to analyze the impact on the suspension system when the injection voltage amplitude is different, Figure 14 shows the *d*-axis current with an injection voltage of 50 V and 100 V. It can be seen that the *d*-axis current fluctuation became significantly smaller when the injection voltage was 50 V. From the analysis in Section 2, the *d*-axis current fluctuation was related to the normal force fluctuation. Smaller *d*-axis current fluctuation can reduce the fluctuation of normal force, thereby reducing the impact on the suspension system. Under the condition of ensuring observation accuracy, the FIR filter design method proposed in this paper can reduce the amplitude of injection voltage compared with the filterless method. Therefore, it can bring a smaller fluctuation of the *d*-axis current and normal force.

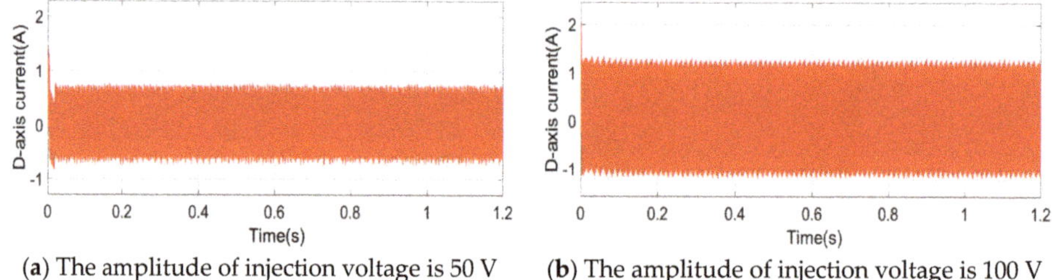

(**a**) The amplitude of injection voltage is 50 V   (**b**) The amplitude of injection voltage is 100 V

Figure 14. The *d*-axis current at different amplitudes of injection voltage.

5. Conclusions

In the high-frequency signal injection method, digital filters were used to extract high-frequency response currents. This paper presented a design method of an FIR filter based on the frequency spectrum of motor current and the frequency response characteristics of the FIR filter. Through theoretical analysis, simulation, and experiment, the following conclusions can be drawn:

1. The time delay caused by the FIR filter was proportional to the filter order. It was necessary to reduce the order of the filter when the requirements are met;
2. By using the designed method proposed in this paper, a low-order FIR filter can be obtained. Its order was as low as five which was much lower than that of the FIR filter designed by the traditional method;
3. The low-order FIR filter designed in this study can complete the magnetic pole position estimation. Compared with the estimation system without filter, the application of this new filter can improve the position estimation accuracy without increasing injection voltage amplitude.

Author Contributions: Conceptualization, H.Z.; Formal analysis, H.Z. and L.Z.; Funding acquisition, L.Z. and C.Z.; Investigation, H.Z.; Methodology, C.Z.; Project administration, L.Z.; Resources, J.L.; Supervision, L.Z.; Validation, C.Z.; Writing—original draft, H.Z.; Writing—review and editing, J.C. and L.S. All authors have read and agreed to the published version of the manuscript.

Funding: The project presented in this article is supported by National Key R&D Program of China (2016YFB1200601-B22).

Conflicts of Interest: The authors declare no conflict of interest.

Nomenclature

Abbreviation or Variable	Definition
PMLSM	permanent magnet linear synchronous motor
PMSM	permanent magnet synchronous motor
IIR	infinite impulse-response
FIR	finite impulse-response
SNR	signal-to-noise ratio
SRF	synchronization reference frame
i_{qf}	fundamental current

i_{qh}	high-frequency response current
i_{qhm}	harmonic current
V_h	injection voltage amplitude
$\Delta\theta$	error between the estimated position and the actual position
ΔL	half of the difference of d–q axis inductance
$\sum L$	half of the sum of d–q axis inductance
τ	pole pitch
p	differential operator
ψ_f	flux linkage of the permanent magnet
$\hat{\theta}$	estimated position
v	motor speed
M	filter order
b_k	filter coefficient

Appendix A

Table A1. Parameters of PMLSM.

Parameter	Value
Resistance (Ω)	1.3
Inductance (mH)	7.8
Continuous current (A)	7.5
Pole pitch (m)	0.018
Thrust constant (N/A)	65.5
Mover mass (kg)	5.8

References

1. Song, X.; Dou, F.; Dai, C. Modeling and Design of the Speed and Location System for Low Speed Maglev Vehicle. In Proceedings of the Fifth ICICTA, Zhangjiajie, China, 12–14 January 2012; pp. 205–209.
2. Zhang, D.; Long, Z.; Dai, C. Design and realization of a novel position-and-speed measurement system with communication function for the low-speed maglev train. *Sens. Actuators A Phys.* **2013**, *203*, 261–271. [CrossRef]
3. Dai, C.; Huang, C.; Tan, L. Comparison and Analysis of Signal Processing Methods for Induction Loop-Cable Position and Speed Detection System. In Proceedings of the 2018 Chinese Automation Congress (CAC), Xi'an, China, 30 November–2 December 2018; pp. 2873–2878.
4. Maeda, T.; Watanabe, K.; Ono, M. Train position detecting system using radio millimeter-waves. In Proceedings of the 10th International Conference on Computer System Design and Operation in the Railway and Other Transit Systems, Prague, Czech Republic, 10–12 July 2006; pp. 469–476.
5. Yufeng, L.; Tefang, C.; Shu, C. Processing and Fusion of Velocity and Position Signal of Medium and High Speed Maglev Train. In Proceedings of the 13th World Congress on Intelligent Control and Automation (WCICA), Changsha, China, 4–8 July 2018; pp. 1578–1583.
6. Wang, H.; Ge, X.; Liu, Y. Second-Order Sliding-Mode MRAS Observer-Based Sensorless Vector Control of Linear Induction Motor Drives for Medium-Low Speed Maglev Applications. *IEEE Trans. Ind. Electron.* **2018**, *65*, 9938–9952. [CrossRef]
7. Shen, J.X.; Zhu, Z.Q.; Howe, D. Improved speed estimation in sensorless PM brushless AC drives. *IEEE Trans. Ind. Appl.* **2002**, *38*, 1072–1080. [CrossRef]
8. Li, Y.; Zhu, Z.Q.; Howe, D.; Bingham, C.M. Improved Rotor Position Estimation in Extended Back-EMF Based Sensorless PM Brushless AC Drives with Magnetic Saliency. In Proceedings of the 2007 IEEE International Electric Machines & Drives Conference, Antalya, Turkey, 3–5 May 2007; pp. 214–219.
9. Kim, H.; Son, J.; Lee, J. A High-Speed Sliding-Mode Observer for the Sensorless Speed Control of a PMSM. *IEEE Trans. Ind. Electron.* **2011**, *58*, 4069–4077.
10. Quang, N.K.; Hieu, N.T. FPGA-based sensorless PMSM speed control using reduced-order extended Kalman filters. *IEEE Trans. Ind. Electron.* **2014**, *61*, 6574–6582. [CrossRef]

11. Raca, D.; Garcia, P.; Reigosa, D.; Briz, F.; Lorenz, R. A comparative analysis of pulsating vs. rotating vector carrier signal injection-based sensorless control. In Proceedings of the Twenty-Third Annual IEEE APEC, Austin, TX, USA, 24–28 February 2008; pp. 879–885.

12. Luo, X.; Tang, Q.; Shen, A.; Zhang, Q. PMSM Sensorless Control by Injecting HF Pulsating Carrier Signal Into Estimated Fixed-Frequency Rotating Reference Frame. *IEEE Trans. Ind. Electron.* **2016**, *63*, 2294–2303. [CrossRef]

13. Liu, J.M.; Zhu, Z.Q. Novel Sensorless Control Strategy With Injection of High-Frequency Pulsating Carrier Signal Into Stationary Reference Frame. *IEEE Trans. Ind. Appl.* **2014**, *50*, 2574–2583. [CrossRef]

14. Tang, Q.; Shen, A.; Luo, X.; Xu, J. PMSM Sensorless Control by Injecting HF Pulsating Carrier Signal Into ABC Frame. *IEEE Trans. Power Electr.* **2017**, *32*, 3767–3776. [CrossRef]

15. Tang, Q.; Shen, A.; Luo, X.; Xu, J. IPMSM Sensorless Control by Injecting Bidirectional Rotating HF Carrier Signals. *IEEE Trans. Power Electr.* **2018**, *33*, 10698–10707. [CrossRef]

16. Ni, R.; Xu, D.; Blaabjerg, F.; Lu, K.; Wang, G.; Zhang, G. Square-Wave Voltage Injection Algorithm for PMSM Position Sensorless Control with High Robustness to Voltage Errors. *IEEE Trans. Power Electr.* **2017**, *32*, 5425–5437. [CrossRef]

17. Xu, P.L.; Zhu, Z.Q. Novel Square-Wave Signal Injection Method Using Zero-Sequence Voltage for Sensorless Control of PMSM Drives. *IEEE Trans. Ind. Electron.* **2016**, *63*, 7444–7454. [CrossRef]

18. Bian, Y.J.; Guo, X.H.; Song, X.F.; Wu, Y.F.; Chen, Y. Initial rotor position estimation of PMSM based on high-frequency signal injection. In Proceedings of the 2014 ITEC Asia-Pacific, Beijing, China, 31 August–3 September 2014; pp. 1–4.

19. Li, M.Q.; Wang, L. An Improved Low Speed Sensorless Control Strategy for Permanent Magnet Synchronous Motor. *Trans. China Electrotech. Soc.* **2018**, *33*, 1967–1974.

20. Li, Q.; Li, C.F.; Liu, K.; Huang, Y.Q. Digital filters design for sensorless control of PMSM. *Electr. Mach. Control* **2018**, *8*, 47–53.

21. Ji, W.; Shi, G.D.; Xu, B. An improved rotating HF signal injection method based on FIR filters for state estimation of BPMSM sensorless control. *Adv. Mech. Eng.* **2017**, *9*, 1–10. [CrossRef]

22. Li, H.; Zhang, X.; Xu, C.; Hong, J. Sensorless Control of IPMSM Using Moving-Average-Filter Based PLL on HF Pulsating Signal Injection Method. *IEEE Trans. Energy. Convers.* **2020**, *35*, 43–52. [CrossRef]

23. Kim, S.H.; Park, N.C. Simple sensorless algorithm for interior permanent magnet synchronous motors based on high-frequency voltage injection method. *IET Electr. Power Appl.* **2014**, *8*, 68–75.

24. Zhang, G.Q.; Wang, G.L.; Xu, D.G. Filterless Square-Wave Injection Based Initial Position Detection for Permanent Magnet Synchronous Machines. *Trans. China Electrotech. Soc.* **2017**, *32*, 162–168.

25. Zhang, X.; Li, H.; Yang, S.; Ma, M. Improved Initial Rotor Position Estimation for PMSM Drives Based on HF Pulsating Voltage Signal Injection. *IEEE Trans. Ind. Electron.* **2018**, *65*, 4702–4713. [CrossRef]

26. Kim, S.; Im, J.; Song, E.; Kim, R. A New Rotor Position Estimation Method of IPMSM Using All-Pass Filter on High-Frequency Rotating Voltage Signal Injection. *IEEE Trans. Ind. Electron.* **2016**, *63*, 6499–6509. [CrossRef]

27. Chen, S.; Chiang, H.; Liu, T.; Chang, C. Precision Motion Control of Permanent Magnet Linear Synchronous Motors Using Adaptive Fuzzy Fractional-Order Sliding-Mode Control. *IEEE ASME Trans. Mech.* **2019**, *24*, 741–752. [CrossRef]

28. Yang, S.; Yang, S.; Hu, J. Design Consideration on the Square-Wave Voltage Injection for Sensorless Drive of Interior Permanent-Magnet Machines. *IEEE Trans. Ind. Electron.* **2017**, *64*, 159–168. [CrossRef]

29. Lan, Y.P.; Hu, X.C.; Chen, Q.L. Finite Element Analysis of Electromagnetic Characteristics of Controllable Excitation Magnetic Suspension Feed Platform. *CJME* **2017**, *53*, 184–189. [CrossRef]

30. Yoon, Y.; Sul, S. Sensorless Control for Induction Machines Based on Square-Wave Voltage Injection. *IEEE Trans. Power Electr.* **2014**, *29*, 3637–3645. [CrossRef]

31. Liang, W.; Wang, J.; Luk, P.C.; Fang, W.; Fei, W. Analytical Modeling of Current Harmonic Components in PMSM Drive with Voltage-Source Inverter by SVPWM Technique. *IEEE Trans. Energy Convers.* **2014**, *29*, 673–680. [CrossRef]

Article

Benchmark of Rotor Position Sensor Technologies for Application in Automotive Electric Drive Trains

Christoph Datlinger * and Mario Hirz

Institute of Automotive Engineering, Graz University of Technology, 8010 Graz, Austria; mario.hirz@tugraz.at
* Correspondence: christoph.datlinger@tugraz.at; Tel.: +43-316-873-35264

Received: 29 May 2020; Accepted: 14 June 2020; Published: 28 June 2020

Abstract: Rotor shaft position sensors are required to ensure the efficient and reliable control of Permanent Magnet Synchronous Machines (PMSM), which are often applied as traction motors in electrified automotive powertrains. In general, various sensor principles are available, e.g., resolvers and inductive- or magnetoresistive sensors. Each technology is characterized by strengths and weaknesses in terms of measurement accuracy, space demands, disturbing factors and costs, etc. Since the most frequently applied technology, the resolver, shows some weaknesses and is relatively costly, alternative technologies have been introduced during the past years. This paper investigates state-of-the-art position sensor technologies and compares their potentials for use in PMSM in automotive powertrain systems. The corresponding evaluation criteria are defined according to the typical requirements of automotive electric powertrains, and include the provided sensor accuracy under the influence of mechanical tolerances and deviations, integration size, and different electrical- and signal processing-related parameters. The study presents a mapping of the potentials of different rotor position sensor technologies with the target to support the selection of suitable sensor technologies for specified powertrain control applications, addressing both system design and components development.

Keywords: automotive electric powertrain; permanent magnet synchronous motor; rotor position sensor; resolver; inductive position sensor; eddy current position sensor; Hall sensor; magnetoresistive position sensor

---

1. Introduction

The electrification of vehicles is becoming increasingly widespread in order to reduce greenhouse gas emissions and fulfill the corresponding exhaust emission legislations. Regardless of the electric powertrain architecture, e.g., Hybrid Electric Vehicle (HEV) or Battery Electric Vehicle (BEV), there are two main types of traction motors used today: Induction Motors (IM) and Permanent Magnet Synchronous Motors (PMSM). The control strategies differ significantly according to the selected motor type [1]. Thus, feedback for controlling the electric machine, which is delivered by a rotor shaft sensor, can be separated into two signal types: the rotor speed signal when applying an IM and the rotor position information when utilizing a PMSM. Sensorless control, as it is often found in the literature [2,3], is not used in automotive powertrains due to high demands on control reliability and is therefore not discussed further in detail here.

In general, the automotive industry sets high demands on the control of electrical drive trains, which are required for the correct electronically controlled commutation of these motor types. Accurate rotor speed and position information is vital to obtain precise torque and speed control. This enables the best motor efficiency, resulting in increased driving ranges, increased comfort (by reducing torque ripple) and maintaining Functional Safety (FUSA)-relevant issues, e.g., according to the ISO 26262 [4–6]. In addition, real-time information about rotor speed and position is required for a defined starting direction from vehicle standstill and the prevention of the powertrain from unintended blocking. As an example, Automotive Safety Integrity Level-D (ASIL level-D) applications, which refer to the highest classification of injury risk and the most stringent level of safety measures, define a maximum position sensing angle error of 2° [7].

Contrary to the IM, the PMSM shows better efficiency and is therefore in widespread use in state-of-the-art automotive propulsion systems. Owing to this fact, the present paper focusses on the analysis of rotor position measurement technologies for PMSM-based automotive drivetrains. In general, different rotor shaft position sensor technologies are available for PMSM, whereby the selection of a suitable system is influenced by several factors, such as accuracy demands; sensibility to mechanical, magnetic and electrical disturbance; integration size; costs; etc. Investigations from the course of this research show that currently applied rotor position sensors in PMSM are resolver-types, inductive/eddy current-based sensors, magnetoresistive sensors or encoders. However, the latter type is not applied in automotive powertrains, since a high accuracy can only be achieved with high (expensive) effort.

In today's electric powertrain development processes, the sensor specifications delivered by sensor manufacturers are taken into account [8], but these do not provide sufficient details about the sensor characteristics. This often leads to a challenge for the automotive industry when selecting an unbiased sensor. In this context, the present work delivers detailed investigations of typical rotor position sensor technologies to provide a comparison of the respective sensor systems. A comprehensive evaluation of the sensor characteristics was performed regarding the angle accuracy under a variety of mechanical installation tolerances, operating temperatures and driving profiles (i.e., rotary speed variances). The aforementioned research activities are performed with the aid of a unique sensor test bench, which enables a highly accurate gauging of objects under test. Furthermore, important selection criteria for the drivetrain development, such as the sensor housing size and sensitivity to magnetic fields, costs, electrical supply effort and signal processing effort, were evaluated. Consequently, a holistic overview of state-of-the-art rotor position sensor system potentials is provided, including an analysis of the provoked error characteristics. The results of the study support the selection of sensor technology in terms of the described influencing factors for a given drivetrain application.

The paper is organized as follows. Section 2 provides a general overview of different state-of-the-art rotor position sensors for both automotive electric powertrains and auxiliaries, including a listing of general strengths and weaknesses of each sensor technology. In conjunction with the description of each sensor principle, a representative automotive application was selected to provide boundary conditions for the application of the different sensor technologies to be evaluated. In Section 3, a sensor test bench and measurement plan are introduced to deliver an overview of both the performed testing procedure and the data acquisition technologies. The measurement results of the different sensor technologies are presented in Section 4. The evaluation results are discussed in Section 5, which also includes a holistic comparison of the measurement results and an assessment of the potentials of the different sensor principles. Finally, conclusions are given in Section 6.

2. Automotive Rotor Position Sensors

Rotor position sensors are used in various fields of application in the automotive industry, e.g., in electric powertrain systems, electric cooling fans or pumps. As stated in [9], position measurement in automotive applications can be performed by different approaches. In general, contacting and non-contacting technologies come into use. An exemplary contacting-based angular measurement principle represents the electronic accelerator pedal (E-GAS), which is comprised of a potentiometer using wipers sliding on copper tracks. Such sensor architectures suffer from wear in long term use, and do not enable angle determination greater than 360°. Thus, this principle is not appropriate for the focused investigation object, which is an electric powertrain that requires non-contacting rotor position measurement. In what follows, state-of-the-art automotive non-contacting sensor principles are introduced and discussed, which are investigated in the scope of this work. These technologies include resolver-, inductive-/eddy current-, Hall effect-based- and magnetoresistive-position sensors. Figure 1 represents a graphical overview of the four main types.

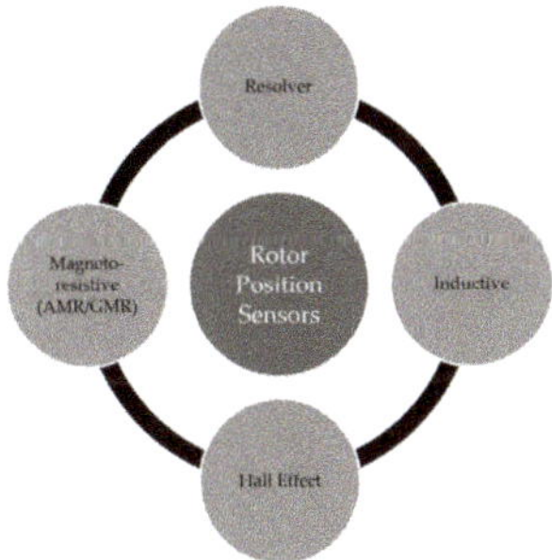

Figure 1. Considered automotive non-contacting rotor position sensor principles.

2.1. Principle of the Resolver Position Sensor

The state-of-the-art position sensing technology in automotive traction systems is the resolver, due to a great reliability in terms of rough conditions, for example contamination, temperature, shock and vibrations. The resolver appears in two types: Wound Field (WF) and Variable Reluctance (VR) resolvers. The difference lies in the design of the sensor's rotor. The WF resolver has a sinusoidal distributed excitation winding with a constant Air Gap (AG) [10]. In contrast, the VR resolver has a uniform distributed excitation winding, in combination with a sinusoidal-shaped AG rotor design [11]. Since the VR resolver is the most established resolver type in automotive drivetrain architectures, the present work focusses on this type. The discussed VR resolver consists of three coils: one excitation coil and two sensing coils. The excitation coil operates as a transmitter. Therefore, the resolver can be considered as a rotating electric transformer. The excitation signal itself is a high frequency sinusoidal voltage U_{exc} at typically $f = 10$ kHz. The generated electromagnetic flux is coupled through the shape of the rotor, which consists of a stamped ferromagnetic material. The sensing coils induce an electrical voltage comprising the excitation signal and a rotor position-dependent magnitude. Note that both sensing coils are mechanically displaced to deliver both a sine- and a cosine-shaped output signal while the rotor is in motion. Consequently, the sensor provides two Amplitude Modulated (AM) signals, which are shifted by 90°. To determine the angular information θ, the excitation term must be removed. This process is performed in a specific Resolver-to-Digital Converter (RDC). Comprehensive study of this sensor system has been performed in previous works, see [12–14]. Figure 2 depicts the operating principle of a VR resolver including an exemplary sensor, which is utilized in automotive drivetrains of BEVs.

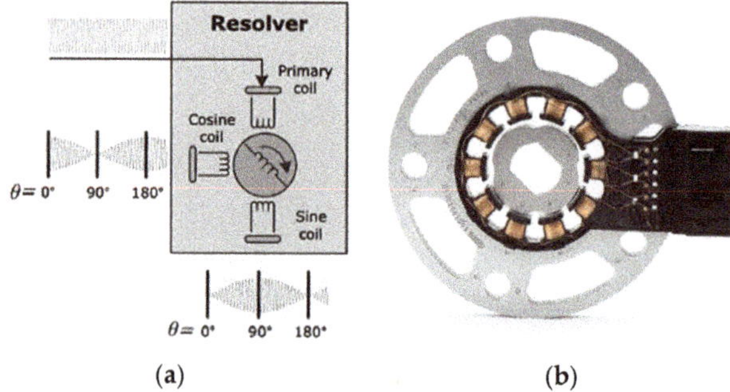

Figure 2. Variable Reluctance (VR) resolver position sensor. (a) Operating principle [13]; (b) Exemplary sensor for Battery Electric Vehicles (BEVs), reproduced from [15]. Copyright 2000–2020 MinebeaMitsumi Inc.

Hereafter, the principle is formulated mathematically by Equations (1)–(4). The utilized variables are described as follows. A represents the amplitude of the respective voltage. The resolver can be considered as a rotating transformer; thus, the feedback voltages U_{sin} and U_{cos} are reduced by a certain factor, which is denoted as transformation ratio k. The electrical rotor shaft position θ is determined by an arctangent function of the demodulated signals $V_{sin,demod}$ and $V_{cos,demod}$, which results in an extraction of the excitation term.

$$V_{exc} = A \times \sin(\omega t) \tag{1}$$

$$V_{sin} = kA \times \sin(\omega t) \times \sin(\theta) \tag{2}$$

$$V_{cos} = kA \times \sin(\omega t) \times \cos(\theta) \tag{3}$$

$$\theta = \arctan\left(\frac{V_{sin,demod}}{V_{cos,demod}}\right) \tag{4}$$

The resulting electrical angle is usually multiplied by a factor, which describes the number of poles. For electric powertrains, the number of poles of the resolver is related to the number of poles of the traction motor. A higher number of resolver poles increases the accuracy of the sensor, but also increases the costs, since a higher number of windings is required. These multi-speed resolvers are preferred for applications with high requirements on sensor system performance, e.g., electric drivetrains.

In the automotive sector, a resolver position sensor system is most commonly used for controlling PMSM-based powertrains. This is mainly because of the robustness and high accuracy of the technology. In the following, the state-of-the-art architecture of a resolver-based rotor position system and the corresponding signal processing are described. As shown in Figure 3, it consists of the resolver itself, a carrier generation unit, which comprises a digital stage and an analog amplification circuit, an Analog-to-Digital Converter (ADC) for digitizing the resolver feedback signals, a demodulation stage and an angle computation algorithm.

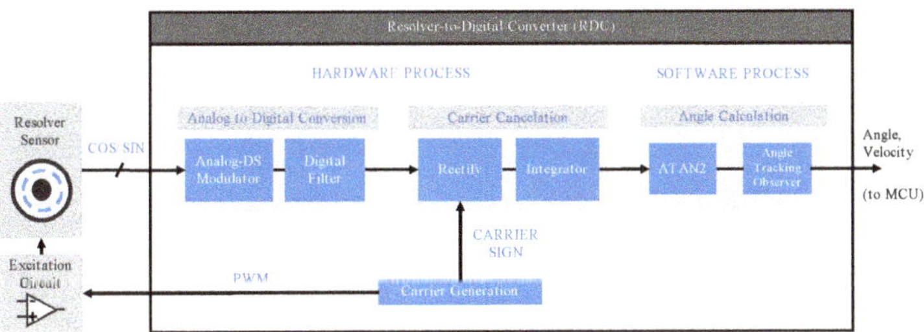

Figure 3. Exemplary illustration of a resolver rotor position signal processing system.

All signal processing stages are integrated in the RDC, which can be either a separate module or be integrated in the powertrain's ECU. The sensor is electrically connected differentially for both the excitation and the feedback signals using Twisted Pair (TP) cables to eliminated external interferences. The angular information is passed on by an interface, e.g., a Transistor Transistor Logic (TTL) encoder emulation, an Inter-Integrated Circuit (I^2C) bus when the RDC unit is placed closely to the subsequent Motor Control Unit (MCU) on a single PCB, or a CAN bus in case of a distributed system architecture. A detailed functional description of the individual processing stages can be found in [12,14]. Table 1 outlines the strengths and weaknesses of state-of-the-art VR resolver position sensors.

Table 1. Property summery of VR resolvers.

	Attributes
Advantages	High accuracy and resolution
	Reliable
	Robust
Disadvantages	High costs
	High weight and inertia (speed limitation)
	Relatively large installation space
	High power consumption due to excitation
	Complex signal processing required for angle determination (i.e., RDC)
	Sensitiveness to mechanical tolerances e.g., eccentricity
	Not immune against stray fields

2.2. Inductive Position Sensor

An alternative to the resolver position sensor is the inductive sensor principle, also denoted as an eddy current position sensor [8]. This sensor principle comprises two main components; a metallic rotor that can be of copper or aluminum and a coil design, which is printed on a two-sided Printed Circuit Board (PCB). Angular position determination, including signal processing, is computed on a dedicated sensor chip, which is also mounted on the same PCB [16]. Figure 4 depicts an exemplary inductive sensor application, including relevant components.

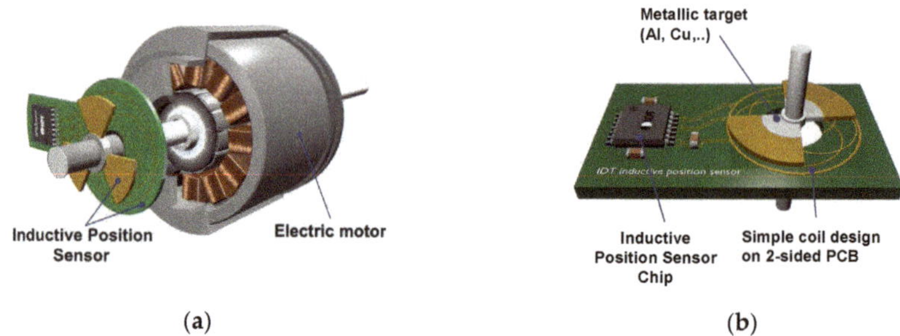

(**a**)　　　　　　　　　　　　　　　　　　(**b**)

Figure 4. Inductive position sensor, reproduced from [16]. Copyright 2019, IDT. (**a**) Illustration of a through-shaft integration; (**b**) Relevant sensor components scheme.

As evident in the figure above, the inductive sensor can be integrated as a through-shaft design in the same manner as the formerly introduced resolver. It should be mentioned that, for the inductive sensor, other designs are feasible, e.g., end-of-shaft (EoS) or side-shaft (i.e., arc design). This makes this sensor technology flexible for a wide range of applications in electric vehicles and auxiliaries.

The working principle of inductive sensors is based on two fundamental physical principles: Michael Faraday's law of induction, and the effects of eddy currents, which have been discovered by Léon Foucault [16,17] (Michael Faraday (1791–1867): scientist in the field of electromagnetism and electrochemistry [18]; Léon Foucault (1819–1868): French physicist who discovered eddy currents and the effect of the earth's rotation [19]). The stator of the sensor (i.e., the PCB) comprises one transmitter coil and two receiver coils. These coils are implemented as copper traces, as depicted in Figure 4b. The first receiving trace is geometrically designed in a sinusoidal shape. The second coil is mechanically displaced by 90°, leading to a cosine-shaped feedback signal. The transmitter trace is excited by a high frequency sinusoidal current, typically between 2 MHz and 6 MHz [20], which is generated by the sensor chip on the PCB. This planar excitation coil builds an LC resonant circuit in combination with additional capacitors. Note that the inductance L is given by the coil trace itself. Electrical interferences can be reduced by ohmic resistors, where one pin is connected to the printed transmitter trace and the other pin is bonded to the capacitors, which are connected with the electrical Ground (GND). The LC resonant circuit generates a magnetic induction field and induces eddy currents in the rotating metallic target. According to the rotor position, a reaction field shields the excitation field in the area beneath the rotating target and induces voltages in the receiving traces. Consequently, inductive position sensors measure the disturbance of a magnetic field by a conductive target.

The sensor accuracy strongly depends on the selected coil design; hence, an ideal receiver coil geometry leads to an ideal sine- and cosine feedback voltage at the open circuit-designed coils. In operation, when the rotating target is facing the sensing area, the magnetic field induces the mentioned eddy currents into the target surface. As a result, these eddy currents produce a counter magnetic field, which leads to a reduction of the flux density beneath the target. This non-uniform flux density generates an electrical voltage at the receiver coil terminals according to Faraday's law of induction. The rotation of the target leads to a change in amplitude and polarity. A mathematical description of the inductive sensor principle is given by Equation (5). The variables are described as follows. The alternating field, which is created by the excitation, is given by $b_e(t)$. The alternating magnetic field, which opposes $b_e(t)$, is described by $b_s(t)$. The surface area of the receiving coil is represented by A.

$$VI = \frac{d\phi}{dt} = -\frac{d \int [b_e(t,x,y) + b_s(t,x,y)] dA}{dt} = \frac{d}{dt} \int b_e(t,x,y) dA + \int b_s(t,x,y)\, dA \qquad (5)$$

For signal processing, the feedback signals are amplified, rectified and filtered before being converted in the digital domain by an integrated ADC. Especially for this sensor type, the signal processing procedure is highly dependent on the application. Consequently, the sensor chip can provide the absolute angular position of the rotor shaft in degrees, based on a trigonometric-based arctan function [21]. If this operation is performed in a subsequent module, e.g., a vehicle's Electronic Control Unit (ECU), only the demodulated feedback signals of the receiver coils are available, which leads to a reduction of signal processing complexity in the sensor, and therefore to lower costs [16,22].

In summary, inductive position sensors offer miscellaneous utilization for both automotive and non-automotive applications. The sensor's principle benefits are high accuracy, immunity against stray fields, an absolute rotor position determination from the electrical start up, and low costs due to both the low amount of copper wires and lack of magnetic material. The latter benefit makes the sensor more eligible for automotive applications compared to other sensor principles, such as magnetoresistive sensors, which will be described in the further context section. Another major advantage over other sensor technologies is that, in safety-critical applications, multiple sensing areas can be placed on the stator PCB. This enables it to combine multiple position sensors in one housing. In addition to the strengths mentioned, this sensor principle has some weaknesses, which are discussed below. Although the housing of the sensor requires only a small mounting area, a certain space is required for the PCB. In addition, the sensor is sensitive to AG variations in terms of mechanical misalignment or target eccentricity and wobbling.

Hereafter, a typical automotive application of an inductive position sensor system is presented. Since the inductive sensor is fully integrated on a single PCB, a small size can be realized, e.g., for the use in the electronic throttle valves of Internal Combustion Engines (ICEs), as shown in Figure 5.

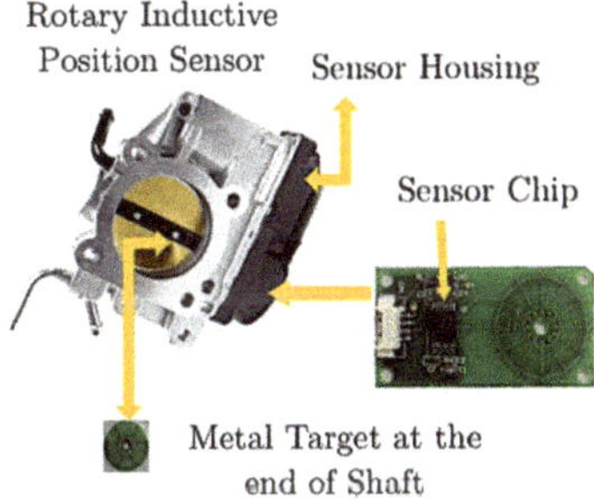

Figure 5. State-of-the-art inductive-based E-Gas actuator position sensor system. Reproduced from [23]. Copyright 1998-2020, Microchip Technology Inc.

In this example, the inductive position sensor acts as a feedback system for throttle valve control. The driver's request is delivered by the electronic accelerator pedal. The MCU determines the flap position and operates a Brushless Direct Current (BLDC) motor. To provide exact information about the flap position for the proper control of the actuator, an inductive position sensor is placed at the moving flap shaft.

Today, different electric motor manufacturers search for alternative technologies to replace the resolver position sensors with smaller and potentially more cost-efficient inductive sensor systems. Table 2 lists the strengths and weaknesses of the inductive position sensor principle.

Table 2. Summary of pros and cons of the inductive position sensor principle.

	Attributes
Advantages	High accuracy and resolution Robust Immune to stray fields Requires no magnetic material Small installation space compared to the resolver technology Adaptable coil design for both linear and rotary movements Low system costs Redundant implementation possible
Disadvantages	Temperature-depending accuracy Thermal limitations of the PCB Sensitiveness to AG variations

2.3. Hall Effect Position Sensor

The fundamental principle of this sensor is based on the eponymous effect discovered by Edwin Hall in 1879 [24] (Edwin Hall (1855–1938): American physicist and innovator of the Hall effect [25]). Hall effect sensors are widely utilized in various fields due to their main advantage of miniaturization and contactless measurement. Applications range from low-cost household products to complex industrial, automotive and aviation machines. Typical examples include BLDC motor commutation sensors, and crankshaft and level measurement. Figure 6 depicts a state-of-the-art Hall effect sensor.

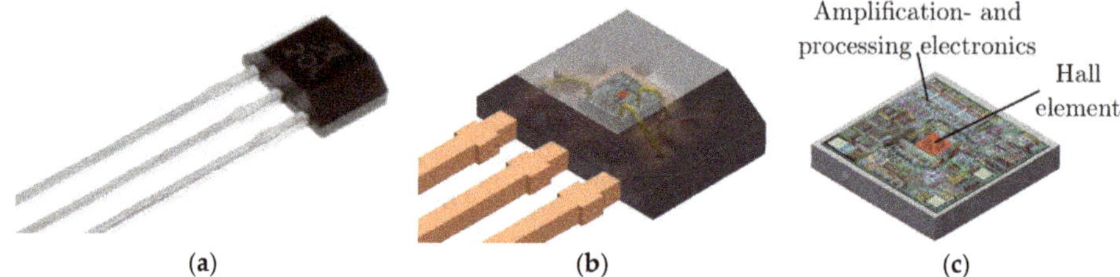

Figure 6. Hall effect sensor. (**a**) Electrical sensor component, reproduced from [26]. Copyright 2009-2013, Allegro MicroSystems, LLC; (**b**) Inner view of a Hall effect sensor comprising the mounted die and the bond wire connections to the external pins for supply, output signal and GND (from left to right) [27]; (**c**) Hall Integrated Circuit (IC) chip, reproduced from [27]. Copyright 2009–2013, Allegro MicroSystems, LLC.

The effect of the Hall sensor is based on the Lorentz-force, which acts on moving charge carriers. In this way, an electrical voltage is applied to the Hall plate, which results in an electrical current flow. When the current-carrying Hall plate is exposed perpendicularly to magnetic induction, e.g., by the use of a magnet, the charge carriers are deflected by a repulsion force to a certain angle, which is orthogonal to the applied field. This deviation is due to the aforementioned Lorentz-force. Accordingly, a Hall-voltage V_H, which is directed transverse to the current direction, can be picked-off between two opposite points on the Hall plate. The voltage is proportional to the magnetic field and the current. The Hall-voltage can be described as shown in Equation (6), where I is the current, B is the applied magnetic field, q_0 defines the charge of an electron, N states the carrier density, d represents the Hall plate thickness and R_H specifies the Hall-coefficient [9,27]. The physical process is illustrated in Figure 7.

$$V_H \frac{IB}{q_0 Nd} = R_H I \left(\frac{B}{d} \right) \tag{6}$$

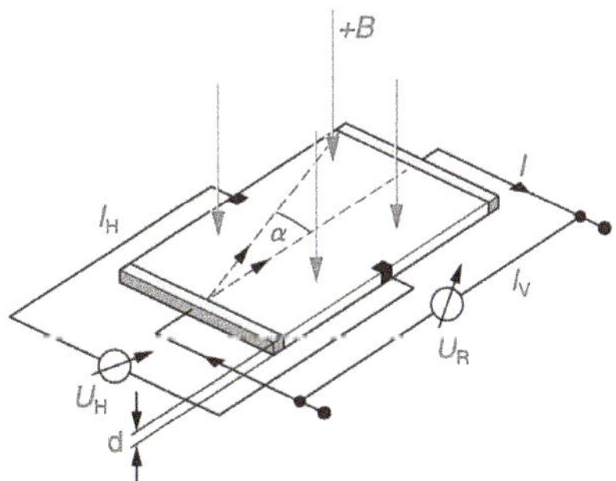

Figure 7. Illustration of the Hall effect principle, reproduced from [9]. Copyright 2007, Robert Bosch GmbH.

There are two main types of Hall effect sensors: linear devices and threshold types. Linear Hall sensors provide a proportional analog output voltage according to the magnetic field strength of the applied magnet. The output voltage moves into the supply direction (e.g., at a south pole) or GND (e.g., in case of a north pole). This can be achieved by a differential Operational Amplifier (OPAMP). As a result, the output voltage is dependent on the orientation of the magnet. When no magnet is detected at all, a bias voltage is present, which is also described as a Null voltage. Note that the fixed offset introduced is to avoid putting two power supplies in the processing circuit, which reduces the overall system costs [28]. A linear Hall sensor can be found, for example, in automotive throttle pedals.

The threshold type provides a digital state at the output (i.e., on and off) in case the field strength of the approached magnet obtains a certain magnetic amplitude or polarity. In terms of signal processing, this sensor configuration comprises a Schmitt-Trigger, which enables a solid, binary output voltage. In addition, these sensors can be configured as latching devices. Here, the sensor output is in the on-state when a south pole is detected (i.e., the magnetic operating point), and it turns off when a north pole appears (i.e., the magnetic release point). This characteristic is often used for rotary speed and position determination [28]. One example is the wheel speed sensor, which is used for an exemplary description of this sensor system in the following.

The measurement of the vehicle wheel speeds is i.a. used for safety and assistance systems, e.g., Anti-lock Braking System (ABS) and Electronic Stability Control (ESC). The design of wheel speed sensor systems is simple due to robustness. Since the Hall sensor is highly reliable, it can be placed close to the bearings of the wheels. A rotary ring includes permanent magnets for speed determination, as depicted in Figure 8.

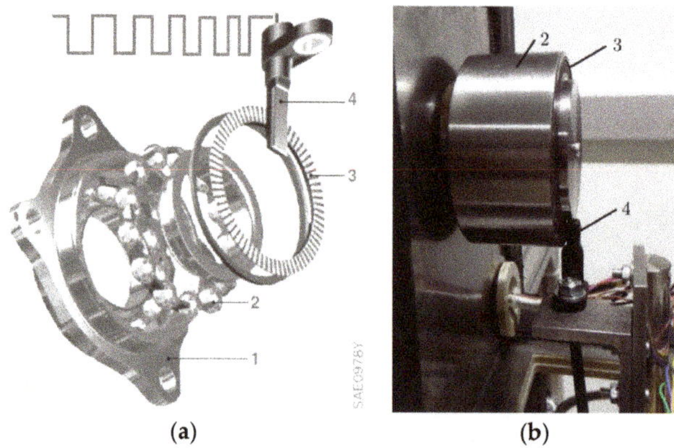

(a) (b)

Figure 8. Hall-based wheel speed sensor. (1) wheel hub, (2) ball bearing, (3) permanent magnet ring, (4) wheel speed sensor. (**a**) Exploded assembly drawing, reproduced from [9]. Copyright 2007, Robert Bosch GmbH.; (**b**) Exemplary sensor on the sensor test bench.

Hall sensors provide a TTL signal to the vehicle ECU, which counts the pulses that correspond to the magnets passing the Hall element. When removing one magnet from the ring, a zero-mark can be obtained to enable absolute position detection.

In conclusion, Hall effect sensors are at their most versatile used when it comes to the measurement of position or rotary speed, since they can withstand rough environmental conditions (e.g., high temperature, shock, dust, contamination), and offer low system costs and miniaturization. However, Hall effect-based sensors are not found in powertrains, since a large amount of both Hall elements and magnets are required to achieve a sufficient accuracy, which increases the costs. Table 3 outlines the strengths and weaknesses of Hall sensors.

Table 3. Advantages and disadvantages of the Hall effect sensor technology.

	Attributes
Advantages	Reliable due to contactless measurement Highly repeatable Fast reaction Broad temperature range Low system costs
Disadvantages	External power supply required Short range detection only in terms of AG

2.4. Magnetoresistive Position Sensor

Magnetoresistive position sensors rely on a similar principle to the previously introduced Hall sensors, because they interact with an external magnetic field. A classification can be made in two types of magnetoresistive position sensors: the Anisotropic Magnetoresistive (AMR) type and the Giant Magnetoresistive type. The physical principle of both is based on Lorentz force [9,29]. A thin-film nickel-iron (NiFe) alloy coating, mostly the permalloy $Ni_{81}Fe_{19}$, with a thickness of 30 to 50 nanometers has an electromagnetically anisotropic characteristic. Accordingly, the material's electrical resistance changes depending on the direction of an external magnetization [9,29,30]. William Thomson discovered this effect in 1857, which describes electrons with different spin orientations possessing variable energy levels, resulting in a magnetic field-dependent conductivity when an external magnetic field is present [29,30] (William Thomson (1824–1907): Scottish engineer, mathematician and physicist [31]). Equation (7) shows the electrical resistance as a function of the magnetization direction.

$$R(\theta) = R_0 + R\cos^2\theta = R_0\left(1 + \frac{\Delta R}{R}\right)\cos^2\theta \tag{7}$$

The term $\Delta R/R$ describes the maximum possible variation of resistance. The squared cosine term in Equation (7) depicts that the AMR sensor generates two sine-shaped curves over one full mechanical revolution. For the determination of the angular position, four magnet-depending resistors are usually combined into a Wheatstone bridge, where the second AMR resistor bridge is mechanically rotated by 45° on the substrate. This enables an additional signal, as a cosine function, to determine the angular position by an arctan computation. In other words, one bridge delivers a sine output signal, the other bridge a cosine signal. The calculation of the resulting voltage per bridge is given by Equation (8). Consequently, the position can be determined according to Equation (9).

$$V_{Bridge\ 1} = A(T)\sin(2\alpha);\ V_{Bridge\ 2} = A(T)\cos(2\alpha) \tag{8}$$

$$\theta = \frac{1}{2}\arctan\left(\frac{V_{Bridge\ 1}}{V_{Bridge\ 2}}\right) \tag{9}$$

Note that, in practical applications, both bridges are placed in one system by one manufacturing process to compensate for resistive temperature influence leading to a nearly equal amplitude per output signal. As a result, high-quality angular information without temperature influences (i.e., drifts) is achievable. Figure 9 demonstrates the principle and layout of an AMR angle sensor for a rotary measurement application in the range of 360°, as it is used in automotive applications.

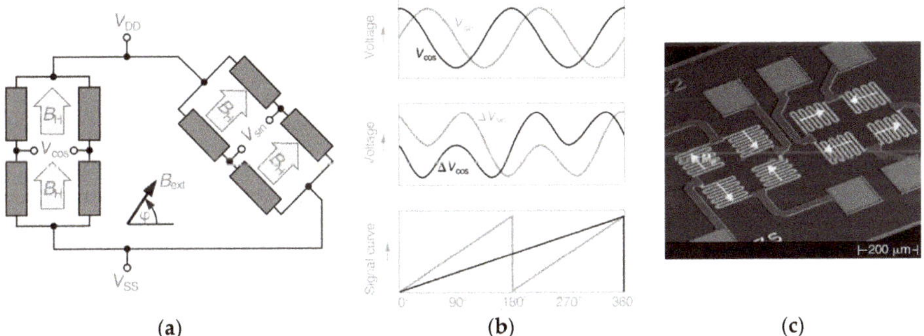

(a)   (b)   (c)

Figure 9. AMR sensor principle. (**a**) Sensor bridge circuit, where B_H is the control induction, U_{sin} and U_{cos} are the measurement voltages, U_{DD} and U_{SS} are the supply voltages and B_{ext} is the external magnetic field with the angle of rotation φ of the magnet; (**b**) Resulting sensor signals; (**c**) Electro microscopical exposure of a 360° AMR sensor layout, reproduced from [9]. Copyright 2007, Robert Bosch GmbH.

AMR are typically used in BLDC-driven auxiliary units, e.g., in electric water/oil pumps or cooling fans. In case of the 360° variant, an additional planar coil is added, which is placed above the AMR resistors. When an electrical current is applied there, an auxiliary field is generated, which evokes changes in the output signals of each AMR bridge to differentiate the measurement range. With this sensor principle, a very high measurement accuracy of less than 1° can be achieved. However, the accuracy is strongly dependent on the utilized magnet and its central placement vertical to the sensing area.

The other sensor principle is based on the GMR effect, which was discovered in 1988. This sensing effect occurs in multilayer structures of interchangeable ferromagnetic and non-ferromagnetic materials.

In a simplified view, two alignment characteristics of this multilayer structure exist; the anti-parallel and the parallel alignment. In case of a lack of an applied external magnetic field, the structure is aligned in an anti-parallel order. This state is also called the initial state. When the multilayers are exposed to an external magnetic field (e.g., by a permanent magnet), the ferromagnetic layers are forced to align to the applied field direction. As a result of this alignment, the resistance of the multilayer structure decreases dependent on the applied magnetic field strength. The minimum resistance of the multilayers is reached when the applied magnetic field surpasses the anti-ferromagnetic coupling, so that all layers are oriented parallel in respect to the external field (saturation point). In conclusion, the GMR layers show a minimum resistance in parallel aligned magnetizations and a maximum resistance in counter-directed magnetization, as shown in Figure 10 [9,29,30,32,33].

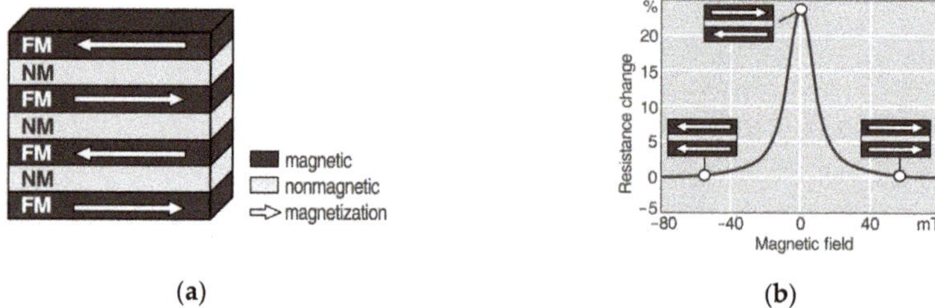

(a)   (b)

Figure 10. GMR sensor principle. (**a**) GMR multilayer structure with FerroMagnetic (FM) and Non FerroMagnetic (NM) layers; (**b**) Resistance characteristic according to an applied magnetic field, reproduced from [9]. Copyright 2007, Robert Bosch GmbH.

The characteristic of a GMR multilayer sensor can be described according to Equation (10).

$$\text{GMR} = \frac{R_{\min} - R_{\max}}{R_{\min}} = \frac{R_{\text{saturated}} - R_0}{R_{\text{saturated}}} \tag{10}$$

Since this effect is uniaxial, the resistance is proportional to the cosine of the angle θ of the applied external field according to Equation (11).

$$R = R_0 - \Delta R \cos(\theta) \tag{11}$$

Focusing on rotational GMR spin valve sensors for automotive applications, two Wheatstone bridges are used; the same as for the AMR principle to determine the position of the rotating shaft, as shown in Figure 11.

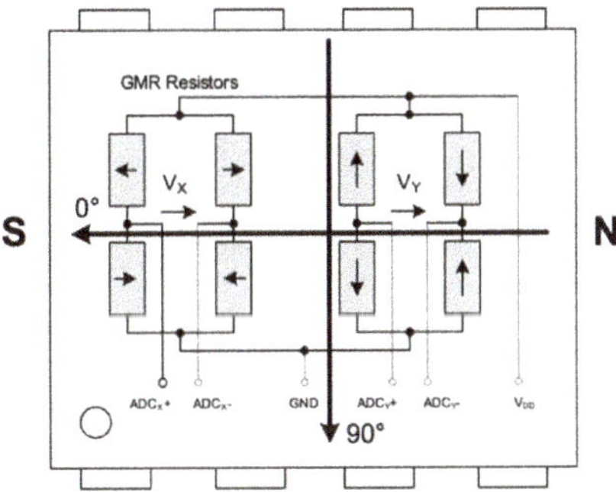

Figure 11. Exemplary GMR sensor die in a magnetic field (i.e., north pole (N) and south pole (S)) containing the sensitive GMR bridges, where V_x and V_y are the sinusoidal output voltages for further angle determination accessible via differential ADC outputs (i.e., ADC$_{x+}$, ADC$_{x-}$, ADC$_{y+}$, ADC$_{y-}$), reproduced from [34]. Copyright 2020, Infineon Technologies AG.

Considering one bridge for explanation, the output voltage of the bridge (i.e., V_X and V_Y) resolves according to Equation (12) [35].

$$V_{X,Y} = \frac{\Delta R(1 - \alpha \Delta T)}{2R_0(1 + \alpha \Delta T) + \Delta R(1 - \alpha \Delta T)} \times V_{DD}; \quad \Delta R = \frac{1}{2}\left(\frac{\Delta R}{R}\right)R_s \frac{iW}{d} \cos\left(\Theta_p - \Theta_f\right) \tag{12}$$

The parameters in Equation (12) are described as follows. The basic GMR and bridge resistance is given by R_0. ΔR is the variable resistance that depends on the direction of the magnetic field, and V_{DD} represents the bridge biasing voltage. In addition, the sensor is temperature dependant. For that reason, α_0 describes the absolute temperature coefficient of R_0, α_Δ represents the absolute temperature coefficient of ΔR and T the relative temperature to a reference temperature [35]. Owing to the fact that the resistance is dependent upon the GMR material, the width is described by W and the thickness of the plate is given by d. The parameters Θ_p and Θ_f are the angles of magnetization of both the pinned and the free layers [36]. As a result, the orientation of the magnetic field—and thus the position of the magnet—can be determined by way of the resistance-dependent voltage. This analog voltage (i.e., V_X and V_Y) is digitized by an ADC for further processing.

In contrast to the formerly introduced AMR position sensors, GMR sensors offer an inherent complete angular measurement range of 360°, where the calculated arctan position information repeats after exceeding 180°. Thus, no additional measures are required, contrary to the AMR sensor, where an additional planar coil is necessary. In addition, GMR sensors are very robust in terms of temperature, which is important for automotive applications (−40 °C to +150 °C). Even if the accuracy of GMR is in the same range as that of AMR, higher measurement signals, and magnets with lower field strength and therefore lower costs can be utilized. Unlike the AMR sensors, the GMR sensors do not only consist of one magnetic functional layer, but rather of a complex layer system. Additional descriptions regarding the differences of both introduced magnetoresistive sensor principles can be found in [29].

Concerning both types, AMR and GMR, the material of the rotary permanent magnet is vital to achieve a good angular measurement performance. Usually, magnets with a diametrically magnetized characteristic are selected. A scheme of the sensor setup is shown in Figure 12, where the permanent magnet is mounted at the end of a rotating shaft, the position of which is the information of interest. In general, the magnet needs to be placed very close to the sensing area of the AMR/GMR sensor, in an orthogonal way. As it can be seen, this design makes the sensor and therefore the accuracy vulnerable to mechanical deviations such as axial displacements and AG variations.

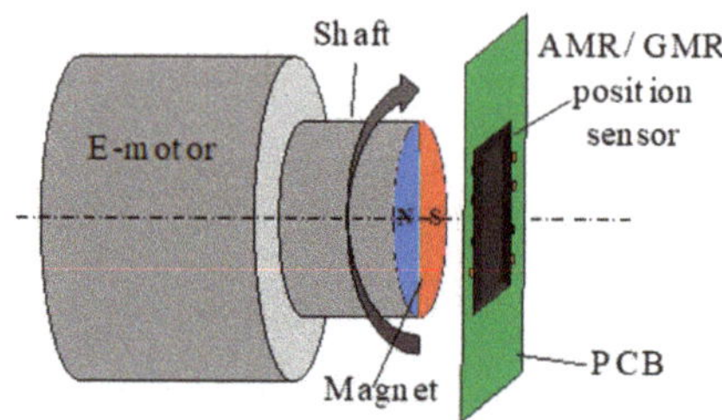

Figure 12. Typical setup of an AMR/GMR sensor.

Table 4 concludes both the strengths and weaknesses of the AMR/GMR position sensor technology.

Table 4. Listing of the strengths and weaknesses of magnetoresistive sensors.

	Attributes
Advantages	Small installation space High accuracy Broad temperature range Fast operation High rotary speed range Low system costs Different data interfaces possible (e.g., analog, digital, bus)
Disadvantages	External power supply required Short range detection only in terms of small AG Sensitive to mechanical tolerances and eccentricity

3. Experimental Characterization Method

In order to benchmark the performance of the previously introduced sensor technologies, an experimental method was applied by use of a rotor position sensor test bench. The corresponding sensor test bench has been specifically designed for rotor position sensor system error characterizations [12–14].

For analysis, the Device Under Test (DUT) is driven by an induction motor, which can reach a maximum speed of n = 24,000 rpm by a maximum angular acceleration of a = 10,000 rpm/s. This motor simulates the driving profiles of an electric powertrain. The speed is controlled by a highly accurate Hall-based position sensor with 8192 increments per rotation [37], which fulfills two tasks. Since the reference sensor provides an encoder emulation by the use of three TTL-based signals (i.e., A-, B- and Z-track), the speed is derived by the zero-position mark, which is the Z-track TTL signal for controlling the test bench drive comprising both the power electronics plus the control algorithm unit. In addition, the reference position information is used to evaluate the DUT angular accuracy in terms of potential angle error. The Hall-based reference sensor is doubly designed, and provides an angular resolution of $\theta < 0.1°$ at the maximum speed. Both angular data of the DUT position and the reference position are captured by a state-of-the-art 5 mega-samples per second (MS/s) Field Programmable Gate Array (FPGA)-based data acquisition unit [38]. This setup enables a universal analysis of various rotation position sensors with different interfaces (e.g., analog, digital, CAN-bus). In addition, a comprehensive sensor system characterization can be performed by integrating the corresponding DUT-ECUs [13].

The test bench not only provides high measurement accuracy, data acquisition rate and speed, but also automated adjustment of mechanical displacements of the DUT, as well as variation of sensor supply voltage and sensor operating temperature. For the investigation of the DUT sensitiveness regarding mechanical displacements (caused, for example, by production tolerances), the DUT stator is mounted on precisely displaceable motorized positioning actuators. This allows for the misalignment of the sensor stator relative to the rotor in four axes: x, y, z and the vertical tilt. The variation of the sensor supply voltage to simulate voltage fluctuations in the vehicle is performed by a controlled power supply. In addition, the ambient operation temperature can be varied by the use of a controlled climatic chamber surrounding the DUT during testing. Figure 13 shows the schematic test bench setup and a photo of the test facility.

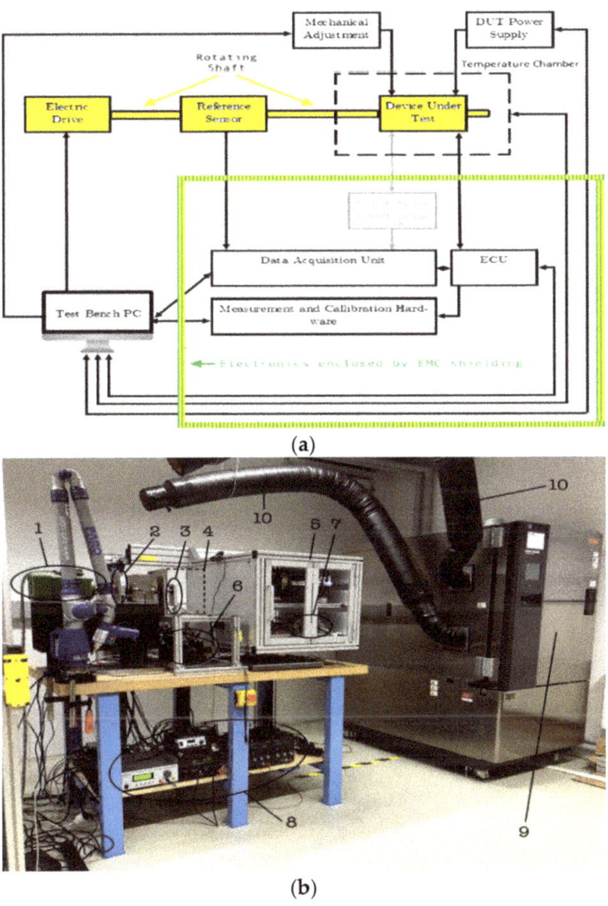

Figure 13. Sensor test bench (**a**) sensor test bench system overview; (**b**) sensor test bench and relevant components: 1, electric drive; 2, reference sensor; 3, device under test (DUT); 4, specific temperature chamber (removed for demonstration—see dashed box); 5, data acquisition housing (incl. electromagnetic compatibility (EMC) shielding); 6, motorized positioning actuators; 7, electronic control unit (ECU); 8, controlled power supply; 9, temperature conditioning system; 10, thermal tubes (for thermal transfer) [13].

In the present work, all investigated position sensors were evaluated in the same procedure. The resolvers are both powered and captured by a generic RDC evaluation board from Texas Instruments [39]. The angular information was provided by a TTL-based encoder emulation interface to the universal data acquisition card of the test bench to enable precise angle error determination. For the inductive sensors, the demodulated sine and cosine signals were captured by the analog inputs of the test bench data acquisition card. The angular position data of the magnetoresistive sensors were transferred by a CAN-bus interface onto the universal data acquisition card.

To enable both flexible and precise mounting of the different investigated sensor types, a universal adapter was used. The adapter system joins the rotary part of the DUT with the rotating drive shaft of the test bench by use of a high-precision clamping. To prevent vibrations at high speed testing, a specific balancing system was designed in the rotary adapter. Figure 14a depicts the universal sensor adapter with an exemplary rotating part of a resolver. Figure 14b shows the highly flexible stationary adapter, holding the stator of a resolver rotor position sensor as an example.

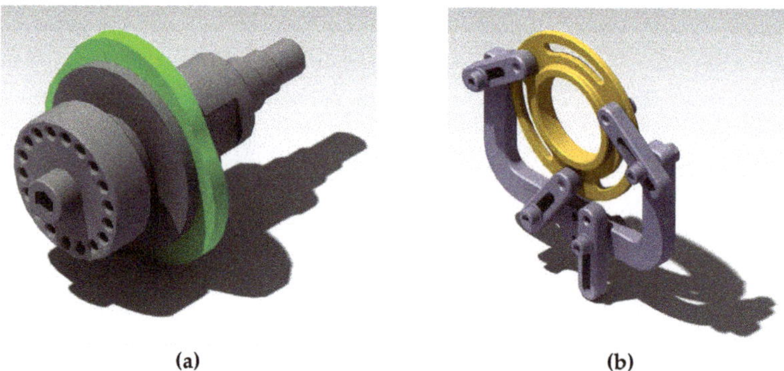

(a) **(b)**

Figure 14. Rendering of the universal sensor adapter system. (**a**) Rotary part with an exemplary mounted resolver rotor (green part); (**b**) Stator adapter strap with an exemplary resolver stator, shown in yellow.

In the following section, the results of the comprehensive rotor position sensor investigations are presented. Figure 15 shows the definitions of reference coordinate system, tilt angle $\Delta\Phi$ and rotation speed Δn used during the different test cases.

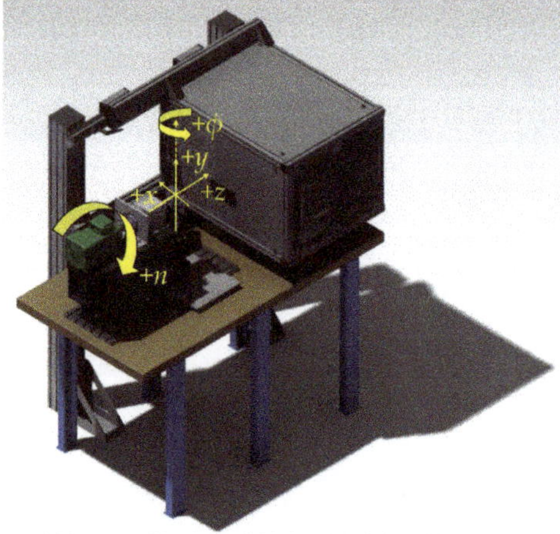

Figure 15. Coordinate system, tilt angle and rotational speed applied during the sensor system tests. The speed is defined as positive clockwise from +z view and the tilt angle positive clockwise from bottom view (+y).

4. Results of a Comprehensive Rotor Position Sensor Technology Evaluation

The investigations focus on the angular accuracy of the previously introduced position sensor technologies (see Section 2). In total, 14 rotor position sensor types have been analyzed by the specified measurement series on the sensor test bench: six resolver-, seven inductive-/eddy current- and one AMR-position sensor. The considered sensors are from different manufacturers and differ in terms of their technical characteristics, e.g., number of pole pairs, mechanical sizes (i.e., rotor diameter) and the AG. Table 5 provides an overview of main data of the tested sensors. Observe that all sensors with an around shaft design are tested with an axial nominal AG of $z = 0$ mm, since the rotor is surrounded by the circular stator. However, a radial offset between stator and rotor still exists due to the rotor geometry.

Table 5. Technical details of the tested sensors.

Resolver Position Sensor						
	Sample #1	Sample #2	Sample #3	Sample #4	Sample #5	Sample #6
Number of Pole Pairs	3	4	3	3	4	10
Maximal Operating Speed [rpm]	17,550	15,000	17,550	22,000	18,000	9000
Operating Temperature Range [°C]	−40 to +150	−40 to +180	−40 to +150	−40 to +150	−40 to +150	−40 to +150
Design	Around Shaft	Around Shaft	Around Shaft	Around Shaft	Around Shaft	Around Shaft
Nominal AG [mm]	0	0	0	0	0	0

Inductive Position Sensor							
	Sample #1	Sample #2	Sample #3	Sample #4	Sample #5	Sample #6	Sample #7
Number of Pole Pairs	3	3	3	2	4	4	2
Maximal Operating Speed [rpm]	24,000	24,000	24,000	20,000	20,000	20,000	20,000
Operating Temperature Range [°C]	−40 to +150	−40 to +150	−40 to +150	−40 to +150	−40 to +150	−40 to +150	−40 to +150
Design	Full Radial EoS	Around Shaft	Full Radial EoS	Arc 180° EoS	Arc 90° EoS	Arc 90° EoS	Arc 90° EoS
Nominal AG [mm]	1	0	4.9	3	3	2	3

AMR	
	Sample #1
Maximal Operating Speed [rpm]	7000
Operating Temperature Range [°C]	−40 to +150
Design	Full Radial
Nominal AG [mm]	4

The evaluation results are based on a variety of test cases in terms of mechanical- and temperature-based variation, considering both varying sensor geometries and operational range. The sensor systems' error characteristics are compared in terms of the angular average peak-to-peak error. Since the peak-to-peak error is the worst-case scenario, the resulting value reveals, from the mean of the collected results, the angular peak-to-peak error of every single measured test case. Observe that the angular errors shown correspond to the mechanical error, as measured in relation to the drive shaft. The peak-to-peak error of every single measured test case N is defined as

$$E_{\text{pp}}(N) = \max[\Delta\theta(t)] - \min[\Delta\theta(t)] \tag{13}$$

where $\Delta\theta(t)$ is the determined angular error:

$$\Delta\theta(t) = \theta_{\text{DUT}}(t) - \theta_{\text{Reference}}(t); \text{ with } \theta_{\text{DUT}}(t) = \arctan(V_{\sin}(t)/V_{\cos}(t)). \tag{14}$$

As a result, the utilized average peak-to-peak error can be determined according to (14).

$$E_{\text{pp, avg}} = \frac{1}{N} \int_0^N |E_{\text{pp}}(N)| dN \tag{15}$$

4.1. Speed-Dependant Angular Error Evaluation

Figure 16 illustrates the angle error characterization of the investigated sensor technologies in the mechanical initial point. The initial point is defined as the point at which the sensor is centrally adjusted according to the data sheet. This implies that all mechanical variation parameters are nulled (i.e., $\Delta\phi = \Delta x = \Delta y = \Delta z = 0$ [°/mm]). Note that, for both the inductive sensors and the AMR sensor, the nominal AG, which was defined by the manufacturer (see Table 5), is applied. The measurements have been performed at a temperature of 25 °C, which is defined as Room Temperature (RT) in further context. The only parameter varied in this test case was the rotational speed of the drive axle according to the maximal speed specification of the manufacturer (e.g., $n = 17{,}550$ rpm, see Table 5-resolver sample #1). The minimum rotational speed, which is described in Figure 16 below, was chosen in a way such that a sufficient number of angular periods was available. The step size of the speed was chosen in hundreds of steps and, respectively, thousands of steps.

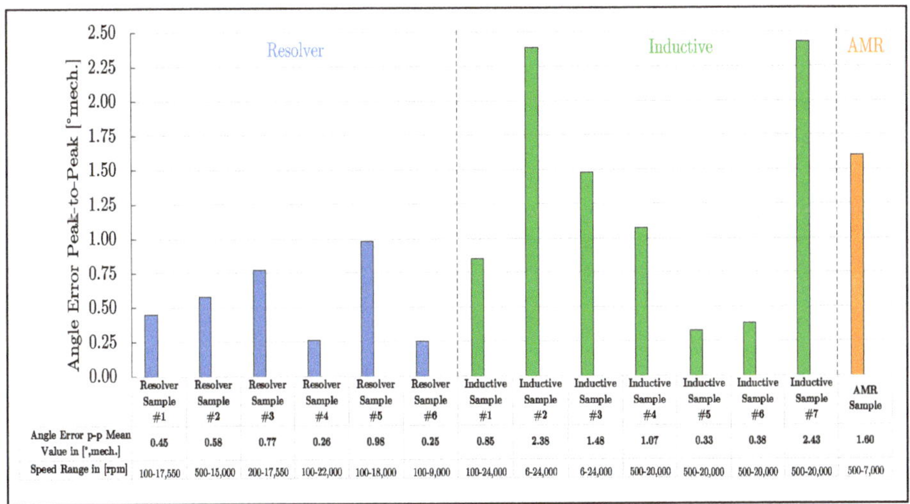

	Resolver Sample #1	Resolver Sample #2	Resolver Sample #3	Resolver Sample #4	Resolver Sample #5	Resolver Sample #6	Inductive Sample #1	Inductive Sample #2	Inductive Sample #3	Inductive Sample #4	Inductive Sample #5	Inductive Sample #6	Inductive Sample #7	AMR Sample
Angle Error p-p Mean Value in [°,mech.]	0.45	0.58	0.77	0.26	0.98	0.25	0.85	2.38	1.48	1.07	0.33	0.38	2.43	1.60
Speed Range in [rpm]	100-17,550	500-15,000	200-17,550	100-22,000	100-18,000	100-9,000	100-24,000	6-24,000	6-24,000	500-20,000	500-20,000	500-20,000	500-20,000	500-7,000

Figure 16. Experimental comparison of the resolver-, inductive- and AMR-position sensor technology, based on a variable applied speed and both constant temperature ($T = 25\,°C = RT$) and constant centric sensor placement.

Considering the error characteristics of resolvers shown in Figure 16, it can be summarized that this technology offers a high accuracy. This is one of the reasons why this sensor type is the preferred choice in electric powertrain architectures today. Basically, it can be argued that a higher number of pole pairs results in better accuracy. However, this is not always the case, as a comparison of the different types shows (e.g., Resolver Sample #4 vs. Sample #5 and #6).

In general, inductive position sensors can reach a similar level of accuracy as resolvers (see Inductive Sample #1, #5, #6). Full radial sensor designs with a circumference of 360° deliver higher accuracy in principal, but other parameters, e.g., the AG, also have a considerable influence; compare Inductive Sample #1 and #3, with AG = z = 1 mm at sample #1 and AG = z = 4.9 mm at sample #3. Inductive Sample #2 with an around shaft design shows a lower accuracy, which might be caused by additional eddy currents induced by the steel-made drive shaft. Observing the arc designs of the inductive position sensors, the angular accuracy is comparably good to the resolvers and other inductive sensor designs. A resolver can be replaced by an inductive position sensor with two pole pairs and an arc design measuring a half circle (i.e., 180°); see Figure 16, Inductive Sample #4. The best accuracy is provided by two different 90° arc design inductive position sensors with four pole pairs. For example, a ten-pole pair resolver shows a similar accuracy to a four-pole pair arc 90° inductive sensor, while the inductive sensor has lower installation space requirements and, probably, lower costs. However, the number of pole pairs should be taken into account, since a lower amount of pole pairs (e.g., Inductive Sample #7) shows a comparatively high angle error compared to a sample with similar specifications and nominal AG (e.g., Inductive Sample #5).

Concerning the AMR technology, the advantage of miniaturization is diminished by the sensitivity of mechanical impacts. This can be seen in the comparatively low accuracy performance, which is in the middle range of the analyzed inductive sensors.

Besides a comparison of the different sensor technologies, Figure 16 shows that the sensor design, in terms of pole pairs and implementation (e.g., arc versus full circle in the case of the inductive sensor principle), strongly influences the accuracy. In the following subsection, mechanical displacements are also considered, in addition to the speed variation, in order to investigate the sensitivity of the sensors with regard to mechanical installation tolerances.

4.2. Multi-Mechanical Parameter Angular Error Characterization

The results of the multi-mechanical parameter variation tests, considering both geometrical misalignment and speed variations, are represented in Figure 17. The Figures A1–A3 in the Appendix A provide an overview of the conducted mechanical parameter variation range of each sensor, where the darker color represents the selected point. The angle error determination performed is based on Equations (13)–(15) and the average peak-to-peak error of each measurement series per parameter variation is considered.

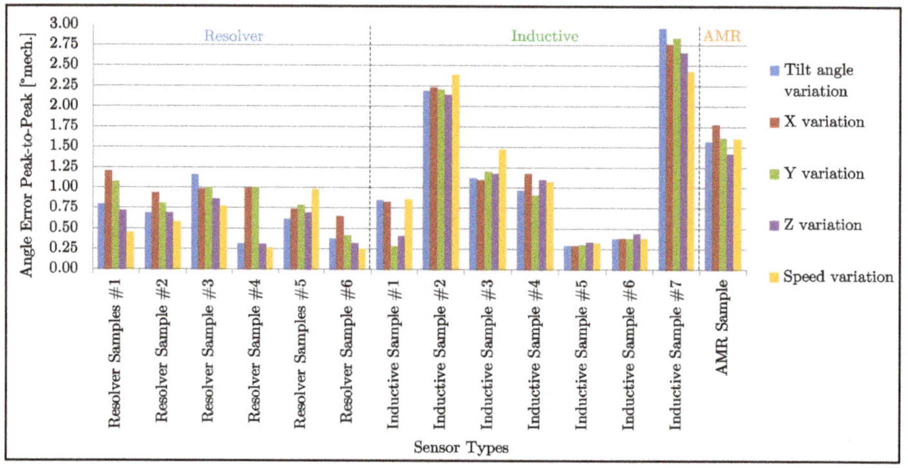

Figure 17. Benchmark of both mechanical and rotor speed parameter variations at RT = 25°, by use of the average angular peak-to-peak error from each test case, which is one parameter variation.

The inductive sensor principle is capable of reaching the characteristics of the resolver technology in terms of mechanical deviations and speed variation (e.g., Inductive Sample #5 & #6). The design of the inductive position sensor is highly responsible for the angular accuracy, e.g., arc design vs. 360° design.

The magnetoresistive sensor principle is suitable for miniaturized applications, and the angular accuracy performance is in the range of the other two sensor types. Furthermore, the robustness in terms of mechanical misalignment in all axes shows that some sensors offer a very good performance (e.g., the Resolver Sample #3), while other sensors are more sensitive to mechanical deviations (e.g., Inductive Sample #1). In general, it can be concluded that the greater the difference between the respective bars of a specific sensor in the figure above, the more sensitive it is to x-, y-, z- and tilt-orientation-related tolerances.

4.3. Temperature Variation-Based Angular Accuracy of Non-Resolver Position Sensors

The error response at different operating ambient temperature conditions was investigated over both the speed range and the mechanical initial position. Since it is known that resolver position sensors are very robust against temperature influences [40], only the inductive- and AMR-based sensors were examined here. Different thermal conditions can lead to angle error distortion for these sensor types, owing to the temperature coefficient of the PCB material (i.e., the inductive sensor) or the magnetic material (i.e., the magnetoresistive position sensor). Figure 18 depicts the error characteristics based on the thermal influences, which were provoked on the sensor system test bench by use of the climatic conditioning system. The angular error for each applied speed at a certain temperature was evaluated based on Equations (13)–(15).

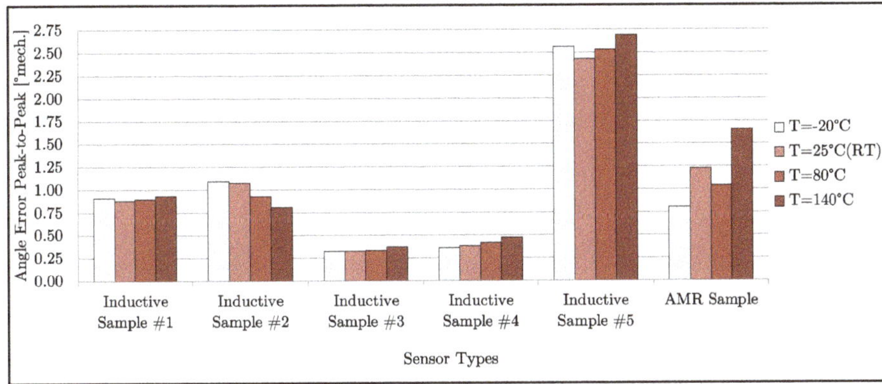

Figure 18. Inductive- and AMR-based sensor accuracy characteristics in the sensor systems' sweet spots at different operational ambient temperatures.

Besides the fact that the sensor systems' accuracy strongly depends on the design of both the target (i.e., the rotor) and the stator coil structure, temperature plays an important role for accurate operation. Aside from Inductive Sample #2, with its around shaft design, all the other evaluated devices show the highest angular error at maximum temperature. It can also be seen that thermal changes do not influence the accuracy very significantly when considering the inductive sensors. However, the AMR sensor principle, which is based on the magnetoresistive effect, shows the best performance at the lowest temperature and worst performance at maximum temperature. The selected temperature of T = 80 °C is a common operational condition in drivetrain applications, and T = 140 °C represents high load conditions, which typically occur for a short duration. It can be concluded that the angular accuracy varies in the range of a maximum of 25% for inductive sensors and about 120% for the AMR. Either way, the absolute angular deviation is about 0.25° for inductive sensors and about 0.90° for the AMR.

4.4. Validation of the Presented Results

This subsection provides a validation of the previously presented results. To summarize, the experimentally-based evaluation method consisted of three main phases. In the first phase, the DUT was mounted on the test bench on both the motorized stator positioning system (Figure 13) and the rotating shaft (Figure 14). In addition, the electrical installation was conducted, i.e., the sine and cosine feedback signals were connected with the data acquisition unit of the test bench. The experimental characterization was performed in phase two of the investigation process, which started with loading the test cases that are defined for each parameter variation (e.g., speed n, temperature T, other mechanical parameters x, y, z, Φ). In the following, the test procedure was started by conditioning the influencing parameters on the test bench. The process of the determination of the angular information of both the reference sensor and the DUT was repeated for every test case. This process included signal acquisition and data storage. After completing every test case, the data was processed in phase three. This comprised the conversion of the mechanical reference angle to the electrical angle value by consideration of the number of pole pairs of the DUT. Furthermore, both electrical angles were compensated in terms of the angular offset. As a result, each position started from 0°. This process was repeated after every overflow, i.e., exceeding 360°. Based on these data, the electrical angle error was subsequently determined, including both the computation and storage of the angular peak-to-peak error for each test case number. To achieve the final result for each parameter variation, which is presented in the figures above, the mean peak-to-peak error of every single parameter variation was conducted. The previously described process is graphically summarized in Figure 19.

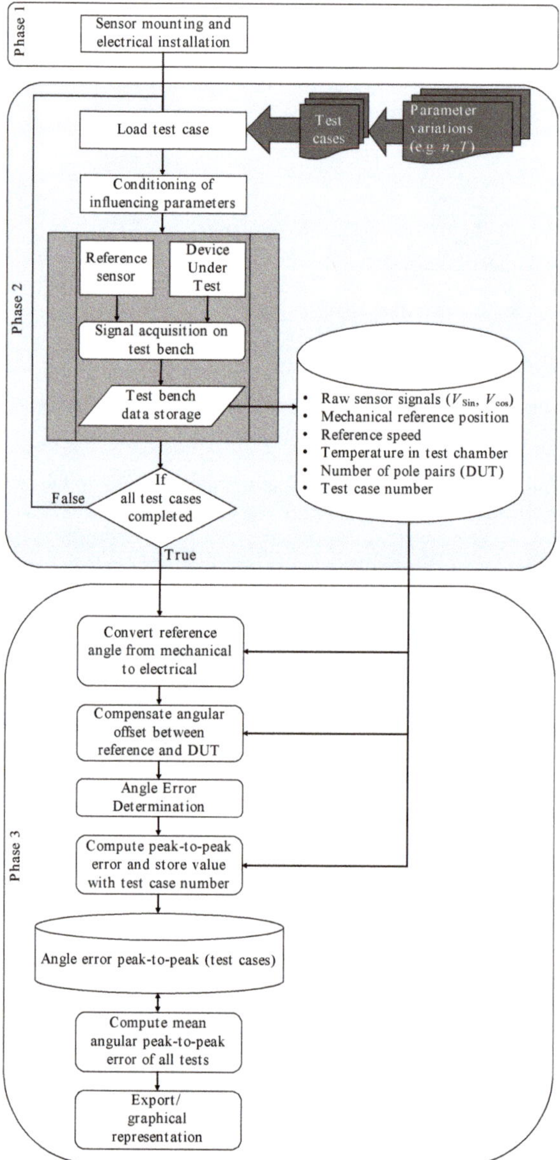

Figure 19. Illustration of the sensor characterization process.

Sensor signal data processing and validation are exemplarily shown for the Inductive Sample #4 at a constant speed of $n = 4000$ rpm under RT, which includes both the speed variation and temperature variation test. The resulting signals can be seen in Figure 20.

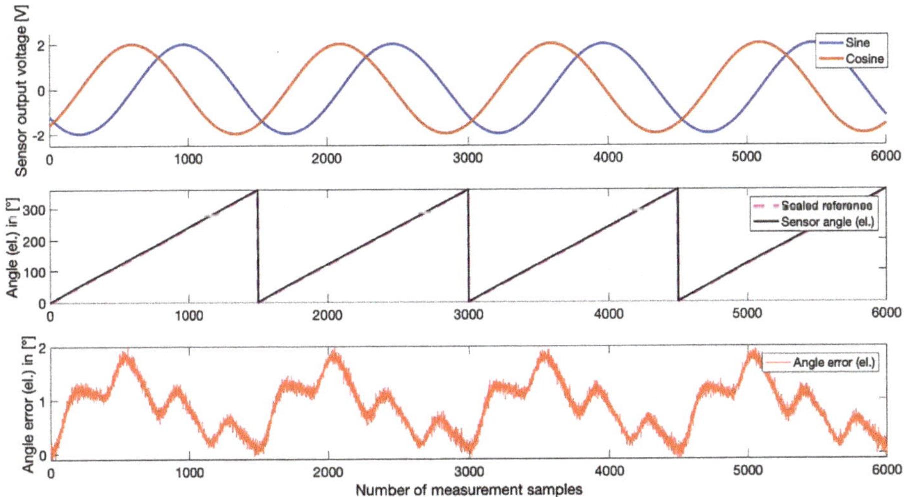

Figure 20. Measurement and processed signals of the Inductive Sample #4 at $n = 4000$ rpm and RT.

The sensor delivers two sinusoidal signals, i.e., sine and cosine (first sublot, Figure 20). The resulting electrical position of the sensor is illustrated in the second subplot of Figure 20. Based on the characterization method, the scaled and offset compensated reference angle can also be found in this graph. The resulting angle error of four periods, i.e., twice the number of pole pairs, is shown in the third subplot in Figure 20. The periodical error can be seen at a glance. Observe that the evaluation was performed over the recorded samples. Consequently, the peak-to-peak error is 2.11° electrically and 1.06° mechanically, since the pole pairs of this sample are two. Table 6 outlines the results for all measurements of the mentioned parameter variation, including the exemplary test, which is presented above. In addition, the mean peak-to-peak error for this specific test series is stated, as considered in Figure 16.

Table 6. Results of the speed variation test of Inductive Sample #4.

Speed n [rpm]	Temperature T [°C]	Angle Error Peak-to-peak [°, el.]	Angle Error Peak-to-peak [°, mech.]
1000	25	2.09	1.05
2000	25	2.02	1.01
4000	25	2.11	1.06
6000	25	2.03	1.01
8000	25	2.04	1.02
10,000	25	2.27	1.13
20,000	25	2.57	1.29
-	-	Mean peak-to-peak error [°, mech.]	1.08

5. Discussion

The present research provides a comprehensive investigation of the angular accuracy of different rotor position sensor technologies in terms of a variation of multi-mechanical and thermal parameters. The considered sensor types in this work include resolver-, inductive-/eddy current-, Hall- and magnetoresistive-based position sensors.

The results show that the inductive sensors are able to reach the performance of resolver position sensors. It was found that the accuracy of the inductive sensor technology strongly depends on the sensor design. As an example, a full circle design has the potential to perform in a comparable accuracy-range to a resolver; the arc design can even achieve equal or better accuracy characteristics. It has to be considered that both parameters, the number of pole pairs and the nominal AG, are important influencing factors. The AMR sensor principle comes with a great advantage for miniaturized applications, but its angular accuracy is heavily reliant on mechanical tolerances.

When operating the position sensor in a powertrain, the accuracy in terms of thermal deviations is an important criterion. Since resolver sensors show nearly constant accuracy over the holistic temperature range [38], non-resolver-based position sensors were investigated according to their angular accuracy under different automotive ambient temperatures. The tested inductive- and magnetoresistive- position sensors show higher sensitivity to temperature changes, because the PCB material is more sensitive in terms of geometrical form changes at induction sensors, and the magnetic performance varies as a function of the ambient temperature in the case of magnetoresistive sensors. In general, the accuracy of the sensor deteriorates at higher temperatures.

Figure 21 shows the potentials of the different sensor technologies under consideration of selected performance indicators, which is relevant for the application in automotive powertrain systems. Based on the results of the study, it can be stated that the resolver is one of the most accurate and, at the same time, robust sensors for powertrain applications. However, complex signal processing (leading to relatively high costs), and comparatively high weight and mechanical size are disadvantages of this sensing principle.

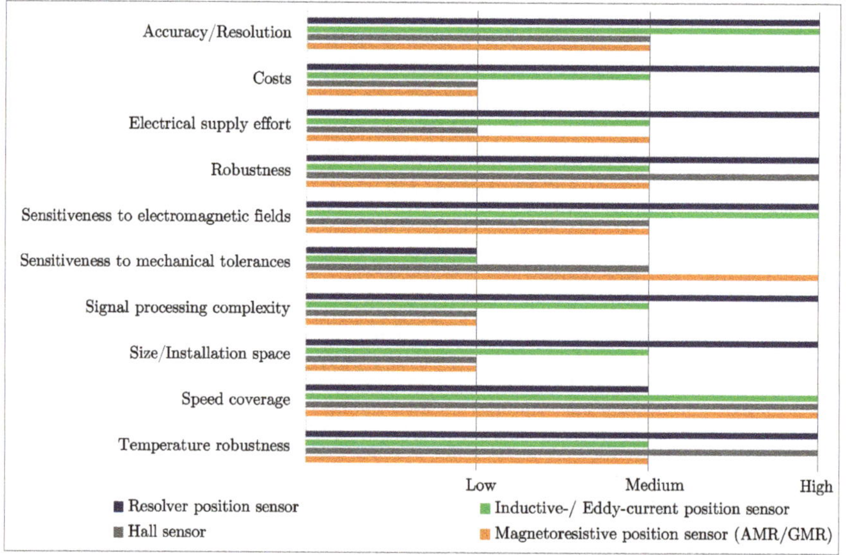

Figure 21. Comparison of the different sensor technologies related to selected performance indicators.

Inductive sensors are capable of fulfilling key requirements to a high level, comparable to resolver performance. An important characteristic is the design (i.e., full 360° vs. arc design) of the sensor, since it has an influence on the accuracy. Furthermore, the thermal sensitivity of the PCB carrier material influences the angular error in the temperature-sensitive application of traction motors. For AMR and GMR sensors, current state-of-the-art semiconductor sensor manufacturing technology enables a very precise determination of the measured rotating shaft. Inductive- and eddy-current position sensors offer less rotor mass and a smaller sensor housing. However, the present research deals with increasing the sensor performance to achieve equal fidelity compared to the commonly-used resolver technology, which is accomplishable by improving the coil design. Hall sensors can be used in large numbers to measure the rotor position quite precisely (e.g., as a reference sensor utilized on the test bench). When applying this sensing method, the larger magnet size needs to be taken into account compared to magnetoresistive sensors. Larger magnets including a higher number of Hall elements (required for high accuracy) also increase the housing size, which might restrict the usability for automotive applications with high demands for compact design. The investigated AMR/GMR sensors offer higher resolution and, at the same time, a small installation space. Therefore, when a precise position determination of the rotating shaft is demanded, e.g., at small BLDC motors, magnetoresistive sensors are preferable. In the case of low requirements on the position information, reduced numbers of Hall elements, e.g., three elements mechanically displaced by 120°, might be sufficient.

Evaluating the sensor costs, the resolver is the most high-priced system. This is due to the fact that a high amount of copper wires and a complex winding process is required, similar to an electric motor. Also, signal processing is effortful, demanding a performant microcontroller (μC), raising the system costs. Inductive position sensors are cheaper compared to resolvers, since all sensing components can be placed on a low-cost PCB. Only the rotor needs to be made of a ferromagnetic material. Especially in mass production, the inductive sensor principle can play to its benefit in comparison to the resolver. Hall-based and magnetoresistive sensors can also be produced relatively cheaply; hence, semiconductor mass production allows low cost manufacturing. The only matter of expense in these sensor systems is the required magnet(s), the costs of which depend mainly on the magnet size and material.

The electrical supply of the sensor plays a key role when it comes to efficiency in automotive applications. Figure 21 shows that the resolver has the highest electric power demand because of its relatively high excitation currents and voltage. Thus, a powerful driver stage is inevitable, increasing the costs and influencing other electrical components in terms of Electromagnetic Compatibility (EMC). Inductive and magnetoresistive sensors do not need such high power for sensor operation. The Hall-effect sensor only requires low constant DC-Voltage to detect a magnet and change the output stage.

Resolvers are used in various powertrains of BEVs and HEVs offering robust rotary measurements today. Even in aviation and military practice, this sensor principle can be found because of its high durability and fail safety behavior. The same characteristics are applicable to the Hall sensor. Regarding inductive and magnetoresistive sensor principles, the overall robustness can be considered equal in case that a solid housing is designed. There are rough conditions for accurate position sensing in electric powertrains, especially when the sensor is mounted close to the electric machine, which produces strong electromagnetic fields. Resolver and inductive position sensors are more sensitive to electromagnetic influences than Hall- and magnetoresistive sensors. Both resolvers and inductive sensors rely on a similar principle, using a ferromagnetic rotor material to determine the position of the rotating shaft. An external field can influence the feedback signals and so distort the angular information. Today, different research activities are ongoing for inductive position sensors to improve their performance in terms of EMC. All magnet-based rotor position sensors show a certain sensitivity against stray fields. For that reason, they are rated as "medium", compared to resolver and inductive sensor technology.

Sensitiveness to mechanical tolerances is one of the investigations in this work, so this attribute is rated based on each sensor working principle. Mechanical tolerances can occur during installation processes, resulting in mechanical offset from the ideal-considered centric position, leading to an angular measurement error. For resolvers and inductive position sensors, this sensitivity is quite low. Resolvers usually operate under certain mechanical boundaries (i.e., the inner side of the stator), and are mainly sensitive to vertical tilt and axial offsets. Inductive position sensors show quite similar good characteristics, because the coil design can be optimized to the operating point. However, mechanical deviations of the rotating target in respect to the stator can cause biased angular information. Hall elements sense only in a relatively small distance. Larger AG ranges between sensor element and magnet or fluctuations (e.g., wobble) distort the measurement result. AMR/GMR position sensors are typical EoS applications, and show high sensitivity regarding mechanical tolerances. Minor axial- or radial displacements cause relatively high angular errors.

In terms of signal processing complexity, the resolver technology requires more complex signal processing, comprising digitalization, filtering, demodulation of the AM feedback signals, trigonometric-based angular calculation and optional observer algorithms to process the rotor position. All these steps, performed in both hardware and software, usually on a multi-core μC, are integrated in a so-called Resolver-To-Digital Converter (RDC), also denoted as an R2D. The high complexity of signal processing is cost-intensive, and can cause additional angular errors, which can be seen as potential drawback of this technology. In inductive position sensors, the feedback signals on the sine- and cosine traces are also AM signals, which need to be demodulated. Typical sensors using this principle offer signal processing as a powerful one-chip solution directly on the PCB. In this way, the resulting angular information can be read in directly by the MCU without drawing computational power, which reduces the effort of angle determination compared to resolver technology. Hall sensors and magnetoresistive sensors are less complex in terms of signal processing effort. Hall sensors provide either an analog or a digital output signal to the subsequent control unit (e.g., ECU, MCU), where, in the case of an analog signal, the information only needs to be digitalized and interpreted. For most applications, digital signals are used in form of TTL pulses. The position is then calculated by counting the pulses, enabling a simple signal processing complexity. Magnetoresistive sensors contain Wheatstone Bridges, delivering (magnet) position dependant sinusoidal voltages at the output as a result of resistance changes. These electrical circuits are relatively low in complexity, and are designed similarly to analog Hall sensors. In this case, the analog voltage is interpreted by an external controlling unit to determine the absolute rotary position.

Resolvers tend to be relatively large radial sensors with considerable weight and rotor inertia. Inductive position sensors only require space for the PCB-based stator and a thin rotary target with distinctly less inertia compared to the resolver. Hall and magnetoresistive sensors are very small components with a chip size of a few millimeters. However, in the case of Hall-sensors, a higher number of elements might be necessary to provide the specified accuracy. Miniaturization can be enabled due to semiconductor manufacturing; consequently, these sensor types need less installation space than resolvers and inductive position sensors. One parameter to be considered is the size of the magnet that is used. All sensors, apart from resolvers, can operate at very high speeds ($n \geq 20$ k rpm), due to the relatively small rotor inertia.

The findings of the temperature-related investigations show that resolvers cover the full automotive range due to their thermally robust sensor design. Here, temperature-biased errors stem only from the thermal resistance change of the copper windings. Hall sensors can also withstand high and low temperatures. AMR/GMR sensors can be designed to be temperature-compensating to reduce temperature-biased angular errors. For both Hall and magnetoresistive sensors, the only limiting criterion is the Curie Temperature T_C of the magnet. Inductive position sensors are able to operate within the automotive temperature range, but the thermal properties of the PCB substrates need to be taken into account.

6. Conclusions

The present research provides a detailed benchmark of modern rotor position sensor technologies for application in automotive electric drive trains, considering different influencing factors. The investigated sensor systems were evaluated considering the typical performance indicators of automotive powertrain applications, including sensor accuracy, sensitiveness to mechanical tolerances and electromagnetic fields, effort of signal processing and temperature robustness, as well as the required installation space and cost aspects. For this purpose, a specific sensor test bench, including sophisticated measurement equipment, mechanical and electrical misalignment equipment, as well as a thermal conditioning system, was used. The measurement results are displayed in diagrams and discussed to elaborate the strengths and weaknesses of the different investigated sensor technologies.

For application in modern electric powertrain systems, resolver (as the most used technology) and inductive sensors show the best potentials due to their high accuracy and robustness against mechanical, thermal and electrical disturbances. For non-traction applications, e.g., auxiliaries driven by BLDC-motors, magnetoresistive and Hall-based sensors seem to be more suitable, due to lower cost and installation space. Nevertheless, the position sensor technology has to be selected according to the requirements of the specific application. In this context, the present work contributes to the selection of suitable sensor technology in the development of new electric propulsion systems.

Author Contributions: Conceptualization, C.D. and M.H.; methodology, C.D.; software, C.D.; validation, C.D.; formal analysis, C.D.; investigations, C.D. and M.H.; resources, C.D. and M.H.; data curation, C.D.; writing—original draft preparation, C.D.; writing—review, M.H.; visualization, C.D.; project supervision, M.H. All authors have read and agreed to the published version of the manuscript.

Funding: Open access funding provided by Graz University of Technology.

Conflicts of Interest: The authors declare no conflict of interest.

Appendix A

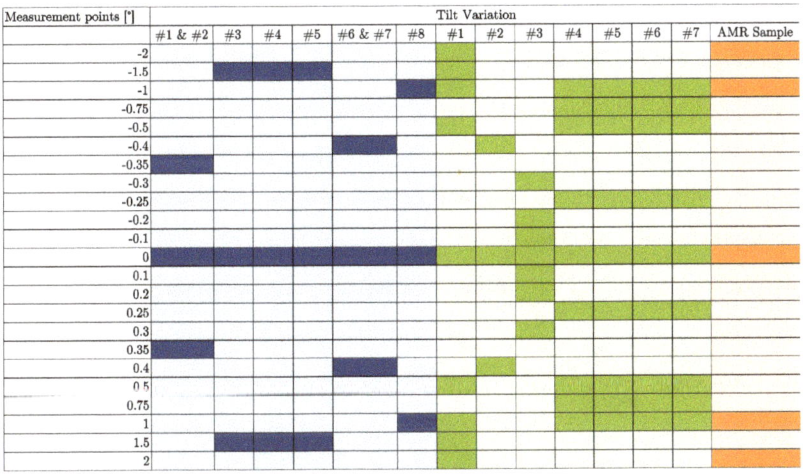

Measurement points [°]	Tilt Variation													
	#1 & #2	#3	#4	#5	#6 & #7	#8	#1	#2	#3	#4	#5	#6	#7	AMR Sample
-2														
-1.5														
-1														
-0.75														
-0.5														
-0.4														
-0.35														
-0.3														
-0.25														
-0.2														
-0.1														
0														
0.1														
0.2														
0.25														
0.3														
0.35														
0.4														
0.5														
0.75														
1														
1.5														
2														

Figure A1. Tilt variation measurement plan for all evaluated sensors; blue, resolvers; green, inductive sensors; orange, AMR sensor.

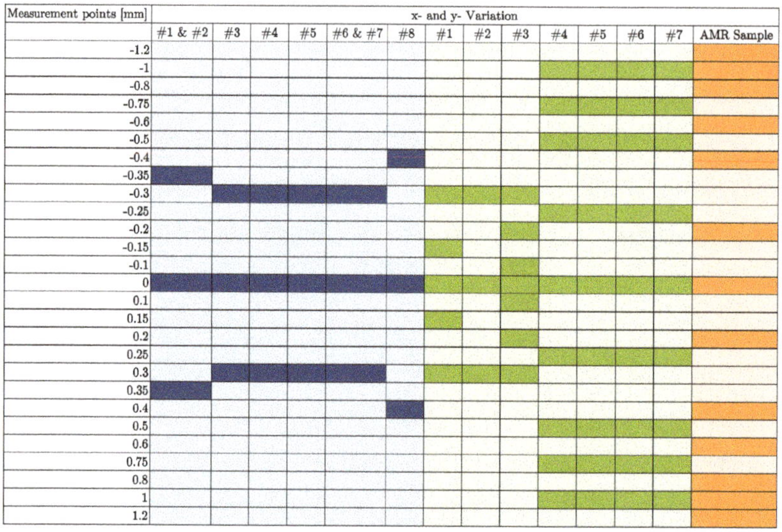

Measurement points [mm]	x- and y- Variation													
	#1 & #2	#3	#4	#5	#6 & #7	#8	#1	#2	#3	#4	#5	#6	#7	AMR Sample
-1.2														
-1														
-0.8														
-0.75														
-0.6														
-0.5														
-0.4														
-0.35														
-0.3														
-0.25														
-0.2														
-0.15														
-0.1														
0														
0.1														
0.15														
0.2														
0.25														
0.3														
0.35														
0.4														
0.5														
0.6														
0.75														
0.8														
1														
1.2														

Figure A2. Test case plan for the x and y variation to evaluate all available sensors; blue, resolvers; green, inductive sensors; orange, AMR sensor.

Measurement points [mm]	z- Variation													
	#1 & #2	#3	#4	#5	#6 & #7	#8	#1	#2	#3	#4	#5	#6	#7	AMR Sample
-2														
-1.6														
-1.5														
-1.4														
-1.2														
-1														
-0.832														
-0.8														
-0.7														
-0.6														
-0.5														
-0.4														
-0.35														
-0.3														
-0.2														
-0.1														
0														
0.1														
0.2														
0.3														
0.35														
0.4														
0.5														
0.6														
0.7														
0.8														
0.832														
1														
1.2														
1.4														
1.5														
1.6														
2														
2.5														
3														

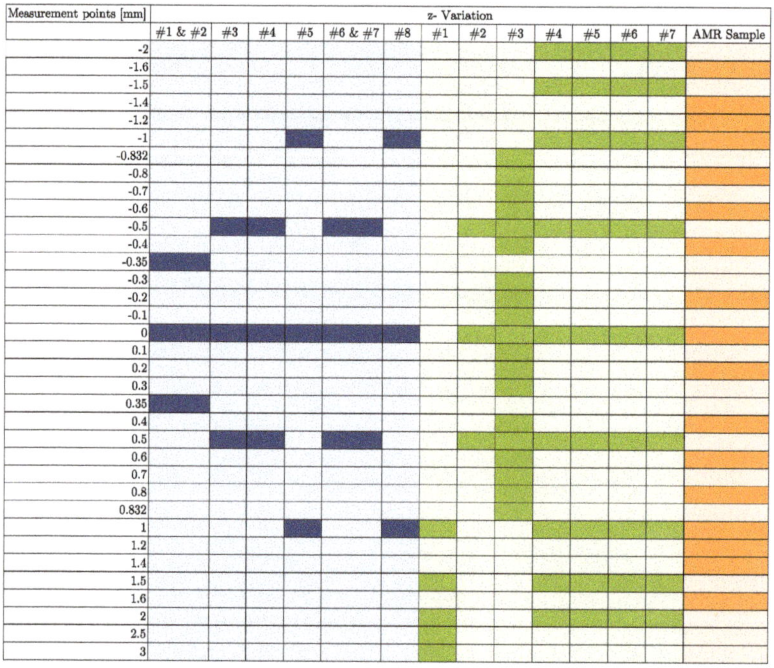

Figure A3. Measurement plan for the axial evaluation of all rotor position sensors; blue, resolvers; green, inductive sensors; orange, AMR sensor.

References

1. Quang, N.P.; Dittrich, J.-A. *Vector Control of Three-Phase AC Machines: System Development in the Practice*; Springer: Berlin/Heidelberg, Germany, 2008; ISBN 9783540790297.
2. Lashkevich, M.; Anuchin, A.; Aliamkin, D.; Briz, F. Investigation of self-sensing rotor position estimation methods for synchronous homopolar motor in traction applications. In *IECON 2017—43rd Annual Conference of the IEEE Industrial Electronics Society*; IEEE: Beijing, China, 2017; pp. 8225–8229. [CrossRef]
3. Foo, G.; Xinan, Z.; Maskell, D.L.; Vilathgamuwa, D.M. Sensorless direct torque control of interior permanent magnet synchronous motor drives using the current derivative approach. In Proceedings of the IEEE Symposium on Sensorless Control for Electrical Drives (SLED), Sydney, Australia, 7–8 June 2015; pp. 1–6. [CrossRef]
4. International Organization for Standardization (ISO). ISO 26262 Road Vehicles-Functional Safety. Available online: https://www.iso.org/standard/68383.html (accessed on 7 January 2020).
5. Birolini, A. *Reliability Engineering: Theory and Practice*, 8th ed.; Springer: Berlin, Germany, 2017.
6. Ross, H.-L. *Funktionale Sicherheit im Automobil: ISO 26262, Systemengineering Auf Basis Eines Sicherheitslebenszyklus und bewährter Managementsysteme*; Hanser: Munich, Germany, 2014.
7. Tille, T. *Automobil-Sensorik*; Springer: Berlin/Heidelberg, Germany, 2016; ISBN 978-3-662-48943-7.
8. Brunel, O.; Moller, R. *Rotor Position Sensor for Hybrid Drives and Electric Drives New Generation Eddy Current Position Sensor*; SAE World Congress & Exhibition: Detroit, MI, USA, 2011. [CrossRef]
9. Robert Bosch GmbH. *Bosch Automotive Electrics and Automotive Electronics*, 5th ed.; Springer Vieweg: Wiesbaden, Germany, 2014.
10. Alipour-Sarabi, R.; Nasiri-Gheidari, Z.; Tootoonchian, F.; Oraee, H. Performance analysis of concentrated wound-rotor resolver for its applications in high pole number permanent magnet motors. *IEEE Sens. J.* **2017**, *17*, 7877–7885. [CrossRef]
11. Xiao, L.; Li, Z.; Bi, C. An optimization approach to variable reluctance resolver. *IEEE Trans. Magn.* **2020**, *56*, 1–5. [CrossRef]
12. Datlinger, C.; Hirz, M. Investigations of rotor shaft position sensor signal processing in electric drive train systems. In Proceedings of the IEEE Transportation Electrification Conference and Expo, Asia-Pacific (ITEC Asia-Pacific), Bangkok, Thailand, 6–9 June 2018; pp. 1–5. [CrossRef]
13. Datlinger, C.; Hirz, M. An extended approach for validation and optimization of position sensor signal processing in electric drive trains. *Electronics* **2019**, *8*, 77. [CrossRef]
14. Datlinger, C.; Hirz, M.; Anuchin, A. Holistic consideration and optimization of signal processing on a resolver-based rotor position sensor systems in electric drive trains. In Proceedings of the 27th International Workshop on Electric Drives: MPEI Department of Electric Drives 90th Anniversary (IWED), Moscow, Russia, 27–30 January 2020; pp. 1–7. [CrossRef]
15. Rotation Angle Sensor (Resolver) for EV/HEV Drive Motor—MinebeaMitsumi. Available online: https://www.minebeamitsumi.com/english/strengths/column/resolver/ (accessed on 4 October 2019).
16. Inductive Position Sensors—IDT: A Renesas Company. 2019. Available online: https://www.sensor-test.de/assets/de/Aussteller/Aussteller-Aktionsprogramm/Fachforen/Manuskripte/JJanisch-IDT-RENESAS.pdf (accessed on 21 December 2019).
17. Hartl, H.; Krasser, E.; Soeser, P.; Winkler, G. *Elektronische Schaltungstechnik*, 2nd ed.; Pearson: Munich, Germany, 2019.
18. Michael Faraday—TUGraz. Available online: http://history.tugraz.at/besonderheiten/groessen_der_technik/faraday.php. (accessed on 20 May 2020).
19. Encyclopedia Britannica: Léon Foucault. Available online: https://www.britannica.com/biography/Leon-Foucault (accessed on 20 May 2020).
20. Passarotto, M.; Qama, G.; Specogna, R. A fast and efficient simulation method for inductive position sensors design. In Proceedings of the IEEE SENSORS, Montreal, QC, Canada, 27–30 October 2019; pp. 1–4. [CrossRef]
21. Volder, J.E. The CORDIC trigonometric computing technique. *IRE Trans. Electron. Comput.* **1959**, *8*, 330–334. [CrossRef]
22. Aschenbrenner, B.; Zagar, B. Analysis and validation of planar high-frequency contactless absolute inductive position sensor. *IEEE Trans. Instrum. Meas.* **2015**, *64*, 768–775. [CrossRef]

23. Microchip Technology Inc.: Inductive Position Sensors. Available online: https://www.microchip.com/design-centers/sensors/inductive-position-sensors (accessed on 23 March 2020).

24. Ebbesson, C. Faculty of Engineering, Lund University, Lund, Sweden: Rotary Position Sensors—Comparative Study of Different Rotary Position Sensors for Electrical Machines Used in a Hybrid Electric Vehicle Application. Available online: http://lup.lub.lu.se/student-papers/record/3632648 (accessed on 10 January 2020).

25. Bridgman, P.W. National Academy of Sciences—Biographical Memoir of Edwin Herbert Hall. Available online: http://www.nasonline.org/publications/biographical-memoirs/memoir-pdfs/hall-edwin.pdf (accessed on 10 January 2020).

26. RS Components: Allegro Microsystems A1120LUA-T Hall-Effect-Sensor Unipolar, 50 Gs, SIP 3-Pin. Available online: https://at.rs-online.com/web/p/hall-effekt-sensor-ics/1709108/ (accessed on 11 February 2020).

27. Milano Shaun—Allegro MicroSystems: Allegro Hall-Effect Sensor ICs—Product Information. 2013. Available online: https://www.allegromicro.com/en/Insights-and-Innovations/Technical-Documents/Hall-Effect-Sensor-IC-Publications/Allegro-Hall-Effect-Sensor-ICs.aspx (accessed on 13 June 2018).

28. Honeywell Inc.: Hall Effect Sensing and Application. Available online: https://sensing.honeywell.com/honeywell-sensing-sensors-magnetoresistive-hall-effect-applications-005715-2-en2.pdf (accessed on 12 February 2020).

29. Jogschies, L.; Klaas, D.; Kruppe, R.; Rittinger, J.; Taptimthong, P.; Wienecke, A.; Rissing, L.; Wurz, M.C. Recent developments of magnetoresistive sensors for industrial applications. *Sensors* **2015**, *15*, 28665–28689. [CrossRef] [PubMed]

30. Hesse, J.; Gardner, J.W.; Goepel, W. *Sensors Applications. Volume 1: Sensors in Manufacturing, Volume 2: Sensors in Intelligent Buildings, Volume 3: Sensors in Medicine and Health Care, Volume 4: Sensors for Automotive Applications, Volume 5: Sensors in Household Appliances*; Wiley-VCH: Weinheim, Germany, 2005.

31. Encyclopedia Britannica: William Thomson. Available online: https://www.britannica.com/biography/William-Thomson-Baron-Kelvin (accessed on 20 May 2020).

32. Hauser, H.; Kloibhofer, R.; Stangl, G. Magnetoresistive sensoren. *Elektrotech. Inftech.* **1998**, *115*, 382–390. [CrossRef]

33. Tumanski, S. *Thin Film Magnetoresistive Sensors*; Institute of Physics Publishing: Bristol, UK, 2001.

34. Infineon Technologies AG, Datasheet: TLE5309D Dual GMR/AMR Angle Sensor. 2017. Available online: https://media.digikey.com/pdf/Data%20Sheets/Infineon%20PDFs/TLE5309D.pdf (accessed on 20 May 2020).

35. Lopez-Martin, A.J.; Carlosena, A. Performance tradeoffs of three novel GMR contactless angle detectors. *IEEE Sens. J.* **2009**, *9*, 191–198. [CrossRef]

36. Djamal, M.; Ramli. Development of sensors based on giant magnetoresistance material. *Procedia Eng.* **2012**, *32*, 60–68. [CrossRef]

37. Baumer Hüebner HMC 16 (M) + HEAG 158 V (M) Incremental Encoder. Available online: www.baumerhuebner.com/uploads/media/hmc16heag158.pdf (accessed on 21 June 2017).

38. Dewetron Ltd.: DEWE-ORION-0816-5Mx Integrated Data Acquisition AD Board. Available online: www.dewetron.com/product/dewe-orion-0816-5mx (accessed on 21 June 2017).

39. Texas Instruments Inc., Texas Instruments PGA411-Q1 Evaluation Module. Available online: http://www.ti.com/tool/PGA411Q1EVM (accessed on 13 August 2019).

40. Datlinger, C.; Hirz, M. An advanced electrical approach to gauge rotor position sensors for optimizing electric drive train systems. *J. Adv. Technol. Eng. Res.* **2019**, *5*. [CrossRef]

Article

Hall-Sensor-Based Position Detection for Quick Reversal of Speed Control in a BLDC Motor Drive System for Industrial Applications

Mohanraj Nandakumar [1,*], Sankaran Ramalingam [1], Subashini Nallusamy [1] and Shriram Srinivasarangan Rangarajan [1,2,*]

[1] Department of EEE, SASTRA Deemed to be University, Thanjavur, Tamil Nadu 613401, India; rs@eee.sastra.ac.in (S.R.); nmnsi@eee.sastra.ac.in (S.N.)

[2] Department of Electrical and computer Engineering, Clemson University, Clemson, SC 29634, USA

* Correspondence: snehammohan@eee.sastra.edu (M.N.); shriras@g.clemson.edu (S.S.R.)

Received: 17 June 2020; Accepted: 14 July 2020; Published: 16 July 2020

Abstract: This paper proposes the novel idea of eliminating the front-end converters used indirect current (DC) bus voltage variation, thereby allowing for control of the speed of the brushless direct current (BLDC) motors in the two-quadrant operation of a permanent magnet brushless direct current (PMBLDC) motor, which is required for multiple bi-directional hot roughing steel rolling mills. The first phase of steel rolling, the manufacture of plates, strips etc., using hot slabs from the continuous casting stage, is carried out for thickness reduction, before the same is sent to the finishing mill for further mechanical processing. The hot roughing process involves applying high, compressive pressure, using a hydraulically operated mechanism, through a pair of backup rolls and work rolls for rolling. Overall, the processes consist of multiple passes of forward and reverse rolling at increasing roll speeds. The rolling process was modeled, taking into account parameters like roller dimensions, angle and length of contact, and rolling force, at various temperatures, using actual data obtained from a steel mill. From this data, speed and torque profiles at the motor shaft, covering the entire rolling process, were created. A profile-based feedback controller is proposed for setting the six-pulse inverter frequency and parameters of the pulse width modulated (PWM) waveform for current control, based on Hall sensor position, and the same is implemented for closed loop operation of the brushless direct current motor drive system. The performance enhancement of the two different controllers was also evaluated, during the rolling of 1005 hot rolled (HR) steel, and was taken into consideration in the research analysis. The entire process was simulated in the MATLAB/Simulink platform, and the results verify the suitability of an entire-drive system for industrial steel rolling applications.

Keywords: Hall sensors; brushless direct current motor drive system; power electronics; industrial application

---

1. Introduction

Steel manufacturing is a highly power-intensive process, wherein about 8% of the total energy consumption is absorbed in hot steel rolling. During the beginning stage of the rough rolling mill, steel slabs from the continuous casting stage are rolled to produce products like plates or strips. Subsequent stages use the above intermediate products for production of thin sheets and similar items. In order to eliminate oxide layers, the hot steel slab specimen has to undergo surface cleaning by means of scrubbing. With the deployment of high-powered electric motors using a gear train arrangement, the set of work and backup rolls are initiated. The physical parameters of the material, like the dimensions of the roller, the work temperature, and the metallurgical properties, are considered during the rolling process, while the torque and rolling force need to be calculated during the multi-pass operation. Profiles involving forward–reverse rolling based on the above calculations facilitate fixing the speed of the shaft and the torque that constitute the operating points of the motor that is driving.

The choice of electric motors includes induction-based motors, synchronous motors, and direct current (DC) motors. Among the various motors, a specially designed motor called the Brushless direct current motor is preferred, based on inherent features like low maintenance, owing to the absence of brushes, no sparking, and less wear and tear.

The modeling of the closed loop drive system, along with a controller using speed and current feedback, integrated with equations governing the steel rolling process, is presented in this paper. A Simulink [1] schematic of the drive system was created to meet the requirements of the work profile. The performance characteristics of the drive motor were ascertained for ten passes of steel rolling. The characteristics involving the back EMF (electromotive force), such as the current of the stator, the speed of the shaft, and the electromagnetic torque are clearly manifested in this paper by means of simulation.

2. Model of a Rolling Mill

Figure 1 shows a simplified schematic of a typical rough rolling mill for handling hot steel slabs for thickness reduction. The Brushless Direct Current (BLDC) motor [2] drive operates the system of backup rolls, the pair of work rolls, and a gear train [3] mechanism. The above drive system is made up of a four-pole BLDC motor [4] and a six-pulse inverter fed from a DC source. The control configuration senses the speed error signal, which is processed through a Proportional plus Integral controller to generate reference torque. The actual stator current is compared with the generated reference current for producing logic and timing signals required for gate control of the six-pulse IGBT inverter.

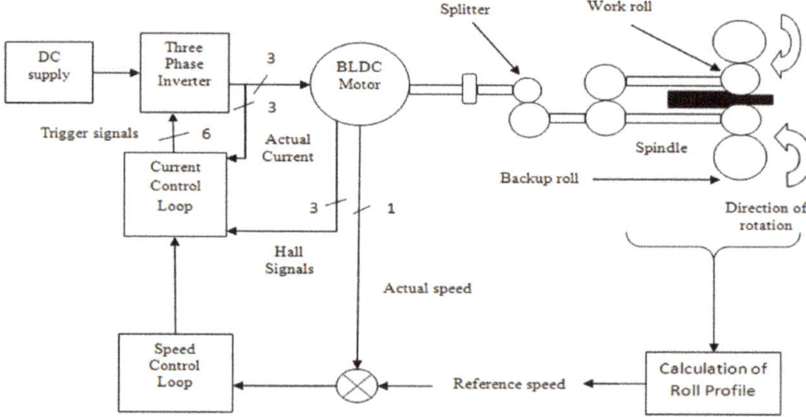

Figure 1. Direct current (DC) motor drive system coupled to a hot rough rolling mill.

2.1. Formulae Governing the Hot Rolling Process

The deformation due to the rolling of a hot slab of steel (which is known by the term draft D) leading to the reduction of thickness of the material [5,6], under specified operating conditions, is governed by the following equation:

$$D = T_I - T_F \tag{1}$$

where D = draft in mm; T_I = starting thickness in mm; and T_F = final thickness in mm.

The work piece under consideration is rolled based on thickness, with ε being the true strain whose equation, before and after the rolling of the work material, is given by

$$\varepsilon = \ln\!\left(\frac{T_I}{T_F}\right).$$ (2)

The following equation is used to compute the average flow stress $(\overline{Y_f})$ on the work material during flat rolling, where K is the strength coefficient of the material in MPa, and n is the strain-hardening exponent obtained as in Figures 2 and 3 for 1005 Hot Rolled steel material:

$$\overline{Y_f} = \frac{K\varepsilon^n}{1+n}$$ (3)

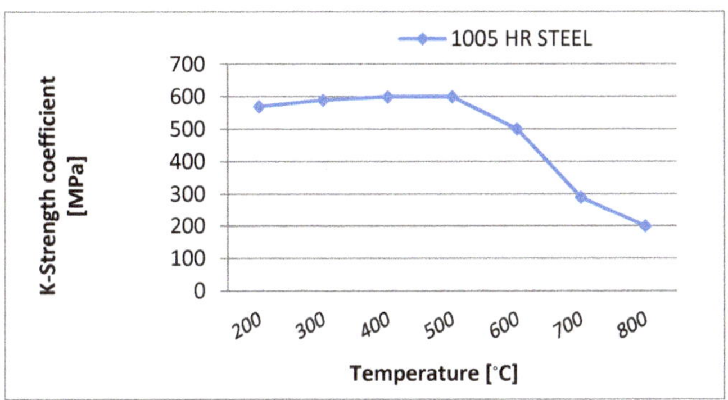

Figure 2. Coefficient versus temperature for a 1005 hot rolled steel specimen.

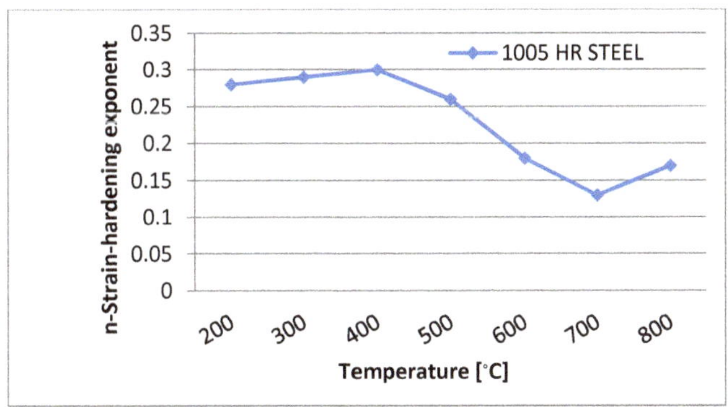

Figure 3. Exponent versus temperature for a 1005 hot rolled steel specimen.

The contact length (C_L) is given by

$$C_L = \left[R_R\;(T_I - T_F)\right]^{0.5}$$ (4)

where R_R = Radius of the work roll in mm.

Rolling force (F) is given by

$$F = \overline{Y_f}w\,C_L$$ (5)

where w = steel slab width in mm.

Load torque (T_L) is given by which is related to force as follows:

$$T_L = FC_L \tag{6}$$

2.2. Rolling Process Parameters

For the low carbon specimen of 1005 HR steel [7,8], the metallurgical data and parameters [9] were obtained from the standard ASM Metals Handbook and shown in Table 1. Figure 2 shows the variation of the strength coefficient (K) and Figure 3 the strain hardening coefficient (n) with respect to temperature in °C.

Table 1. Mechanical properties of 1005 HR Steel.

Properties	1005 HR Steel
Density (gr/cm^3)	7.872
Modulus of Elasticity (GPa)	210
Yield strength (MPa)	>280
Poisson's ratio	0.27 to 0.3
Break Elongation (%)	28 (in 80 mm)
Hardness, Vickers	105
Tensile strength (MPa)	325

The actual rolling of the steel slab [10] was carried out in a forward–reverse manner, cyclically, over 10 passes with successive thickness reductions and increases in roll speed, as shown in Table 2.

Table 2. Rolling parameters for ten passes.

Pass	Temperature, °C	K, MPa	n
1	800	200	0.18
2	795	205	0.177
3	790	210	0.174
4	785	215	0.171
5	780	220	0.168
6	775	225	0.165
7	770	230	0.162
8	765	235	0.159
9	760	240	0.156
10	750	245	0.153

3. Creation of Speed and Torque Values

The speed and torque values were deduced from the equations formed from the hot rolling process by employing various values from the graphs of the strength coefficient versus temperature, and the strain hardening exponent versus temperature. It is desirable for the thickness reduction to be performed at a higher speed, provided that the operation is quick and utmost care is taken, as a slight reduction in temperature causes hardening of the hot specimen. Hence, several passes are suggested for rolling operations and the direction of rotation of the work roll reverses at the end of the work specimen, which also makes the steel slab move in the reverse direction. This action is repeated several times to realize multiple-roll operation. Alternate positive and negative values were acquired from both the torque and drive speed, corresponding to the 1st and 3rd quadrant operation of the BLDC motor [11–13]. Table 3 indicates the values of roller speed, motor speed, shaft torque, and thickness reduction for the total time duration of 50 s.

Table 3. Profile values of speed and torque (operation: 10 passes).

Roller Speed (RPM)	Motor Speed (RPM)	Shaft Torque (N-m)	Reduced Thickness (cm)	Time Duration (s)
20	800	423	9.7	5
−20	−800	−432	9.4	5
22.5	900	449	9.1	5
−22.5	−900	−466	8.8	5
25	1000	483	8.5	5
−25	−1000	−500	8.2	5
27.5	1100	518	7.9	5
−27.5	−1100	−536	7.6	5
30	1200	555	7.3	5
−30	−1200	−575	7.0	5

4. Feedback Controller Design

The profile values from Table 3 are the specific values of torque and speed for the steel rolling operation of the mill. The BLDC drive should confine its operation to the required stator current and inverter frequency. This was effected by means of a suitable controller, and a suitable inner and outer control loop. A PI controller was used in the loops. Speed error was processed in the outer loop of the controller, which produced a reference for the torque, in order to generate the three phase currents that were used as a further reference in the control mechanism as follows:

$$\omega_e(t) = \omega^*(t) - \omega_{act}(t) \tag{7}$$

$$T^*(t) = K_p \omega_e(t) + K_i \int_0^t \omega_e(t) \tag{8}$$

$$I^*(t) = \frac{T^*(t)}{K_t} \tag{9}$$

$$I_a^*(t) = I^*(t)0° \tag{10}$$

$$I_b^*(t) = I^*(t) - 120° \tag{11}$$

$$I_c^*(t) = I^*(t) - 240°. \tag{12}$$

The actual values of the phase currents belonging to the motor were compared with the above generated reference currents, involving a bang-bang control for limiting the signal within the hysteresis band, for further generation of six gating pulses for the three-phase inverter [2]. The first and third quadrant operation of the brushless DC motor, for both forward and reverse operation was governed by profile values of speed and torque, with polarity reversal in the rolling operation. Suitable parameters for the controller were chosen for the smooth operation of the entire system.

5. Schematic of the Modeled System

The MATLAB/Simulink simulation platform was employed to carry out the modeling of the hot roughing steel rolling mill, as depicted in Figure 4. Along with the various parameters of the process model, torque–speed profile values at the motor shaft over ten passes, as shown in Table 3, were specified. These profile data [14–16] correspond to the operating conditions during multi-cyclic, forward–reverse runs of rolling being carried out [17,18].

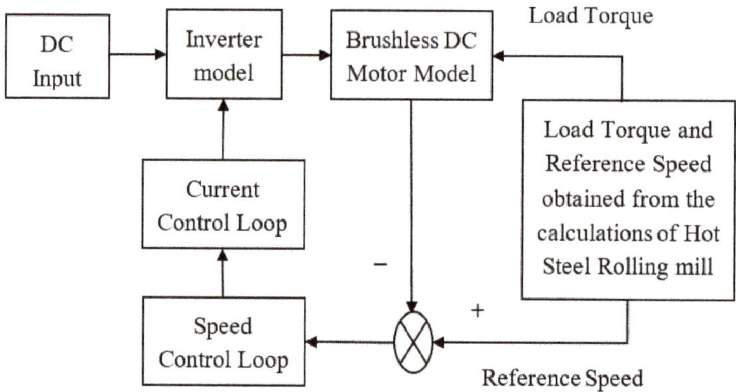

Figure 4. The entire closed loop system in Simulink.

Simulation Results

A simulation run was captured for a period of 50 s, portraying the stator current, back EMF, electromagnetic torque, and speed of the rotor.

Figure 5 shows the simulation results of the rolling operation over an entire period of 50 s for ten passes, depicting the first and third quadrant operation of the BLDC drive system following the speed profile.

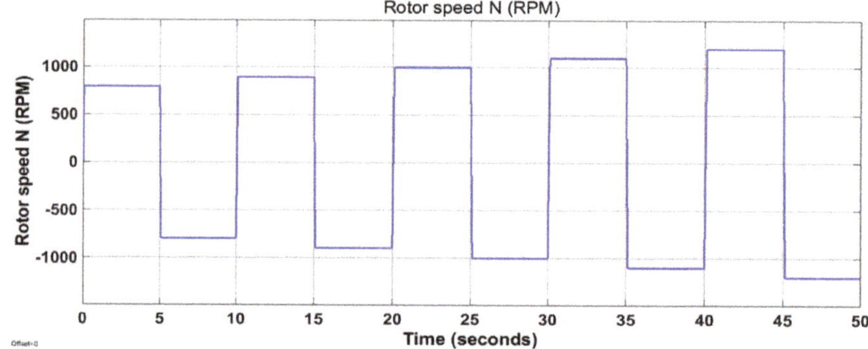

Figure 5. Speed over a period of 50 s.

In Figure 6, the back-EMF waveform is shown for 50 s over ten passes, depicting the first and third quadrant operation of the BLDC drive system. Its expanded view, which is trapezoidal, is shown over a particular period in Figure 7.

The waveform shown in Figure 8 depicts the simulated output of the electromagnetic torque over an entire period of 50 s and for the ten passes required for the first and third quadrant operation of the BLDC drive system [19,20] following the torque profile.

The efficacious nature of the controller is clearly demonstrated in the rapid reversal involved, with respect to the torque and speed characteristics, needed during the rolling process, as shown in Figures 5 and 8. As the rolling process progresses, the quick response of the drive system for bidirectional rolling clearly presents a series of stepped increases of all variables over successive periods.

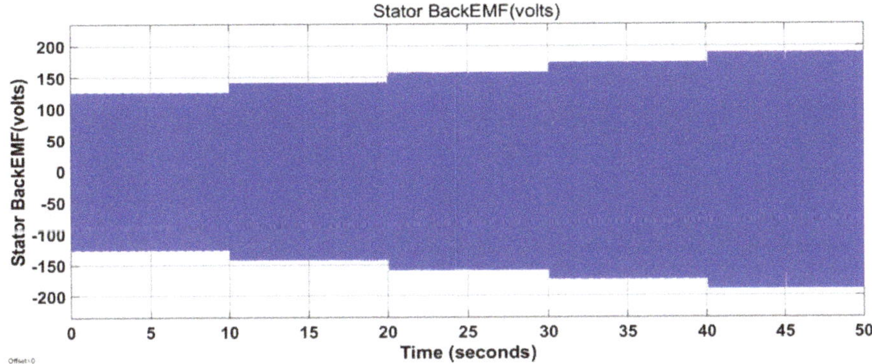

Figure 6. Back-Electro Motive Force waveform in phase A over 50 s.

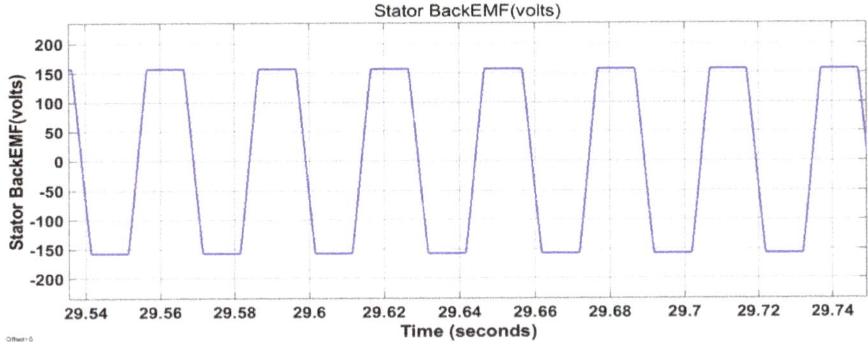

Figure 7. View of the motor's back electromotive force waveform.

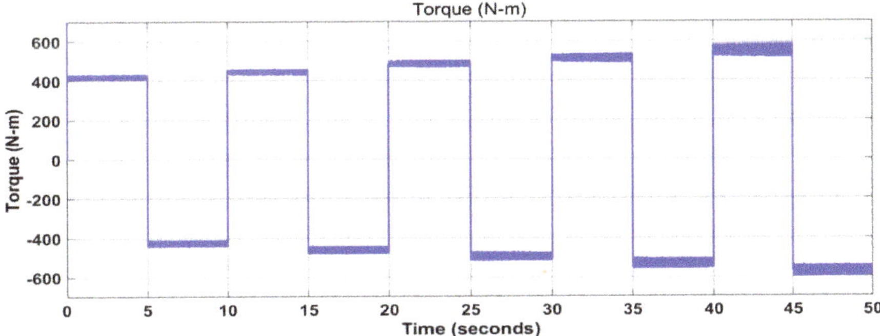

Figure 8. Torque over 50 s.

The stator current waveform, shown in Figure 9, depicts the simulation results for an entire period of 50 s, over ten passes, and its expanded view is also shown in Figure 10, over a particular period, which is rectangular in shape.

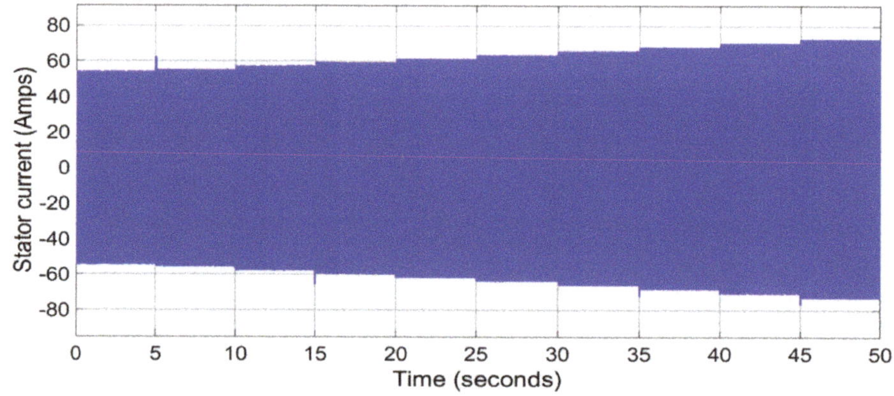

Figure 9. Current waveform over 50 s.

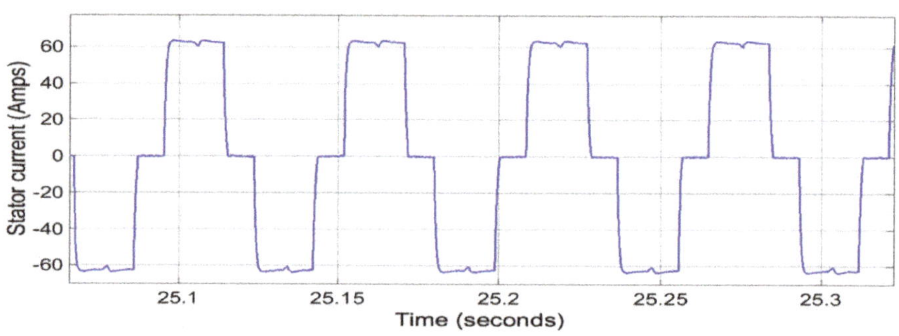

Figure 10. View of the stator's current waveform.

A profile-based feedback controller is proposed for setting the six-pulse inverter frequency and parameters of the PWM waveform for current control based on Hall sensor position, and the same is implemented for the closed loop operation of the brushless direct current motor drive system. The speed response of the two different controllers, controller 1 and controller 2, are shown in Figure 11, and its expanded view depicts that controller 2 provided a smoother response compared with controller 1. In contrast, controller 1 has a higher amount of overshoot, which is quite evident in the expanded view. Moreover, controller 1 and controller 2 had different gain parameters selected for obtaining better performance.

Table 4 shows the ratings and parameters of the BLDC motor used in this paper.

Table 4. Brushless direct current (BLDC) motor parameters.

Back EMF Motor Rating	Trapezoidal62 HP
Voltage	3-phase, 440 V
Rated Speed	1500 RPM
Resistance (Phase)	0.21 Ω
Inductance (Phase)	8.5 mH
Number of pole pairs	2

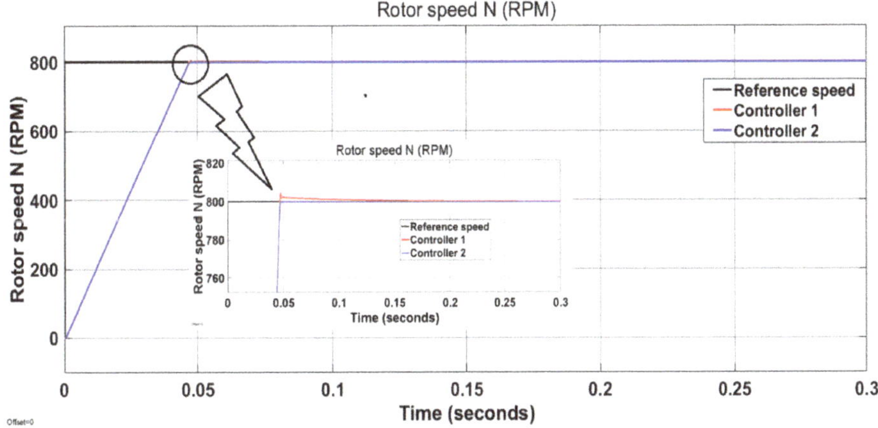

Figure 11. Rotor speed of the two different controllers following the speed profile, and its expanded view.

6. Hardware Setup

Figure 12 illustrates the hardware setup of the drive system employed for the research, in a laboratory-based environment. It consisted of an IGBT-based intelligent power module (IPM), the brushless direct current (BLDC) motor belt coupled to a direct current (DC) generator, and the load arrangements. For the drive system to operate in a closed-loop mechanism, an ARM development board with speed sensors and interfaces was employed. A three-phase alternating current (AC) supply fed the bridge rectifier for obtaining the DC supply to drive the BLDC motor. The generated reference parameters were compared with the actual parameters, as discussed earlier, by means of a suitable controller mechanism. A CPU that was interfaced with the BLDC motor drive system was employed for monitoring the actual and reference speeds. This can be clearly observed from the screen display in Figure 12.

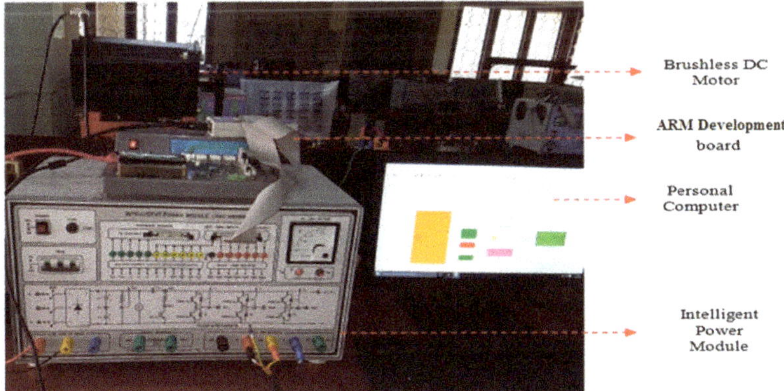

Figure 12. Brushless DC motor drive hardware arrangements.

The actual rotor speed of the BLDC motor, following the set or reference speed is shown in Figure 13. The reference speed ranges from 900 RPM to 1300 RPM, with incremental steps of 100 RPM. A red-colored line indicates the reference or set speed, which was set by the user and accordingly the motor speed tracked the set speed indicated in a blue color.

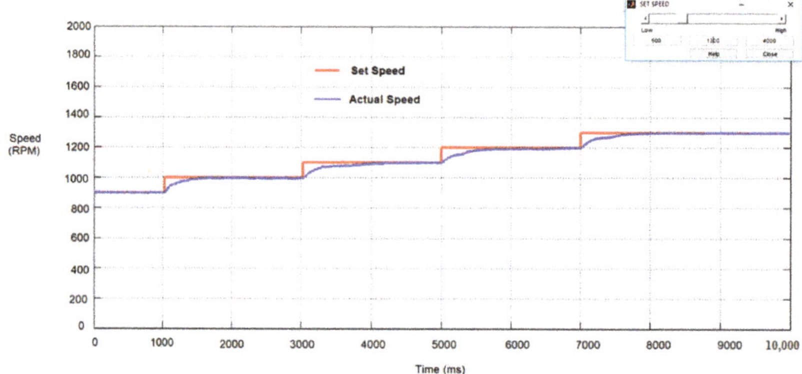

Figure 13. Actual speed of brushless DC motor tracks increased values of set speed viewed in a Graphical User Interface (GUI).

The forward motion of the Brushless DC motor tracked the set or reference speed with decreasing values spanning from 1300 RPM to 900 RPM in steps of 100 RPM, as shown in Figure 14.

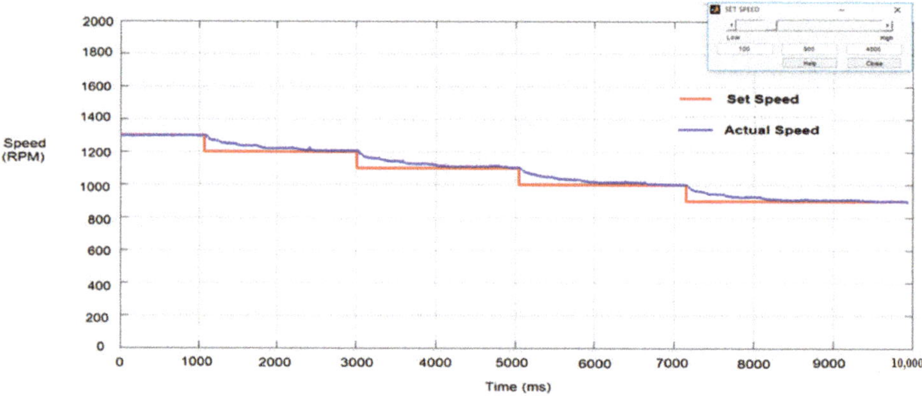

Figure 14. Actual speed of the brushless DC motor tracked decreasing values of set speed, as viewed in the GUI.

7. Conclusions

A total system for hot rough steel rolling, consisting of a BLDC motor drive, coupled to a set of work rolls and backup rolls, along with a closed loop controller, with a multi-loop configuration has been presented in this paper. The set of mathematical equations covering the steel rolling process, the BLDC motor model, and the controller model have been developed for simulation. The simulation results, corresponding to the first and third quadrant operation of the BLDC motor drive system, for a multi-pass steel roughing mill of a 1005 hot rolled steel feed specimen, have been obtained. A profile-based controller has been successfully employed for the smooth operation of the entire system. Furthermore, the performance of the drive system proved it to be an efficient mechanism for fast hot rolling. Hence, it is concluded that a BLDC-motor-based drive system, as presented in this paper, can serve as a powerful replacement for the traditional induction motor drive systems that have been employed for steel rolling, thereby eliminating the need for front-end converters, which in turn eliminates the cost of equipping the components used in them, and, thus, also improving the conversion efficiency.

Author Contributions: M.N. performed the work under the guidance and direction of S.R., S.N. and S.S.R. All authors analyzed the data and contributed towards the paper. M.N. wrote the paper and executed the experiments under the guidance and supervision of, and with suggestions from, S.R., S.N. and S.S.R. All authors also have contributed towards writing and editing the paper. All authors have read and agreed to the published version of the manuscript.

Funding: This research received no external funding.

Acknowledgments: The authors thank the SASTRA Deemed University for providing all of the facilities in the research lab of Power Electronics & Drives that were used to carry out the research.

Conflicts of Interest: The authors declare no conflict of interest.

References

1. Math Works. The Math Works—MATLAB and Simulink for Technical Computing. 2016. Available online: http://www.mathworks.com (accessed on 7 September 2016).

2. Gamazo-Real, J.C.; Vázquez-Sánchez, E.; Gómez-Gil, J. Position and Speed Control of Brushless DC Motors Using Sensorless Techniques and Application Trends. *Sensors* **2010**, *10*, 6901–6947. [CrossRef] [PubMed]

3. Mahfouf, M.; Yang, Y.Y.; Gama, M.A.; Linkens, D.A. Roll speed and Roll Gap control with Neural Network Compensation. *ISIJ Int.* **2005**, *45*, 841–850. [CrossRef]

4. Masood, R.J.; Wang, D.B.; Ali, Z.A.; Khan, B. DDC Control Techniques for Three-Phase BLDC Motor Position Control. *Algorithms* **2017**, *10*, 110. [CrossRef]

5. Chattopadhyay, A.K. Alternating Current Drives in the Steel Industry, Advancements in the last 30 years. *Ind. Electron. Mag.* **2010**, *4*, 30–42. [CrossRef]

6. Kun, E.; Szemmelveisz, T. Energy efficiency enhancement in the Hot Rolling Mill. *Mater. Sci. Eng.* **2014**, *3*, 4350.

7. Rodríguez, J.R.; Pontt, J.; Newman, P.; Musalem, R.; Miranda, H.; Morán, L.; Alzamora, G. Technical Evaluation and Practical Experience of High-Power Grinding Mill Drives in Mining Applications. *IEEE Trans. Ind. Appl.* **2005**, *41*, 866–873. [CrossRef]

8. Lotter, U.; Schmitz, H.-P.; Zhang, L. Application of the Metallurgically Oriented Simulation System "TKS-StripCam" to predict the properties of Hot Strip Steels from the Rolling Conditions. *Adv. Eng. Mater.* **2002**, *4*, 207–213. [CrossRef]

9. Yamada, F.; Sekiguchi, K.; Tsugeno, M.; Anbe, Y.; Andoh, Y.; Forse, C.; Guernier, M.; Coleman, T. Hot Strip Mill Mathematical models and SetUp Calculation. *IEEE Trans. Ind. Appl.* **1991**, *27*, 131–139. [CrossRef]

10. Orcajo, G.A.; Rodriguez, J.; Ardura, P.; Cano, J.M.; Norniella, J.G.; Llera, R.; Cifrián, D. Dynamic Estimation of Electrical Demand in Hot Rolling Mills. In Proceedings of the IEEE Industry Applications Society Annual Meeting 2015, Dallas, TX, USA, 18–22 October 2015.

11. Gieras, J.F. *Permanent Magnet Motor Technology, Design and Applications*; Second Edition Revised and Expanded; CRC Press: Boca Raton, FL, USA, 2002.

12. Niknejad, P.; Agarwal, T.; Barzegaran, M.R. Utilizing Sequential Action Control Method in GaN-Based High-Speed Drive for BLDC Motor. *Machines* **2017**, *5*, 28. [CrossRef]

13. Miller, T.J.E. *Brushless Permanent Magnet & Reluctance Motor Drives*; Clarendon Press: Oxford, UK, 1989; Volume 2, pp. 192–199.

14. ASM International. *ASM Metals Handbook, Vol.14, Forming and Forging*; ASM International: Novelty, OH, USA, 1996; pp. 351–355.

15. Groover, M.P. *Fundamentals of Modern Manufacturing. Materials, Processes and Systems*, 4th ed.; John Wiley & Sons: Hoboken, NJ, USA, 2010; pp. 395–403.

16. *AISI/SAE Standard (American Iron and Steel Institute/Society of Automotive Engineers)*; SAE J403 1050; SAE: Warrendale, PA, USA, 2014.

17. Cunkas, M.; Aydoğdu, O. Realization of Fuzzy Logic Controlled Brushless DC Motor Drives Using Matlab/Simulink. *Math. Comput. Appl.* **2010**, *15*, 218–229.

18. Singh, B.; Bist, V. An improved power quality bridgeless Cuk converter fed BLDC motor drive for air conditioning system. *IET Power Electron.* **2013**, *6*, 902–913. [CrossRef]

19. Pillay, P.; Krishnan, R. Modeling, simulation, and analysis of permanent- magnet motor drives, part II: The brushless dc motor drive. *IEEE Trans. Ind. Appl.* **1989**, *IA-25*, 274–279. [CrossRef]
20. Nandakumar, M.; Ramalingam, S.; Nallusamy, S.; Rangarajan, S.S. Novel efficacious utilization of Fuzzy Logic Controller based Two-quadrant operation of PMBLDC Motor Drive System for Multi-Pass hot steel rolling Process. *Electronics* **2020**, *9*, 1008. [CrossRef]

Article

Research on Harmonic Torque Reduction Strategy for Integrated Electric Drive System in Pure Electric Vehicle

JIanJun Hu [1,*], Ying Yang [1], Meixia Jia [1], Yongjie Guan [1], Chunyun Fu [1] and Shuiping Liao [2]

[1] State Key Laboratory of Mechanical Transmissions, School of Automotive Engineering, Chongqing University, Chongqing 400044, China; yangying2020cqu@163.com (Y.Y.); jiameixia23@163.com (M.J.); guanyongjiecqu@163.com (Y.G.); fuchunyun@cqu.edu.cn (C.F.)

[2] Chongqing Changan Automobile Co., Ltd., Chongqing 400023, China; liaosp@changan.com.cn

* Correspondence: hujianjun@cqu.edu.cn

Received: 9 July 2020; Accepted: 30 July 2020; Published: 1 August 2020

Abstract: In order to study the influence of harmonic torque on the performance of the integrated electric drive system (permanent magnet synchronous motor + reducer gear pair) in a pure electric vehicle (PEV), the electromechanical coupling dynamic model of a PEV was established by considering the dead-time effect and voltage drop effect of an inverter and the nonlinear characteristics of the transmission system. Based on the model, the dynamic characteristics of an integrated electric drive system (IEDS) are studied, and the interaction between the mechanical system and electrical system is analyzed. On this basis, a harmonic torque reduction strategy for an IEDS is proposed in this paper. The simulation results show that the proposed strategy can effectively reduce the harmonic torque of the motor and reduce the speed fluctuation and dynamic load of the system components, which can improve the stability of the IEDS and prolong the life of the mechanical components.

Keywords: integrated electric drive system; permanent magnet synchronous motor; electromechanical coupling; harmonic torque reduction strategy

---

1. Introduction

1.1. Motivation

Electric drive assembly is the key component of a pure electric vehicle (PEV), which plays a key role in the driving experience. Recent studies have shown that the integration of a drive motor and reducer can greatly improve the efficiency, power density, and reliability of the drive system, and it can also effectively reduce the volume, weight, and production cost of a PEV [1,2]. The research and development of high speed, high efficiency, high power density, lightweight integrated motor-reducer assembly has become a hotspot. The general structure of an integrated electric drive system (IEDS) is shown in Figure 1, which mainly includes a permanent magnet synchronous motor (PMSM), motor control system, helical gear pair reducer, and output shaft. The IEDS includes the driving motor part and the mechanical transmission part.

1.2. Literature Review

Due to the dead-time effect of an inverter, the voltage drop effect, and the structural factors of the motor itself, electromagnetic torque generated by the motor contains harmonic torque [3], which will cause adverse effects on the mechanical transmission system. At the same time, due to the nonlinear factors such as gear time-varying stiffness and meshing error, the mechanical transmission system will also affect the stability of the electrical system and finally significantly affect the performance of the

PEV. Therefore, scholars have paid much attention to the research of this problem, mainly including two categories: (1) motor and controller research and (2) gear transmission system research.

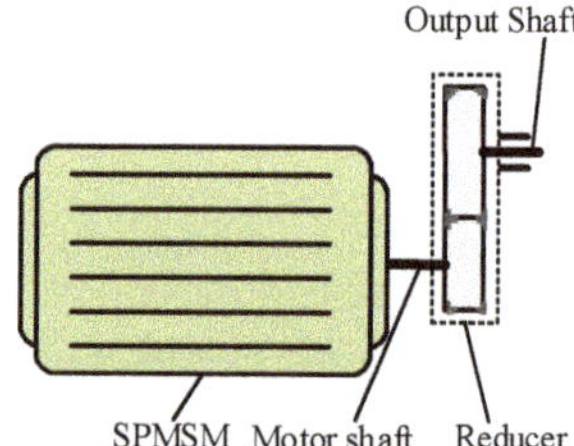

Figure 1. General structure diagram of an integrated electric drive system.

In order to reduce the motor torque ripple caused by the nonlinear characteristics of the inverter, scholars have done a lot of research on the optimization and improvement of the electrical system control strategy, mainly including the following two aspects: (1) inverter optimization research and (2) motor control strategy. From the aspect of inverter optimization, scholars have proposed various inverter topologies, which can greatly reduce the torque ripple [4–6], but this method has high requirements for development cost and it cannot be popularized in a short time. So, scholars have done a lot of research on the switching control strategy of an inverter such as hybrid space vector modulation (SVM) strategies in [7,8] to reduce torque ripple. In [9], a minimum root mean square (RMS) torque ripple-remote-state pulse-width modulation (MTR-RSPWM) technique was proposed for minimizing the RMS torque ripple under the reduced common-mode voltage condition of three-phase voltage source inverter-fed brushless alternating current motor drives. Besides, a new modulation method, modified trapezoidal modulation (MTM), was proposed for an inverter–PMSM drive in [10], which can increase torque and reduce torque ripple simultaneously. [11] studied the theoretical distortion index of a multilevel motor drive considering control sensitivity. By calculating the distortion index, the optimal equivalent carrier frequency to minimize the torque ripple was obtained. In addition to optimizing the inverter, some scholars studied the motor control strategy. In [12–14], the duty cycle of direct torque control (DTC) was optimized to reduce the torque and flux ripple at low switching frequency. Furthermore, in order to reduce the torque and flux ripple under all operating conditions, the three-level direct torque control (3L-DTC) based on constant switching frequency, which was suitable for low and constant switching frequency operation [15], and generalized direct torque control (GDTC) strategy, which was suitable for any voltage level inverter [16], were studied. Refs [17,18] proposed a seven-level torque comparator and a multi band torque hysteresis controller respectively, in which the voltage vectors were optimized to reduce the torque ripple at different speeds. Some scholars also adopted predictive torque control to reduce torque, flux ripple, and switching frequency [19]. In terms of motors for robots, the modified distributed control framework and on-line tuning fuzzy proportional-derivative (PD) controller of controlling of 5 degree-of-freedom (DOF) robot manipulators based on the equivalent errors method were used in [20,21] to improve the dynamic performance of robots under large disturbance and high frequency. In recent years, the harmonic injection method [22–26] has attracted the attention of many scholars. By adding a harmonic current feedback loop and harmonic voltage compensation loop to the traditional motor double closed-loop control system, and then using the method of injecting harmonic voltage [22] and harmonic current [23–26], the electromagnetic torque ripple of the motor was significantly reduced.

From single-stage gear transmission research to multi-stage gear transmission research, the field of gear system dynamics has formed a relatively mature theoretical system with the joint efforts of scholars [27]. Many scholars have studied the influence law of external load [28], time-varying meshing stiffness, backlash, tooth surface wear [29,30], meshing frequency, eccentricity [31], error, position error, bearing stiffness, and other internal and external factors on the gear transmission system, and they have conducted in-depth research on the nonlinear dynamic characteristics of the gear system. Among them, gear noise and dynamic load caused by vibration are areas of major concern [32]. Due to the rise of IEDS, scholars have done some research on the electromechanical coupling characteristics of IEDS in recent years. A kind of trajectory-based stability preserving dimension reduction (TSPDR) methodology was proposed to investigate the nonlinear dynamic characteristics of the gear–motor system in the literature [33], which revealed the relationship between the stability and resonance of the gear motor system combined with modal analysis. Combining the nonlinear permeance network model of a squirrel cage induction motor (IM) and the bending torsion coupling dynamic model of a planetary gear rotor system and considering external excitation such as load mutation and voltage transient, [34] analyzed the dynamic characteristics of electromechanical coupling of a motor–gear system, and the author further provided an effective method for detecting the asymmetric voltage sag condition [35]. In [36], nonlinear damping characteristics, the time-varying meshing stiffness of gears, the wheel rail contact relationship, and other nonlinear factors were considered to reveal the dynamic performance of the motor car–track system model.

From the above analysis, it can be seen that scholars all over the world have conducted in-depth research on torque ripple within motor systems and the dynamic load of gear transmission systems, but little attention has been paid to the influence of coupling between a mechanical system and an electrical system on the dynamic performance of an IEDS. A small number of electromechanical coupling studies only focus on the electromechanical coupling characteristics of the IEDS, rarely conducting in-depth study on the overall torque ripple of the IEDS, and they do not propose methods to improve the system performance, which limits the ride comfort and NVH (noise, vibration, harshness) performance improvement of a PEV equipped with an IEDS.

1.3. Original Contributions of This Paper

In order to explore the influence of motor harmonic torque on the stability and dynamic load of an IEDS, the electromechanical coupling dynamic model of a PEV equipped with an IEDS is established in this paper. The simulation results show that the mechanical nonlinear factors such as time-varying meshing stiffness and meshing error will lead to the fluctuation of the motor shaft speed. The dead-time effect and voltage drop effect of the inverter will cause the harmonic torque of the motor and increase the dynamic load of the mechanical system. To reduce the influence of motor torque on the IEDS, a harmonic torque reduction strategy is proposed in this paper. Harmonic voltage is injected into the traditional field-oriented control (FOC) to reduce the harmonic torque of the IEDS, and it can ensure the system's stability and a response equal to the traditional FOC Furthermore, the simulation results show that the harmonic torque reduction strategy proposed in this paper can effectively reduce the speed fluctuation and dynamic load of the system components and improve the stability of the IEDS.

2. Dynamics Modeling of PEV Equipped with IEDS

The structure of a PEV equipped with an IEDS is shown in Figure 2. Its power transmission system mainly includes an IEDS, final drive, differential, axle shaft, and wheel. In order to study the influence of harmonic torque on the working performance of an IEDS, the transient models of the electrical system and mechanical transmission system of an IEDS are established in this paper. On this basis, the dynamic model of the PEV equipped with an IEDS is established.

2.1. Harmonic Torque Mathematical Model of PMSM

Combined with Clarke transform and Park transform, the mathematical model of a three-phase PMSM in a d-q rotating coordinate system can be obtained. The stator voltage equation is as follows:

$$\begin{cases} u_d = Ri_d + L_d \frac{d}{dt}i_d - \omega_e L_s i_q \\ u_q = Ri_q + L_q \frac{d}{dt}i_q + \omega_e(L_s i_d + \psi_f) \end{cases} \tag{1}$$

where u_d and u_q are the voltages of the d-axis and q-axis respectively; i_d and i_q are the currents of the d-axis and q-axis respectively; L_d and L_q are the stator inductance of the d-axis and q-axis respectively; R is the stator resistance of the motor; and ω_e is the rotor angular velocity.

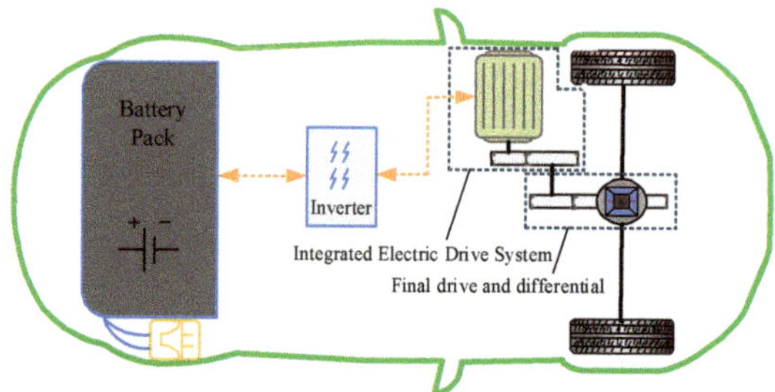

Figure 2. Structure diagram of a pure electric vehicle (PEV) power transmission system equipped with an integrated electric drive system (IEDS).

The torque equation of a surface-mounted permanent magnet synchronous motor (SPMSM) is as follows:

$$T_m = \frac{3}{2}P_n i_q \left[i_d(L_d - L_q) + \psi_f \right]. \tag{2}$$

Stated thus, through the transformation from an a-b-c coordinate system to a d-q rotating coordinate system, the electromagnetic torque of PMSM is only related to its q-axis current. This means that by controlling the q-axis current of the motor, the electromagnetic torque control of the motor can be realized. The parameters of PMSM used in this paper are shown in Table 1.

Table 1. Parameters of the permanent magnet synchronous motor (PMSM) in this paper.

Parameter	Value	Unit
Peak power	60	kW
Peak torque	250	Nm
Maximum speed	7000	rpm
Base speed	2292	rpm
DC voltage	650	V
Stator resistance	0.153	Ω
Stator inductance	1.8	mH
Permanent magnet flux linkage ψ_f	0.2778	Wb
Pole number P_n	4	-
Phase current limit	150	A

The inverter circuit of the PMSM for an electric vehicle is shown in Figure 3. Sa, Sb, Sc, S'a, S'b, and S'c are defined as the switch states of the inverter. The working principle of the inverter is as follows: when Sa/Sb/Sc is in 1 state, S'a/S'b/S'c is in 0 state. At this time, the three switching devices of the upper bridge arm of the inverter circuit are on, while the three switching devices of the lower bridge arm of the inverter circuit are turned off.

Figure 4 shows a structural diagram of the A-phase bridge arm of the inverter, in which the switch device is an insulated gate bipolar transistor (IGBT), and the motor stator winding is equivalent to a resistance inductance circuit, forming a series structure with the voltage source. As shown in Figure 4, when the inverter switching device Sa is on, S'a must be turned off. If the speed of Sa is turning on faster than that of S'a is turning off, the bridge arm will be short-circuited. Therefore, in order to ensure the safe operation of the inverter, dead time T_{dead} should be added to the process of switching between the two switches. During the dead time, it can be equivalent to the switch of the upper or lower leg of the inverter being on. There is a deviation between the actual switch signal and the ideal switch driving signal, which leads to the deviation between the actual output voltage and the demand voltage.

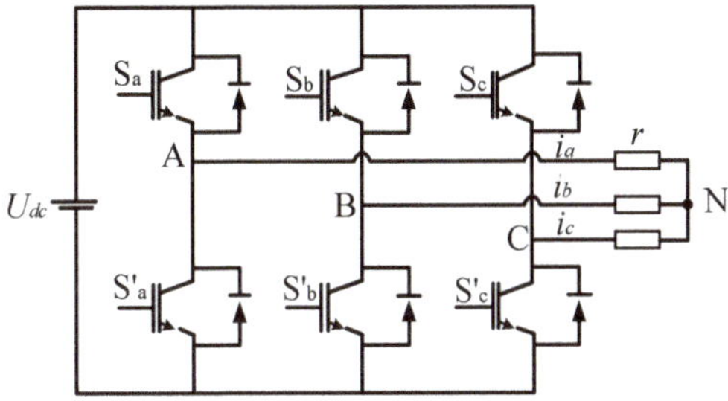

Figure 3. Inverter circuit of the PMSM.

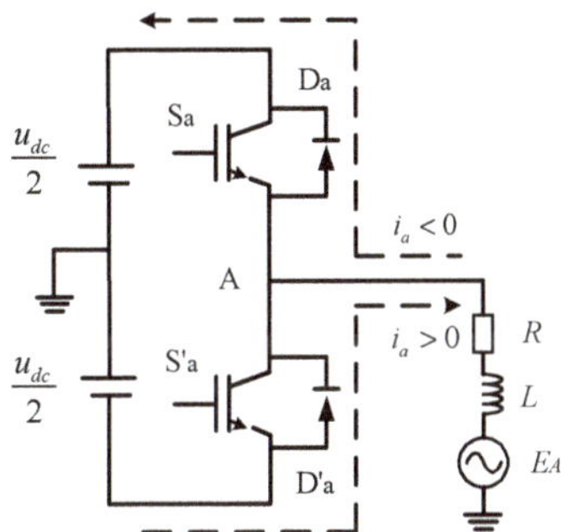

Figure 4. Structure diagram of the A-phase bridge arm of the inverter.

The dead time in the actual output voltage waveform of the inverter can be expressed as follows:

$$T_d = t_d + t_{on} - t_{off} \tag{3}$$

where t_{on} and t_{off} represent the on time and off time of the IGBT respectively; while t_d represents the preset dead time in the switch driving signal.

The above content is the error voltage brought by the dead time of the inverter to the system. The switch tube voltage drop and freewheeling diode conduction voltage drop will also bring an error voltage to the system. As shown in Figure 5, the current direction and path in the inverter bridge arm are different under different switching states. The error voltage U_{err} considering dead time and tube voltage drop is obtained by making a difference, as shown in Figure 5. v_t is the conduction voltage drop of the switch tube, and v_d is the conduction voltage drop of the freewheeling diode. Supposing that T_{s1} and T_{s2} are the real turn-on times of S_a and S'_a of the upper and lower switches in the A-phase of the inverter respectively during a pulse width modulation (PWM) cycle, the average error voltage Δu can be obtained by averaging the error voltage in a period of time according to the area equivalent principle. After averaging the error voltage over a period of time, a square wave signal can be obtained, as shown in Figure 6.

Through Fourier decomposition of the square wave signal, the mathematical expression of the average error voltage of the inverter can be obtained as follows:

$$u_{err} = \frac{4\Delta u}{\pi}\left(\sin \omega t + \frac{1}{3}\sin 3\omega t + \frac{1}{5}\sin 5\omega t + \frac{1}{7}\sin 7\omega t + \cdots\right). \tag{4}$$

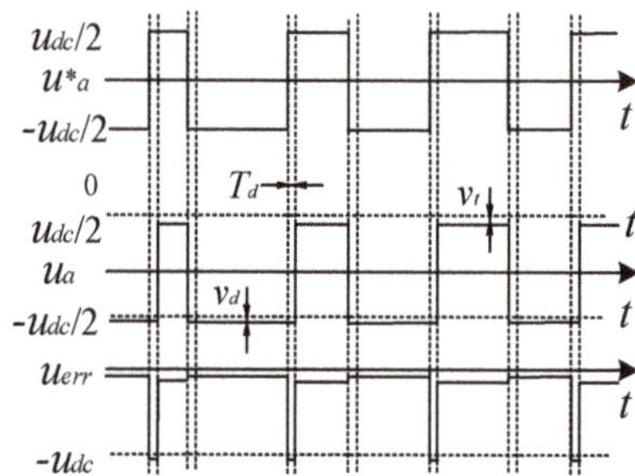

Figure 5. Voltage waveform of the A-phase bridge arm.

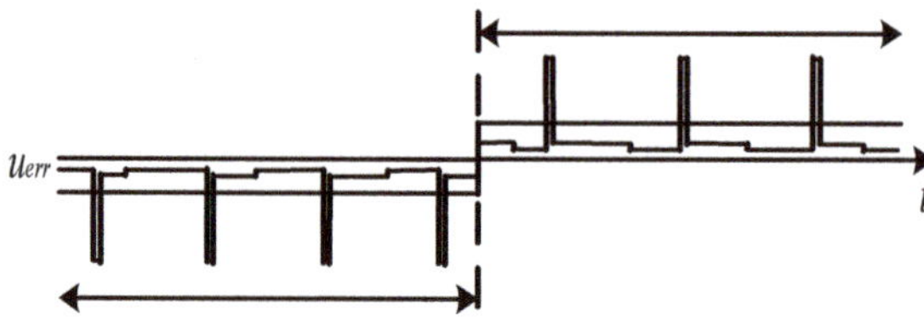

Figure 6. Error voltage waveform generated by inverter nonlinear characteristics.

From the results of Fourier decomposition, it can be seen that the nonlinear characteristics such as the dead-time effect and voltage drop of the inverter will introduce a large number of harmonics into the motor voltage, then causing the motor to produce a harmonic current. Since the stator winding of the PMSM mostly adopts star connection mode, the 3rd and integral multiple harmonic components of the stator current can not be circulated. Therefore, the actual system mainly includes 5th, 7th, 11th, and 13th harmonic components [37]. The specific parameters of the inverter used in this paper are shown in Table 2.

Table 2. Parameters of the inverter in this paper. IGBT: insulated gate bipolar transistor.

Parameter	Valve	Unit
DC bus voltage u_{dc}	650	V
Modulation carrier period T_{PWM}	100	μs
IGBT turn on time t_{on}	1	μs
Voltage drop of IGBT switch tube u_t	3	V
Modulation carrier frequency f_{PWM}	10	kHz
Dead time t_d	4	μs
IGBT turn off time t_{off}	2	μs
Conduction voltage drop of freewheeling diode u_d	2	V

In the a-b-c coordinate system, the rotation speed of the 5th harmonic voltage vector in the motor stator winding is -5ω, and the rotation speed of the 7th harmonic voltage vector in the motor stator winding is 7ω [38]. Therefore, the stator voltage equation in the three-phase static coordinate system can be expressed as follows:

$$\begin{cases} u_a = u_1 \sin(\omega t + \theta_1) + u_5 \sin(-5\omega t + \theta_5) + u_7 \sin(7\omega t + \theta_7) + \cdots \\ u_b = u_1 \sin\left(\omega t + \theta_1 - \frac{2}{3}\pi\right) + u_5 \sin\left(-5\omega t + \theta_5 - \frac{2}{3}\pi\right) + u_7 \sin\left(7\omega t + \theta_7 - \frac{2}{3}\pi\right) + \cdots \\ u_c = u_1 \sin\left(\omega t + \theta_1 + \frac{2}{3}\pi\right) + u_5 \sin\left(-5\omega t + \theta_5 + \frac{2}{3}\pi\right) + u_7 \sin\left(7\omega t + \theta_7 + \frac{2}{3}\pi\right) + \cdots \end{cases} \tag{5}$$

where θ_1 is the initial phase angle of the stator fundamental voltage; θ_5 and θ_7 are the initial phase angles of the 5th and 7th harmonic voltages of the stator, respectively; u_1 is the amplitude of the stator fundamental voltage; and u_5 and u_7 are the amplitude of the 5th and 7th harmonic voltages of the stator, respectively.

Adopting the amplitude invariable transformation, the stator voltage equation expressed in Equation (5) is transformed into a d-q rotating coordinate system, and the converted stator voltage equation of the d-axis and q-axis is obtained as follows:

$$\begin{cases} u_d = u_{d1} + u_5 \cos(-6\omega t + \theta_5) + u_7 \cos(6\omega t + \theta_7) \\ u_q = u_{q1} + u_5 \sin(-6\omega t + \theta_5) + u_7 \sin(6\omega t + \theta_7) \end{cases} \tag{6}$$

where u_{d1} and u_{q1} are the d-axis and q-axis components of fundamental voltage in the d-q rotating coordinate system, respectively.

It can be seen from Equation (6) that in the d-q rotating coordinate system, the 5th and 7th harmonic components on the stator of the motor in the original three-phase static coordinate system are shown as the 6th harmonic component, with the rotation directions of them being opposite. If there is harmonic voltage in the stator voltage, there will be a corresponding harmonic current. In the same way, transforming the three-phase current of the motor into the d-q rotating coordinate system by means of amplitude invariance, the stator current equation in the d-q rotating coordinate system is obtained as follows:

$$\begin{cases} i_d = i_{d1} + i_5 \cos(-6\omega t + \theta'_5) + i_7 \cos(6\omega t + \theta'_7) \\ i_q = i_{q1} + i_5 \sin(-6\omega t + \theta'_5) + i_7 \sin(6\omega t + \theta'_7) \end{cases} \tag{7}$$

where i_{d1} and i_{q1} represent the d-axis and q-axis components of the stator fundamental current in the d-q rotating coordinate system; i_5 and θ_5 represent the amplitude and initial phase angle of the 5th harmonic current in the d-q rotating coordinate system; and i_7 and θ_7 represent the amplitude and initial phase angle of the 7th harmonic current of the motor in the d-q rotating coordinate system.

Ignoring the harmonic caused by the motor itself, there is no harmonic flux component in the permanent magnet flux linkage. By substituting Equation (7) into Equation (1), the stator voltage equation with harmonic component is obtained as follows:

$$\begin{cases} u_d = & -\omega_e L_s i_{q1} + R i_{d1} + 5\omega_e L_s i_5 \sin(-6\omega t + \theta'_5) + R i_5 \cos(-6\omega t + \theta'_5) \\ & -7\omega_e L_s i_7 \sin(6\omega t + \theta'_7) + R i_7 \cos(6\omega t + \theta'_7) \\ u_q = & \omega_e \psi_f + \omega_e L_s i_{d1} + R i_{q1} - 5\omega_e L_s i_5 \cos(-6\omega t + \theta'_5) + R i_5 \sin(-6\omega t + \theta'_5) \\ & +7\omega_e L_s i_7 \cos(6\omega t + \theta'_7) + R i_7 \sin(6\omega t + \theta'_7) \end{cases} \tag{8}$$

Projecting the fundamental current and harmonic current of the motor into the d-q rotating coordinate system, the current in the d-q axis is obtained as follows:

$$\begin{cases} i_d = i_{d1} + i'_{d5} + i'_{d7} + i'_{d11} + i'_{d13} \\ i_q = i_{q1} + i'_{q5} + i'_{q7} + i'_{q11} + i'_{q13} \end{cases} \tag{9}$$

where i_{d1} is the d-axis component of the stator fundamental current; i'_{d5}, i'_{d7}, i'_{d11}, and i'_{d1} are the d-axis components of the stator harmonic current of 5th, 7th, 11th, and 13th harmonic currents, respectively; i_{q1} is the q-axis component of the stator fundamental current; and i'_{q5}, i'_{q7}, i'_{q11}, and i'_{q13} are the q-axis components of the stator harmonic current of the 5th, 7th, 11th, and 13th harmonic currents, respectively.

By introducing Equation (9) into the electromagnetic torque equation of the motor, the harmonic equation of the torque is obtained as follows:

$$\begin{aligned} T_m &= \tfrac{3}{2} P_n \psi_f i_q \\ &= \tfrac{3}{2} P_n \psi_f \left[i_{q1} + \left(i'_{q5} + i'_{q7} \right) + \left(i'_{q11} + i'_{q13} \right) \right] \\ &= T_{e0} + T_{e6} + T_{e12} \end{aligned} \tag{10}$$

where T_{e0} is the constant component of electromagnetic torque; and T_{e6} and T_{e12} are the 6th and 12th harmonics of the electromagnetic torque respectively.

It can be seen from Equation (10) that the constant component of electromagnetic torque is generated by the interaction of fundamental current and permanent magnet flux linkage under the premise of ignoring the harmonics introduced by the motor. The 6th harmonic of electromagnetic torque is mainly generated by the interaction of the rotor flux and stator 5th and 7th harmonic currents. Similarly, the 12th harmonic of electromagnetic torque is mainly generated by the interaction of the rotor flux and stator 11th and 13th harmonic currents. Generally, the larger the harmonic number, the smaller the corresponding harmonic torque amplitude becomes.

2.2. Transmission System Model of IEDS

The mechanical transmission system of an IEDS mainly includes a wheel gear reducer and output shaft. The dynamic model of a helical gear transmission system is established as below.

The meshing displacement of the gear tooth along the meshing line is:

$$\delta_y = R_1\theta_1 - R_2\theta_2 - e \tag{11}$$

where θ_1 is the angular displacement of gear 1; θ_2 is the angular displacement of driven gear 2; R_1 is the radius of gear 1; R_2 is the radius of gear 2; and e is the meshing error.

The normal elastic deformation of the fear tooth along the contact line is as follows:

$$\delta_i = \delta_y \cos \beta_b \tag{12}$$

where β_b is the helix angle of the base circle.

The meshing force of the gear pair can be expressed as follows:

$$F_m = c_m\dot{\delta}_i + k_m\delta_i = \left[c_m(R_1\dot{\theta}_1 - R_2\dot{\theta}_2 - \dot{e}) + k_m(R_1\theta_1 - R_2\theta_2 - e)\right]\cos \beta_b \tag{13}$$

where c_m is the gear meshing damping and k_m is the gear meshing stiffness. The calculation methods of the two will be described in detail later.

The y-direction meshing force of the gear pair can be expressed as follows:

$$F_y = F_m \cos \beta_b = \left[c_m(R_1\dot{\theta}_1 - R_2\dot{\theta}_2 - \dot{e}) + k_m(R_1\theta_1 - R_2\theta_2 - e)\right]\cos^2 \beta_b. \tag{14}$$

The torsional vibration model of a parallel shaft helical cylindrical gear pair is as follows:

$$\begin{cases} I_1\ddot{\theta}_1 = T_1 - F_yR_1 \\ I_2\ddot{\theta}_2 = -T_2 + F_yR_2 \end{cases} \tag{15}$$

where I_1 is the moment of inertia of gear 1; I_2 is the moment of inertia of gear 2; T_1 is the active torque acting on gear 1; and T_2 is the load torque acting on gear 2.

The specific parameters of helical gears involved in this paper are shown in Table A1 of Appendix A. By calculating the length change of the meshing line in the process of gear meshing and describing the time-varying meshing stiffness of the helical gear by the angular displacement of the driving gear, the relationship between the stiffness and angular displacement of the driving gear is obtained, as shown in Figure 7.

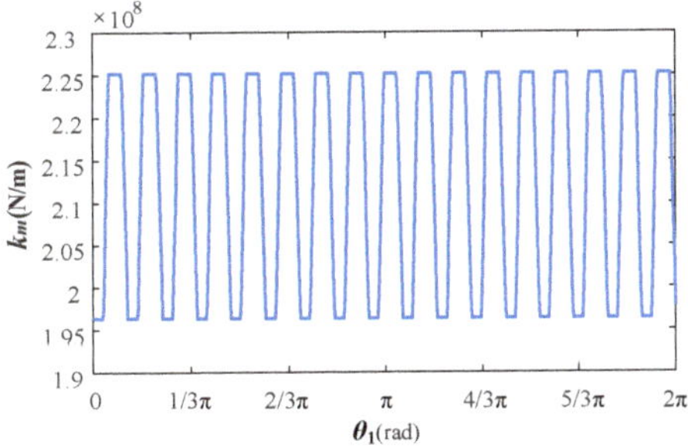

Figure 7. Time-varying meshing stiffness of gear pair.

The meshing damping of the gear pair can be expressed as follows:

$$c_m = 2\xi_m \sqrt{\bar{k}_{12} \frac{m_1 m_2}{m_1 + m_2}} \tag{16}$$

where ξ_m is the meshing damping ratio in the range of 0.03–0.17; and m_i is the mass of gear i.

The mathematical expression of the gear meshing error is as follows:

$$e = e_m + e_r \sin(Z_1 \theta_1 + \delta) \tag{17}$$

where e_m is the constant value of the meshing error, and e_r is the amplitude of the meshing error, both of which are related to the manufacturing accuracy of the gear; and δ is the initial phase of the meshing error.

2.3. Vehicle Powertrain Model

The moment of inertia of the whole vehicle is equivalent to the wheel in this paper, and the torsional vibration model of the whole vehicle powertrain is obtained, as shown in Figure 8. Ignoring the translational vibration of the system, a 6-DOF dynamic model of the power transmission system is established by using the lumped parameter method, as shown in Equation (18).

$$\begin{cases} I_m \ddot{\theta}_m + c_{s1}(\dot{\theta}_m - \dot{\theta}_1) + k_{s1}(\theta_m - \theta_1) = T_m \\ I_1 \ddot{\theta}_1 - c_{s1}(\dot{\theta}_m - \dot{\theta}_1) - k_{s1}(\theta_m - \theta_1) + R_1 F_y = 0 \\ I_2 \ddot{\theta}_2 + c_{s2}(\dot{\theta}_2 - \dot{\theta}_g) + k_{s2}(\theta_2 - \theta_g) - R_2 F_y = 0 \\ I_g \ddot{\theta}_g - c_{s2}(\dot{\theta}_2 - \dot{\theta}_g) - k_{s2}(\theta_2 - \theta_g) + 2/i_0 \cdot c_a(\dot{\theta}_g/i_0 - \dot{\theta}_w) + 2/i_0 \cdot k_a(\theta_g/i_0 - \theta_w) = 0 \\ I_w \ddot{\theta}_w - c_a(\dot{\theta}_g/i_0 - \dot{\theta}_w) - k_a(\theta_g/i_0 - \theta_w) + c_w(\dot{\theta}_w - \dot{\theta}_v) + k_w(\theta_w - \theta_v) = 0 \\ I_v \ddot{\theta}_v - 2c_w(\dot{\theta}_w - \dot{\theta}_v) - 2k_w(\theta_w - \theta_v) = -T_L \end{cases} \tag{18}$$

where T_m is the electromagnetic driving torque generated by PMSM; and T_L is the external resistance load when the vehicle is working. The parameters and their values represented by other symbols are shown in Table A2 of Appendix A.

Among them, the resistance load of the vehicle can be expressed as follows:

$$T_L = \left(mgf \cos \alpha + \frac{C_D A}{21.15} V^2 + mg \sin \alpha\right) \cdot r \tag{19}$$

where r is the wheel radius of the car.

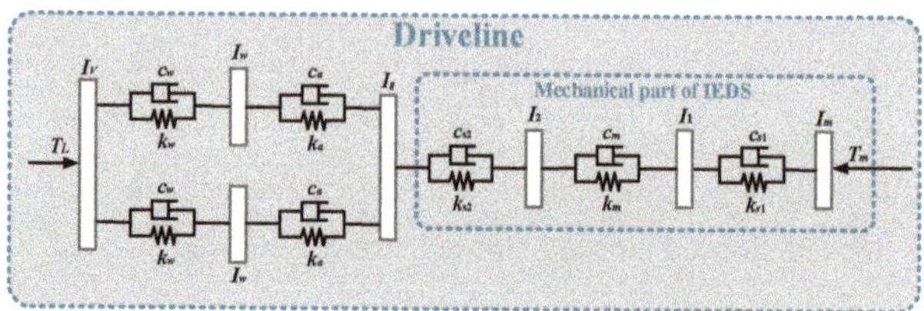

Figure 8. 6-DOF torsional vibration model of a power transmission system.

3. Research on Electromechanical Coupling Characteristics of IEDS

In order to study the electromechanical coupling effect of an IEDS, the transmission system dynamic model, PMSM, and PMSM control system model are built in this paper. The speed and torque of the motor shaft are used as common variables to transfer data between the electrical system and mechanical system in real time. Figure 9 shows the electromechanical coupling model of a PEV in motor torque mode. It should be noted that the simulation model of the IEDS is built on the MATLAB/Simulink 2018b simulation platform and the computer processor is Intel(R) Core(TM) i7-8700 CPU@ 3.20 Ghz.

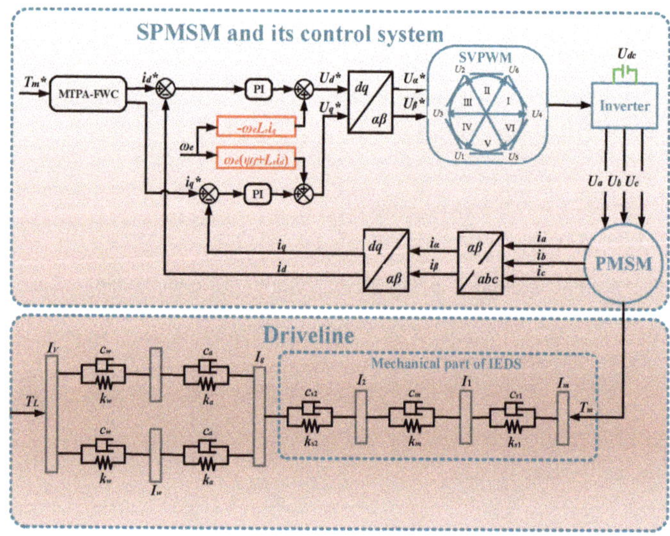

Figure 9. Electromechanical coupling dynamic model of a PEV.

In order to explore the electromechanical coupling characteristics of an IEDS under the condition of uniform speed, the simulation conditions are set as: the speed of PMSM is 1000 rpm; the vehicle load is 100 Nm; and the motor working mode is torque control mode. When the speed of the PMSM is 1000 rpm (the rotation frequency is 16.67 Hz), the electric angular frequency of the motor is 66.67 Hz, and the gear meshing frequency of the reducer of the IEDS is 283.33 Hz. The electromagnetic torque and its frequency spectrum of the PMSM are shown as Figure 10a,b, respectively. Under the influence of nonlinear factors such as inverter dead-time effect and voltage drop effect, the electromagnetic torque shows the characteristics of pulsation. Moreover, the fluctuation range of harmonic torque is about 11 Nm, and the fluctuation frequencies of harmonic torque are mainly 400 Hz, 800 Hz, and 1200 Hz, which are 6 times, 12 times, and 18 times the angular frequency, respectively. Figure 11a shows the three-phase current of the PMSM, and it can be seen that the current waveform in the time domain is not completely sinusoidal. The frequency domain analysis of the A phase current in three-phase current by fast Fourier transform (FFT) shows that in addition to the fundamental frequency (67 Hz), the current also contains harmonic currents, among which the obvious components are 333 Hz, 467 Hz, 733 Hz, and 867 Hz, which are 5 times, 7 times, 11 times, and 13 times the angular frequency, respectively, as shown in Figure 11b.

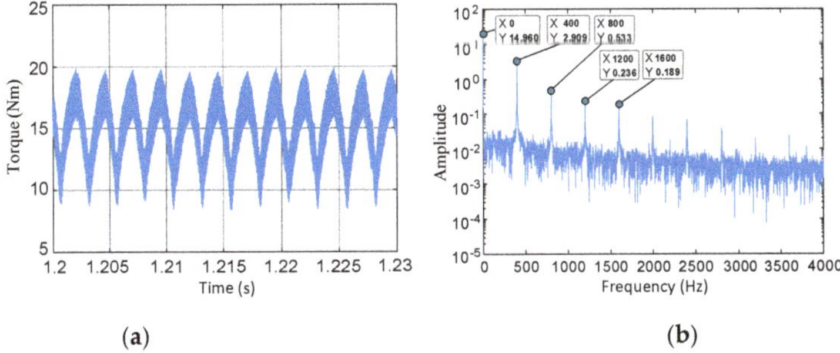

Figure 10. Electromagnetic torque of PMSM in an IEDS: (**a**) Electromagnetic torque in time domain; (**b**) Frequency spectrum of electromagnetic torque.

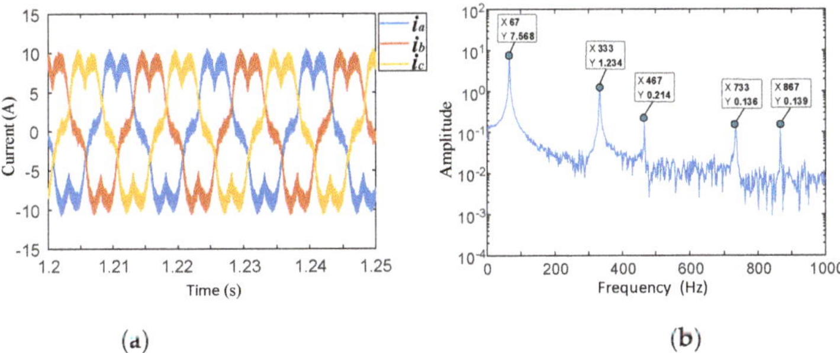

Figure 11. Three-phase current of PMSM in an IEDS: (**a**) Three-phase current in time domain; (**b**) Frequency spectrum of A phase current.

The shaft speed and the frequency spectrum of the PMSM are shown as Figure 12a,b, respectively. Under the influence of harmonic torque, gear time-varying stiffness, gear meshing error, and other factors, the motor shaft speed fluctuates continuously, and the fluctuation amplitude is about 2 rpm. The main components of the motor shaft speed fluctuation are gear meshing frequency (283 Hz), 2nd gear meshing frequency (567 Hz), motor 6th harmonic torque (400 Hz), and 12th harmonic torque (800 Hz). The gear pair speed and the frequency spectrum of the reducer in the IEDS are shown in Figure 13, which shows that the speed of the driving and driven gears fluctuates constantly, and the main frequency component of fluctuation is the same as that of the motor shaft. In conclusion, under the influence of mechanical transmission system factors such as gear time-varying stiffness and gear meshing error, the rotational speed of each component of the IEDS presents the characteristics of fluctuation. Meanwhile, the harmonic torque of the motor introduces more harmonics to the speed of each component, which intensifies the speed fluctuation of each component.

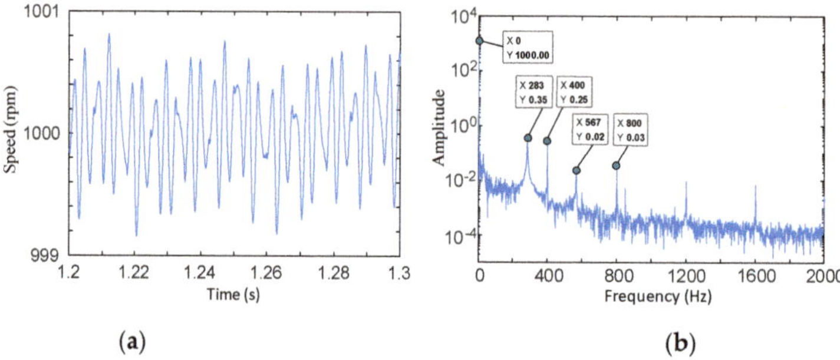

Figure 12. Shaft speed of the PMSM in the IEDS: (**a**) Shaft speed in time domain; (**b**) Frequency spectrum of shaft speed.

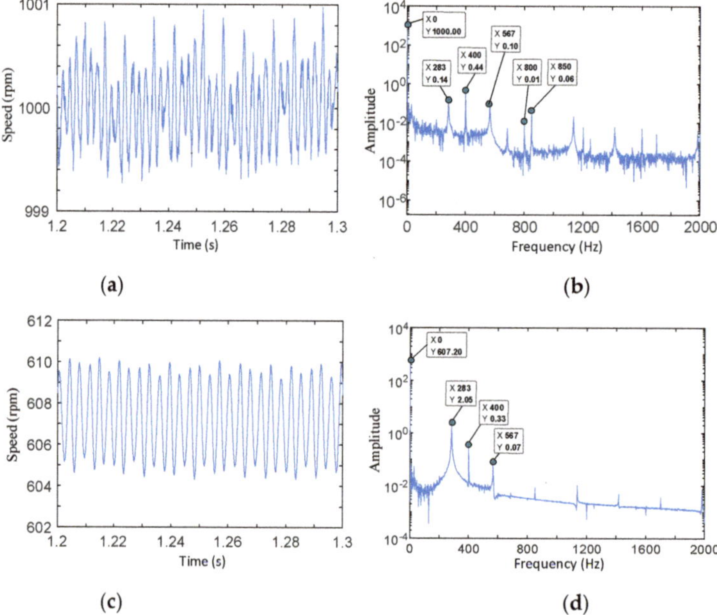

Figure 13. Gear pair speed of reducer in the IEDS: (**a**) Driving gear speed in time domain; (**b**) Frequency spectrum of driving gear speed; (**c**) Driven gear speed in time domain; (**d**) Frequency spectrum of driven gear speed.

The meshing force and the frequency spectrum of the gear pair of the reducer in the IEDS is shown in Figure 14, and the meshing displacement and its frequency spectrum is shown in Figure 15. Under the influence of time-varying meshing stiffness and meshing error, the meshing force fluctuates around the theoretical meshing force. The frequency domain analysis of the gear meshing force and meshing displacement not only includes the gear meshing frequency (283 Hz) and its 2nd harmonic frequency (567 Hz), but it also includes the 6th harmonic torque (400 Hz) and the 12th harmonic torque (800 Hz). The dynamic transfer torque and the frequency spectrum of the motor shaft are shown in Figure 16, and that of the output shaft of the IEDS is shown in Figure 17. The dynamic torque transmitted by the two shafts also fluctuates continuously. It can be seen from the frequency domain analysis diagram that the main component of the fluctuation is the same as the meshing force of the gear.

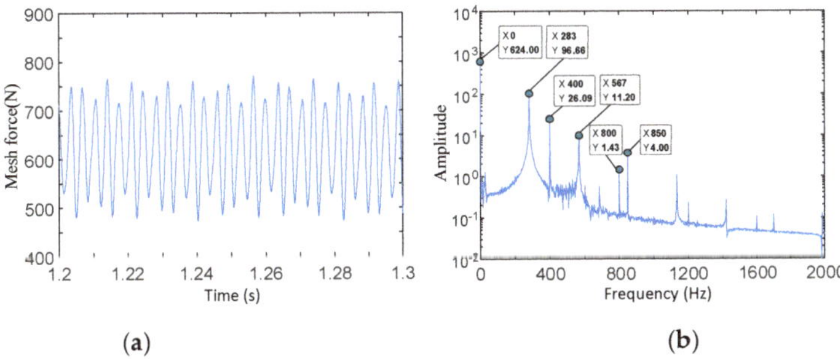

Figure 14. Gear pair meshing force of the reducer in the IEDS: (**a**) Meshing force in the time domain; (**b**) Frequency spectrum of the meshing force.

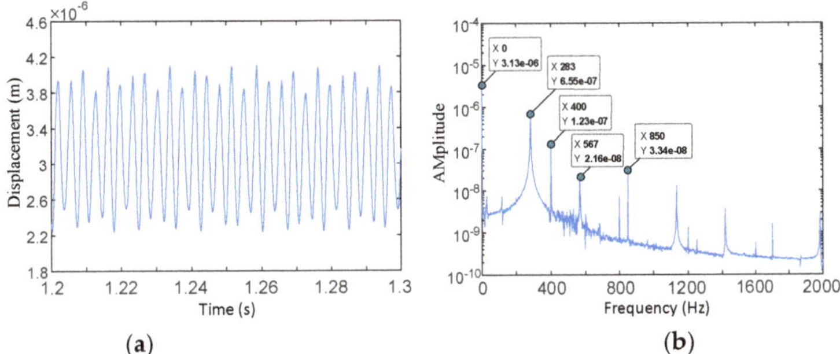

Figure 15. Gear pair meshing displacement of the reducer in the IEDS: (**a**) Meshing displacement in the time domain; (**b**) Frequency spectrum of the meshing displacement.

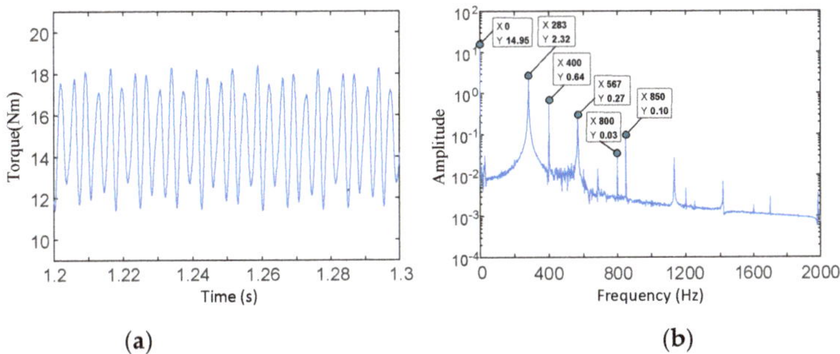

Figure 16. Dynamic transfer torque of the motor shaft in the IEDS: (**a**) Dynamic transfer torque in the time domain; (**b**) Frequency spectrum of dynamic transfer torque.

Through the above simulation and analysis of the electromechanical coupling characteristics of the IEDS under uniform speed conditions, it is found that the electrical system and mechanical system of the IEDS will affect each other. Gear meshing frequency is the main component of motor shaft speed fluctuation, which means that mechanical nonlinear factors such as time-varying meshing stiffness and meshing error are the main causes of motor shaft speed fluctuation. Besides, the electrical system will also affect the operation of mechanical system, because the dead-time effect and voltage drop effect of the inverter make the output electromagnetic torque of the PMSM contain 6th and 12th harmonics. The harmonic torque frequency of the motor is included in the meshing force of the gear and the transmission torque of the shaft, which means that the harmonic torque of the motor will increase the dynamic load of the mechanical system.

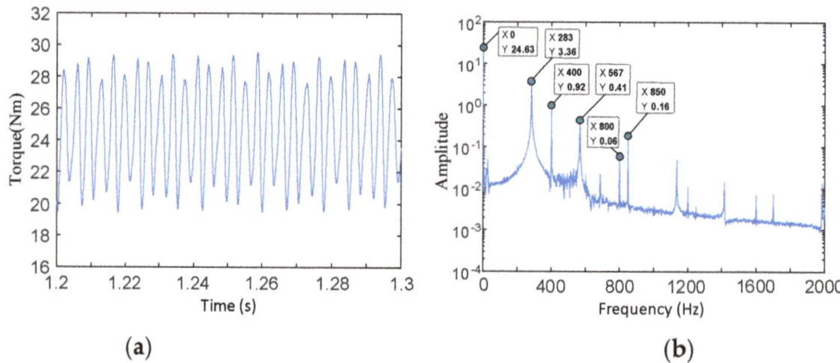

(a)   (b)

Figure 17. Dynamic transfer torque of the output shaft in the IEDS: (**a**) Dynamic transfer torque in the time domain; (**b**) Frequency spectrum of the dynamic transfer torque.

It should be noted that when designing an IEDS, if the 6th harmonic torque of the motor is consistent with the gear meshing frequency, the amplitude of the motor speed fluctuation may be larger after superposition, which is not conducive to the smooth operation of the motor. Moreover, the dynamic load amplitude of the mechanical system is larger and the service life of the mechanical system is reduced. So, in order to avoid it, the gear meshing frequency of the reducer should not be equal to 6 times that of the electric angular frequency of the motor, namely $Z_{p1} \neq 6P_n$ (where Z_{p1} is the number of driving gear teeth of the reducer gear pair; and P_n is the number of magnetic pole pairs).

4. Harmonic Torque Reduction Strategy for IEDS

According to the analysis above, the harmonic torque will cause additional load fluctuation in an IEDS, which will aggravate the speed fluctuation of each component of the IEDS, and that is not conducive to the stable and efficient operation of the system. To solve the above problems, a harmonic torque reduction strategy to reduce the adverse effects of motor harmonic torque in an IEDS is proposed this section.

4.1. Design of Harmonic Torque Reduction Strategy

The stator of the motor mainly contains the 5th and 7th harmonic currents. In the three-phase static coordinate system, the rotation speed of the 5th harmonic current is -5ω, and the rotation speed of the 5th harmonic current is 7ω. To better control the 5th and 7th harmonic currents of the motor, the 5th and 7th rotation coordinate systems are established in this section. Through coordinate transformation, the 5th harmonic voltage and current are converted into DC flow in the 5th rotation coordinate system, while the 7th harmonic voltage and current are converted into DC flow in the 7th rotation coordinate system. In a relative manner, other current components are converted into AC flow.

Consequently, at this time, a low-pass filter can be used to separate the 5th and 7th harmonics in the three-phase current of PMSM, and then the synchronous rotation a proportional-integral (PI) controller can be used to make the actual d-q axis current follow the reference current command to realize the injection of harmonic voltage to eliminate the torque harmonic component.

The 5th and 7th synchronous rotation coordinate systems used in Figure 18 are shown in Figure 19. It should be noted that the transformation matrix between the coordinate systems is shown in Appendix B.

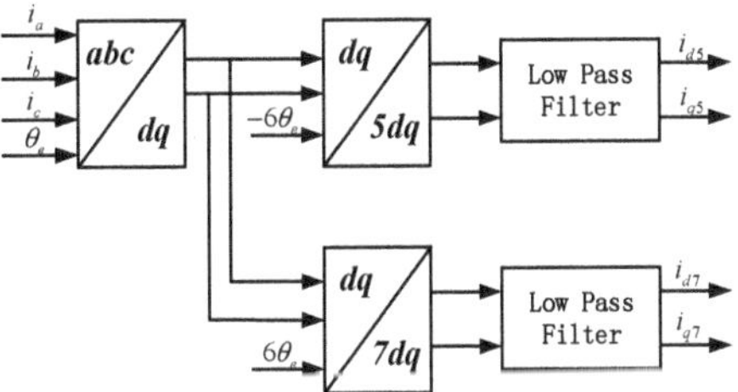

Figure 18. Separation strategy of 5th and 7th harmonic currents of a PMSM.

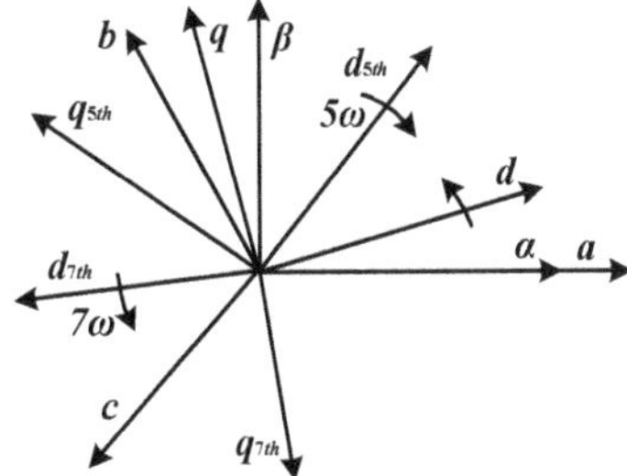

Figure 19. 5th and 7th synchronous rotation coordinate systems of a motor.

Based on the above coordinate transformation, Equation (8) is transformed into a 5th d-q rotating coordinate system:

$$
\begin{cases}
u_{d5} = & \omega_e\psi_f\sin(-6\omega t+\theta_0)-\omega_e L_s i_1\sin(6\omega t+\theta_1'')+Ri_1\cos(6\omega t+\theta_1'')\\
& +5\omega_e L_s i_{q5}+Ri_{d5}-7\omega_e L_s i_7\sin(12\omega t+\theta_7')+Ri_7\cos(12\omega t+\theta_7')\\
u_{q5} = & \omega_e\psi_f\cos(-6\omega t+\theta_0)+\omega_e L_s i_{d1}\cos(6\omega t+\theta_1'')+Ri_1\sin(6\omega t+\theta_1'')\\
& -5\omega_e L_s i_{d5}+Ri_{q5}+7\omega_e L_s i_7\cos(12\omega t+\theta_7')+Ri_7\sin(12\omega t+\theta_7')
\end{cases}
\tag{20}
$$

where i_{d5} and i_{q5} are the d-axis and q-axis DC current components in the 5th d-q rotation coordinate system.

The harmonic steady-state voltage equation in the 5th d-q rotating coordinate system is obtained by omitting the AC quantities contained in Equation (20), which is written as:

$$
\begin{cases}
u_{d5} = 5\omega L_q i_{q5}+Ri_{d5}\\
u_{q5} = -5\omega L_q i_{q5}+Ri_{d5}
\end{cases}
\tag{21}
$$

Similarly, Equation (8) is transformed into the 7th d-q rotating coordinate system:

$$
\begin{cases}
u_{d7} = & \omega_e\psi_f\sin(6\omega t+\theta_0)-\omega_e L_s i_1\sin(-6\omega t+\theta_1'')+Ri_1\cos(-6\omega t+\theta_1'')\\
& +5\omega_e L_s i_5\sin(-12\omega t+\theta_5')+Ri_5\cos(-12\omega t+\theta_5')-7\omega_e L_s i_{q7}+Ri_{d7}\\
u_{q7} = & \omega_e\psi_f\cos(6\omega t+\theta_0)+\omega_e L_s i_1\cos(-6\omega t+\theta_1'')+Ri_1\sin(-6\omega t+\theta_1'')\\
& -5\omega_e L_s i_5\cos(-12\omega t+\theta_5')+Ri_5\sin(-12\omega t+\theta_5')+7\omega_e L_s i_{d7}+Ri_{q7}
\end{cases}
\tag{22}
$$

where i_{d7} and i_{q7} are the d-axis and q-axis DC current components in the 7th d-q rotation coordinate system.

The harmonic steady-state voltage equation in the 7th d-q rotating coordinate system is obtained by omitting the AC quantities contained in Equation (22), which is written as:

$$
\begin{cases}
u_{d7} = -7\omega L_q i_{q7}+Ri_{d7}\\
u_{q7} = 7\omega L_q i_{q7}+Ri_{d7}
\end{cases}
\tag{23}
$$

Using a PI controller, combined with the 5th and 7th harmonic steady-state voltage equations, the harmonic current loop control strategy is obtained, as shown in Figure 20a,b. Among them, the 5th and 7th harmonic steady-state voltage and current are coupled with each other. In order to better control the harmonic current, this paper realizes the decoupling of stator harmonic voltage and current by adding compensation terms. Superimposing the voltage generated by the harmonic voltage steady-state equation and the harmonic current PI controller, the required injection voltage of each harmonic current in the rotating coordinate system can be obtained.

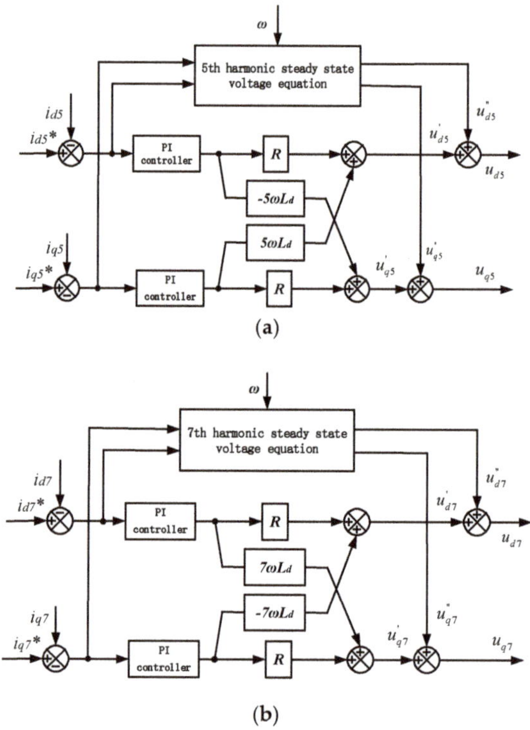

Figure 20. Control block diagram of motor harmonic current decoupling: (**a**) 5th harmonic current; (**b**) 7th harmonic current.

As shown in Figure 21, after coordinate transformation, injecting the harmonic output from the harmonic current controller into voltage signal U'_α and U'_β and then adding them to the required voltage U^*_α and U^*_β of the motor itself, harmonic voltage injection is realized, which constitutes the new reference voltage signals U_β and U_β of the motor as the inverter demand voltages.

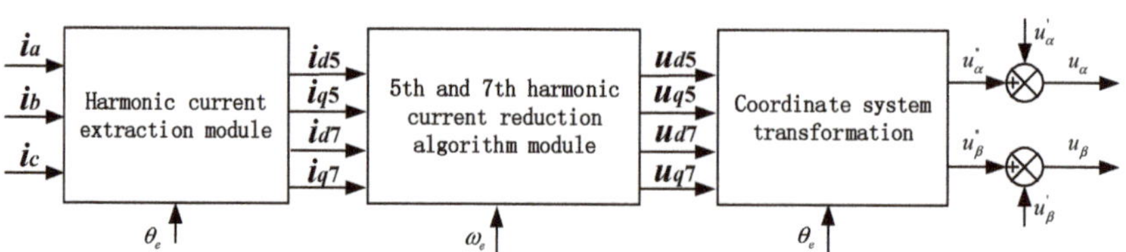

Figure 21. Harmonic torque reduction control strategy.

4.2. Simulation Analysis of Harmonic Torque Reduction Control Strategy Effectiveness

The electromechanical coupling dynamic model of a PEV with a harmonic torque reduction strategy is shown in Figure 22. In order to verify the effect of the harmonic torque reduction strategy, the simulation conditions are set as: the speed of the PMSM is 1000 rpm; the load of the vehicle is 100 Nm; and the motor is in torque control mode.

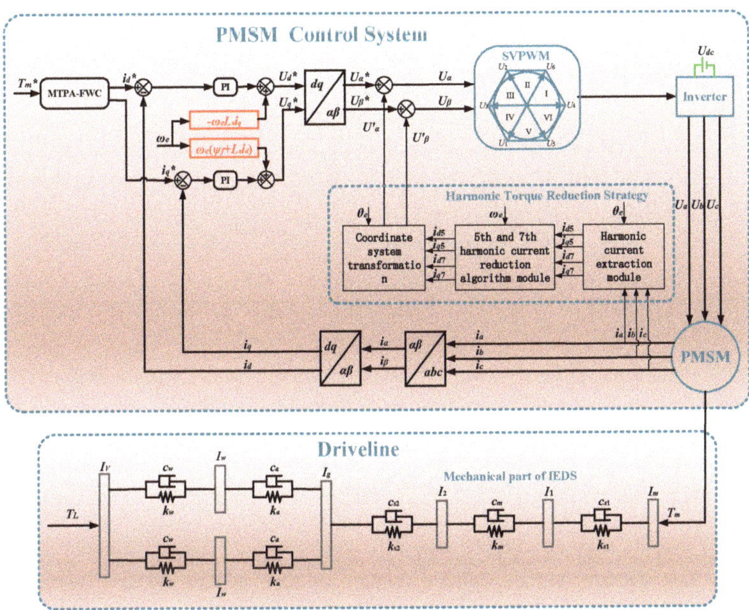

Figure 22. Electromechanical coupling dynamic model of a PEV considering harmonic torque.

Figure 23a,b show the three-phase current before and after adding the harmonic reduction strategy, respectively, while Figure 23c,d show the frequency domain analysis of the three-phase current before and after adding the harmonic reduction strategy, respectively. Moreover, Figure 24a,b show the electromagnetic torque of the motor before and after adding the harmonic reduction strategy, respectively, while Figure 24c,d show the frequency domain analysis of the motor electromagnetic torque before and after adding the harmonic reduction strategy, respectively. Under the effect of the harmonic reduction strategy, the 5th and 7th harmonics in the current are significantly reduced. Furthermore, the sinusoidal degree of the current is significantly improved, and the 6th harmonic torque of the motor is effectively reduced. It can be found that the amplitude of the 5th harmonic current is reduced from 1.234 A to 0.06 A, and that of the 7th harmonic current is reduced from 0.214 A to 0.006 A. Meanwhile, the total fluctuation amplitude of electromagnetic torque is reduced by 50% from 10 Nm to 5 Nm. The 6th harmonic torque amplitude of the motor is reduced from 2.909 Nm to 0.060 Nm. From the analysis of the above results, it can be seen that the harmonic current content can be effectively reduced by harmonic voltage injection, and then the harmonic torque can be reduced.

Figure 25a compares the motor shaft speed before and after adding the harmonic reduction strategy while Figure 25b,c show the frequency domain analysis of the motor shaft speed before and after adding the harmonic reduction strategy, respectively. The fluctuation amplitude of the motor shaft speed caused by the 6th harmonic torque of motor is reduced from 0.25 to 0.01 rpm, which means that the overall fluctuation amplitude of the motor speed is effectively reduced.

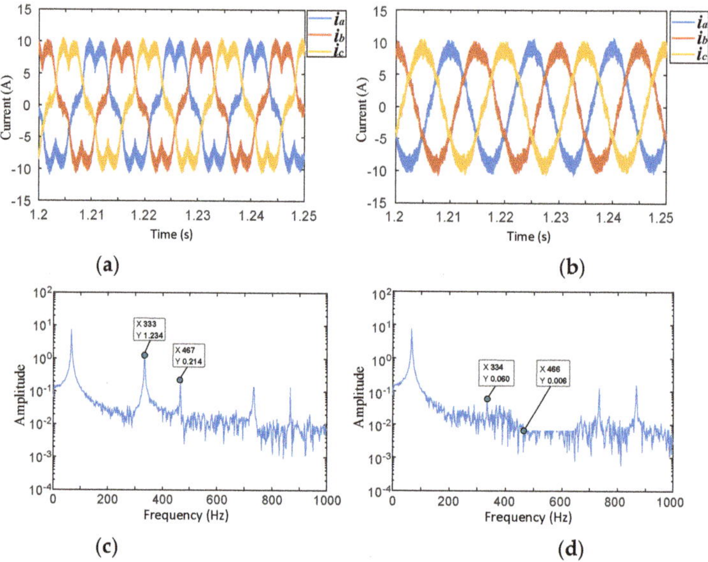

Figure 23. Three-phase current of a PMSM in an IEDS: (**a**) Three-phase current of the stator in the time domain before adding the harmonic reduction strategy; (**b**) Three-phase current of the stator in the time domain after adding the harmonic reduction strategy; (**c**) Frequency spectrum of the A-phase current before adding the harmonic reduction strategy; (**d**) Frequency spectrum of the A-phase current after adding the harmonic reduction strategy.

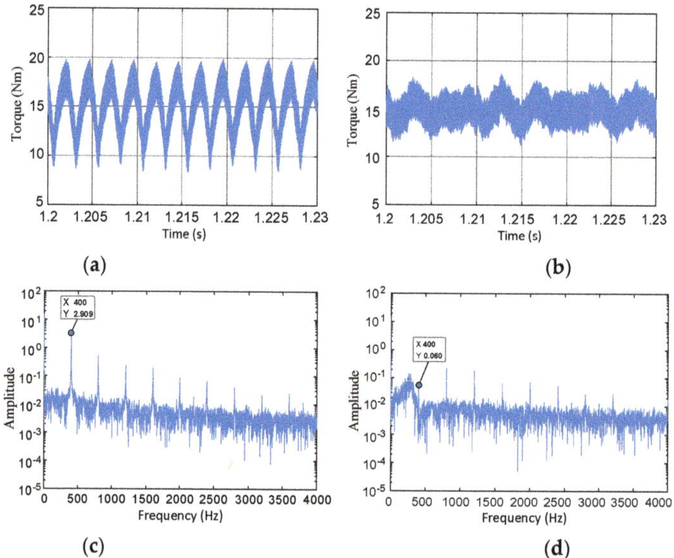

Figure 24. Electromagnetic torque of a PMSM in an IEDS: (**a**) Electromagnetic torque in the time domain before adding the harmonic reduction strategy; (**b**) Electromagnetic torque in the time domain after adding the harmonic reduction strategy; (**c**) Frequency spectrum of electromagnetic torque before adding the harmonic reduction strategy; (**d**) Frequency spectrum of electromagnetic torque after adding the harmonic reduction strategy.

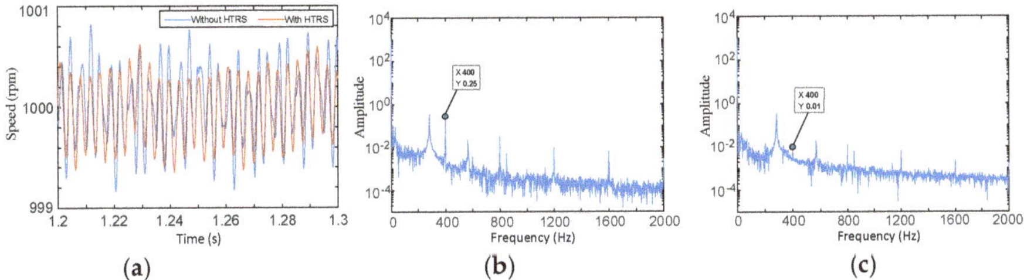

Figure 25. Shaft speed of a PMSM in an IEDS: (**a**) Shaft speed in the time domain; (**b**) Frequency spectrum of the shaft speed before adding the harmonic reduction strategy; (**c**) Frequency spectrum of the shaft speed after adding the harmonic reduction strategy.

Figure 26a compares the driving gear speed of the reducer of the IEDS before and after adding the harmonic reduction strategy, while Figure 26b,c shows the frequency domain analysis of the reducer driving gear speed before and after adding the harmonic reduction strategy, respectively. The results show that the fluctuation amplitude of the driving gear speed caused by the 6th harmonic torque of the motor is reduced from 0.44 to 0.01 rpm.

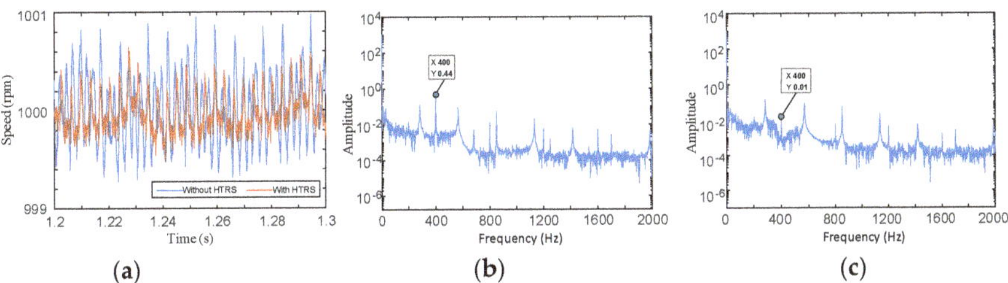

Figure 26. Driving gear speed of the reducer in an IEDS: (**a**) Driving gear speed in the time domain; (**b**) Frequency spectrum of the driving gear speed before adding the harmonic reduction strategy; (**c**) Frequency spectrum of the driving gear speed after adding the harmonic reduction strategy.

Moreover, Figure 27a compares the driving gear speed of the reducer of IEDS after adding the harmonic reduction strategy, while Figure 27b,c respectively show the frequency domain analysis of the driven gear speed of the reducer before and after adding the harmonic reduction strategy. It can be seen that the fluctuation amplitude of the driven gear speed caused by the 6th harmonic torque of the motor decreases from 0.33 to 0.01 rpm, and the overall fluctuation amplitude of the driven gear speed decreases slightly. Under the effect of the harmonic reduction strategy, the 6th harmonic torque component of the motor speed harmonics of each component of the IEDS is significantly reduced, and the fluctuation amplitudes of the speed of each component are reduced to a certain extent, which is conducive to the smooth operation of the system.

The meshing force of the reducer gear pair of the IEDS before and after adding the harmonic reduction strategy is shown in Figure 28a, while the frequency domain analysis of the meshing force of the gear pair before and after adding the harmonic reduction strategy is shown in Figure 28b,c respectively. It can be found that the amplitude of the gear pair meshing force caused by the 6th harmonic torque of motor is reduced from 26.09 to 0.67 Nm. Figure 29a compares the dynamic transmission torque of the motor shaft of the IEDS before and after adding the harmonic reduction strategy; then, Figure 29b,c show the frequency domain analysis of the motor shaft dynamic transmission torque before and after adding the harmonic reduction strategy, respectively. The amplitude of the motor shaft dynamic transmission torque caused by the 6th harmonic torque of the motor decreases from 0.64 to 0.02 Nm. Figure 30a compares the dynamic transmission torque of the output shaft of the

IEDS before and after adding the harmonic reduction strategy. Then, Figure 30b,c show the frequency domain analysis of the output shaft dynamic transmission torque before and after adding the harmonic reduction strategy. The dynamic transmission torque of the output shaft caused by the 6th harmonic torque of the motor is reduced from 0.92 to 0.02 Nm. That is to say, under the effect of the harmonic reduction strategy, the 6th harmonic torque component of the dynamic load of each component of the IEDS is significantly reduced, which is conducive to improving the reliability of the system and prolonging the service life of the mechanical system.

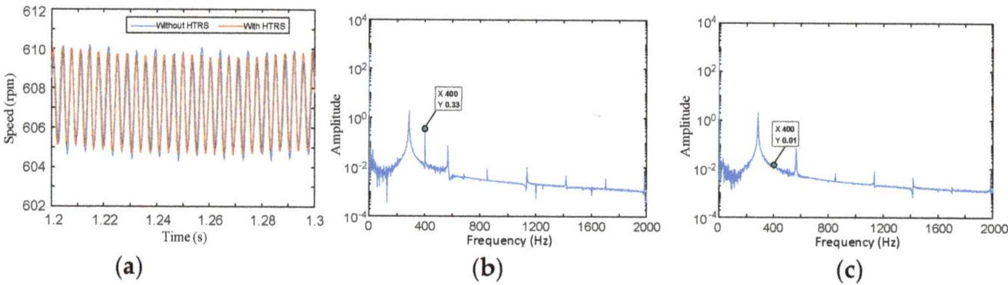

Figure 27. Driven gear speed of the reducer in an IEDS: (**a**) Driven gear speed in the time domain; (**b**) Frequency spectrum of the driven gear speed before adding the harmonic reduction strategy; (**c**) Frequency spectrum of the driven gear speed after adding the harmonic reduction strategy.

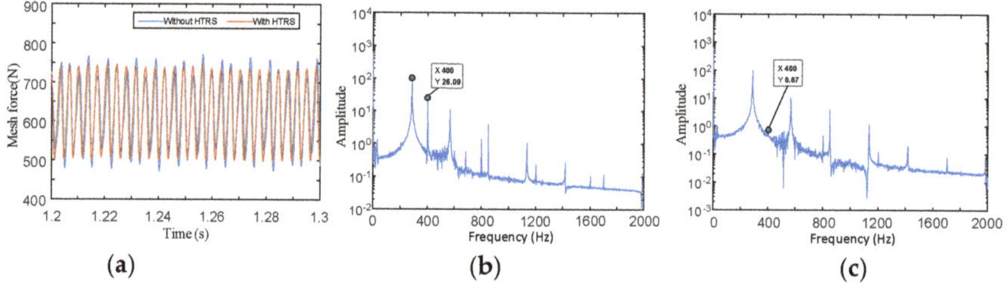

Figure 28. Driven meshing force of the reducer gear pair in an IEDS: (**a**) Reducer gear pair meshing force in the time domain; (**b**) Frequency spectrum of the meshing force before adding the harmonic reduction strategy; (**c**) Frequency spectrum of the meshing force after adding the harmonic reduction strategy.

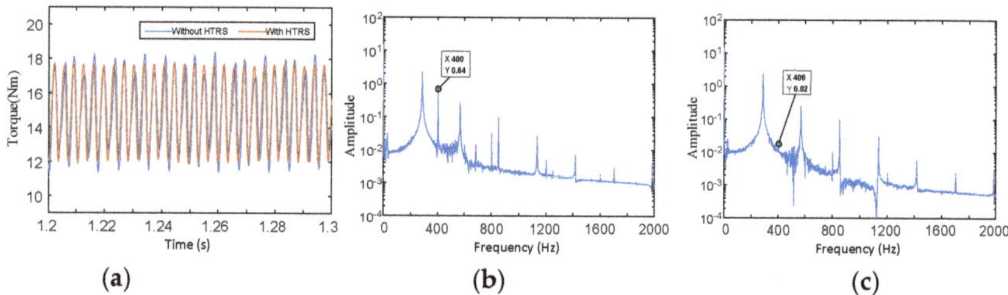

Figure 29. Dynamic torque of the motor shaft in an IEDS: (**a**) Motor shaft dynamic torque in the time domain; (**b**) Frequency spectrum of dynamic torque before adding the harmonic reduction strategy; (**c**) Frequency spectrum of dynamic torque after adding the harmonic reduction strategy.

The detailed data comparison of IEDS before and after the harmonic torque reduction strategy is shown in Table 3. The simulation results show that the 5th and 7th harmonic currents of the PMSM are obviously reduced under the action of the harmonic torque reduction strategy, which makes the 6th harmonic torque successfully reduced, thus reducing the speed fluctuation of each component of the

IEDS and making the system run more stably. At the same time, the reduction of harmonic current also helps reduce the motor heating and improve the efficiency of the motor.

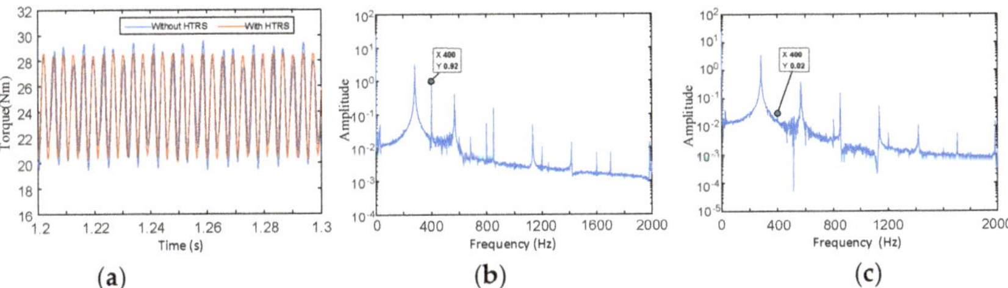

Figure 30. Dynamic torque of the output shaft in an IEDS: (**a**) Output shaft dynamic torque in the time domain; (**b**) Frequency spectrum of dynamic torque before adding the harmonic reduction strategy; (**c**) Frequency spectrum of dynamic torque after adding the harmonic reduction strategy.

Table 3. Data comparison of the IEDS before and after adding the harmonic torque reduction strategy.

Parameter	Before Adding Harmonic Torque Reduction Strategy	After Adding Harmonic Torque Reduction Strategy	Optimization Effect (%)
5th harmonic current amplitude (A)	1.234	0.06	95.1
7th harmonic current amplitude (A)	0.214	0.006	97.2
6th harmonic torque (Nm)	2.909	0.06	97.9
Overall fluctuation amplitude of motor torque (Nm)	10	5	50.0
Fluctuation amplitude of motor shaft speed (rpm)	0.25	0.01	96.0
Amplitude of driving gear speed fluctuation caused by 6th harmonic torque (rpm)	0.44	0.01	97.7
Fluctuation amplitude of driven gear speed caused by 6th harmonic torque (rpm)	0.33	0.01	97.0
Amplitude of meshing force fluctuation of gear pair caused by 6th harmonic torque (N)	26.09	0.67	97.4
Amplitude of shaft dynamic load fluctuation caused by 6th harmonic torque (Nm)	0.64	0.02	96.9
Amplitude of output shaft dynamic load fluctuation caused by 6th harmonic torque (Nm)	0.92	0.02	97.8

5. Conclusions

In this paper, the electromechanical coupling model of an electric vehicle equipped with an IEDS is established, and the electromechanical coupling characteristics of the IEDS are simulated and analyzed. On this basis, the method to suppress the harmonic torque of a PMSM is studied. There are two points that should be noted:

1. The electrical system and mechanical system of the IEDS will interact with each other. Mechanical nonlinear factors such as time-varying meshing stiffness and the meshing error

of the gears can lead to a speed fluctuation of the motor shaft. Meanwhile, the dead-time effect and voltage drop effect of the inverter will cause the 6th harmonic torque and 12th harmonic torque of the motor, which will increase the dynamic load of the mechanical system. When designing an IEDS, the gear meshing frequency of the reducer should not be equal to 6 times that of the electric angular frequency of the motor, namely $Z_{p1} \neq 6P_n$. In order to avoid that when the 6th harmonic torque of the motor is always consistent with the gear meshing frequency, the superposition may lead to a more serious speed fluctuation of the system, resulting in a greater dynamic load amplitude of the mechanical system and reducing the service life of the mechanical system.

2. By injecting harmonic voltage, a harmonic torque reduction strategy is proposed for an IEDS in this paper. Under the effect of the harmonic torque reduction strategy, the 5th and 7th harmonic currents are effectively reduced, and the total fluctuation amplitude of the electromagnetic torque is reduced by 50%. The simulation results show that the harmonic torque reduction strategy proposed in this paper can effectively reduce the harmonic torque of the IEDS, thus reducing the speed fluctuation and dynamic load of each component of the system and improving the stability of the IEDS.

Author Contributions: Conceptualization, J.H.; Data curation, M.J. and Y.G.; Formal analysis, Y.G.; Funding acquisition, J.H. and S.L.; Investigation, M.J.; Methodology, J.H. and Y.Y.; Project administration, J.H. and C.F.; Resources, S.L.; Software, Y.Y.; Supervision, C.F.; Validation, M.J., Y.G. and C.F.; Visualization, Y.Y.; Writing—original draft, Y.Y.; Writing—review and editing, Y.Y. All authors have read and agreed to the published version of the manuscript.

Funding: This research was funded by the National Key R&D Program of China under Grant No. 2018YFB0106100 and the Chongqing Artificial Intelligence Technology Innovation Major Special Project under Grant No. cstc2018jszx-cyztzx0047.

Conflicts of Interest: The authors declare no conflict of interest.

Appendix A

Table A1. Parameters of the helical gear pair.

Parameter	Driving Gear	Driven Gear	Unit
Teeth number	$Z_1 = 17$	$Z_2 = 28$	-
Normal modulus m_n	3	3	mm
Modulus of end face m_t	3.19	3.19	mm
Tooth surface width b	23	23	mm
Normal pressure angle α_n	20	20	deg
Transverse pressure angle α_t	21.17	21.17	deg
Pitch circle helix angle β	20	20	deg

Table A2. Parameters of the transmission system.

Parameter	Value	Unit
Numerical motor rotor and motor shaft moment of inertia I_m	0.035	kg·m^2
Rotational inertia of driving gear of reducer I_1	1.67×10^{-4}	kg·m^2
Rotational inertia of driven gear of reducer I_2	1.2×10^{-3}	kg·m^2
Equivalent moment of inertia of final drive and differential I_g	8×10^{-3}	kg·m^2
Wheel moment of inertia I_w	0.915	kg·m^2
Body equivalent moment of inertia I_V	139.8	kg·m^2
Torsional stiffness of motor shaft k_{s1}	8×10^4	Nm/rad
Normal meshing stiffness per unit length of gear pair k_u	6×10^9	Nm/mm
Torsional stiffness of output shaft k_{s2}	2×10^5	Nm/rad
Half shaft torsional stiffness k_a	8×10^3	Nm/rad
Wheel torsional stiffness k_V	4.5×10^3	Nm/rad
Torsional damping ratio of motor shaft c_{s1}	2	Nm·s/rad
Gear meshing damping c_m	800	N/(m/s)
Torsional damping ratio of output shaft c_{s2}	2	Nm·s/rad

Electronics 2020, 9, 1241

Appendix B

$$
C_{\frac{3s}{5r}} = \frac{2}{3}
\begin{bmatrix}
\cos(-5\theta_e) & \cos\left(-5\theta_e - \frac{2}{3}\pi\right) & \cos\left(-5\theta_e - \frac{4}{3}\pi\right) \\
-\sin(-5\theta_e) & -\sin\left(-5\theta_e - \frac{2}{3}\pi\right) & -\sin\left(-5\theta_e - \frac{4}{3}\pi\right) \\
\sqrt{\frac{1}{2}} & \sqrt{\frac{1}{2}} & \sqrt{\frac{1}{2}}
\end{bmatrix};
$$

$$
C_{\frac{5r}{3s}} = \frac{2}{3}
\begin{bmatrix}
\cos(-5\theta_e) & -\sin(-5\theta_e) & \sqrt{\frac{1}{2}} \\
\cos\left(-5\theta_e - \frac{2}{3}\pi\right) & -\sin\left(-5\theta_e - \frac{2}{3}\pi\right) & \sqrt{\frac{1}{2}} \\
\cos\left(-5\theta_e - \frac{4}{3}\pi\right) & -\sin\left(-5\theta_e - \frac{4}{3}\pi\right) & \sqrt{\frac{1}{2}}
\end{bmatrix}
$$

$$
C_{\frac{3s}{7r}} = \frac{2}{3}
\begin{bmatrix}
\cos(7\theta_e) & \cos\left(7\theta_e - \frac{2}{3}\pi\right) & \cos\left(7\theta_e - \frac{4}{3}\pi\right) \\
-\sin(7\theta_e) & -\sin\left(7\theta_e - \frac{2}{3}\pi\right) & -\sin\left(7\theta_e - \frac{4}{3}\pi\right) \\
\sqrt{\frac{1}{2}} & \sqrt{\frac{1}{2}} & \sqrt{\frac{1}{2}}
\end{bmatrix};
\quad
C_{\frac{7r}{3s}} = \frac{2}{3}
\begin{bmatrix}
\cos(7\theta_e) & -\sin(7\theta_e) & \sqrt{\frac{1}{2}} \\
\cos\left(7\theta_e - \frac{2}{3}\pi\right) & -\sin\left(7\theta_e - \frac{2}{3}\pi\right) & \sqrt{\frac{1}{2}} \\
\cos\left(7\theta_e - \frac{4}{3}\pi\right) & -\sin\left(7\theta_e - \frac{4}{3}\pi\right) & \sqrt{\frac{1}{2}}
\end{bmatrix}
$$

$$
C_{\frac{2r}{5r}} =
\begin{bmatrix}
\cos(-6\theta_e) & \sin(-6\theta_e) \\
-\sin(-6\theta_e) & \cos(-6\theta_e)
\end{bmatrix}
\quad
C_{\frac{5r}{2r}} =
\begin{bmatrix}
\cos(-6\theta_e) & -\sin(-6\theta_e) \\
\sin(-6\theta_e) & \cos(-6\theta_e)
\end{bmatrix}
$$

$$
C_{\frac{2r}{7r}} =
\begin{bmatrix}
\cos(6\theta_e) & \sin(6\theta_e) \\
-\sin(6\theta_e) & \cos(6\theta_e)
\end{bmatrix}
\quad
C_{\frac{5r}{2r}} =
\begin{bmatrix}
\cos(6\theta_e) & -\sin(6\theta_e) \\
\sin(6\theta_e) & \cos(6\theta_e)
\end{bmatrix}
$$

References

1. Wu, Y.C.; Sun, Z.H. Design and Analysis of a Novel Speed-Changing Wheel Hub with an Integrated Electric Motor for Electric Bicycles. *Math. Probl. Eng.* **2013**, *2013*, 8. [CrossRef]
2. Burkhardt, Y.; Spagnolo, A.; Lucas, P.; Zavesky, M.; Brockerhoff, P.; IEEE. Design and analysis of a highly integrated 9-phase drivetrain for EV applications. In Proceedings of the 2014 International Conference on Electrical Machines, Berlin, Germany, 2–5 September 2014; pp. 450–456.
3. Zhu, C.; Zeng, Z.; Zhao, R. Torque ripple elimination based on inverter voltage drop compensation for a three-phase four-switch inverter-fed PMSM drive under low speeds. *IET Power Electron.* **2017**, *10*, 1430–1437. [CrossRef]
4. Viswanathan, V.; Seenithangom, J. Commutation Torque Ripple Reduction in the BLDC Motor Using Modified SEPIC and Three-Level NPC Inverter. *IEEE Trans. Power Electron.* **2018**, *33*, 535–546. [CrossRef]
5. Viswanathan, V.; Jeevananthan, S. Hybrid converter topology for reducing torque ripple of BLDC motor. *IET Power Electron.* **2017**, *10*, 1572–1587. [CrossRef]
6. Viswanathan, V.; Jeevananthan, S. Reducing torque ripple of BLDC motor by integrating dc-dc converter with three-level neutral-point-clamped inverter. *Compel. Int. J. Comput. Math. Electr. Electron. Eng.* **2016**, *35*, 959–981. [CrossRef]
7. Zeng, Z.; Zhu, C.; Jin, X.; Shi, W.; Zhao, R. Hybrid Space Vector Modulation Strategy for Torque Ripple Minimization in Three-Phase Four-Switch Inverter-Fed PMSM Drives. *IEEE Trans. Ind. Electron.* **2017**, *64*, 2122–2134. [CrossRef]
8. Zhu, C.; Zeng, Z.; Zhao, R. Comprehensive Analysis and Reduction of Torque Ripples in Three-Phase Four-Switch Inverter-Fed PMSM Drives Using Space Vector Pulse-Width Modulation. *IEEE Trans. Power Electron.* **2017**, *32*, 5411–5424. [CrossRef]
9. Baik, J.; Yun, S.; Kim, D.; Kwon, C.; Yoo, J. Remote-State PWM with Minimum RMS Torque Ripple and Reduced Common-Mode Voltage for Three-Phase VSI-Fed BLAC Motor Drives. *Electronics* **2020**, *9*, 586. [CrossRef]
10. Joryo, S.; Tatsumi, K.; Morizane, T.; Taniguchi, K.; Kimura, N.; Omori, H.; IEEE. Study of Torque ripple reduction and Torque boost by Modified Trapezoidal Modulation. In Proceedings of the 2018 International Power Electronics Conference, Niigata, Japan, 20–24 May 2018; pp. 1202–1205.

11. Shimmyo, S.; Takeuchi, K.; Takahashi, N.; Matsushita, M.; Ohnishi, K. Multi-level Motor Drives for Torque Ripple Suppression Taking Control Sensitivity into Account. *IEEJ J. Ind. Appl.* **2016**, *5*, 69–77. [CrossRef]

12. Mohan, D.; Zhang, X.; Foo, G.H.B. Three-Level Inverter-Fed Direct Torque Control of IPMSM with Torque and Capacitor Voltage Ripple Reduction. *IEEE Trans. Energy Convers.* **2016**, *31*, 1559–1569. [CrossRef]

13. Mohan, D.; Zhang, X.; Foo, G.H.B. A Simple Duty Cycle Control Strategy to Reduce Torque Ripples and Improve Low-Speed Performance of a Three-Level Inverter Fed DTC IPMSM Drive. *IEEE Trans. Ind. Electron.* **2017**, *64*, 2709–2721. [CrossRef]

14. Wang, X.; Zhou, Y.; Yang, D.; Shi, X.; IEEE. Direct Torque Control of Three-Level Inverter-Fed PMSM Based on Zero Voltage Vector Distribution for Torque Ripple Reduction. In Proceedings of the 2017 29th Chinese Control and Decision Conference, Chongqing, China, 28–30 May 2017; pp. 7776–7781.

15. Mohan, D.; Zhang, X.N.; Foo, G.H.B. Three-Level Inverter-Fed Direct Torque Control of IPMSM With Constant Switching Frequency and Torque Ripple Reduction. *IEEE Trans. Ind. Electron.* **2016**, *63*, 7908–7918. [CrossRef]

16. Mohan, D.; Zhang, X.N.; Foo, G.H.B. Generalized DTC Strategy for Multilevel Inverter Fed IPMSMs With Constant Inverter Switching Frequency and Reduced Torque Ripples. *IEEE Trans. Energy Convers.* **2017**, *32*, 1031–1041. [CrossRef]

17. Tatte, Y.N.; Aware, M.V. Torque Ripple and Harmonic Current Reduction in a Three-Level Inverter-Fed Direct-Torque-Controlled Five-Phase Induction Motor. *IEEE Trans. Ind. Electron.* **2017**, *64*, 5265–5275. [CrossRef]

18. Dhiman, S.; Hussain, A.; Kumar, V.T. An Effective Voltage Switching State Algorithm for Direct Torque Controlled Five-Phase Induction Motor Drive to Reduce Torque Ripple. In Proceedings of the 2016 IEEE Students' Conference on Electrical, Electronics and Computer Science, Bhopal, India, 5–6 March 2016.

19. Zhang, G.Z.; Chen, C.; Gu, X.; Wang, Z.A.; Li, X.M. An Improved Model Predictive Torque Control for a Two-Level Inverter Fed Interior Permanent Magnet Synchronous Motor. *Electronics* **2019**, *8*, 769. [CrossRef]

20. Dong, Y.; Nuchkrua, T.; Shen, T. Asymptotical stability contouring control of dual-arm robot with holonomic constraints: Modified distributed control framework. *IET Control Theory Appl.* **2019**, *13*, 2877–2885. [CrossRef]

21. Li, K.L.; Boonto, S.; Nuchkrua, T. On-line Self Tuning of Contouring Control for High Accuracy Robot Manipulators under Various Operations. *Int. J. Control Autom. Syst.* **2020**, *18*, 1818–1828. [CrossRef]

22. Lu, J.; Yang, J.G.; Ma, Y.C.; IEEE. Research on Harmonic Compensation for Flux and Current of Permanent Magnet Synchronous Motor. In Proceedings of the 2015 IEEE International Conference on Advanced Intelligent Mechatronics, Busan, Korea, 7–11 July 2015; pp. 589–594.

23. Boroujeni, M.S.; Markadeh, G.R.A.; Soltani, J. Torque ripple reduction of brushless DC motor based on adaptive input-output feedback linearization. *ISA Trans.* **2017**, *70*, 502–511. [CrossRef]

24. Boroujeni, M.S.; Markadeh, G.A.; Soltani, J.; IEEE. Adaptive Input-Output Feedback Linearization Control of Brushless DC Motor with Arbitrary Current Reference using Voltage Source Inverter. In Proceedings of the 2017 8th Power Electronics, Drive Systems & Technologies Conference, Mashhad, Iran, 14–16 February 2017; pp. 537–542.

25. Jedryczka, C.; Danielczyk, D.; Szelag, W. Torque Ripple Minimization of the Permanent Magnet Synchronous Machine by Modulation of the Phase Currents. *Sensors* **2020**, *20*, 2406. [CrossRef]

26. Boroujeni, M.S.; Markadeh, G.A.; Soltani, J.; Blaabjerg, F. Torque ripple reduction of brushless DC motor with harmonic current injection based on integral terminal sliding mode control. *IET Electr. Power Appl.* **2018**, *12*, 25–36. [CrossRef]

27. Li, Z.; Peng, Z. Nonlinear dynamic response of a multi-degree of freedom gear system dynamic model coupled with tooth surface characters: A case study on coal cutters. *Nonlinear Dyn.* **2016**, *84*, 271–286. [CrossRef]

28. Chen, Z.G.; Zhai, W.M.; Wang, K.Y. Vibration feature evolution of locomotive with tooth root crack propagation of gear transmission system. *Mech. Syst. Signal Process.* **2019**, *115*, 29–44. [CrossRef]

29. Ding, H.L.; Kahraman, A. Interactions between nonlinear spur gear dynamics and surface wear. *J. Sound Vibr.* **2007**, *307*, 662–679. [CrossRef]

30. Cai, Z.; Lin, C. Dynamic Model and Analysis of Nonlinear Vibration Characteristic of a Curve-Face Gear Drive. *Stroj. Vestn. J. Mech. Eng.* **2017**, *63*, 161–170. [CrossRef]

31. Lin, C.; Liu, Y.; Gu, S.J. Analysis of nonlinear twisting vibration characteristics of orthogonal curve-face gear drive. *J. Braz. Soc. Mech. Sci. Eng.* **2015**, *37*, 1499–1505. [CrossRef]

32. Parker, R.C.; Ambarisha, V.K.; Asme. Nonlinear dynamics of planetary gears using analytical and finite element models. *J. Sound Vib.* **2008**, *302*, 577–595.

33. Liu, C.Z.; Qin, D.T.; Wei, J.; Liao, Y.H. Investigation of nonlinear characteristics of the motor-gear transmission system by trajectory-based stability preserving dimension reduction methodology. *Nonlinear Dyn.* **2018**, *94*, 1835–1850. [CrossRef]

34. Bai, W.; Qin, D.; Wang, Y.; Lim, T.C. Dynamic characteristics of motor-gear system under load saltations and voltage transients. *Mech. Syst. Signal Process.* **2018**, *100*, 1–16. [CrossRef]

35. Bai, W.; Qin, D.; Wang, Y.; Lim, T.C. Dynamic characteristic of electromechanical coupling effects in motor-gear system. *J. Sound Vib.* **2018**, *423*, 50–64. [CrossRef]

36. Wang, Z.W.; Mei, G.M.; Xiong, Q.; Yin, Z.H.; Zhang, W.H. Motor car-track spatial coupled dynamics model of a high-speed train with traction transmission systems. *Mech. Mach. Theory* **2019**, *137*, 386–403. [CrossRef]

37. Wei, J.T.; Kang, J.S.; Cui, Y.H. Analysis and Inhibition of Permanent Magnet Synchronous Motor Torque Ripple. *Mechatronics* **2014**, *12*, 15–19. [CrossRef]

38. Liao, Y.; Zhen, S.; Liu, R.; Yao, J. Torque Ripple Suppression of Permanent Magnet Synchronous Motor by the Harmonic Injection. In Proceedings of the 2018 Asia-Pacific Magnetic Recording Conference, Shanghai, China, 15–17 November 2011; Volume 21, pp. 121–129.

Article

Discrete-Time Neural Control of Quantized Nonlinear Systems with Delays: Applied to a Three-Phase Linear Induction Motor

Alma Y. Alanis [1], Jorge D. Rios [1,*], Javier Gomez-Avila [1], Pavel Zuniga [1] and Francisco Jurado [2]

[1] University Center of Exact Sciences and Engineering, University of Guadalajara, Marcelino Garcia Barragan Blvd. 1421, Guadalajara 44430, Mexico; alma.alanis@academicos.udg.mx (A.Y.A.); jenrique.gomez@academicos.udg.mx (J.G.-A.); pavel.zuniga@academicos.udg.mx (P.Z.)

[2] División de Estudios de Posgrado e Investigación, Tecnológico Nacional de México/I.T. La Laguna, Revolución Blvd. and Instituto Tecnológico de La Laguna Av., Torreón 27000, Mexico; fjurado@itlalaguna.edu.mx

* Correspondence: jorge.rarranaga@academicos.udg.mx

Received: 22 July 2020; Accepted: 5 August 2020; Published: 7 August 2020

Abstract: This work introduces a neural-feedback control scheme for discrete-time quantized nonlinear systems with time delay. Traditionally, a feedback controller is designed under ideal assumptions that are unrealistic for real-work problems. Among these assumptions, they consider a perfect communication channel for controller inputs and outputs; such a perfect channel does not consider delays, or noise introduced by the sensors and actuators even if such undesired phenomena are well-known sources of bad performance in the systems. Moreover, traditional controllers are also designed based on an ideal plant model without considering uncertainties, disturbances, sensors, actuators, and other unmodeled dynamics, which for real-life applications are effects that are constantly present and should be considered. Furthermore, control system design implemented with digital processors implies sampling and holding processes that can affect the performance; considering and compensating quantization effects of measured signals is a problem that has attracted the attention of control system researchers. In this paper, a neural controller is proposed to overcome the problems mentioned above. This controller is designed based on a neural model using an inverse optimal approach. The neural model is obtained from available measurements of the state variables and system outputs; therefore, uncertainties, disturbances, and unmodeled dynamics can be implicitly considered from the available measurements. This paper shows the performance and effectiveness of the proposed controller presenting real-time results obtained on a linear induction motor prototype. Also, this work includes stability proof for the whole scheme using the Lyapunov approach.

Keywords: quantized; nonlinear systems; time delay; lyapunov approach; real-time implementation

---

1. Introduction

Traditionally, a control system is designed based on many assumptions, which are rarely satisfied in real-life systems. One of these assumptions is a mathematical model that perfectly represents the behavior of the system. However, most of the used mathematical models do not consider disturbances, uncertainties, parametric variations, unmodeled dynamics, among other issues. Moreover, there is the assumption that the state measurements and calculated control signals are transmitted through ideal communication channels without the noise of sensors, signal delays, and packet losses, which are phenomena that are present in the day to day systems. These factors are sources that need to be compensated adequately by the controller to have proper performance and avoid stability problems.

In control system design, the existence of delays can severely degrade the performance of the controllers, as well as causing a loss in system stability, among other undesirable effects [1–3]. Ignoring measurement noise when designing a controller can cause undesired performance or system failure, especially for systems with high sensitivity to uncertainties [4]. Furthermore, control system design implemented with digital processors implies sampling and holding processes; both operations can severely affect the performance of the designed controller [5]. Moreover, the design of controllers that consider and compensate quantization effects of measured signals is a problem that has attracted the attention of control system researchers for many years [6], which is highly relevant due to the importance that the investigation of networked control systems has gained recently [7].

Previous research [8,9] show a successful trajectory tracking based on ideal conditions. Also, the use of intelligent system techniques has allowed the control schemes to be designed in a manner inspired by nature. These complex dynamic systems include robotic navigation, synchronization, complex networks, control of complex processes, smart grids, among others, see [10–13]. Mainly, neural networks and fuzzy logic systems have made significant progress in designing solutions for complex control systems. In [14], deterministic learning is presented to identify nonlinear dynamic systems based on sampled data sequences with simulation results applied to chaotic systems using radial basis neural networks. In [15], a backstepping controller based on a fuzzy observer for the control of a nonlinear system with strict second-order feedback is presented. In [16], an adaptive neural controller for uncertain nonlinear systems with state and input restrictions is presented, with simulation-level results, for a single-link robot.

Regarding the design of controllers considering quantization, in [17], a feedback control with static quantization is presented for discrete-time systems. In [18], a feedback optimal neural controller is designed for nonlinear systems quantized in discrete time with input constraints. In [19], the case of the stabilization of nonlinear systems based on three-layer neural networks as a universal approximator of dynamic systems is considered. In [20], a quantized adaptive controller for uncertain nonlinear systems in the presence of external disturbances is presented. Another current approach is described in [21], which shows the design of an adaptive fault-tolerant controller with input prescribed performance for trajectory tracking with quantization and unknown control directions. A similar problem is analyzed in [3], where the use of a Takagi–Sugeno fuzzy controller is proposed for systems in the network under actuator saturation, measurement noise, and quantization. In [22], an inverse optimal neural controller is presented for nonlinear discrete systems considering disturbances; in [2], time delays, as well as disturbances, are considered in neural controller design. Nevertheless, it must be noted that none of the above-referenced works include a real-time implementation of a controller, which considers at the same time disturbances, noise, delays, and quantization for a system with an unknown model.

Please note that all the controllers mentioned above only include simulation-level results due to the difficulty of adequately measuring the effects referred above and the complexity of their corresponding stability analysis.

Motivated by these facts, this paper presents the design and implementation of a discrete-time neural controller for quantized nonlinear systems with delays and measurement noise, while presenting disturbances, uncertainties and unmodeled dynamics, also, without the need for prior knowledge of the system model or the need to estimate its bounds, nor nominal models. Accordingly, the main contributions of this paper can be defined as follows: First, it is designed a neural controller for uncertain nonlinear systems under the inverse optimality approach; second, such controller is implemented in real time and third stability proof of the entire system is included using a Lyapunov approach considering: quantization of the state and input signals, unknown state delays, measurement noise, external disturbances, uncertainties, unknown model, among others.

This paper is organized as follows: Section 2 includes some preliminary concepts and problem statement. Section 3 describes the design of the discrete-time neural controller for uncertain nonlinear systems, including the Lyapunov proof. Section 4 presents real-time results, and in Section 5, conclusions about our proposal are presented.

2. Problem Statement

We propose to solve the problem of designing a discrete-time neural controller for quantized uncertain nonlinear systems under the presence of delays, without the need to have prior knowledge of the system model. First, essential definitions are considered below to define a neural model that considers all the dynamics caused by the above-mentioned phenomena. Based on such a model, a controller is designed to achieve trajectory tracking.

2.1. System Definition

Let us consider a discrete-time unknown nonlinear delayed system that can be written in an affine form as follows:

$$
\begin{aligned}
x\,(k+1) &= f(x\,(k)) + g(x\,(k))u\,(k)\,, \\
y\,(k) &= Cx\,(k)\,,
\end{aligned}
\tag{1}
$$

where $x\,(k) \in \Re^n$ is the state of the system, $u\,(k) \in \Re^m$ is the control input, and $f(x\,(k)) : \Re^n \to \Re^n$ and $g(x\,(k)) : \Re^n \to \Re^{n \times m}$ are smooth maps. Without loss of generality, system (1) is assumed to have an equilibrium point at $x\,(k) = 0$.

Then (1) can be rewritten as:

$$
\begin{aligned}
x\,(k-l) &= \begin{bmatrix} x_1\,(k) & \cdots & x_i\,(k) & \cdots & x_n\,(k) \end{bmatrix}^\mathsf{T}, \\
d\,(k) &= \begin{bmatrix} d_1\,(k) & \cdots & d_i\,(k) & \cdots & d_n\,(k) \end{bmatrix}^\mathsf{T}, \\
x_i\,(k+1) &= f_i(x\,(k)) + g_i(x\,(k))u\,(k)\,, \\
y\,(k) &= Cx\,(k)\,.
\end{aligned}
\tag{2}
$$

2.2. Inverse Optimal Control Methodology

The main idea behind inverse optimal control methodology is to design a control law and then the cost functional [23]. For the state feedback control synthesis, it is necessary to fulfill the following Assumption:

Assumption 1. *The full state x (k) of system (1) is available.*

If the full state vector is not available for measurement, Assumption 1 may be fulfilled using a state estimator considering the discrete-time separation principle for nonlinear systems.

The trajectory tracking problem can be solved as a regulation problem for an error system that should be regulated to the origin. For such transformation, an auxiliary error system should be defined using error equations and its dynamics, in this case, let us consider the trajectory tracking error as:

$$
\psi\,(k) = x\,(k) - x_d\,(k)\,,
\tag{3}
$$

with $x_d\,(k)$ the desired trajectory for $x\,(k)$, it is important to note that the proposed controller only requires the knowledge of $x_d\,(k+1)$ and there is no longer a restriction for x_d, which is an improvement with respect to other controllers. Let us propose a control Lyapunov function (CLF) as

$$
V(\psi\,(k)) = \frac{1}{2}\psi^T\,(k)\,\mathcal{P}\psi\,(k)\,, \quad \mathcal{P} = \mathcal{P}^T > 0,
\tag{4}
$$

to ensure the stability of the trajectory tracking error (3), which will be achieved by defining an appropriate matrix $\mathcal{P}$. Instead of solving the Hamilton-Jacobi-Bellman (HJB) equation, the inverse optimal control synthesis is based on the knowledge of $V(\psi(k))$. The control law $u(k)$ can be defined as:

$$
\begin{aligned}
u(k) &= -\frac{1}{2}\mathcal{R}^{-1}(\psi(k))\,g^T(\psi(k))\,\frac{\partial V(\psi(k+1))}{\partial \psi(k+1)} \\
&= -\frac{1}{2}\left(\mathcal{R}(\psi(k)) + \frac{1}{2}g^T(\psi(k))\,\mathcal{P}g(\psi(k))\right)^{-1} g^T(\psi(k))\,\mathcal{P}f^T(\psi(k)),
\end{aligned}
\tag{5}
$$

where $\mathcal{R}(\psi(k)) = \mathcal{R}^T(\psi(k)) > 0$ is a weighting matrix whose entries can be functions of the system state or can be fixed, i.e., it can be selected as $\mathcal{R}(\psi(k)) = \mathcal{R}$. Since $\mathcal{P}$ and $\mathcal{R}(\psi(k))$ are positive definite and symmetric matrices, the existence of the inverse in (5) is ensured [22,23]. Now, the control synthesis goal is to obtain an appropriate matrix $\mathcal{P}$ for (5) such that $\mathcal{P} = \mathcal{P}^T > 0$, and its values are determined heuristically in order to improve trajectory tracking performance; stability analysis guarantees optimality of the inverse optimal controller. The following theorem establishes a sufficient condition for the matrix $\mathcal{P}$ regarding inverse optimal control.

Theorem 1 ([23]). *Consider the system* (1)*. If there exists a matrix* $\mathcal{P} = \mathcal{P}^T > 0$ *such that the following inequality holds:*

$$
V_f(\psi(k)) - \frac{1}{4}\mathcal{P}_1^T(\psi(k))\,(\mathcal{R}(\psi(k)) + \mathcal{P}_2(\psi(k)))^{-1}\mathcal{P}_1(\psi(k)) \leq -\psi^T(k)\,\mathcal{Q}\psi(k),
$$

where

$$
\begin{aligned}
V_f(\psi(k)) &= \frac{1}{2}f^T(\psi(k))\,\mathcal{P}f(\psi(k)) - V(\psi(k)), \\
\mathcal{P}_1(\psi(k)) &= g^T(\psi(k))\,\mathcal{P}f(\psi(k)), \\
\mathcal{P}_2(\psi(k)) &= \frac{1}{2}g^T(\psi(k))\,\mathcal{P}g(\psi(k)),
\end{aligned}
\tag{6}
$$

$V(\psi(k))$ *as defined in* (4) *and* $\mathcal{Q} = \mathcal{Q}^T > 0$*, then, the equilibrium point* $\psi(k) = 0$ *from the system* (1) *is globally exponentially stabilized by the control law* (5)*. Moreover, this control law is inverse optimal in the sense that it minimizes the meaningful functional given by*

$$
J(k) = \sum_{k=0}^{\infty}\left(l(\psi(k)) + u^T(k)\,\mathcal{R}(\psi(k))\,u(k)\right),
\tag{7}
$$

with $l(\psi(k)) = -V(\psi(k+1)) + V(\psi(k)) - u^{*T}\mathcal{R}(\psi(k))\,u^*(k)$.

Proof. For the detailed proof of Theorem 1, we refer the reader to [23]. $\square$

System (1), is described in an affine form, which is not a trivial requirement for general nonlinear systems, mainly for unknown nonlinear systems. Therefore, in this paper, a neural identifier is used to provide an accurate model for the nonlinear system to be controlled. The model should be presented in a block control form [23].

2.3. Quantizers in a Control Loop

Different kinds of quantizers are considered in the literature. These include static, dynamic, uniform, logarithmic, hysteresis, among other kinds of quantizers. However, in practice, uniform quantizers are typically used, mainly due to technological constraints, although other quantizers may be advantageous for analytical purposes [7]. In our work, a dynamic uniform quantizer is considered.

In this paper, the general quantizer is defined as the one described in [17]. Let us assume that $v \in \Re^s$ and that there exist numbers $M > 0$ and $D > 0$ such that the following condition holds:

$$|q(v) - v| \leq D, \text{ if } |v| \leq M, \tag{8}$$

where v is the signal to be quantized, $q(v)$ is the signal after quantization, M denotes the range of the quantizer, and D stands for the quantization error boundary of the quantizer. In the developed control strategy, we consider the one-parameter family of quantizers that can be defined as:

$$q_\mu(v) = \mu q \left(\frac{v}{\mu} \right), \tag{9}$$

where $\mu > 0$ is the parameter of the quantizer which can be computed independently, therefore $q(v)$ in (8) can be defined as (9), although, in this paper, it is considered unknown. Let us consider the quantized measured state signal and the quantized control input signal in the following form:

$$x_q(k) = q_{\mu_x}(x(k)) = \mu_x q \left(\frac{x(k)}{\mu_x} \right), \tag{10}$$

$$u_q(k) = q_{\mu_u}(u(k)) = \mu_u q \left(\frac{u(k)}{\mu_u} \right), \tag{11}$$

where $q_{\mu_*}(\cdot)$ is a dynamic quantizer defined by (9), which is comprised of a dynamic scaling μ_* and by a static quantizer $q(\cdot)$ defined by (8) with range M_* and error D_* [18].

3. Neural Controller Design

The inverse optimal controller defined in the previous section is defined for an ideal system. Such a system is considered ideal since the calculated control law is directly applied to the system without the actuator limitations and input quantization. Moreover, such a controller considers full access to the ideal value of state variables, i.e., without quantization error in sensors. However, real-time applications include the system to be controlled as well as attached sensors and actuators. These actuators and sensors are part of the entire system. Therefore, their dynamics affect system operation and should be considered for the control design. In this section, a discrete-time neural controller is proposed for quantized nonlinear systems with signal delays. First, it is necessary to obtain an accurate model for the unknown nonlinear system. The use of a recurrent high order neural network is proposed to identify the unknown nonlinear system. Second, based on the obtained neural model, a controller is designed using an inverse optimality approach. It is important to note that the system to be modeled by the neural network includes the system itself, and the dynamics of the actuators and sensors. This is possible since the neural network is trained with physical measurements from sensors, as depicted in Figure 1.

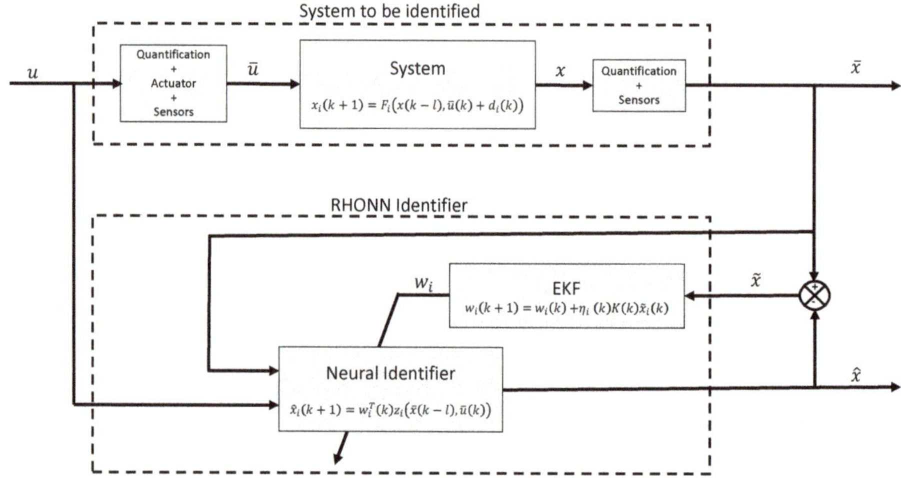

Figure 1. Schematic representation for neural identification of discrete-time unknown nonlinear quantized delayed systems with measurement noise.

3.1. Neural Identification

As previously explained, the quantization error is a source of problems that can be encountered in real-time implementations. In almost all quantized output feedback control systems, quantization effects exist in the communication channel from sensor-to-controller, as well as from controller-to-actuator [17]. From now on, let us consider that the measured system signals and the controller output will be quantized before they are transmitted to the controller and the plant via a communication network.

However, it is necessary to consider that both measurement and actuator errors due to the unmodeled dynamics that appear in real-life problems; in this manner, an identification scheme can be represented as in Figure 1. Moreover, for unknown nonlinear systems with uncertainties, disturbances, and delays, a neural identifier can give an accurate model. System illustrated in Figure 1, can be modeled as:

$$\begin{aligned} x_i(k+1) &= f_i(\bar{x}(k-l)) + g_i(\bar{x}(k-l))u(k) + d_i(k), \quad i = 1, \cdots, n, \\ y(k) &= Cx(k), \end{aligned} \tag{12}$$

where $\bar{x}$ is the quantized state, u represents the ideal calculated control law, l is the unknown delay for the state, and d_i represents unknown disturbances. From Figure 1, it can be seen that a recurrent high order neural network (RHONN) is trained with quantized measured values of the state $\bar{x}$ and calculated values of the input u.

In this paper, the measured values of the state $\bar{x}$ are considered to be an unknown nonlinear function which depends on its quantized error e_s, delay l and time k, as follows:

$$\bar{x}(k-l) = \delta\left(x_q(k), e_s(k), l, k\right) < \bar{\delta},$$

with γ an unknown function assumed bounded by $\bar{\gamma}$ which is also considered unknown; also, x_q is defined as in (10). Then, we propose a neural identifier for the system (12), with the following structure:

$$\hat{x}(k) = \begin{bmatrix} \hat{x}_1(k) & \cdots & \hat{x}_i(k) & \cdots & \hat{x}_n(k) \end{bmatrix}^\top,$$

$$\hat{x}_i(k+1) = w_i^\top z_i(\overline{x}(k-l), u(k)), \tag{13}$$

with $z_i(\overline{x}(k-l), u(k))$ defined as

$$z_i(x(k), \varrho(k)) = \begin{bmatrix} z_{i_1} \\ z_{i_2} \\ \vdots \\ z_{i_{L_i}} \end{bmatrix} = \begin{bmatrix} \Pi_{j \in I_1} \varsigma_{i_j}^{d_{ij}(1)} \\ \Pi_{j \in I_2} \varsigma_{i_j}^{d_{ij}(2)} \\ \vdots \\ \Pi_{j \in I_{L_i}} \varsigma_{i_j}^{d_{ij}(L_i)} \end{bmatrix}, \tag{14}$$

with $d_{i_j}(k)$ being non-negative integers, and

$$\varsigma_i = \begin{bmatrix} \varsigma_{i_1} \\ \vdots \\ \varsigma_{i_n} \\ \varsigma_{i_{n+1}} \\ \vdots \\ \varsigma_{i_{n+m}} \end{bmatrix} = \begin{bmatrix} S(\overline{x}_1) \\ \vdots \\ S(\overline{x}_n) \\ u_1(k) \\ \vdots \\ u_m(k) \end{bmatrix}, \tag{15}$$

where $S(\bullet)$ is defined as any sigmoid function as:

$$S(\varsigma) = \frac{1}{1 + \exp(-\beta\varsigma)}, \quad \beta > 0, \tag{16}$$

and ς is any real valued variable.

As discussed in [24], the general discrete-time nonlinear system (1) can be approximated by a RHONN. In this paper the following discrete-time RHONN has been selected:

$$x(k+1) = w^{*\top} z(\overline{x}(k-l), u(k)) + \epsilon_z, \tag{17}$$

and in the state component-wise form as

$$x_i(k+1) = w_i^{*\top} z_i(\overline{x}(k-l), u(k)) + \epsilon_{z_i}, \quad i = 1, \cdots, n, \tag{18}$$

where $\overline{x}$ is the measured plant state, and ϵ_{z_i} is a bounded approximation error, which can be reduced by increasing the number of the adjustable weights [24]. Let us assume that there exists an optimal weights vector $w_i^* \in \Re^{L_i}$, such that $\|\epsilon_{z_i}\|$ is minimized on a compact set $w_{z_i} \subset \Re^{L_i}$ which is an artificial quantity required only for analytical purposes. In general, it is assumed that this optimal weight vector exists, and it is constant but unknown [24].

Let us define the estimate of w_i^* as w_i; then, the weight estimation error $\tilde{w}_i(k)$ and identification error $\tilde{x}_i(k)$ are defined, respectively as

$$\tilde{w}_i(k) = w_i^* - w_i(k), \tag{19}$$

and

$$\tilde{x}_i(k) = \overline{x}_i(k) - \hat{x}_i(k). \tag{20}$$

The weight vectors are updated online with a decoupled extended Kalman filter (EKF) training algorithm, described by

$$
\begin{aligned}
w_i(k+1) &= w_i(k) + \eta_i(k) K_i(k) \tilde{x}_i(k), & (21)\\
K_i(k) &= P_i(k) H_i(k) M_i(k), & (22)\\
P_i(k+1) &= P_i(k) - K_i(k) H_i^{\top}(k) P_i(k) + Q_i(k), & (23)
\end{aligned}
$$

with

$$
\begin{aligned}
M_i(k) &= \left[R_i(k) + H_i^{\top}(k) P_i(k) H_i(k) \right]^{-1}, & (24)\\
i &= 1,\cdots,n,
\end{aligned}
$$

and the identification error is defined as in (20). The dynamics of $x_i(k+1)$ can be expressed as

$$
\tilde{x}_i(k+1) = x_i(k+1) - \hat{x}_i(k+1) + e_{qx_i}(k), \tag{25}
$$

with

$$
e_{qx_i}(k) = \bar{x}_i(k) - x_i(k), \tag{26}
$$

as the state quantization error. From (10), this last equation can be rewritten as:

$$
e_{qx_i}(k) = \mu_x \left(q\left(\frac{x(k)}{\mu_x} \right) - \frac{x(k)}{\mu_x} \right). \tag{27}
$$

To determine the state quantization error, first consider the property defined in (8), then,

$$
\left| \frac{x(k)}{\mu_x} \right| \le M_x, \tag{28}
$$

and

$$
\left| q\left(\frac{x(k)}{\mu_x} \right) - \frac{x(k)}{\mu_x} \right| \le D_x, \tag{29}
$$

By using the homogeneity property of the Euclidian norm, from (27) the following equation is obtained

$$
\begin{aligned}
\left| e_{qx_i}(k) \right| &= \left| \mu_x \left(q\left(\frac{x(k)}{\mu_x} \right) - \frac{x(k)}{\mu_x} \right) \right| \\
&= \mu_x \left| \left(q\left(\frac{x(k)}{\mu_x} \right) - \frac{x(k)}{\mu_x} \right) \right| \le \mu_x \Delta_x.
\end{aligned} \tag{30}
$$

Considering the quantized system (18) and (13), the dynamics of (25) can be defined as:

$$
\begin{aligned}
\tilde{x}_i(k+1) &= x_i(k+1) - \hat{x}_i(k+1) + e_{qx_i}(k) \\
&= \tilde{w}_i^{T}(k) z_i(\bar{x}(k)) + \epsilon_{z_i} + e_{qx_i}(k).
\end{aligned} \tag{31}
$$

By using (21), the weight estimation error dynamics (19) can be defined as:

$$
\tilde{w}_i(k+1) = \tilde{w}_i(k) - \eta_i(k) K_i(k) \tilde{x}_i(k). \tag{32}
$$

Thus, considering (12), (31) and (32), then, it is possible to establish the following main result for an unknown nonlinear system with nonlinear output.

Theorem 2. *For a quantized delayed system* (12), *the neural identifier* (13), *trained with the EKF-based algorithm* (21), *ensures that the identification error* (25) *and the weight estimation error* (32) *are semi-globally uniformly ultimately bounded (SGUUB).*

Proof of Theorem 2. Step 1. For each element of $V(k)$, defined as $V_i(k)$ for $i = 1, 2, \cdots, n$, let us consider the following Lyapunov function candidate:

$$V_i(k) = \gamma_i (\tilde{x}_i(k))^2 + \tilde{w}_i^T(k) P_i(k) \tilde{w}_i(k). \tag{33}$$

Its first difference can be written as:

$$\begin{aligned}
\Delta V_i(k) &= \gamma_i (\tilde{x}_i(k+1))^2 + \tilde{w}_i^T(k+1) P_i(k+1) \tilde{w}_i(k+1) - \gamma_i (\tilde{x}_i(k))^2 \\
&\quad - \tilde{w}_i^T(k) P_i(k) \tilde{w}_i(k).
\end{aligned} \tag{34}$$

Substituting (31) and (32) in (34) it yields

$$\begin{aligned}
\Delta V_i(k) &= \gamma_i \left(\tilde{w}_i^T(k) z_i (\bar{x}(k-l), u(k)) + \epsilon_{z_i} + e_{qx_i}(k) \right)^2 \\
&\quad + (\tilde{w}_i(k) - \eta_i(k) K_i(k) \tilde{x}_i(k))^T (P_i(k) - A_i(k)) \\
&\quad \times (\tilde{w}_i(k) - \eta_i(k) K_i(k) \tilde{x}_i(k)) - \gamma_i (\tilde{x}_i(k))^2 - \tilde{w}_i^T(k) P_i(k) \tilde{w}_i(k).
\end{aligned}$$

Then,

$$\begin{aligned}
\Delta V_i(k) &= \gamma_i \left(\tilde{w}_i^T(k) z_i (\bar{x}(k-l), u(k)) \right)^2 - \tilde{w}_i^T(k) A_i(k) \tilde{w}_i(k) \\
&\quad + \eta_i^2(k) \tilde{x}_i^T(k) K_i^T(k) P_i(k) K_i(k) \tilde{x}_i(k) - \gamma_i (\tilde{x}_i(k))^2 \\
&\quad - \eta_i(k) \tilde{x}_i^T(k) K_i^T(k) A_i(k) \tilde{x}_i(k) + 2\gamma_i \tilde{w}_i^T(k) z_i (\bar{x}(k-l), u(k)) \epsilon_{z_i} \\
&\quad + 2\gamma_i \tilde{w}_i^T(k) z_i (\bar{x}(k-l), u(k)) e_{qx_i}(k) - \eta_i(k) \tilde{w}_i^T(k) P_i(k) K_i(k) \tilde{x}_i(k) \\
&\quad + \tilde{w}_i^T(k) A_i(k) \tilde{x}_i(k) - \eta_i(k) \tilde{x}_i^T(k) K_i^T(k) P_i(k) \tilde{w}_i(k) \\
&\quad + \eta_i(k) \tilde{x}_i^T(k) K_i^T(k) A_i(k) \tilde{w}_i(k) + \gamma_i e_{z_i}^2 + 2\gamma_i \epsilon_{z_i} e_{qx_i}(k) + \gamma_i e_{qx_i}^2(k).
\end{aligned}$$

with $A_i(k) = K_i(k) H_i^\top(k) P_i(k) - Q_i(k)$. Now, using the following inequalities

$$\begin{aligned}
X^T X + Y^T Y &\geq 2X^T Y, \\
X^T X + Y^T Y &\geq -2X^T Y, \\
-\lambda_{\min}(P)(X^T X) &\geq -X^T P X \geq -\lambda_{\max}(P)(X^T X),
\end{aligned} \tag{35}$$

which are valid $\forall X, Y \in \Re^n$ with $\forall P \in \Re^{n \times n}$, $P = P^T > 0$, using (35), then $\Delta V_i(k)$ can be rewritten as

$$
\begin{aligned}
\Delta V_i(k) \;\leq\;& \gamma_i \left(\tilde{w}_i^T(k) z_i(\overline{x}(k-1), \overline{u}(k)) \right)^2 - \tilde{w}_i^T(k) A_i(k) \tilde{w}_i(k) \\
&+ \eta_i^2(k) \tilde{x}_i^T(k) K_i^T(k) P_i(k) K_i(k) \tilde{x}_i(k) - \gamma_i (\tilde{x}_i(k))^2 \\
&- \eta_i(k) \tilde{x}_i^T(k) K_i^T(k) A_i(k) \tilde{x}_i(k) + 4\gamma_i \left(\tilde{w}_i^T(k) z_i(\overline{x}(k-1), u(k)) \right)^2 + \epsilon_{z_i}^2 \\
&+ 4\gamma_i \left(\tilde{w}_i^T(k) z_i(\overline{x}(k-1), u(k)) \right)^2 + e_{qx_i}^2(k) + \eta_i^2(k) \tilde{x}_i^T(k) K_i^T(k) K_i(k) \tilde{x}_i(k) \\
&+ A_i(k) \tilde{w}_i(k) \tilde{w}_i^T(k) A_i(k) + \eta_i^2(k) \tilde{w}_i^T(k) P_i(k) P_i^T(k) \tilde{w}_i(k) \\
&+ K_i(k) \tilde{x}_i(k) \tilde{x}_i^T(k) K_i^T(k) + \tilde{w}_i^T(k) A_i(k) A_i^T(k) \tilde{w}_i(k) \\
&+ \tilde{x}_i(k) \tilde{x}_i^T(k) + \eta_i^2(k) \tilde{x}_i^T(k) K_i^T(k) K_i(k) \tilde{x}_i(k) \\
&+ P_i(k) \tilde{w}_i(k) \tilde{w}_i^T(k) P_i(k) + \gamma_i \epsilon_{z_i}^2 + 2\gamma_i \epsilon_{z_i} e_{qx_i}(k) + \gamma_i e_{qx_i}^2(k).
\end{aligned}
$$

Then,

$$
\begin{aligned}
\Delta V_i(k) \;\leq\;& \gamma_i \|\tilde{w}_i(k)\|^2 \|z_i(\overline{x}(k-1), u(k))\|^2 - \|\tilde{w}_i(k)\|^2 \lambda_{\min}(A_i(k)) \\
&- \gamma_i \|\tilde{x}_i(k)\|^2 + \eta_i^2(k) \|\tilde{x}_i(k)\|^2 \|K_i(k)\|^2 \lambda_{\max}(P_i(k)) \\
&- \eta_i(k) \|\tilde{x}_i(k)\|^2 \lambda_{\min}(K_i(k)) \lambda_{\min}(A_i(k)) \\
&+ 8\gamma_i^2 \|\tilde{w}_i(k)\|^2 \|z_i(\overline{x}(k-1), u(k))\|^2 + |\epsilon_{z_i}|^2 \\
&+ |e_{qx_i}(k)|^2 + \eta_i^2(k) \|\tilde{x}_i(k)\|^2 \|K_i(k)\|^2 + \|A_i(k)\|^2 \|\tilde{w}_i(k)\|^2 \\
&+ \eta_i^2(k) \|\tilde{w}_i(k)\|^2 \|P_i(k)\|^2 + \|K_i(k)\|^2 \|\tilde{x}_i(k)\|^2 \\
&+ \|\tilde{w}_i(k)\|^2 \|A_i(k)\|^2 + \|\tilde{x}_i(k)\|^2 + \eta_i^2(k) \|\tilde{x}_i(k)\|^2 \|K_i(k)\|^2 \\
&+ \|\tilde{w}_i(k)\|^2 \|P_i(k)\|^2 + \gamma_i |\epsilon_{z_i}|^2 + 2\gamma_i |\epsilon_{z_i}| |e_{qx_i}(k)| + \gamma_i |e_{qx_i}(k)|^2.
\end{aligned} \tag{36}
$$

Defining

$$
\begin{aligned}
E_i(k) \;=\;& -\gamma_i \|z_i(\overline{x}(k-1), u(k))\|^2 + \lambda_{\min}(A_i(k)) - 8\gamma_i^2 \|z_i(\overline{x}(k-1), u(k))\|^2 \\
&- 2\|A_i(k)\|^2 - \eta_i^2(k) \|P_i(k)\|^2 - \|P_i(k)\|^2, \\
F_i(k) \;=\;& -\eta_i^2(k) \|K_i(k)\|^2 \lambda_{\max}(P_i(k)) + \gamma_i + \eta_i(k) \lambda_{\min}(K_i(k)) \lambda_{\min}(A_i(k)) \\
&- 2\eta_i^2(k) \|K_i(k)\|^2 - \|K_i(k)\|^2 - 1,
\end{aligned}
$$

and selecting γ_i, η_i, Q_i and R_i such that $E_i > 0$ and $F_i > 0$, $\forall k$, then (36) can be expressed as:

$$
\begin{aligned}
\Delta V_i(k) \;=\;& - \|\tilde{w}_i(k)\|^2 E_i(k) - \|\tilde{x}_i(k)\|^2 F_i(k) + |\epsilon_{z_i}|^2 + |D_{x_i}|^2 + \gamma_i |\epsilon_{z_i}|^2 \\
&+ 2\gamma_i |\epsilon_{z_i}| |D_{x_i}| + \gamma_i |D_{x_i}|^2,
\end{aligned} \tag{37}
$$

with $\left| e_{qx_i}(k) \right| \leq \mu_x D_x$ from (30). Then, $\Delta V_i(k) < 0$ when

$$
\|\tilde{w}_i(k)\| > \sqrt{ \frac{ |\epsilon_{z_i}|^2 + |\Delta_{x_i}|^2 + \gamma_i |\epsilon_{z_i}|^2 + 2\gamma_i |\epsilon_{z_i}| |\Delta_{x_i}| + \gamma_i |\Delta_{x_i}|^2 }{ E_i(k) } } \equiv \kappa_1(k), \tag{38}
$$

or

$$
\|\tilde{x}_i(k)\| > \sqrt{ \frac{ |\epsilon_{z_i}|^2 + |\Delta_{x_i}|^2 + \gamma_i |\epsilon_{z_i}|^2 + 2\gamma_i |\epsilon_{z_i}| |\Delta_{x_i}| + \gamma_i |\Delta_{x_i}|^2 }{ F_i(k) } } \equiv \kappa_2(k). \tag{39}
$$

Step 2. Now, for $V(k)$, consider the Lyapunov function candidate

$$
V(k) = \sum_{i=1}^{n} \left(\gamma_i (\tilde{x}_i(k))^2 + \tilde{w}_i^T(k) P_i(k) \tilde{w}_i(k) \right). \tag{40}
$$

Its first difference can be written as:

$$
\begin{aligned}
\Delta V(k) \;=\; & \sum_{i=1}^{n} \Big(\gamma_i \left(\tilde{x}_i(k+1) \right)^2 + \tilde{w}_i^T(k+1)\, P_i(k+1)\, \tilde{w}_i(k+1) \\
& - \gamma_i \left(\tilde{x}_i(k) \right)^2 - \tilde{w}_i^T(k)\, P_i(k)\, \tilde{w}_i(k) \Big).
\end{aligned}
\tag{41}
$$

Substituting (31) and (32) in (41) yields

$$
\begin{aligned}
\Delta V(k) \;=\; & \sum_{i=1}^{n} \Big(\gamma_i \left(\tilde{w}_i^T(k)\, z_i\left(\overline{x}(k-1), u(k) \right) + \epsilon_{z_i} + e_{qx_i}(k) \right)^2 \\
& + \left(\tilde{w}_i(k) - \eta_i(k)\, K_i(k)\, \tilde{x}_i(k) \right)^T \left(P_i(k) - A_i(k) \right) \\
& \times \left(\tilde{w}_i(k) - \eta_i(k)\, K_i(k)\, \tilde{x}_i(k) \right) \\
& - \gamma_i \left(\tilde{x}_i(k) \right)^2 - \tilde{w}_i^T(k)\, P_i(k)\, \tilde{w}_i(k) \Big).
\end{aligned}
\tag{42}
$$

Defining

$$
\begin{aligned}
A_i(k) \;=\;& K_i(k)\, H_i^{\top}(k)\, P_i(k) - Q_i(k), \\
E_i(k) \;=\;& -\gamma_i \left\| z_i\left(\overline{x}(k-1), u(k) \right) \right\|^2 + \lambda_{\min}(A_i(k)) - 8\gamma_i^2 \left\| z_i\left(\overline{x}(k-1), u(k) \right) \right\|^2 \\
& -2\left\| A_i(k) \right\|^2 - \eta_i^2(k)\left\| P_i(k) \right\|^2 - \left\| P_i(k) \right\|^2, \\
F_i(k) \;=\;& -\eta_i^2(k)\left\| K_i(k) \right\|^2 \lambda_{\max}(P_i(k)) + \gamma_i + \eta_i(k)\, \lambda_{\min}(K_i(k))\, \lambda_{\min}(A_i(k)) \\
& -2\eta_i^2(k)\left\| K_i(k) \right\|^2 - \left\| K_i(k) \right\|^2 - 1,
\end{aligned}
$$

and selecting γ_i, η_i, Q_i and R_i such that $E_i > 0$ and $F_i > 0$, $\forall k$, then (42) can be expressed as:

$$
\begin{aligned}
\Delta V_i(k) \;=\; & \sum_{i=1}^{n} \Big(-\left\| \tilde{w}_i(k) \right\|^2 E_i(k) - \left\| \tilde{x}_i(k) \right\|^2 F_i(k) + \left| \epsilon_{z_i} \right|^2 + \left| \Delta_{x_i} \right|^2 \\
& + \gamma_i \left| \epsilon_{z_i} \right|^2 + 2\gamma_i \left| \epsilon_{z_i} \right| \left| \Delta_{x_i} \right| + \gamma_i \left| \Delta_{x_i} \right|^2 \Big).
\end{aligned}
\tag{43}
$$

Then, $\Delta V_i(k) < 0$ when (38) and (39) is fulfilled for $i = 1, 2, \cdots, n$. Therefore, considering Step 1 and Step 2 for (40), the solution of (31) and (32) is SGUUB. $\quad\square$

Remark 1. *Considering Theorem 2 and its proof, it can be demonstrated that the result can be extended for systems with multiple delays such as $x(k - l_i)$ with $i = 1, 2, \cdots$ used instead of $x(k - l)$ in (1), and for systems with time-varying delays $x(k - l_i(k))$ with $l_i(k)$ bounded by $l_i(k) \leq l_i$ in such case $x(k - l_i(k))$ can be substituted by $x(k - l_i)$ and can be handled as in previous proof due to the boundedness of RHONN. Also, from (43) it can be seen that RHONN (38) can identify a quantized discrete-time unknown delayed nonlinear system, this neural model can be arbitrarily designed in an affine form in order to deal with control goals.*

3.2. Neural Inverse Optimal Control

In the previous section, a neural identifier has been designed for a quantized discrete-time unknown delayed nonlinear system under disturbances and uncertainties; all these phenomena are now modeled by the neural identifier (13) without the need to measure or estimate or even know their boundaries. In this section, the obtained neural model is used to design a controller for such systems, using the inverse optimality methodology, a controller for the system is designed.

Then, considering Theorem 1 and Theorem 2, it is possible to establish the following main result, for the closed-loop system:

Theorem 3. *Consider a control law* (5) *for a quantized delayed system* (12) *modeled by a RHONN* (13), *trained with an EKF* (21), *ensuring that the identification error* (31) *and the trajectory tracking error* (3) *are SGUUB; moreover, the RHONN weight* (32) *remain bounded.*

Proof. Consider the following augmented Lyapunov function candidate

$$V(k) = V_1(k) + V_2(k),$$ (44)

with

$$V_1(k) = \sum_{i=1}^{n} \left(\Gamma_i \left(\tilde{x}_i(k) \right)^2 + \tilde{w}_i^T(k) P_i(k) \tilde{w}_i(k) \right),$$ (45)

$$V_2(k) = \frac{1}{2} \psi^T(k) \mathcal{P} \psi(k).$$ (46)

The first difference from (44) results

$$\begin{aligned} \Delta V(k) &= V_1(k+1) - V_1(k) + V_2(k+1) - V_2(k) \\ &= \sum_{i=1}^{n} \left(\Gamma_i \left(\tilde{x}_i(k+1) \right)^2 + \tilde{w}_i^T(k+1) P_i(k) \tilde{w}_i(k+1) \right) \\ &\quad - \sum_{i=1}^{n} \left(\Gamma_i \left(\tilde{x}_i(k) \right)^2 + \tilde{w}_i^T(k) P_i(k) \tilde{w}_i(k) \right) \\ &\quad + \frac{1}{2} \psi^T(k+1) \mathcal{P} \psi(k+1) - \frac{1}{2} \psi^T(k) \mathcal{P} \psi(k). \end{aligned}$$ (47)

Case 1. For $\Delta V_1(k)$. From Theorem 2, (47) can be expressed as:

$$\begin{aligned} \Delta V_1(k) &= \sum_{i=1}^{n} \left(- \|\tilde{w}_i(k)\|^2 E_i(k) - \|\tilde{x}_i(k)\|^2 F_i(k) + |\epsilon_{z_{iu}}|^2 + |\Delta_{x_i}|^2 \right. \\ &\quad \left. + \Gamma_i |\epsilon_{z_{iu}}|^2 + 2\Gamma_i |\epsilon_{z_{iu}}| |\Delta_{x_i}| + \Gamma_i |\Delta_{x_i}|^2 \right). \end{aligned}$$ (48)

Then, $\Delta V_1(k) < 0$ when (38) and (39) is fulfilled for $i = 1, 2, \cdots, n$.
Case 2. For $\Delta V_2(k)$. Consider (1) and (3), then

$$\begin{aligned} \Delta V_2(k) &= \frac{1}{2} f^T(\psi(k)) \mathcal{P} f(\psi(k)) + \frac{1}{2} f^T(\psi(k)) \mathcal{P} g(\psi(k)) u(k) \\ &\quad + \frac{1}{2} u^T(k) g^T(\psi(k)) \mathcal{P} f(\psi(k)) + \frac{1}{2} u^T(k) g^T(\psi(k)) \mathcal{P} g(\psi(k)) u(k) \\ &\quad - \frac{1}{2} \psi^T(k) \mathcal{P} \psi(k), \end{aligned}$$ (49)

with (5). From Theorem 1 [23], $\Delta V_2(k) \leq -\psi^T(k) \mathcal{Q} \psi(k)$ with $\mathcal{Q} = \mathcal{Q}^T > 0$. Therefore, considering Case 1 and Case 2, for (44), the solutions of (25), (32) and (3) are SGUUB. $\square$

The entire closed-loop system is depicted in Figure 2. Theorem 3 ensures the stability of the closed-loop system, without the assumption of persistent excitation. It is necessary to point out that the inverse optimal controller is designed for the neural model not for the system model, which is considered unknown. This means that the neural model considers system dynamics, sensor errors, actuators errors, disturbances, uncertainties, and unmodeled dynamics.

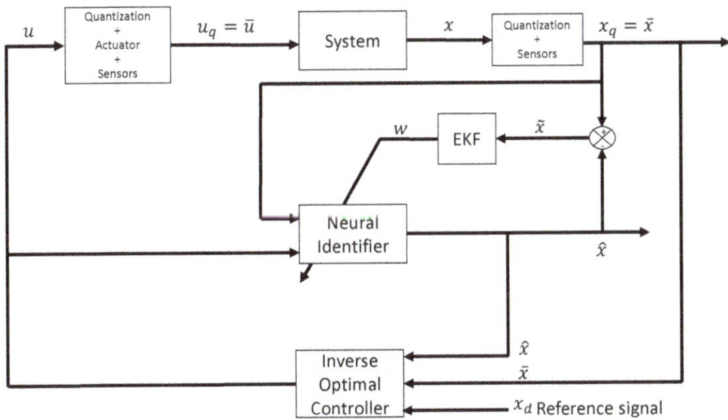

Figure 2. Schematic closed-loop system representation of a discrete-time unknown nonlinear quantized delayed system with measurement noise.

Remark 2. *It is important to note that in this paper, the unknown discrete-time quantized delayed system (12) is modeled by a RHONN (13), trained with an EKF learning algorithm (21), then the controller is designed based on the RHONN model. Therefore, the RHONN model (13) is designed in order that the controllable block form required by the controller (5) can be easily obtained in the required affine form. Furthermore, controllability weights of RHONN model (13) are selected as fixed weights to avoid singularities due to controller feedback [23].*

Remark 3. *As explained in Section 3.2 any nonlinear trajectory tracking problem can be transformed into an error regulation problem [9]; this transformation is displayed in Figure 3 including the design of a neural inverse optimal control (NIOC).*

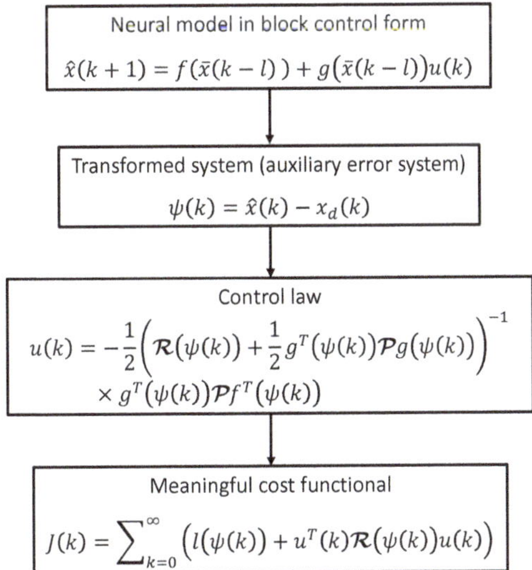

Figure 3. Scheme of a NIOC.

4. Real-Time Results

4.1. Implementation to a LIM

In this section, a NIOC scheme is designed to be implemented in a linear induction motor (LIM) prototype in the presence of all the phenomena, as mentioned earlier. First, a neural identifier is designed for the LIM system, considering an unknown model with disturbances, uncertainties, quantization in measured signals, measurement noise, and delays. Therefore, the obtained neural state model is used to design a controller for the LIM using the inverse optimal methodology considering actuator limitations.

A LIM is an electrical machine in which electrical energy is directly transformed into mechanical energy as a translational movement. LIMs have excellent performance characteristics [25,26], such as high-starting thrust force, elimination of gears between motor and motion devices, reduction of mechanical losses and size of motion devices, high-speed operation, silent operation, among others. The driving principles of a LIM are similar to the ones of a traditional rotary induction motor. However, the LIM mathematical model is more complicated; its control design is also a complex task due to the exceedingly time-varying nature of its parameters. This is due to the change of operating conditions, such as speed, temperature, and rail configuration, to name a few. LIMs have been used for mass public transportation, 3D printing, material transportation, elevators, aircraft carrier, industrial processes, automation, and home appliances, among others [25].

From previous works, it is possible to remark that in [22], a neural inverse optimal controller has been proposed for a LIM; however, such controller only considers disturbances meanwhile delays are disregarded. Also, a neural identifier for a LIM is proposed in [1]. Then, in [2], time delays are considered as well as disturbances, with a sliding mode neural controller for the same system. Nonetheless, none of the above-referenced works include real-time implementation of the controller when simultaneously considering, disturbances, noise, delays, actuator limitations, and quantization for a system with an unknown model. Therefore, in this section, the design of a neural controller for a LIM is proposed. Such a controller considers all the phenomena mentioned earlier while intending to control complex systems with conditions increasingly closer to that of real-time experimentation.

4.2. NIOC Scheme

4.2.1. RHONN Model

The $\alpha - \beta$ mathematical model for a LIM is represented by the state vector (50), which includes position, velocity, flux magnitude, α current, and β current. The control law vector $u(k)$ to be designed using the NIOC algorithm, given by (51), and its entries, namely $u_\alpha(k)$ and $u_\beta(k)$, represent the α and β components, of the input voltage applied to the LIM, respectively.

$$x(k) = \begin{bmatrix} x_1(k) \\ x_2(k) \\ x_3(k) \\ x_4(k) \\ x_5(k) \end{bmatrix} = \begin{bmatrix} p(k) \\ v(k) \\ \psi(k) \\ i_\alpha(k) \\ i_\beta(k) \end{bmatrix}, \tag{50}$$

$$u(k) = \begin{bmatrix} u_\alpha(k) \\ u_\beta(k) \end{bmatrix}. \tag{51}$$

For the design of the proposed RHONN identifier, which is given through (53), the state variables x_1, x_2, x_4 and x_5 are considered as measurable signals, and $\Psi(k) = f_\alpha(k)^2 + f_\beta(k)^2$ where f_α and f_β are the α and β flux components, respectively, estimated by a nonlinear observer [22]. The function $S(v)$ is defined as (54), and the identifier's state variables are defined as:

$$\hat{x}(k) \;=\; \begin{bmatrix} \hat{x}_1\,(k) \\ \hat{x}_2\,(k) \\ \hat{x}_3\,(k) \\ \hat{x}_4\,(k) \\ \hat{x}_5\,(k) \end{bmatrix} \;=\; \begin{bmatrix} \hat{p}\,(k) \\ \hat{v}\,(k) \\ \hat{\psi}\,(k) \\ \hat{\bar{i}}_\alpha\,(k) \\ \hat{\bar{i}}_\beta\,(k) \end{bmatrix}. \tag{52}$$

For the identifier in (53), the quantized fluxes $\overline{f}_\alpha$ and $\overline{f}_\beta$, are not considered since the implemented prototype shown in Figure 4 cannot measure magnetic fluxes. As mentioned before, these variables are estimated by a reduced-order state-observer [22]. Also, the following weights are fixed: $w_f = 0.001$, $w_{12} = 0.001$, $w_{24} = 0.001$, $w_{25} = 0.001$, $w_{45} = 0.05$, $w_{55} = 0.05$ for controllability purposes, these values have been obtained heuristically in order to improve identification performance [27], neural identification accuracy is obtained due to the adaptable weight connections of the neural network another solution is the zero-crossing avoidance studied in [28]; the remaining weight vectors w_i from the identifier in (53) are trained online by the EKF training algorithm in (21), with the parameters shown in Table 1.

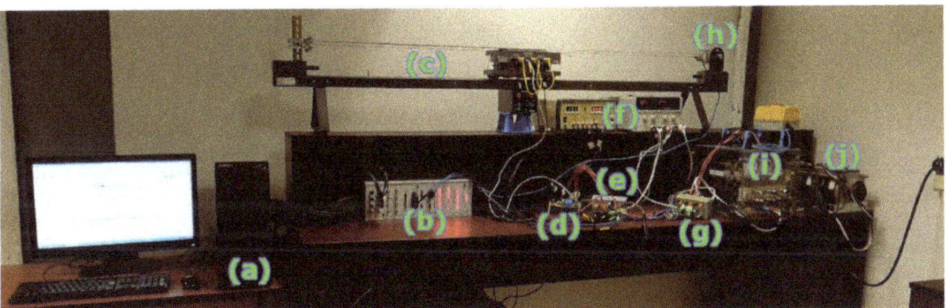

Figure 4. Linear induction motor (LIM) prototype. (a) Computer with dSPACE® software and RTI 1104 board installed, (b) dSPACE® RTI 1104 panel connector, (c) Linear induction motor, (d) Voltage sensor, (e) Current sensor, (f) Power supply, (g) TTL to CMOS converter, (h) encoder, (i) IGBT module (or insulated-gate bipolar transistor), (j) Auto-transformer.

$$
\begin{aligned}
\hat{x}_1\,(k+1) \;&=\; w_{11}\,(k)\,S\left(\overline{p}\,(k-1)\right) + w_{12}\overline{v}\,(k-1), \\[4pt]
\hat{x}_2\,(k+1) \;&=\; w_{21}\,(k)\,S\left(\overline{v}\,(k-1)\right) + w_{22}\,(k)\,S\left(\psi_\alpha\,(k-1)\right) + w_{23}\,(k)\,S\left(\psi_\beta\,(k-1)\right) \\
&\quad - w_{24}\left(S\left(\psi_\alpha\,(k-1)\right) + S\left(\psi_\beta\,(k-1)\right)\right)\overline{i}_\alpha \\
&\quad + w_{25}\left(S\left(\psi_\alpha\,(k-1)\right) + S\left(\psi_\beta\,(k-1)\right)\right)\overline{i}_\beta, \\[4pt]
\hat{x}_3\,(k+1) \;&=\; w_{31}\,(k)\,S\left(\psi_\alpha\,(k-1)\right)^2 + w_{32}\,(k)\,S\left(\psi_\beta\,(k-1)\right)^2 \\
&\quad + w_{33}\,(k)\,w_f S\left(\overline{v}\,(k-1)\right)^2 \\
&\quad + 2w_f\left(w_{31}\,(k)\,S\left(\psi_\alpha\,(k-1)\right) - w_{32}\,(k)\,S\left(\psi_\beta\,(k-1)\right)\right)\overline{i}_\alpha\,(k-1) \\
&\quad + 2w_f\left(w_{31}\,(k)\,S\left(\psi_\alpha\,(k-1)\right) + w_{32}\,(k)\,S\left(\psi_\beta\,(k-1)\right)\right)\overline{i}_\beta\,(k-1), \\[4pt]
\hat{x}_4\,(k+1) \;&=\; w_{41}\,(k)\,S\left(\overline{v}\,(k-1)\right) + w_{42}\,(k)\,S\left(\psi_\alpha\,(k-1)\right) + w_{43}\,(k)\,S\left(\psi_\beta\,(k-1)\right) \\
&\quad + w_{44}\,(k)\,S\left(\overline{i}_\alpha\,(k-1)\right) + w_{45}u_\alpha\,(k), \\[4pt]
\hat{x}_5\,(k+1) \;&=\; w_{51}\,(k)\,S\left(\overline{v}\,(k-1)\right) + w_{52}\,(k)\,S\left(\psi_\alpha\,(k-1)\right) + w_{53}\,(k)\,S\left(\psi_\beta\,(k-1)\right) \\
&\quad + w_{54}\,(k)\,S\left(\overline{i}_\beta\,(k-1)\right) + w_{55}u_\beta\,(k),
\end{aligned}
\tag{53}
$$

$$S\,(v) = \tanh\,(v). \tag{54}$$

Table 1. EKF training parameters.

i	η	P	Q	R
w_1	0.5	2	0.63759	0.0061914
w_2	0.5	$2 \times I_{3\times3}$	$0.0067168 \times I_{3\times3}$	0.0061914
w_3	0.5	$2 \times I_{3\times3}$	$0.0067168 \times I_{3\times3}$	0.0061914
w_4	0.5	$2 \times I_{4\times4}$	$0.0045569 \times I_{4\times4}$	0.0061914
w_5	0.5	$2 \times I_{4\times4}$	$0.0045569 \times I_{4\times4}$	0.0061914

4.2.2. Inverse Optimal Control

It is easy to see that the identifier in (53) is written as an affine system, as given by (1). Using the obtained model for the identifier, the system can be represented by the following blocks

$$\hat{X}_1(k+1) = \hat{x}_1(k+1), \tag{55}$$

$$\hat{X}_2(k+1) = \begin{bmatrix} \hat{x}_2(k+1) \\ \hat{x}_3(k+1) \end{bmatrix}, \tag{56}$$

$$\hat{X}_3(k+1) = \begin{bmatrix} \hat{x}_4(k+1) \\ \hat{x}_5(k+1) \end{bmatrix}. \tag{57}$$

Then, considering the desired trajectory $\hat{X}_{1d}$ for $\hat{X}_1$, the system can be represented as (58)–(60).

$$\psi_1(k+1) = \hat{X}_1(k+1) - X_{1_d}(k+1) = \hat{x}_1(k+1) - x_{1_d}(k+1), \tag{58}$$

$$\psi_2(k+1) = \hat{X}_2(k+1) - X_{2_d}(k+1) = \begin{bmatrix} \hat{x}_2(k+1) - x_{2_d}(k+1) \\ \hat{x}_3(k+1) - x_{3_d}(k+1) \end{bmatrix}, \tag{59}$$

$$\psi_3(k+1) = \hat{X}_3(k+1) - X_{3_d}(k+1) = \begin{bmatrix} \hat{x}_4(k+1) - x_{4_d}(k+1) \\ \hat{x}_5(k+1) - x_{5_d}(k+1) \end{bmatrix}. \tag{60}$$

For this implementation, X_{1_d} is the desired position trajectory, X_{2_d} is defined as $X_{2_d} = \left[x_{2_d}, x_{3_d}\right]$, where x_{2_d} (v reference) is determined as

$$k_1 \psi_1(k) = f_1(\bar{x}(k)) + g_1(\bar{x}(k))\bar{x}_2(k) - X_{1_d}(k+1),$$
$$k_1 \psi_1(k) = w_{f_1}^{\top}(k) z_{f_1}(\bar{x}(k)) + w_{g_1}\bar{x}_2(k) - X_{1_d}(k+1),$$
$$X_{2_d}(x(k)) = w_{g_1}^{-1}\left[-w_{f_1}^{\top}(k) z_{f_1}(\bar{x}(k)) + k_1 \psi_1(k) + X_{1_d}(k+1)\right],$$

and x_{3_d} (Ψ reference) is the given flux magnitude reference; the X_{3_d} reference is defined as $X_{3_d} = \left[x_{4_d}, x_{5_d}\right]$, which gives both the α and β current references and is computed as

$$k_2 \psi_2(k) = f_2(\bar{x}(k)) + g_2(\bar{x}(k)) X_3(k) - X_{2_d}(k+1),$$
$$k_2 \psi_2(k) = w_{f_2}^{\top}(k) z_{f_2}(\bar{x}(k)) + w_{g_2} X_3(k) - X_{2_d}(k+1),$$
$$X_{3_d}(x(k)) = w_{g_2}^{-1}\left[-w_{f_2}^{\top}(k) z_{f_2}(\bar{x}(k)) + k_2 z_2(k) + X_{2_d}(k+1)\right].$$

Finally, the schematic representation for the calculation of u used in (5) is presented in Figure 5.

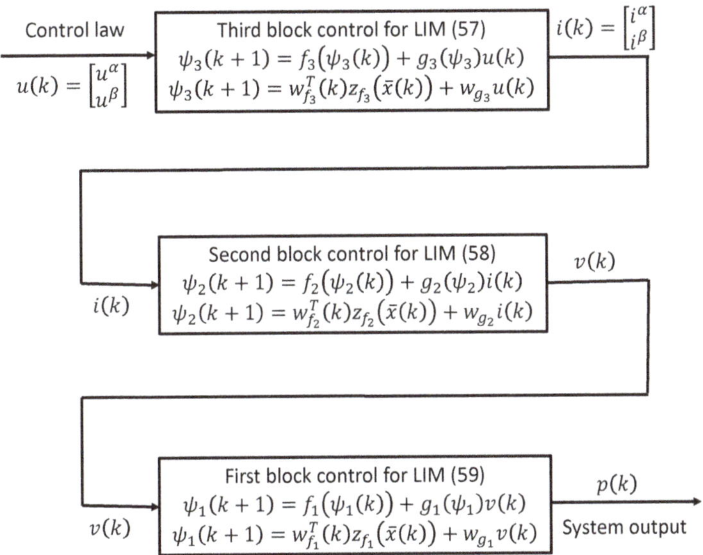

Figure 5. Schematic representation for control law calculation.

Remark 4. *The LIM prototype used to test in real time the proposed controller can be modeled by a RHONN, which can be designed in an affine block controllable form composed of two blocks. The first block contains the two desired signals to be controlled, which are controlled by virtual controllers (controllers that are not physically applied to the LIM). These virtual controllers are designed as a feedback linearization system in a block control form whose last block is controlled using an inverse optimal controller to achieve trajectory tracking, minimizing a functional cost criteria [23].*

4.3. Results

The following results were obtained from a real-time testing prototype using a LIM; the complete setup is shown in Figure 4. The proposed scheme is programmed in Matlab\Simulink®(Matlab and Simulink are registered trademarks of MathWorks), and then loaded to a dSPACE®(dSPACE is a registered trademark of dSPACE GmbH) 1104R&D board, mounted on the PC equipment (Figure 4a), which computes the proposed controller–identifier algorithm using all the information coming from the prototype; all its inputs and outputs go through the connector panel. The dSPACE® ControlDesk®(ControlDesk is a registered trademark of dSPACE GmbH) software is used to visualize and store the input and output signals, as well as managing the LIM prototype.

The test is performed in real time on a LIM prototype, using the dSPACE® 1104R&D controller board with a sampling time equal to 0.3 ms. Blocks to induce delays in the signals during the test are programmed, and they are activated at 4 s, 7 s and 11 s for the position, α current, and β current signals, respectively. These blocks randomly substitute the information of the most recent sample with the information form one of the previous ten samples; moreover, each delaying block performs this operation independently from one another in a way that the number of samples that each signal is delayed is chosen randomly as well.

The control objective of the test is the tracking of a trapezoidal reference and the performance is shown in Figure 6, where two zoomed graphs are also presented with intervals 6.8 s to 7.2 s and 10 s to 10.4 s. Figure 7, shows trajectory tracking error for position. Figure 8, shows norm of weight vectors of identifier (53) confirming their boundedness as stated in Theorem 3. From Figures 6 and 7 it can be observed that the measured signal closely follows the reference, clearly establishing the excellent behavior of the proposed controller, moreover, it can also be seen that the measured and identified waveforms are superimposed to one another, which also verifies the superior performance of the proposed identifier structure.

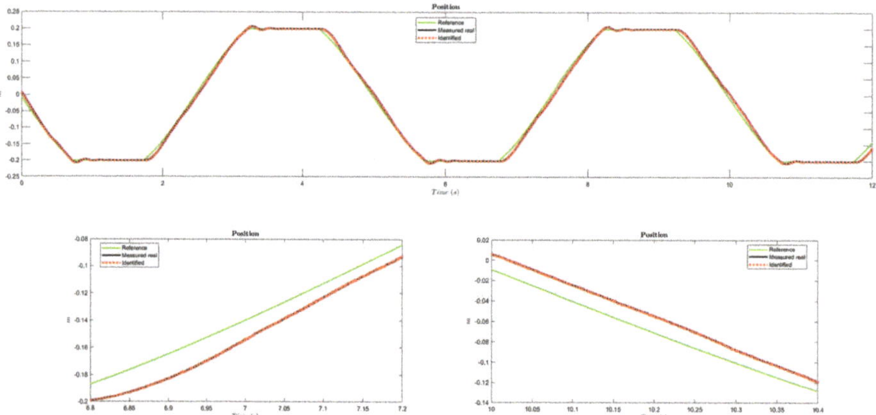

Figure 6. Tracking performance of our proposed neural control scheme for a trapezoid signal and zoomed images at the same time intervals of the zoomed images of the control signals.

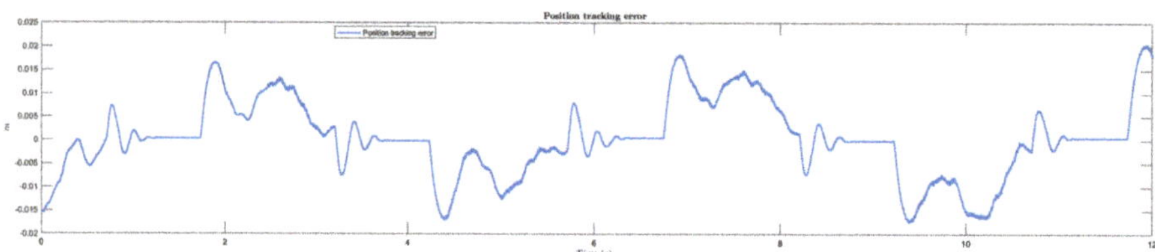

Figure 7. Position tracking error.

From results depicted in Figures 6 and 7, it is easy to see that the obtained trajectory tracking results are SGUUB, due to the boundedness of the trajectory tracking error depends of the unknown bounds of all the error sources, like quantization error, unmodeled dynamics, uncertainties, disturbances, sensor noises, among others. All these unknown quantities contribute to trajectory tracking error, consistent with the result obtained in Theorem 3 regarding SGUUB [23].

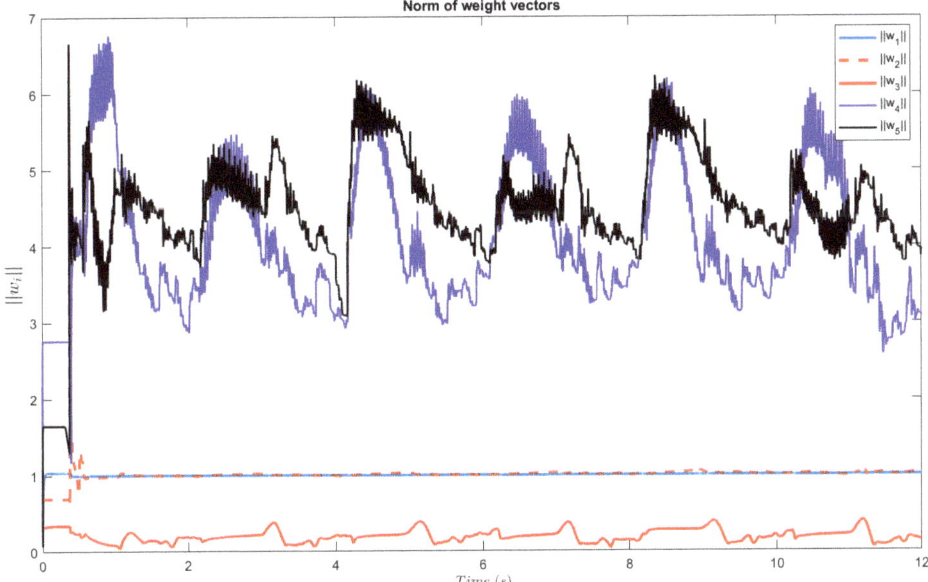

Figure 8. Norm of weight vectors.

In this paper, a stressing profile for the position has been used to show the applicability and relevance of the proposed controller, showing significant results for real-time trajectory tracking of nonlinear systems, which is not easily found in previous works of quantized systems.

The time instances shown in Figure 6 coincide with the ones of the zoomed graphs in Figures 9 and 10, where both the u_α and u_β components computed by the proposed controller are compared against the α and β components of the LIM input voltage, measured in volts (V), at its terminals by voltage sensors. The zoomed images are presented to give an insight into the quantization effects, measurement noise, and delays. It is important to note that the measured signals are converted to $\alpha - \beta$ quantities in the controller board, and then compared with the ones calculated by the proposed controller.

Figure 11 shows two graphs corresponding to the differences between the computed control signals and the ones obtained from measurements of the prototype. It must be noticed that this difference remains bounded and that even though the difference is relatively significant, the performance of our proposed discrete-time neural control scheme is as expected. This can be verified from Figure 7, where the tracking error of the reference trapezoidal signal from Figure 6 is shown to be very small. It is worth noting that the effects considered (quantization from the analog to digital conversion of the measured signals, noise introduced by the measurement sensors, and the induced delay, which can represent a delay of up to ten samples) do not represent conditions that diminish the excellent performance of the proposed controller.

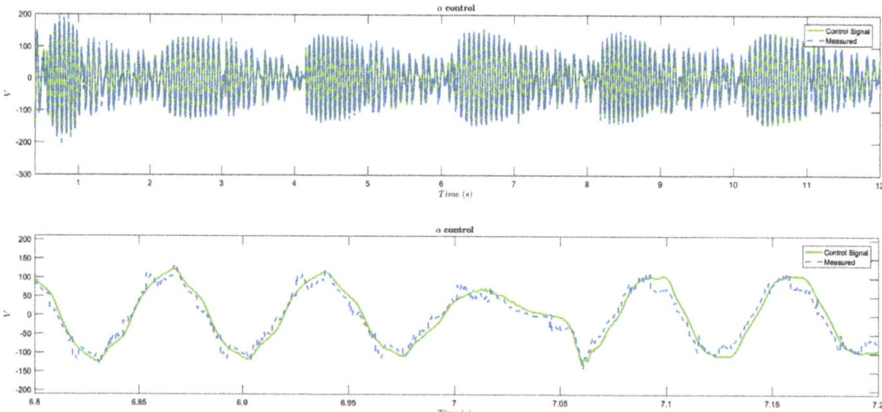

Figure 9. Control signal u_α from the test and the zoomed image for a time interval from 6.8 s to 7.2 s.

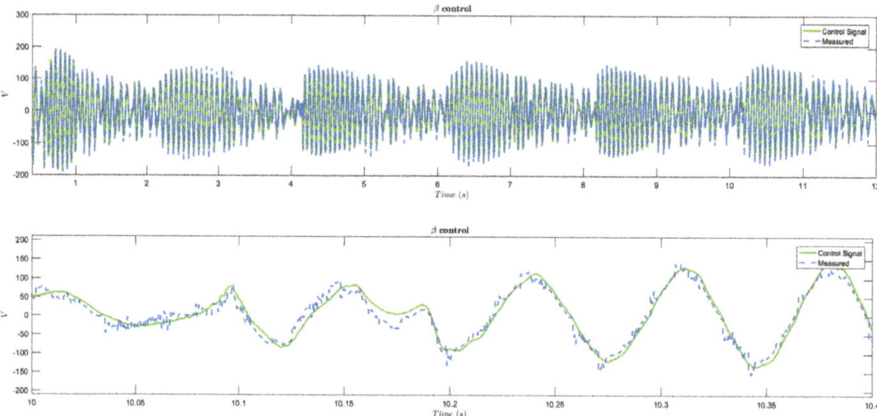

Figure 10. Control signal u_β from the test and the zoomed image for a time interval from 10 s to 10.4 s.

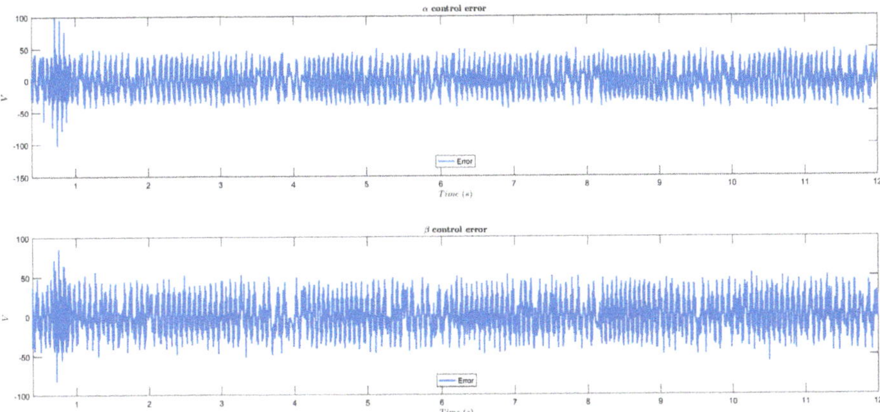

Figure 11. Control error signals between desired control signals and measured signals.

It is important to consider that the proposed identifier has an excellent performance, such that in Figure 6 is almost impossible to note differences between plant state variables and RHONN ones; therefore, to measure such differences, their statistical information is included in Table 2. As can be seen from Table 2, the identification errors are small enough to confirm the validity and accuracy of the RHONN-based identifier model for the LIM, in actual operating conditions and with real-life effects considered. In order to further support this claim, the identification RMSE of the state variables from the LIM system are given in Table 2. These errors correspond to the difference between $\bar{x}_i - \hat{x}_i$; meanwhile, the flux errors are described by the difference between the observed flux signal and the identified flux signal; all RMSE values are calculated for the 12 s window of the test. As can be seen from Table 2, the identification errors are small enough to confirm the validity and accuracy of the RHONN-based identifier model for the LIM, in actual operating conditions and with real-life effects considered.

Table 2. Identification root mean square errors.

State Variable	RMSE
x_1	0.000153 m
x_2	0000607 m/s
x_3	0.004495 Wb^2
x_4	1.88391 A
x_5	1.3332866 A

4.4. Comparative Analysis

To compare the proposed controller against existing ones, Table 3, displays the mean value and standard deviation for speed tracking error, to a comparison between an optimal tracking control (OTC) [29], a conventional discrete-time sliding mode control (SMC) scheme [30], a NIOC with a neural observer (NIOCNO) [22] with respect to the proposed Quantized NIOC. It is shown that the proposed Quantized NIOC presents the best performance.

Table 3. Position tracking error information.

Controller	Mean Value (m/s)	Standard Deviation (m/s)
OTC [29]	0.027393	0.04821
SMC [30]	0.008358	0.09212
NIOCNO [22]	0.002468	0.04703
NIOC	**0.001023**	**0.00763**

4.5. Discussion about Experimental Environment

Real-time implementations involve major challenges caused by unknown dynamics, actuator limitations, quantization errors, unknown delays, and disturbances that can be magnified by the prototype components shown in Figure 11; all these components affect the algorithm. However, all these effects are absorbed and compensated by the proposed identifier-controller approach, as has been displayed by results of Section 4.3 and comparative analysis of Section 4.4. To exemplify this kind of effects, most of them are explained for each of the prototype components:

- Voltage and current sensors are characterized by the presence of noise, measurement error, and quantization error.
- IGBT module has a limited frequency of operation as well as the AC power supply saturates its output. Moreover, the IGBT module has its controllers for IGBTs, which provoke unknown dynamics.
- TTL to CMOS converter, cause noise, uncertainties and unknown dynamics and quantization effects.
- Computer equipment and dSPACE® 1104 board, presents quantization effects mainly due to the signal conversions at input and output and unknown dynamics provoked by inner circuits.
- LIM contains high nonlinear behaviors, including end effects and unknown uncertainties, mainly due to winding heating, unknown dynamics, hysteresis effects, and parasitic currents, among unknown disturbances by mass effects and the presence of friction, among others. Moreover, this kind of system does not exhibit evident delays effects. However, in this experiment, the delays have been externally included through a time-delay MATLAB® block, as explained in Section 4.3.

These effects can be easily seen in Figure 10, which shows control error signals between desired control signals and measured signals. Therefore, this paper presents real-time implementation of the controller when simultaneously considering, disturbances, noise, delays, actuator limitations, and quantization for a system with unknown plant model.

These are common problems for real-life implementations, and their consequences can be observed in the results included above in this section. Figures 6 and 7 show that due to the capabilities of the RHONN as an identifier (53), it can replicate LIM dynamics, with small identification errors, as shown in Table 2. In Figures 9 and 10 depict the $\alpha - \beta$ components of the calculated control law and the components of the actual measured signal. The difference is presented in Figure 11. This notable error is the consequence of all the phenomena mentioned above, among others. However, meaningful capabilities of the proposed NIOC scheme allow us to obtain good performance for trajectory tracking as is shown in the results depicted in Figures 6 and 7, accompanied of the statistical information included in Table 3 for identification RMSEs, even without the need for previous knowledge of a model for the controlled system nor an explicit estimation of its bounds. In order to explain the nature of the above-mentioned effects, in Figure 12 schematic representation of the control signal route is included, from the calculated value of vector $u(k)$ along with its application to the LIM prototype and its measurement for $\alpha - \beta$ components calculation to be depicted in Figures 9–11.

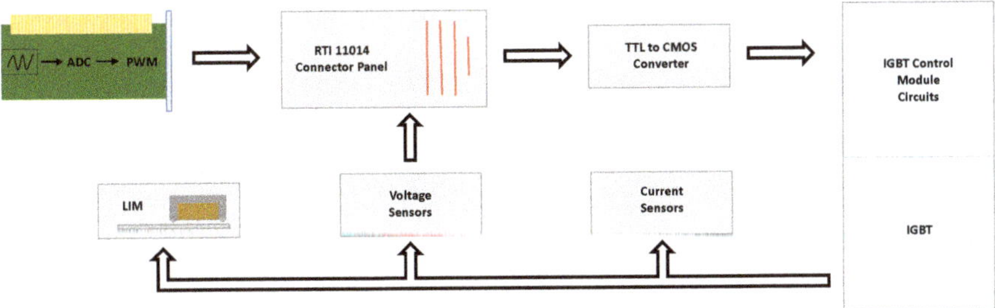

Figure 12. Control signal route: From calculated to applied signal.

5. Conclusions

This work reports a complete neural identifier-control model that deals with quantization error, measurement noise, disturbances, uncertainties, and signal delays. For any such conditions, it is not necessary to know neither the nominal model nor its estimations or bounds. The proposed controller is applied to an actual LIM in a real-time experimental test, which shows that the identification errors for all the state variables and the trajectory tracking errors are small enough. The presented results were obtained working with a LIM without knowledge of its model or parameters.

It is important to note that the experimental test intrinsically considers quantization error, measurement noise, disturbances, uncertainties, and delay. However, an additional delay of up to ten samples is randomly applied to the position, α current, and β current signals. Despite all this, the proposed neural identifier-control scheme shows an excellent real-time performance with a sample time equal to 0.3 ms. As seen from the results, the tracking of the reference signal is effectively achieved even though a trapezoidal waveform represents a difficult task for a LIM.

Moreover, a Lyapunov analysis is included with the proof of the semi-globally uniformly ultimately boundedness (SGUUB) of the proposed scheme. Furthermore, it must be noticed that the stability proof requires neither persistent excitation condition nor separation principle relaxing conditions to achieve boundedness of the entire identifier-control scheme. Proposed results can be extended for systems with hysteresis, backlash, friction, and other interesting problems.

For the experimental results included in this paper, controller parameters have been heuristically selected. However, as future work, authors consider the use of an evolutionary algorithm to define an auto-tuning approach.

Author Contributions: Conceptualization, A.Y.A. and J.D.R.; methodology, A.Y.A.; software, J.D.R. and J.G.-A.; validation, A.Y.A., P.Z. and F.J.; formal analysis, A.Y.A.; investigation, A.Y.A.; data curation, A.Y.A., J.D.R. and J.G.-A.; writing—original draft preparation, A.Y.A., J.D.R., P.Z. and F.J.; visualization, J.D.R.; supervision, A.Y.A.; project administration, A.Y.A. All authors have read and agreed to the published version of the manuscript.

Funding: The authors thank the support of CONACYT Mexico, through Project CB-2015-256769 (Project supported by Fondo Sectorial de Investigación para la Educación.

Acknowledgments: The authors also thank Universidad de Guadalajara and Tecnológico Nacional de México/I.T. La Laguna for their support in this research.

Conflicts of Interest: The authors declare no conflict of interest.

Abbreviations

The following abbreviations are used in this manuscript:

RHONN Recurrent high order neural network
EKF Extended Kalman filter
SGUUB Semi-globally uniformly ultimately bounded
NIOC Neural inverse optimal control
LIM Linear induction motor
RMSE Root mean square error

References

1. Alanis, A.Y.; Rios, J.D.; Arana-Daniel, N.; Lopez-Franco, C. Neural identifier for unknown discrete-time nonlinear delayed systems. *Neural Comput. Appl.* **2016**, *27*, 2453–2464. [CrossRef]
2. Rios, J.D.; Alanis, A.Y.; Lopez-Franco, C.; Arana-Daniel, N. RHONN identifier-control scheme for nonlinear discrete-time systems with unknown time-delays. *J. Frankl. Inst.* **2018**, *355*, 218–249. [CrossRef]
3. Zhang, D.; Han, Q.; Zhang, X. Network-Based Modeling and Proportional–Integral Control for Direct-Drive-Wheel Systems in Wireless Network Environments. *IEEE Trans. Cybern.* **2020**, *50*, 2462–2474. [CrossRef] [PubMed]
4. Zhang, D.; Zhou, Z.; Jia, X. Networked fuzzy output feedback control for discrete-time Takagi–Sugeno fuzzy systems with sensor saturation and measurement noise. *Inf. Sci.* **2018**, *457–458*, 182–194. [CrossRef]
5. Okuyama, Y. *Discrete Control Systems*; Springer: London, UK, 2014.
6. Kalman, R.E. Nonlinear aspects of sampled-data control systems. In *Proceedings of the Symposium on Nonlinear Circuit Analysis VI*; Polytechnic Institute of Brooklyn: Brooklyn, NY, USA, 1956; pp. 273–313.
7. Xia, Y.; Fu, M.; Liu, G. *Analysis and Synthesis of Networked Control Systems*; Lecture Notes in Control and Information Sciences; Springer: Berlin/Heidelberg, Germany, 2011.
8. Isidori, A. *Nonlinear Control Systems*; Communications and Control Engineering; Springer: London, UK, 2013.
9. Khalil, H. *Nonlinear Systems*; Pearson Education, Prentice Hall: Upper Saddle River, NJ, USA, 2002.
10. Liu, Y.; Peng, Y.; Wang, B.; Yao, S.; Liu, Z. Review on cyber-physical systems. *IEEE/CAA J. Autom. Sin.* **2017**, *4*, 27–40. [CrossRef]
11. Chen, G. Pinning control and controllability of complex dynamical networks. *Int. J. Autom. Comput.* **2017**, *14*, 1–9. [CrossRef]
12. Salgado, I.; Ahmed, H.; Camacho, O.; Chairez, I. Adaptive sliding-mode observer for second order discrete-time MIMO nonlinear systems based on recurrent neural-networks. *Int. J. Mach. Learn. Cybern.* **2019**, *10*, 2851–2866. [CrossRef]
13. Li, L.; Zheng, N.; Wang, F. On the Crossroad of Artificial Intelligence: A Revisit to Alan Turing and Norbert Wiener. *IEEE Trans. Cybern.* **2019**, *49*, 3618–3626. [CrossRef]
14. Wu, W.; Wang, C.; Yuan, C. Deterministic learning from sampling data. *Neurocomputing* **2019**, *358*, 456–466. [CrossRef]
15. Liu, C.; Liu, X.; Wang, H.; Zhou, Y.; Lu, S. Observer-based adaptive fuzzy funnel control for strict-feedback nonlinear systems with unknown control coefficients. *Neurocomputing* **2019**, *358*, 467–478. [CrossRef]
16. Xi, C.; Dong, J. Adaptive neural network-based control of uncertain nonlinear systems with time-varying full-state constraints and input constraint. *Neurocomputing* **2019**, *357*, 108–115. [CrossRef]
17. Chang, X.; Xiong, J.; Li, Z.; Park, J.H. Quantized Static Output Feedback Control For Discrete-Time Systems. *IEEE Trans. Ind. Inform.* **2018**, *14*, 3426–3435. [CrossRef]
18. Xu, H.; Zhao, Q.; Jagannathan, S. Finite-Horizon Near-Optimal Output Feedback Neural Network Control of Quantized Nonlinear Discrete-Time Systems With Input Constraint. *IEEE Trans. Neural Netw. Learn. Syst.* **2015**, *26*, 1776–1788. [CrossRef] [PubMed]
19. Wang, Y.; Shen, H.; Duan, D. On Stabilization of Quantized Sampled-Data Neural-Network-Based Control Systems. *IEEE Trans. Cybern.* **2017**, *47*, 3124–3135. [CrossRef] [PubMed]
20. Zhou, J.; Wang, W. Adaptive Control of Quantized Uncertain Nonlinear Systems. *IFAC-PapersOnLine* **2017**, *50*, 10425–10430. [CrossRef]

21. Wang, C.C.; Yang, G.H. Prescribed performance adaptive fault-tolerant tracking control for nonlinear time-delay systems with input quantization and unknown control directions. *Neurocomputing* **2018**, *311*, 333–343. [CrossRef]

22. Lopez, V.G.; Sanchez, E.N.; Alanis, A.Y.; Rios, J.D. Real-time neural inverse optimal control for a linear induction motor. *Int. J. Control.* **2017**, *90*, 800–812. [CrossRef]

23. Sanchez, E.; Ornelas-Tellez, F. *Discrete-Time Inverse Optimal Control for Nonlinear Systems*; CRC Press: Boca Raton, FL, USA, 2017.

24. Rovithakis, G.; Christodoulou, M. *Adaptive Control with Recurrent High-Order Neural Networks: Theory and Industrial Applications*; Advances in Industrial Control; Springer: London, UK, 2012.

25. Boldea, I.; Nasar, S. *Linear Electric Actuators and Generators*; Cambridge University Press: Cambridge, UK, 2005.

26. Takahashi, I.; Ide, Y. Decoupling control of thrust and attractive force of a LIM using a space vector control inverter. *IEEE Trans. Ind. Appl.* **1993**, *29*, 161–167. [CrossRef]

27. Levin, A.U.; Narendra, K.S. Control of nonlinear dynamical systems using neural networks: controllability and stabilization. *IEEE Trans. Neural Netw.* **1993**, *4*, 192–206. [CrossRef]

28. Sanchez, E.; Alanís, A.; Loukianov, A. *Discrete-Time High Order Neural Control: Trained with Kalman Filtering*; Studies in Computational Intelligence; Springer: Berlin/Heidelberg, Germay, 2008.

29. Ornelas-Tellez, F.; Alanis, A.Y.; Rios, J.D.; Graff, M. Reduced-order observer for state-dependent coefficient factorized nonlinear systems. *Asian J. Control.* **2019**, *21*, 1216–1227. [CrossRef]

30. Alanis, A.Y.; Rios, J.D.; Rivera, J.; Arana-Daniel, N.; Lopez-Franco, C. Real-time discrete neural control applied to a Linear Induction Motor. *Neurocomputing* **2015**, *164*, 240–251. [CrossRef]

Article

Realization of the Neural Fuzzy Controller for the Sensorless PMSM Drive Control System

Hung-Khong Hoai [1,2], Seng-Chi Chen [1,*] and Chin-Feng Chang [3]

[1]   Department of Electrical Engineering, Southern Taiwan University of Science and Technology, Tainan City 71005, Taiwan; da62b201@stust.edu.tw

[2]   Faculty of Electrical and Electronics Engineering, Ho Chi Minh City University of Transport, Ho Chi Minh City 70000, Vietnam

[3]   Fukuta Electric & Machinery Co., Ltd., Taichung City 429, Taiwan; gordon.chang@fukuta-motor.com.tw

*   Correspondence: amtfcsg123@stust.edu.tw; Tel.: +886-6-253-3131 (ext. 3324)

Received: 30 July 2020; Accepted: 19 August 2020; Published: 24 August 2020

Abstract: A neural fuzzy controller (NFC)-based speed controller for the sensorless permanent magnet synchronous motor (PMSM) drive control system is realized in this paper. The NFC is a fuzzy logic controller (FLC), which adjusts the RBFNN-based (radial basis function neural network) parameter by adapting the dynamic system characteristics. For sensorless PMSM drive, the integration of sliding mode observer (SMO) and phase-locked loop (PLL) is executed to estimate the rotor position and speed. To eliminate the initial rotor position estimation and overcome the conventional PLL-based position estimation error in the direction reversion transition, the *I-f* control strategy is applied to start up the motor and change the rotational direction effectively. The system performance was verified in various experimental conditions. The simulation and experimental results indicate that the proposed control algorithm is implemented efficiently. The motor starts up with diverse external loads, operates in a wide speed range for both positive and negative directions, and reverses the rotational direction stably. Furthermore, the system presents robustness against disturbance and tracks the command speed properly.

Keywords: permanent magnet synchronous motor; sliding mode observer; sensorless control; neural fuzzy controller; *I-f* control strategy

---

1. Introduction

Recently, permanent magnet synchronous motors (PMSMs) have been applied in industrial applications and electric vehicles because of their outstanding specifications such as high efficiency, high power density, and large torque-to-weight ratio. In these applications, the field-oriented control (FOC) algorithm is executed effectively, requiring precise rotor position and speed, which are usually measured by a mechanical position sensor installed on the motor's shaft such as digital encoder, resolver, and Hall sensor. However, the position sensor installation increases the size and cost for the PMSM drive control system. In some cases, sensor failure makes the PMSM system unstable and reduces control reliability. To solve these shortcomings, sensorless control techniques have been studied. Since the back electromotive force (EMF) provides the information of rotor position and speed, back EMF-based estimators are widely proposed, such as the Luenberger observer [1,2], the extended Kalman filter (EKF) observer [3–5], the model reference adaptive system (MRAS) observer [6,7], or the sliding mode observer (SMO) [8–15]. Among those observers, SMO is the most applicable because it has simple structure, robustness against disturbance, low sensitivity to parameter perturbations, and easy implementation.

Generally, the arctangent function is realized to get the rotor position information based on the estimated back EMF obtained by the SMO solution. However, the back EMF is impacted by the chattering problem and noise. To increase the estimator's accuracy and robustness, several phase-locked loop (PLL) structures have been widely researched [11–13,15–18]. The conventional PLL is suitable for one-direction rotation motor only when the parameters are already configured. The reason is that it is utilized with an estimation error, influenced by the sign of speed in the estimated back EMF, which makes a large error of 180° in rotor position estimation when the direction reversion transition happens. Thus, the conventional PLL is not applied in some applications, requiring reversible motor rotational direction operation. To overcome the disadvantage of the conventional PLL, the tangent function-based estimation error is introduced in [16]. However, it is very sensitive to noise, and the obvious estimation error may be obtained when the back EMF crosses zero. Other research using the back EMF signal reconstruction-based estimation error is presented in [12,18]. Although two abovementioned solutions can solve the limitation of the conventional PLL, the position estimation error is not guaranteed to converge to zero because they have three stable points in the phase portrait. Additionally, the estimation error in three aforementioned structures of PLL is constructed from the estimated back EMF, which is too small at the standstill stage or the low-speed region and is affected by noise significantly. Furthermore, to ensure the estimator works precisely, the motor must be ramped up to the specific speed threshold, where the back EMF is large enough for estimation. Therefore, to overcome the startup and reversal problems, the combination of the *I-f* control strategy and the SMO-PLL estimator is implemented in this paper, where the conventional PLL is modified by a position offset value.

The high-performance motor drive control system requires a good tracking response, robustness against disturbance, adaption to parameter variations, and a wide adjustable speed range. Therefore, many speed controllers, such as the RBFNN (radial basis function neural network)-based self-tuning PID (Proportional-Integral-Derivative) controller [15], the sliding mode controller [19], the fuzzy neural network controller [20], and the adaptive fuzzy controller [21] have been proposed to improve system performance. Among these algorithms, in this paper, the neural fuzzy controller (NFC) researched in [22] with only the simulation results for the sensor-based PMSM drive control system is presented and realized in the real-time hardware platform. The main core of NFC is the fuzzy logic controller (FLC). FLC is one of the most effective approaches for controlling the nonlinear system, converting the linguistic control rules based on experts' knowledge into an inference mechanism to regulate the control signals on the system appropriately. Although several successful industrial applications of FLC have been implemented, proper fuzzy membership functions and fuzzy rules designation without the online learning and adaptive adjustment algorithm are not easy to be executed. Thus, FLC has incorporated a neural network with an adaptive mechanism. This integrated controller inherits the merits of both the FLC and the neural network. It performs the identification and learning ability, the parameter adaptation, and uncertain information handling concurrently. In the structure of NFC, the system is firstly identified by an RBFNN which has a simple neural network structure with fast convergence property. Then, the parameters of FLC are tuned by an adjusting mechanism based on the real-time identified plant information (Jacobian transformation) and the gradient descent method to adapt to the dynamics system characteristics.

Accordingly, in our paper, an NFC-based speed controller is realized to enhance the performance of the sensorless PMSM drive control system, which is based on the combination of the SMO-PLL estimator and the *I-f* control strategy. The control algorithm was first evaluated in simulation and then deployed to a real-time platform, established on the DSP (Digital Signal Processor) F28379D (C2000™ LAUNCHXL, Texas Instruments, Inc., Dallas, TX, USA). The system performance is checked under the different experimental conditions with a dynamic load system. The analyzed results demonstrate that the system starts up with diverse external loads, operates in two directions, and switches the rotational direction stably. Comparing to the PI (Proportional-Integral) speed controller, the NFC-based speed controller presents a better tracking performance, faster response and robustness against disturbance. Additionally, the development time for DSP application is substantially shortened when the motor control algorithm is integrated into MATLAB Simulink (MATLAB 2017a, The Mathworks, Inc., Natick, MA, USA) based on the Embedded Coder Support Package utilities.

The remainder of this paper is arranged as follows. Section 2 presents the sensorless PMSM drive control system, including the sliding mode observer, the PLL's structures, and the *I-f* control strategy. The structure of neural fuzzy controller is described in Section 3, consisting of the FLC, the RBFNN, and the adjusting mechanism. In Section 4, the effectiveness and the correctness of the control algorithm are evaluated by the simulation results. The proposed algorithm is confirmed on the real-time platform in Section 5. Finally, Section 6 gives the conclusion.

2. Description of the Sensorless PMSM Drive Control System

The overall architecture of the NFC-based speed controller for the sensorless PMSM drive control system is presented in Figure 1. As mentioned above, the *I-f* control mode and the sensorless control mode are investigated to control the motor in a wide speed range for both directions, which also includes starting the motor from the standstill stage and the rotational direction reversion. In the low-speed range, the PMSM is controlled in the *I-f* control mode, where the switches are activated in position 1. In the medium- to high-speed range, the motor operates in the sensorless control mode, where the switches are activated in position 2. For the sensorless control, the control algorithm includes two closed control loops—the outer speed control loop and the inner current control loop. The FOC algorithm and two PI controllers are implemented in the current control loop. The speed control loop is realized by an NFC-based speed controller, consisting of an FLC, an RBFNN, an adjusting mechanism, and a PI controller. The detailed control algorithm and related formulation are described as follows.

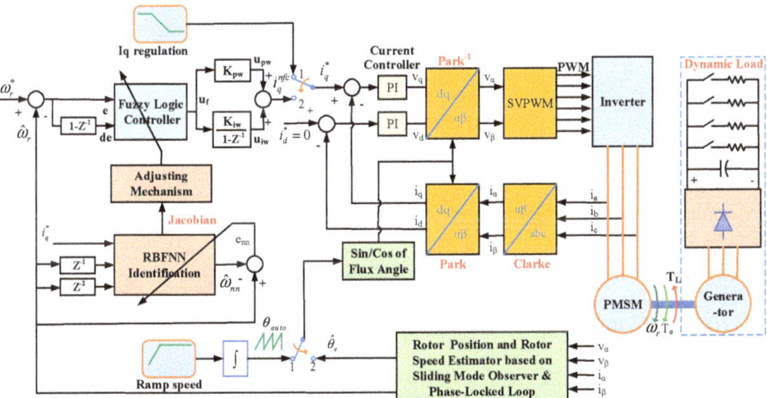

Figure 1. The overall architecture of the neural fuzzy controller (NFC)-based speed controller for the sensorless permanent magnet synchronous motor (PMSM) drive control system.

2.1. Sliding Mode Observer

The modeling of the PMSM was presented in detail previously [15]. In this paper, the main equations for PMSM and the principle of sliding mode observer are summarized as follows.

On the α–β stationary coordinates, the current equation of the surface-mounted PMSM (SPMSM) can be expressed by

$$\begin{cases} \frac{d}{dt}i_\alpha = \frac{1}{L_s}\left(-r_s i_\alpha + v_\alpha - e_\alpha\right) \\ \frac{d}{dt}i_\beta = \frac{1}{L_s}\left(-r_s i_\beta + v_\beta - e_\beta\right) \end{cases} \tag{1}$$

where, $i_\alpha, i_\beta, v_\alpha, v_\beta$ represent the current and the voltage of the α and β axis, respectively; e_α, e_β are the back EMF, satisfying $e_\alpha = -\omega_e \lambda_f sin\theta_e, e_\beta = \omega_e \lambda_f cos\theta_e$; θ_e is the electrical rotor position; λ_f is the permanent magnet flux linkage; ω_e is the rotating speed of magnet flux; and r_s and L_s are the phase resistance and the phase inductance, respectively.

The current observer formulation for SPMSM can be formulated as follows, based on the sliding mode observer theory:

$$\begin{cases} \frac{d}{dt}\hat{i}_\alpha = \frac{1}{L_s}\left(-r_s\hat{i}_\alpha + v_\alpha - kH(\hat{i}_\alpha - i_\alpha)\right) \\ \frac{d}{dt}\hat{i}_\beta = \frac{1}{L_s}\left(-r_s\hat{i}_\beta + v_\beta - kH(\hat{i}_\beta - i_\beta)\right) \end{cases} \tag{2}$$

where $\hat{i}_\alpha, \hat{i}_\beta$ represent the estimated current of the α and β axis, $H(x)$ is the sigmoid function, and k is the observer gain value. Additionally, formulated in the dynamic equation, the estimated current error can be expressed by subtracting Equation (1) from Equation (2):

$$\begin{cases} \frac{d}{dt}\widetilde{i}_\alpha = \frac{1}{L_s}\left(-r_s\widetilde{i}_\alpha + e_\alpha - kH(\widetilde{i}_\alpha)\right) \\ \frac{d}{dt}\widetilde{i}_\beta = \frac{1}{L_s}\left(-r_s\widetilde{i}_\beta + e_\beta - kH(\widetilde{i}_\beta)\right) \end{cases} \tag{3}$$

where $\widetilde{i}_\alpha, \widetilde{i}_\beta$ represent the estimated current error of the α and β axis, defined as: $\widetilde{i}_\alpha = \hat{i}_\alpha - i_\alpha$ and $\widetilde{i}_\beta = \hat{i}_\beta - i_\beta$.

Furthermore, the asymptotic stability of the SMO based on the Lyapunov theory is guaranteed when the observer gain k is designed to be a sufficiently large positive value, $k > max(|e_\alpha|, |e_\beta|)$. Therefore, the estimated current will converge to the actual current. The estimated back EMF $\hat{e}_\alpha, \hat{e}_\beta$ can be obtained as:

$$\begin{cases} \hat{e}_\alpha = kH(\hat{i}_\alpha - i_\alpha) = kH(\widetilde{i}_\alpha) \\ \hat{e}_\beta = kH(\hat{i}_\beta - i_\beta) = kH(\widetilde{i}_\beta) \end{cases} \tag{4}$$

2.2. Speed and Position Estimation Method

Based on the estimated back EMF, the rotor position is directly determined by the arctangent function in the traditional method

$$\hat{\theta}_e = arctan\left(\frac{\hat{e}_\alpha}{\hat{e}_\beta}\right) \tag{5}$$

and the electrical rotor speed can be obtained by $\hat{\omega}_e = \frac{d\hat{\theta}_e}{dt}$. However, due to using the arctangent function, the estimated position and speed are vulnerable to noise and harmonics. Particularly, the apparent estimation errors can be obtained when $\hat{e}_\beta$ crosses zero. Therefore, the phase-locked loop algorithm is executed to mitigate the adverse effect. The structure of the PLL is shown in Figure 2.

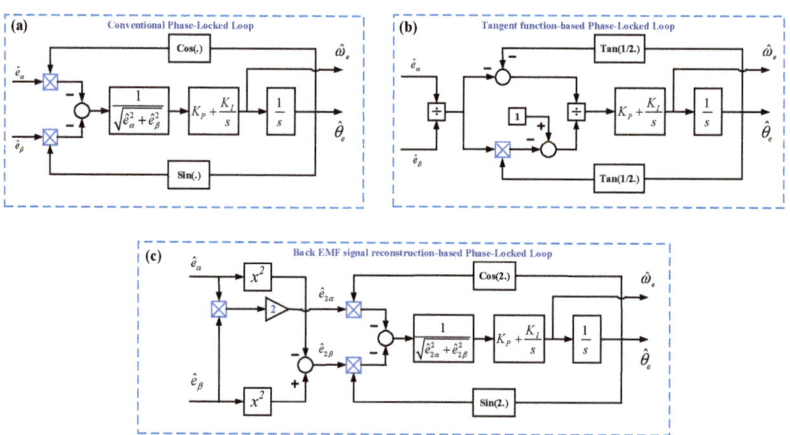

Figure 2. The structures of the PLL (phase-locked loop).

In the conventional PLL [15], the estimation error of rotor position is expressed by

$$\varepsilon = \frac{1}{\sqrt{\hat{e}_\alpha^2 + \hat{e}_\beta^2}}\left[-\hat{e}_\alpha cos(\hat{\theta}_e) - \hat{e}_\beta sin(\hat{\theta}_e)\right]$$
$$= -\hat{e}_{\alpha n}cos(\hat{\theta}_e) - \hat{e}_{\beta n}sin(\hat{\theta}_e) \qquad (6)$$
$$= sin(\theta_e)cos(\hat{\theta}_e) - cos(\theta_e)sin(\hat{\theta}_e)$$
$$= sin(\theta_e - \hat{\theta}_e) \approx \theta_e - \hat{\theta}_e$$

In the tangent function-based PLL [16], the estimation error of rotor position is expressed by

$$\varepsilon = \frac{-\frac{\hat{e}_\alpha}{\hat{e}_\beta} - tan\left(\frac{\hat{\theta}_e}{2}\right)}{1 - \frac{\hat{e}_\alpha}{\hat{e}_\beta}tan\left(\frac{\hat{\theta}_e}{2}\right)} = \frac{tan(\theta_e) - tan\left(\frac{\hat{\theta}_e}{2}\right)}{1 + tan(\theta_e)tan\left(\frac{\hat{\theta}_e}{2}\right)} = tan\left(\theta_e - \frac{\hat{\theta}_e}{2}\right) . \qquad (7)$$

In the back EMF signal reconstruction-based PLL [12,18], the estimation error of rotor position is expressed by

$$\varepsilon = \frac{1}{\sqrt{\hat{e}_{2\alpha}^2 + \hat{e}_{2\beta}^2}}\left[-\hat{e}_{2\alpha}cos(2\hat{\theta}_e) - \hat{e}_{2\beta}sin(2\hat{\theta}_e)\right]$$
$$= sin\left[2\left(\theta_e - \hat{\theta}_e\right)\right] \qquad (8)$$

where $\hat{e}_{2\alpha}$, $\hat{e}_{2\beta}$ are the modified back EMF signal, defined as: $\hat{e}_{2\alpha} = 2\hat{e}_\alpha\hat{e}_\beta = -\left(\omega_e\lambda_f\right)^2 sin(2\theta_e)$ and $\hat{e}_{2\beta} = \hat{e}_\beta^2 - \hat{e}_\alpha^2 = \left(\omega_e\lambda_f\right)^2 cos(2\theta_e)$.

The phase portrait of the PLLs is presented in Figure 3. In the tangent function-based PLL and back EMF signal reconstruction-based PLL, there are three stable points for both positive and negative rotor speed, which are (0,0), (π,0) and ($-\pi$,0). When the PI regulator's parameters are tuned properly, the rotor position estimation error can approach to the origin point. However, in practical application, when the motor changes its direction, the back EMF is very small and is affected by noise significantly in the low-speed region, which causes a large estimation error. Moreover, depending on the initial rotor position and speed estimation error, the trajectories can converge to the stable point (0,0) or ($\pm\pi$,0) when the motor starts up from the standstill stage. Especially, in the case of the *I-f* startup strategy without initial rotor position estimation, the estimation error uncertainly converges to the origin point. It implies that the position estimation error can be obtained of 0 or π (rad). Otherwise, the cycle of rotor position is 2π; thus, the conventional PLL is considered to always have only one stable point in each direction. In the conventional PLL, when the motor reverses its direction from the positive rotor speed to the negative rotor speed, the stable point (0,0) changes into the saddle point while the saddle points ($\pm\pi$,0) become stable points. This shows that, when the speed reversion transition occurs, the trajectories in the phase portrait of the conventional PLL system will deviate from (0,0) to reach ($\pm\pi$,0), and cause a rotor position estimation error of π (rad). To solve this problem, the PI regulator's parameters should be reset. However, it is not easy to adjust to the real-time control system. To overcome the abovementioned problems, during the direction reversion transition, the *I-f* control strategy in the combination with the modified conventional PLL is studied to make sure that the motor can switch the rotational direction stably, and the conventional PLL can work effectively for both directions. That also rejects the effect of the back EMF estimation error in the low-speed region.

Therefore, a position offset is considered to modify the estimated rotor position. The estimation error of rotor position for the conventional PLL is rewritten by

$$\varepsilon = \frac{1}{\sqrt{\hat{e}_\alpha^2 + \hat{e}_\beta^2}}\left[-\hat{e}_\alpha cos(\hat{\theta}_e + \theta_{offset}) - \hat{e}_\beta sin(\hat{\theta}_e + \theta_{offset})\right]$$

$$with \ \theta_{offset} = \begin{cases} 0 \ \ if \ \hat{\omega}_e \geq 0 \\ \pi \ \ if \ \hat{\omega}_e < 0 \end{cases}$$

(9)

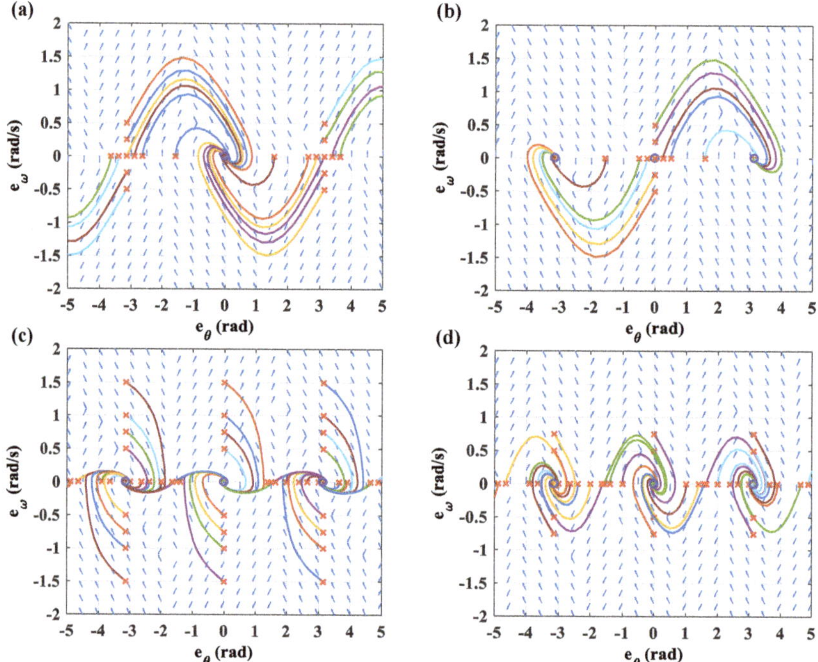

Figure 3. The phase portrait of: (**a**) the conventional PLL for positive rotor speed; (**b**) the conventional PLL for negative rotor speed; (**c**) the tangent-based PLL for both positive and negative rotor speed; and (**d**) the back EMF (electromotive force) signal reconstruction-based PLL for both positive and negative rotor speed.

If the estimation error is attracted to zero by a PI regulator, the rotor position and speed are estimated by

$$\begin{cases} \hat{\omega}_e = K_P\varepsilon + K_i \int \varepsilon dt \\ \hat{\theta}_e = \int \hat{\omega}_e dt \end{cases}.$$

(10)

2.3. The I-f Control Strategy for the Direction Reversion Transition

The *I-f* startup strategy was described and implemented previously [15,23], which smooths the torque and speed transition during the inter-mode transition from the startup stage to the sensorless control stage. Furthermore, for the direction reversion transition, the *I-f* control strategy is extended to make sure that the motor can switch the rotational direction stably. This procedure is presented in Figure 4 and portrayed as the following steps:

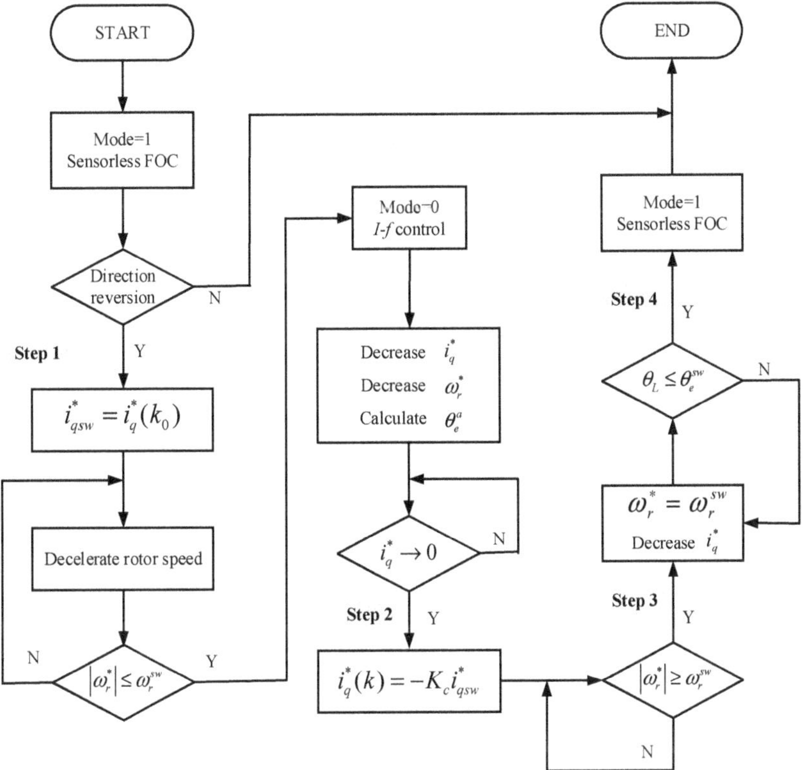

Figure 4. The *I-f* control strategy flowchart for the direction reversion transition.

Step 1: When the speed command is set up to reverse the rotational direction, the system captures the reference current i^*_q and decelerates the rotor speed until the specific speed threshold value for switching condition. When the rotor speed is lower than the abovementioned value, the back-EMF signal is not large enough and significantly affected by noise, which causes the rotor position and speed information to be obtained by the SMO-PLL estimator inaccurately. Thus, the system switches to the *I-f* control mode. The speed command is decreased by a ramp function and i^*_q is regulated properly. The reference current i^*_q and reference speed ω^*_r are expressed by

$$\begin{cases} i^*_q(k+1) = i^*_q(k) - K_a T_s i^*_{qsw} \\ \omega^*_r(k+1) = \omega^*_r(k) \mp K_r T_s \text{ with } |\omega^*_r| < \omega^{sw}_r \qquad \text{for } k > k_0 \\ \theta^a_e(k+1) = \theta^a_e(k) + \omega^*_r T_s \text{ with } \theta^a_e(k_0) = \hat{\theta}_e(k_0), \end{cases} \qquad (11)$$

where the symbol "−" denotes the positive–negative direction transition, and the symbol "+" denotes the negative–positive direction transition; i^*_{qsw} is the reference current at the time the direction reversion command is set up; K_a is the proportional gain for decreasing the current i^*_q; θ^a_e is the auto-generated rotor position; ω^{sw}_r is the specific speed threshold value for switching to the sensorless control mode; and K_r is the proportional gain of the ramp function for the command speed.

Step 2: When the reference current i^*_q approaches zero, it is reset to the certain value to speed up the motor, while the reference speed is still changed by the ramp function.

$$i^*_q(k) = -K_c i^*_{qsw} \qquad (12)$$

where K_c is the proportional gain of the reference current i_q^* in the new direction.

Step 3: When the reference speed approaches ω_r^{sw}, the reference current i_q^* is decreased, similar to the procedure in the *I-f* startup strategy.

$$\begin{cases} i_q^*(k+1) = i_q^*(k) \mp K_a T_s \\ \omega_r^*(k+1) = \omega_r^*(k) = \omega_r^{sw} \end{cases} \quad while \; \theta_L = \hat{\theta}_e - \theta_e^a > \theta_e^{sw} \tag{13}$$

where θ_e^{sw} is the angular threshold value of the switching condition.

Step 4: The motor is switched to operate in the sensorless control mode.

3. The Neural Fuzzy Controller Design for PMSM Drive Control System

3.1. Fuzzy Logic Controller

In Figure 1, there are two inputs for FLC, the rotor speed error e and the rotor speed error change de, which are expressed by

$$\begin{cases} e(k) = \omega_r^*(k) - \hat{\omega}_r(k) \\ de(k) = e(k) - e(k-1) \end{cases} \tag{14}$$

and the output of the FLC is presented as u_f.

According to the fuzzy control theory, the design procedure of the FLC algorithm in this paper is depicted as follows. Firstly, the input variables are selected, where their linguistic variables are presented by E and dE. There are seven linguistic values for each input variables, labeled as $\{A_1, A_2, A_3, A_4, A_5, A_6, A_7\}$ and $\{B_1, B_2, B_3, B_4, B_5, B_6, B_7\}$, respectively. The membership functions are constructed on the symmetrical triangular type, as shown in Figure 5. Secondly, the membership degrees for e and de are computed. For any input value, there are only two active linguistic values and the membership degrees are obtained by

$$\begin{cases} \mu_{A_i}(e) = \dfrac{e_{i+1} - e}{e_{i+1} - e_i} \; and \; \mu_{A_{i+1}}(e) = 1 - \mu_{A_i}(e) \\ \mu_{B_j}(de) = \dfrac{de_{j+1} - de}{de_{j+1} - de_j} \; and \; \mu_{B_{j+1}}(de) = 1 - \mu_{B_j}(de) \end{cases} \tag{15}$$

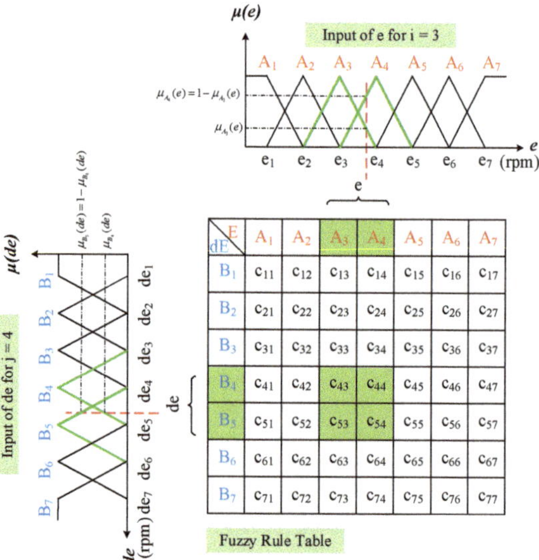

E \ dE	A₁	A₂	A₃	A₄	A₅	A₆	A₇
B₁	c_{11}	c_{12}	c_{13}	c_{14}	c_{15}	c_{16}	c_{17}
B₂	c_{21}	c_{22}	c_{23}	c_{24}	c_{25}	c_{26}	c_{27}
B₃	c_{31}	c_{32}	c_{33}	c_{34}	c_{35}	c_{36}	c_{37}
B₄	c_{41}	c_{42}	c_{43}	c_{44}	c_{45}	c_{46}	c_{47}
B₅	c_{51}	c_{52}	c_{53}	c_{54}	c_{55}	c_{56}	c_{57}
B₆	c_{61}	c_{62}	c_{63}	c_{64}	c_{65}	c_{66}	c_{67}
B₇	c_{71}	c_{72}	c_{73}	c_{74}	c_{75}	c_{76}	c_{77}

Figure 5. The membership function of e, de and fuzzy rule table.

Thirdly, the initial FLC rules are designed by referring to the dynamic response characteristics, such as

$$\text{IF } e \text{ is } A_i \text{ and } de \text{ is } B_j \text{ THEN } u_f \text{ is } c_{j,i} \tag{16}$$

where $i, j = 1, 2 \ldots 7$ and $c_{j,i}$ is the real number. There are 49 fuzzy rules in the fuzzy rule table. Finally, the product-inference rule, singleton fuzzifier, and central average defuzzification technique are implemented to infer the output $u_f(e, de)$ of the fuzzy system. For any input value (e, de), the value of (i,j) is determined, and only four fuzzy rules result in a non-zero output. Therefore, four linguistic values A_i, A_{i+1}, B_j, and B_{j+1} and the corresponding consequent values $c_{j,i}$, $c_{j+1,i}$, $c_{j,i+1}$, and $c_{j+1,i+1}$ are active, and the output of the fuzzy system can be obtained by

$$u_f(e, de) = \frac{\sum_{n=i}^{i+1} \sum_{m=j}^{j+1} c_{m,n} \lfloor \mu_{A_n}(e) \times \mu_{B_m}(de) \rfloor}{\sum_{n=i}^{i+1} \sum_{m=j}^{j+1} \mu_{A_n}(e) \times \mu_{B_m}(de)} = \sum_{n=i}^{i+1} \sum_{m=j}^{j+1} c_{m,n} \times d_{m,n} \tag{17}$$

where $d_{m,n} = \mu_{A_n}(e) \times \mu_{B_m}(de)$, and the value of the $\sum_{n=i}^{i+1} \sum_{m=j}^{j+1} d_{m,n} = 1$ in Equation (17) can be easily derived by using Equation (15). Moreover, the $c_{m,n}$ are adjustable parameters of the FLC.

Additionally, in Figure 1, for the sensorless control mode, the reference current i_q^* is expressed by the output of the fuzzy controller u_f as the following equation:

$$i_q^*(k) = u_{iw}(k-1) + \left(K_{pw} + K_{iw}\right) \times u_f(k) \tag{18}$$

where K_{pw} and K_{iw} are the proportional and integral gain of PI controller in the speed control loop, respectively.

3.2. Radial Basis Function Neural Network

The parameters of FLC in the NFC-based speed controller should be adjusted effectively. That needs the Jacobian transformation obtained from the dynamic system characteristics. Therefore, the PMSM drive control system must be identified. In Figure 6, an RBFNN is designed to acquire that information. It is a feedforward neural network architecture comprised of an input layer, a hidden layer, and a single output layer.

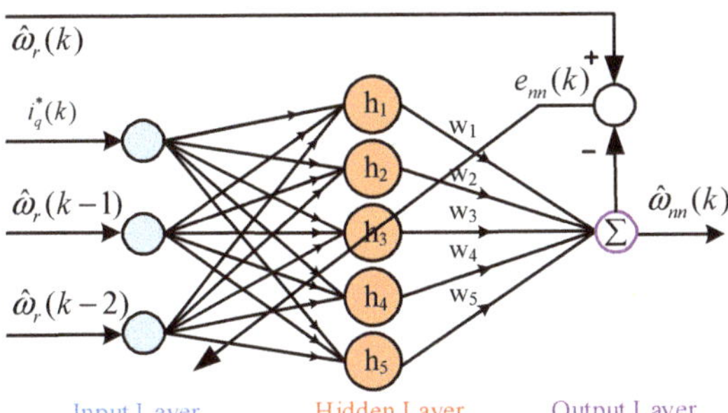

Figure 6. The structure of RBFNN (radial basis function neural network) in the NFC-based speed controller.

There are three inputs in the input layer, namely $i_q^*(k)$, $\hat{\omega}_r(k-1)$, $\hat{\omega}_r(k-2)$, and the input vector form is expressed by

$$x(k) = \left[i_q^*(k) \; \hat{\omega}_r(k-1) \; \hat{\omega}_r(k-2) \right]^T \tag{19}$$

In the hidden layer, there are five neurons and the Gaussian function is used as the activation function, which is presented by the following equation

$$h_l(k) = exp\left(-\frac{\|x(k) - C_l(k)\|^2}{2b_l^2(k)}\right) \tag{20}$$

where $C_l(k) = [c_{1l}(k)\ c_{2l}(k)\ c_{3l}(k)]^T$ and c_{rl}, b_l, are the center and width of the Gaussian function.

In the output layer, the output is formulated by a linearly weighted sum of hidden nodes:

$$\hat{\omega}_{nn}(k) = \sum_{l=1}^{5} w_l(k)h_l(k) \tag{21}$$

Additionally, to adjust the parameters of RBFNN by the online learning law, the instantaneous cost function is presented by

$$J(k) = \frac{1}{2}(\hat{\omega}_r(k) - \hat{\omega}_{nn}(k))^2 = \frac{1}{2}e_{nn}^2(k) \tag{22}$$

According to the stochastic gradient descent (SGD) method, the weights, node widths, and node centers can be calculated and updated by the learning algorithm as below equations [24]:

$$\begin{cases} \Delta\omega_l(k) = -\eta\frac{\partial J(k)}{\partial\omega_l(k)} = \eta e_{nn}(k)h_l(k) \\ \omega_l(k) = \omega_l(k-1) + \Delta\omega_l(k) + \alpha(\omega_l(k-1) - \omega_l(k-2)) \end{cases} \tag{23}$$

$$\begin{cases} \Delta b_l(k) = -\eta\frac{\partial J(k)}{\partial b_l(k)} = \eta e_{nn}(k)\omega_l(k)h_l(k)\frac{\|x(k)-C_l(k)\|^2}{b_l^3(k)} \\ b_l(k) = b_l(k-1) + \Delta b_l(k) + \alpha(b_l(k-1) - b_l(k-2)) \end{cases} \tag{24}$$

$$\begin{cases} \Delta c_{rl}(k) = -\eta\frac{\partial J(k)}{\partial c_{rl}(k)} = \eta e_{nn}(k)\omega_l(k)h_l(k)\frac{x_r(k)-c_{rl}(k)}{b_l^2(k)} \\ c_{rl}(k) = c_{rl}(k-1) + \Delta c_{rl}(k) + \alpha(c_{rl}(k-1) - c_{rl}(k-2)) \end{cases} \tag{25}$$

where α is the momentum factor; η is the learning rate, and $r = 1,2,3$; $l = 1,2\ldots5$.

Additionally, the Jacobian transformation can be expressed by

$$\frac{\partial\hat{\omega}_r}{\partial i_q^*} \approx \frac{\partial\hat{\omega}_{nn}}{\partial i_q^*} = \sum_{l=1}^{5} w_l(k)h_l(k)\frac{c_{1l}(k) - i_q^*(k)}{b_l^2(k)} \tag{26}$$

3.3. Adjusting Mechanism of Fuzzy Logic Controller

To achieve the smallest squared error between the reference speed and the estimated rotor speed, the FLC parameters are adjusted online in the closed-loop control. Hence, the cost function can be firstly defined by

$$J_e(k) = \frac{1}{2}(\omega_r^*(k) - \hat{\omega}_r(k))^2 = \frac{1}{2}e^2(k) \tag{27}$$

Then, the NFC learning law is derived based on the gradient descent method. In the fuzzy rule table (Figure 5), parameters of $c_{m,n}$ can be updated and optimally adjusted according to

$$\Delta c_{m,n} = -\gamma\frac{\partial J_e}{\partial c_{m,n}} \tag{28}$$

where γ is the adaptive rate. The chain rule is executed, and the partial differential equation for J_e in Equation (28) can be presented by

$$\frac{\partial J_e}{\partial c_{m,n}} = -\frac{\partial J_e}{\partial \hat{\omega}_r}\frac{\partial \hat{\omega}_r}{\partial u_f}\frac{\partial u_f}{\partial c_{m,n}} \tag{29}$$

Furthermore, from Equation (15) and using the Jacobian transformation from Equation (26), the following equations are obtained:

$$\frac{\partial u_f}{\partial c_{m,n}} = d_{m,n} \tag{30}$$

and

$$\frac{\partial \hat{\omega}_r}{\partial u_f} = \frac{\partial \hat{\omega}_r}{\partial i_q^*}\frac{\partial i_q^*}{\partial u_f} \approx \left(K_{pw} + K_{iw}\right)\frac{\partial \hat{\omega}_r}{\partial i_q^*} = \left(K_{pw} + K_{iw}\right)\sum_{l=1}^{5} w_l(k)h_l(k)\frac{c_{1l}(k) - i_q^*(k)}{b_l^2(k)} \tag{31}$$

Substituting Equations (30) and (31) into Equation (29), the parameters $c_{m,n}$ of FLC defined in Equation (17) can be adjusted by the following equation:

$$\Delta c_{m,n}(k) = \gamma e(k)\left(K_{pw} + K_{iw}\right)d_{m,n}\sum_{l=1}^{5} w_l(k)h_l(k)\frac{c_{1l}(k) - i_q^*(k)}{b_l^2(k)} \tag{32}$$

with $m = j, j+1$ and $n = i, i+1$.

4. Simulation Results

To demonstrate the correctness of the proposed control algorithm for the PMSM drive system, the system performance was tested in both a simulation and real-time hardware within different experimental conditions. The first case is controlling the motor for two directions in the wide speed range to validate the tracking response and the estimator's correction. The estimated values were compared to the actual values, which were measured by the digital incremental encoder. The second case is the direction reversion transition to make sure that the system can switch the direction stably. The third case is starting up the motor with the different initial external load and comparing the system performance between the proposed NFC-based speed controller and the PI speed controller. That confirms the executed ability of the *I-f* startup strategy and the system's robustness against disturbance. All instances are inspected on the PMSM control system and dynamic load system, where the PMSM's parameters are listed as rated power of 750 W, rated current of 4.24 A, rated speed of 2000 rpm, phase resistance of 1.326 Ω, phase inductance of 2.952 mH, back-EMF constant of 56.5 $V_{L\text{-}L}$/krpm, torque constant of 0.86 Nm/A, inertia of 3.63 Kg·cm^2, and pole pairs of 4.

Furthermore, the PI speed controller has fixed parameters and not flexible to the PMSM drive control system, which is a dynamic, multivariable, and nonlinear system. Especially, the PMSM drive control system is usually operated under various conditions or dynamics load. Therefore, the PI controller only works effectively at a specific operating condition, where the PI parameters are designed in correspondence to the system characteristics. For a fair comparison between the NFC-based speed controller and the PI speed controller, the system performances were analyzed with the same PI's parameters. In the third experimental condition, the PI speed controller is firstly set for the case of an external load of 50 Ω, satisfying the performance criteria such as the transient response without the overshoot or undershoot, and with the short settling time and the zero steady-state error. Then, the operating condition is varied by applying two different external load values on the system to evaluate the system performance for both the PI and NFC-based speed controllers.

The NFC-based speed controller for the sensorless PMSM drive control system was completely designed in MATLAB Simulink, and the overall system is shown in Figure 7. The sampling frequency for real-time platform modeling (block A) is 20 kHz. In the motor control algorithm (block B), the sampling frequency for speed loop control is 1 kHz, while the sampling frequency for the estimator and the current control loop is 20 kHz. The simulation data (block C) are acquired and monitored with the sampling frequency of 20 kHz. The PI controllers' parameters are set in the current control loop as $K_{Pq} = 0.25$, $K_{Iq} = 0.025$, $K_{Pd} = 0.25$, $K_{Id} = 0.025$. The NFC-based speed controller is realized in the speed control loop. For the FLC, the membership function's values are set as $E = [\,e_1\ e_2\ e_3\ e_4\ e_5\ e_6\ e_7\,]$ $= [-225\ -150\ -75\ 0\ 75\ 150\ 225]$ and $dE = [de_1\ de_2\ de_3\ de_4\ de_5\ de_6\ de_7] = [-187.5\ -125\ -62.5\ 0\ 62.5\ 125\ 187.5]$. The initial fuzzy rule table values are set as in Table 1. These parameters are adjusted during the operating time with the adaptive rate of 0.25. For the neural network, the initial neuron parameters are set as follows: node centers $C_l = [-0.5\ -0.25\ 0.0\ 0.25\ 0.5]$, node widths $b = [0.25\ 0.25\ 0.25\ 0.25\ 0.25]$, connective weights $w = [0.00625\ 0.00625\ 0.00625\ 0.00625\ 0.00625]$, momentum factor $\alpha = 0.95$, and the learning rate $\eta = 0.475$. Moreover, the PI's parameters of the PI and NFC-based speed controller are set as $K_{pw} = 1.725$ and $K_{iw} = 0.030$. Additionally, the dynamic load system is modeled by a generator and the electrical load. The electrical load is comprised of a rectifier, a capacitor of 470 µF, and the resistor load bank. In the first and second experimental conditions, the motor is operated with a total resistance load of 100 Ω.

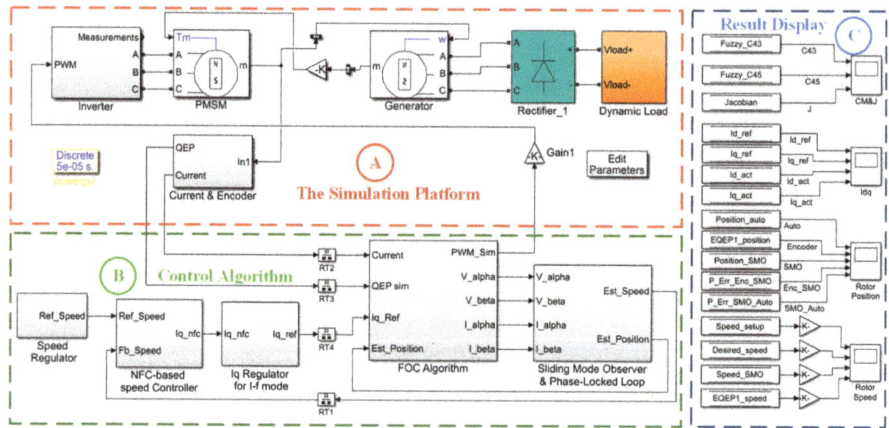

Figure 7. The overall simulation structure of the sensorless PMSM drive control system with a dynamic load system in MATLAB Simulink.

Table 1. The initial fuzzy rule table of fuzzy logic controller (FLC).

dE \ E	A_1	A_2	A_3	A_4	A_5	A_6	A_7
B_1	−0.324	−0.324	−0.324	−0.324	−0.216	−0.108	0
B_2	−0.324	−0.324	−0.324	−0.216	−0.108	0	0.108
B_3	−0.324	−0.324	−0.216	−0.108	0	0.108	0.216
B_4	−0.324	−0.216	−0.108	0	0.108	0.216	0.324
B_5	−0.216	−0.108	0	0.108	0.216	0.324	0.324
B_6	−0.108	0	0.108	0.216	0.324	0.324	0.324
B_7	0	0.108	0.216	0.324	0.324	0.324	0.324

Figure 8 presents the motor's performance in the positive direction for the wide speed range, including: (1) the startup from the standstill; (2) stepping up to the rated speed; and (3) slowing down. The command speed is varied in a sequence of $0 \rightarrow 200 \rightarrow 500 \rightarrow 1000 \rightarrow 1500 \rightarrow 2000 \rightarrow 1600 \rightarrow 1200 \rightarrow 800 \rightarrow 1000$ rpm. Figure 8a indicates that the estimated rotor speed overlaps the actual value and closely tracks the command speed properly. Figure 8b implies that the current i_q is regulated proportionally to the command speed while the current i_d almost is equal to zero. Figure 8c–f illustrates the estimated position, actual position, and the estimation errors at the speeds of 500, 1000,

1500, and 2000 rpm in a period of 0.15 s. There are 5, 10, 15, and 20 position cycles, corresponding to the rotation frequencies of 33.33, 66.67, 100, and 133.33 Hz, respectively. These values are proper to the motor with four pole pairs. The actual and estimated positions come close to each other; thus, the estimation error is equal to zero.

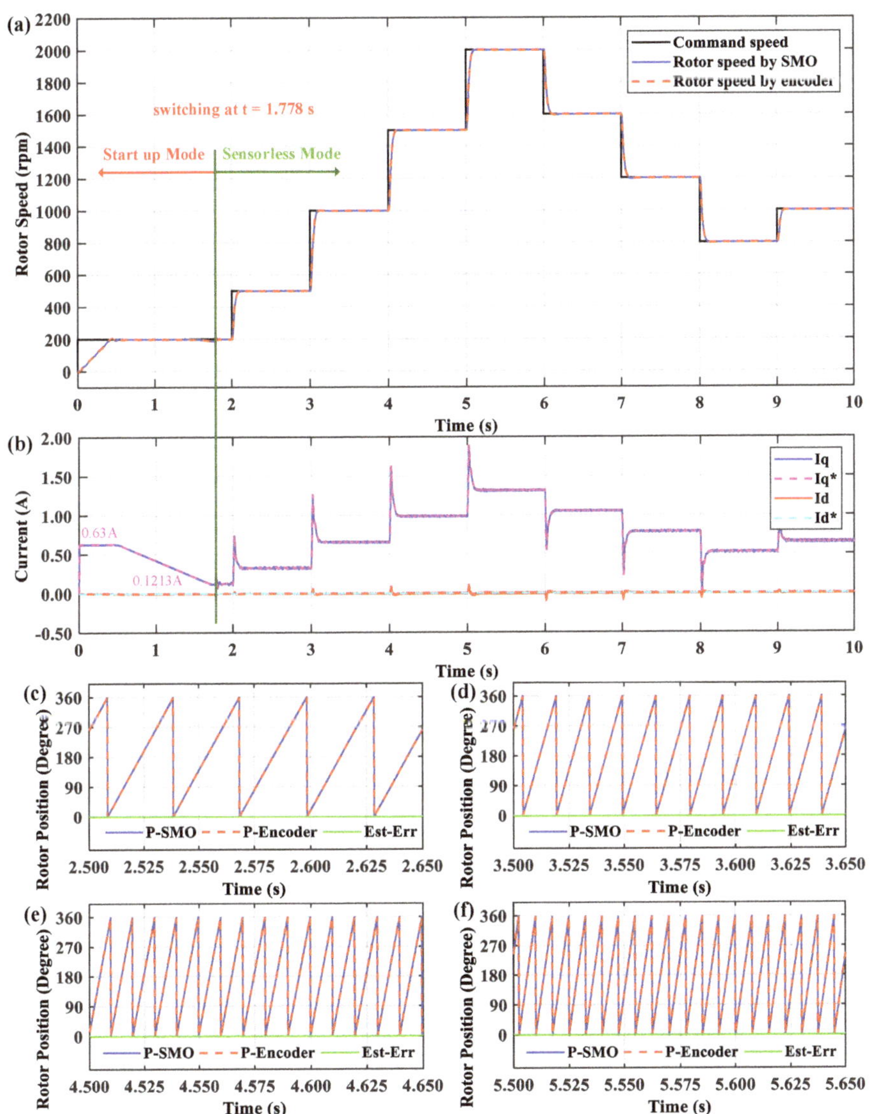

Figure 8. Simulation results in the wide speed range in the positive direction for: (**a**) speed response; (**b**) current response; and rotor position response at (**c**) 500; (**d**) 1000; (**e**) 1500; and (**f**) 2000 rpm.

The motor's performance in the negative direction for the wide speed range is illustrated in Figure 9. The waveform of the command speed is the same as in the positive direction, only the speed has the opposite value. Comparing to Figure 8, the speed response, the current, and the rotor position in the negative direction are just the reverse of their value. In the positive direction, the rotor position is increased from 0° to 360° for one electrical cycle, while the rotor position is decreased from 360° to 0° in the negative direction. Therefore, Figures 8 and 9 refer that the motor control system has worked for two directions with the same quality. In both rotational directions, the motor startups from the standstill stage with the initial reference current i_q^* of 0.63 A. The ratio for decreasing the reference current i_q^* is set up at the value of 0.42 A/s. The motor control algorithm is switched to the sensorless control mode at t = 1.788 s, where the reference current i_q^* is decreased to 0.1213 A.

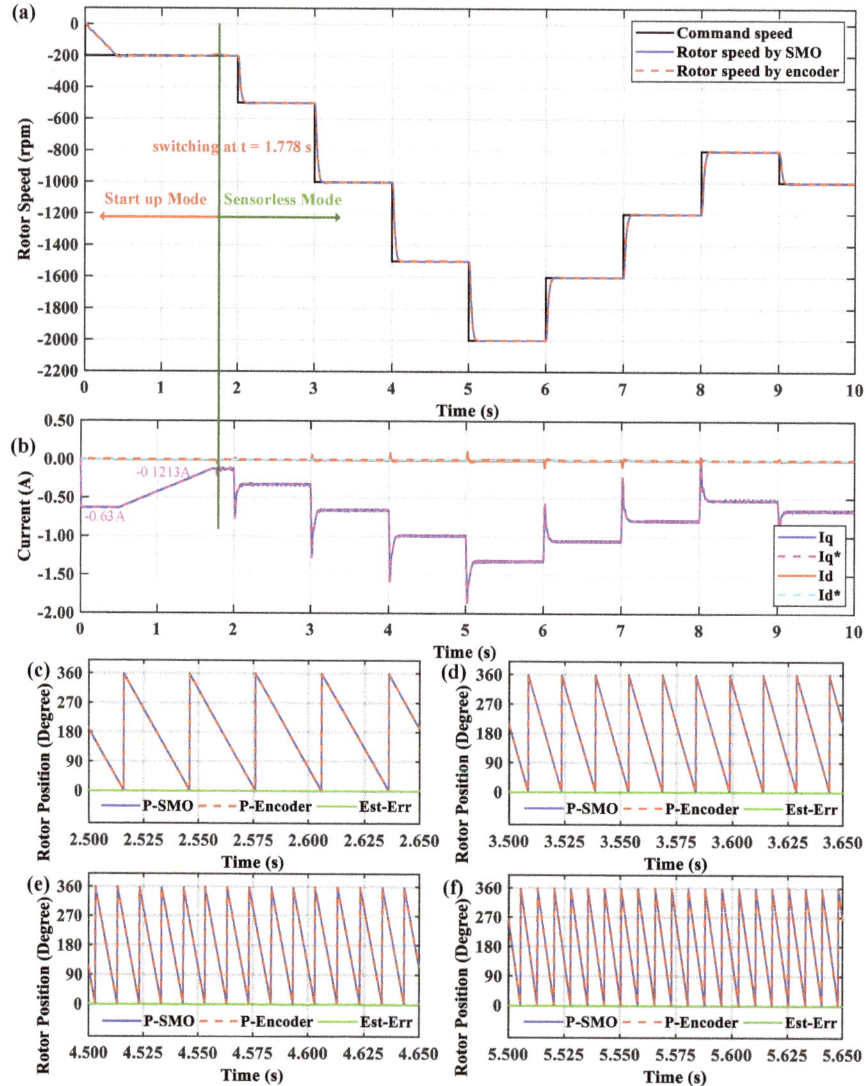

Figure 9. Simulation results in the wide speed range in the negative direction for: (**a**) speed response; (**b**) current response; and rotor position response at (**c**) 500; (**d**) 1000; (**e**) 1500; and (**f**) 2000 rpm.

The motor's speed performance in the reversal operation is presented in Figure 10. Firstly, the motor is started in the positive direction, and then accelerated and decelerated, following the sequence: 0 → 200 → 700 → 200 rpm. Secondly, the command speed is set up to reverse the rotational direction from 200 to −200 rpm at t = 4 s. The motor control algorithm is transferred to the *I-f* control mode from the sensorless control mode. When the motor has already operated in the negative direction, the motor control algorithm is switched to the sensorless control mode again and the command speed is varied by the sequence: −200 → −700 → −200 rpm. Thirdly, a similar procedure is implemented when the command speed is changed to reverse the rotational direction from −200 to 200 rpm at t = 7 s. During the operation of the motor, the estimated rotor speed almost tracks the reference speed and approximates the actual rotor speed for both the sensorless control mode and the *I-f* control mode. Moreover, the current response in Figure 10b indicates that the reference current i_q^* is regulated differently for the sensorless control mode and the *I-f* control mode in the rotational direction reversion. Therefore, the rising time and settling time are different in the speed transition between the rotational direction reversion and the varied speed in one direction. It takes a larger time to reverse the direction. Figure 10c illustrates the rotor position when the motor changes its direction to the negative direction; there is a larger estimation error and the maximum value is −38.67° at t = 4.077 s. Additionally, the maximum rotor position estimation error is 45.25° at t = 7.078 s in the case of changing the direction from the negative value to the positive value in Figure 10d. However, these estimation errors could be accepted because the direction reversion time only takes about 0.111 s. It is the period that the system is operated in the *I-f* control mode. Therefore, Figure 10 implies that the *I-f* control mode makes the motor change the direction stably and the modified conventional PLL works effectively. The estimation error still equals to zero after changing the rotational direction.

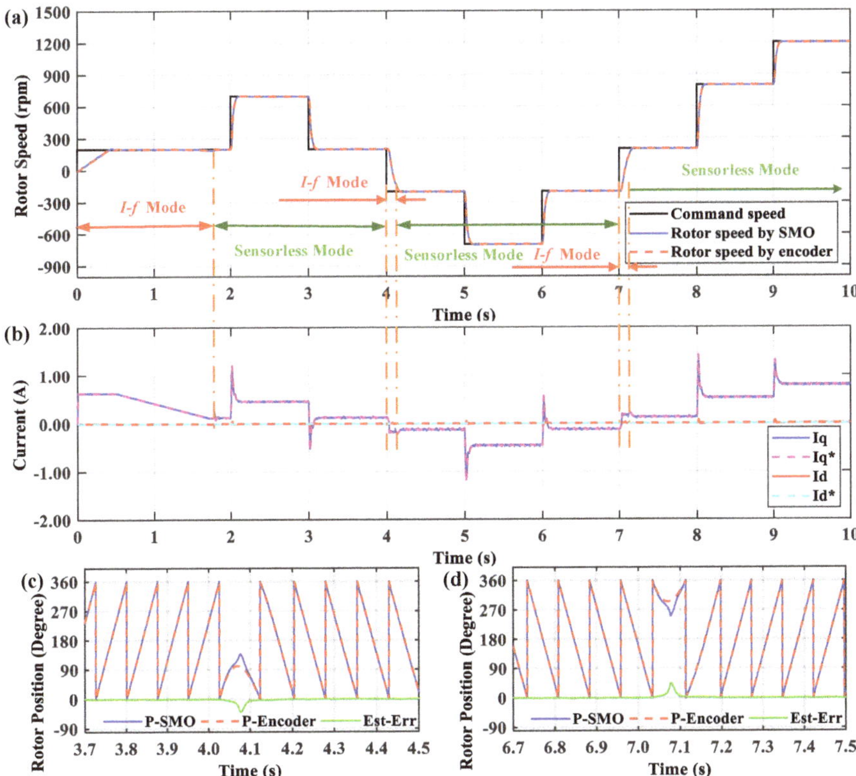

Figure 10. Simulation results in the case of the varied rotational direction for: (**a**) speed response; (**b**) current response; (**c**) rotor position (positive–negative); and (**d**) rotor position (negative–positive).

The system performance for the PI speed controller and the NFC-based speed controller is presented in Figure 11 when the motor startups with the initial resistance load of 100 Ω. The rotor speed follows the sequence of 0 → 200 → 500 → 1000 rpm until t = 4 s. Then, the command speed is regulated as a square wave with a period of 1 s and the speed variation from 1000 to 1500 rpm. At t = 5.5 s, more resistance load is added to the total value of 50 Ω. It implies that the larger external load is suddenly applied to the motor, resulting in a drop of rotor speed briefly because it takes some time to raise the energy supplied to the system. In the PI controller, the rotor speed decreases to the minimum of 950 rpm at t = 5.531 s and steadies at 1000 rpm again at t = 5.751 s. The recovery time is 0.220 s with a speed reduction of 50 rpm. In the NFC controller, the rotor speed decreases to the minimum of 964 rpm at t = 5.523 s and steadies at 1000 rpm again at t = 5.679 s. The recovery time is 0.156 s with a speed reduction of 36 rpm.

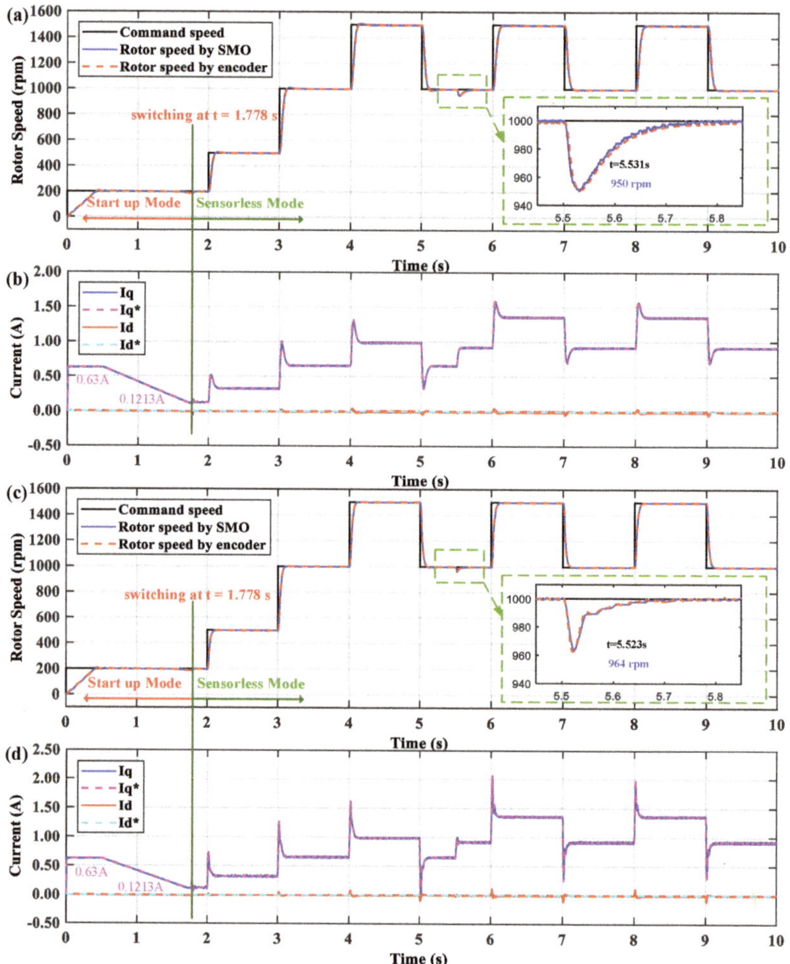

Figure 11. Simulation results for PI (Proportional-Integral) controller: (**a**) speed response; and (**b**) current response; and for NFC controller: (**c**) speed response; and (**d**) current response in the case of $R_L = 100 \rightarrow 50\ \Omega$.

Figure 12 shows the system performance when the motor startups with the initial resistance load of 50 Ω. The speed command is also varied as the same as its waveform, as shown in Figure 11. Because the larger external load is applied in the startup mode, before the earlier sensorless control mode switching at t = 1.702 s, the reference current i_q^* is reduced to the value of 0.1818 A (Figure 12), which is higher than the case shown in Figure 11. Those corresponding values in Figure 11 are 1.778 s and 0.1213 A, respectively. At t = 5.5 s, the resistance load is varied to obtain the total value of 100 Ω. It implies that less external load is applied to the system. In the PI controller, the motor increases the speed until reaching the maximum speed of 1051 rpm, at t = 5.526 s, and then stabilizes at 1000 rpm again at t = 5.724 s. The recovery time is 0.198 s with the speed increment of 51 rpm. In the NFC controller, the rotor speed is increased to the maximum of 1025 rpm, at t = 5.519 s, and stabilizes at 1000 rpm again at t = 5.687 s. The recovery time is 0.168 s with the speed increment of 25 rpm.

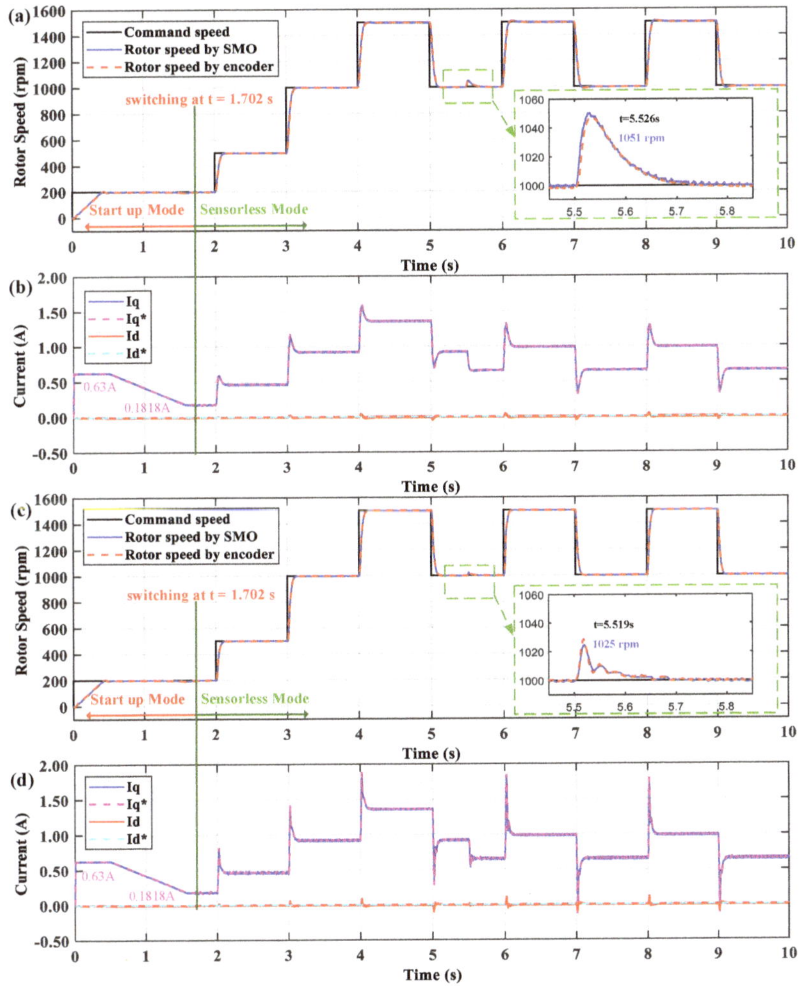

Figure 12. Simulation results for PI controller: (**a**) speed response; and (**b**) current response; and for NFC controller: (**c**) speed response; and (**d**) current response in the case of $R_L = 50 \rightarrow 100$ Ω.

The comparison of the different speed performances for the PI speed controller and the NFC-based speed controller is presented in Figures 13 and 14. It is easy to see that the system has a better performance in the NFC-based speed controller. The motor speed tracks the command speed perfectly, without the overshoots or undershoots for two cases of external load. It also has a faster response, with the settling time of 0.149 s. However, there is a little difference in the speed performance for two cases of external load in the PI speed controller. In the case of $R_L = 100$ Ω, the system has an overshoot and an undershoot, at about +10 rpm, and the settling time of 0.267 s. The system has better performance, without overshoots or undershoots, and the settling time of 0.164 s in the case of $R_L = 50$ Ω, where the PI's parameters are set appropriately. Figures 13 and 14 demonstrate that the PI speed controller can only work properly at the defined condition ($R_L = 50$ Ω), while the proposed NFC-based speed controller works effectively for both cases to obtain a good performance because it has a mechanism to adjust its parameters, adapted to the dynamic system characteristic.

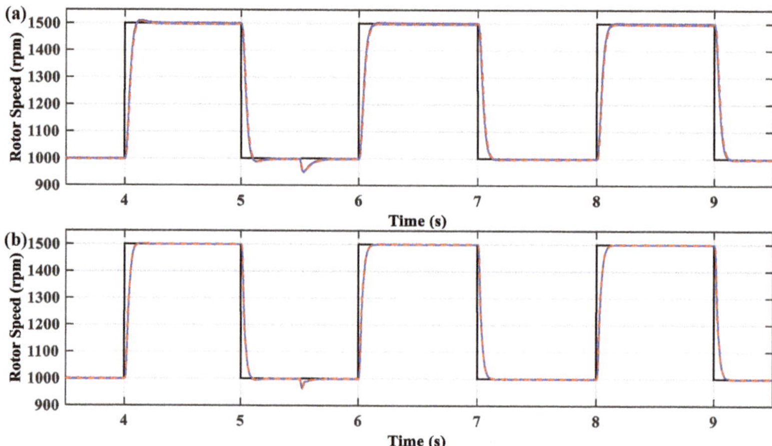

Figure 13. Comparison of the speed response of simulation results in the case of the varied external load for $R_L = 100 \to 50$ Ω: (**a**) PI controller; and (**b**) NFC controller.

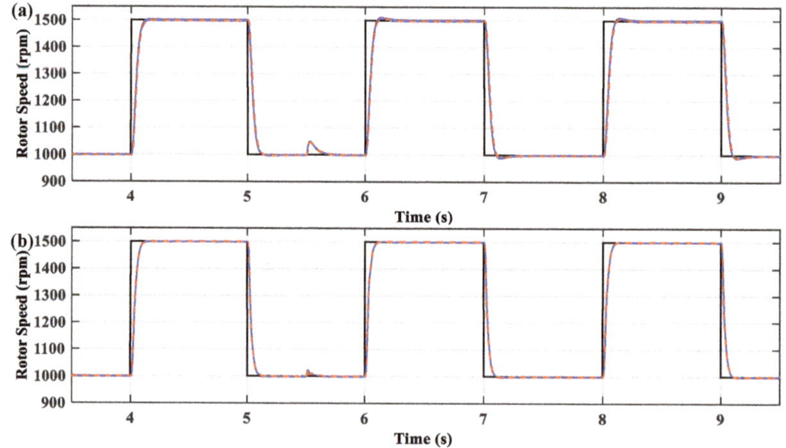

Figure 14. Comparison of the speed response of simulation results in the case of the varied external load for $R_L = 50 \to 100$ Ω: (**a**) PI controller; and (**b**) NFC controller.

In summary, the motor control system's performance is analyzed in Figures 8–14. Several command speed waveforms and the varied dynamic load conditions are applied. Those results imply that the estimator works successfully. The estimated values approach the actual values for both rotor speed and rotor position. The NFC-based speed controller improves the system performance effectively and presents the robustness against disturbance. The rotor speed almost tracks the command perfectly, with a zero steady-state error, without overshoot or undershoot, and having shorter settling time, in comparison to the PI speed controller. Additionally, the system switches the control mode smoothly and reverses the rotational direction stably when the *I-f* control strategy is implemented. The simulation results have already confirmed that the proposed control algorithm for the sensorless PMSM drive control system is correct and effective.

5. Experimental Verification and Results

After evaluating the system performance by simulation, the motor control algorithm was compiled and deployed to the experimental PMSM drive control system, as presented in Figure 15. Moreover, to analyze the system performance, the system data were transmitted to MATLAB Simulink by the SCI function, integrated into the DSP with the sampling frequency of 2 kHz. The experimental system involves a PMSM coupled to a generator, an inverter, a control circuit, and a DSP F28379D. The DSP F28379D is equipped with 200 MHz dual C28xCPUs and dual CLAs, 1 MB Flash, 16-bit/12-bit ADCs, comparators, 12-bit DACs, HRPWMs, eCAPs, eQEPs, CANs, etc. The isolation and protection functions are integrated into the control circuit, which helps lock the PWM control signal in the case of overcurrent problem. The electrical load is comprised of a rectifier, a capacitor of 470 µF and 450 V, and the resistor load bank.

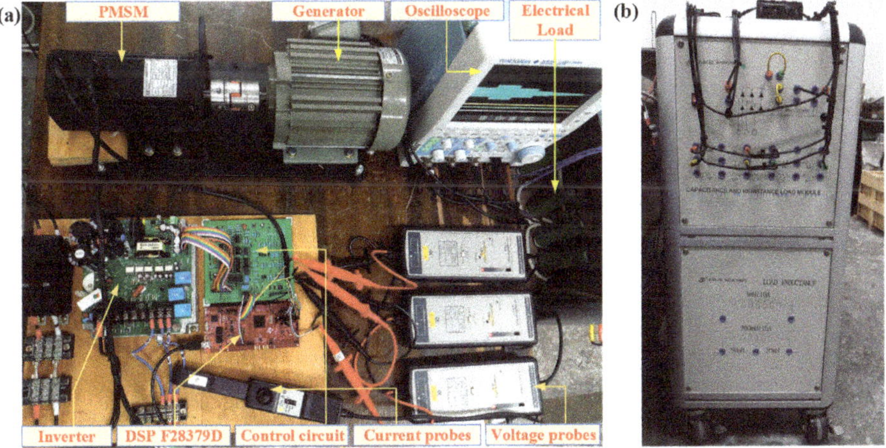

Figure 15. The experimental system with: (**a**) real platform; and (**b**) electrical load.

The sampling frequencies of the speed control loop, the current control loop and the SMO-PLL estimator are 1, 20, and 20 kHz, respectively. The inverter's switching frequency is set at 15 kHz. The PI controllers' parameters are set as $K_{Pq} = 0.25$, $K_{Iq} = 0.025$, $K_{Pd} = 0.25$, $K_{Id} = 0.025$, $K_{pw} = 0.735$, and $K_{iw} = 0.00918$. The membership function's values are set the same as the simulation configuration. The fuzzy rule table values are initialized with a ratio of 0.833 to the values in Table 1 and adjusted with the adaptive rate of 0.25. The neural network's parameters are set as follows: node centers $C_l = [-2.5\ -1.25\ 0.0\ 1.25\ 2.5]$, node widths $b = [2.5\ 2.5\ 2.5\ 2.5\ 2.5]$, connective weights $w = [0.025\ 0.025\ 0.025\ 0.025\ 0.025]$, momentum factor $\alpha = 0.75$, and learning rate $\eta = 0.435$. In the first and second experimental conditions, the motor is operated with the total resistance load of 100 Ω and 400 W.

Figure 16 presents the motor's performance in the positive direction for the wide speed range. The motor startups from the standstill stage at t = 1.193 s and switches to the sensorless control mode at t = 9.213 s. The rotor speed is accelerated to the rated speed and then decelerated. The command speed is varied in a sequence of $0 \rightarrow 300 \rightarrow 500 \rightarrow 1000 \rightarrow 1500 \rightarrow 2000 \rightarrow 1600 \rightarrow 1200 \rightarrow 800 \rightarrow 1000$ rpm for each period of 5 s. Figure 16a refers that the estimated rotor speed overlaps the actual rotor speed and closely tracks the command speed perfectly. In Figure 16b, although the current i_q fluctuates around the reference current i_q^*, its average value is still regulated proportionally to the command speed while the current i_d almost oscillates around the zero value. Figure 8c–f illustrates the estimated position, actual position, and the estimation errors at the speeds of 500, 1000, 1500, and 2000 rpm in a period of 0.15 s. There are 5, 10, 15 and 20 position cycles corresponding to the rotation frequencies of 33.33, 66.67, 100, and 133.33 Hz, respectively. These values are proper to the motor with four pole pairs. The actual and estimated positions come close to each other; thus, the estimation error approximates zero.

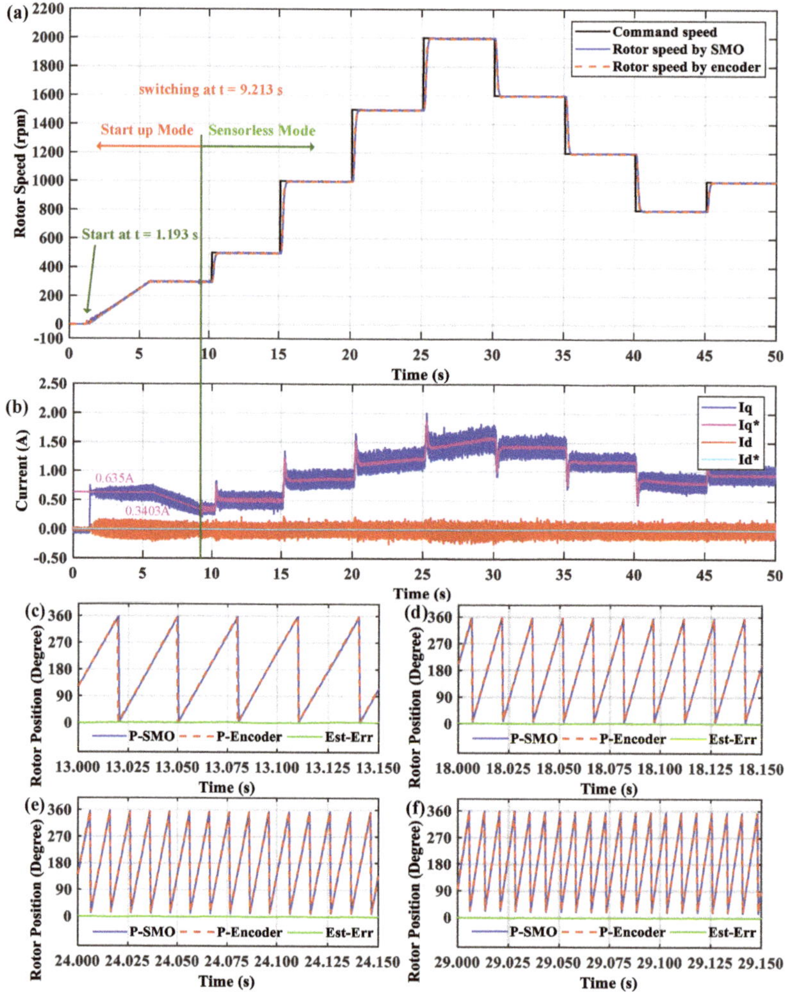

Figure 16. Experimental results in the wide speed range in the positive direction: (**a**) speed response; (**b**) current response; and rotor position response at: (**c**) 500; (**d**) 1000; (**e**) 1500; and (**f**) 2000 rpm.

Figure 17 illustrates the motor's performance in the negative direction for the wide speed range. The waveform of the command speed is the same as in the positive direction, only the speed has the opposite value. The motor starts up at t = 1.035 s and switches to the sensorless control mode at t = 9.201 s. Comparing to Figure 16, in the positive direction, the rotor position is increased from 0° to 360° for one electrical cycle and the estimation error is close to zero, while the rotor position is decreased from 360° to 0° in the negative direction and the estimation error is not completely equal to zero; the maximum error values can reach to 12°. Therefore, the reference current i_q^* is regulated with a little larger value, in comparison to the case of the positive direction. However, the rotor position still tracks the command speed very well. In both rotational directions, the motor starts up from the standstill stage with the initial reference current i_q^* of 0.635 A. The ratio for decreasing the reference current i_q^* is set up at the value of 0.085 A/s.

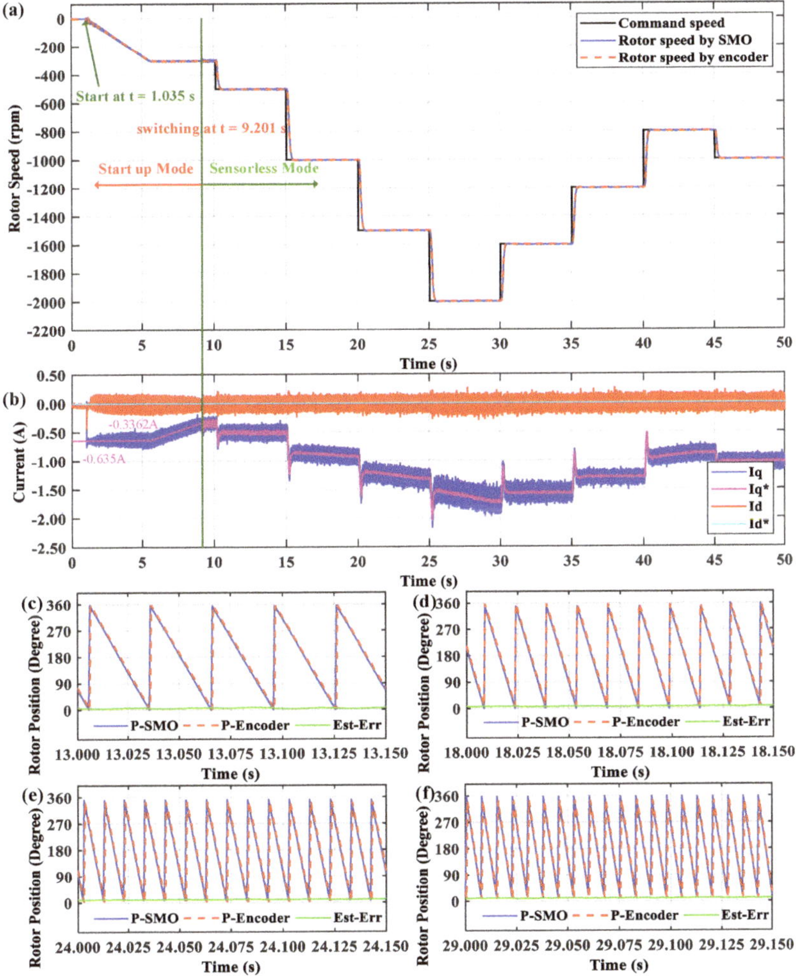

Figure 17. Experimental results in the wide speed range in the negative direction: (**a**) speed response; (**b**) current response; and rotor position response at: (**c**) 500; (**d**) 1000; (**e**) 1500; and (**f**) 2000 rpm.

The motor's speed performance in the reversal operation is presented in Figure 18. Firstly, the motor rotates in the negative direction, and the rotor speed is varied in the sequence of −300 →

−700 → −300 rpm. Secondly, the command speed is set up to reverse the rotational direction from −300 to +300 rpm at t = 11 s. The motor control algorithm is transferred to the *I-f* control mode from the sensorless control mode. The motor is decelerated to zero and accelerated to the command speed, following a ramp function with a ratio of 266.67 rpm/s. The actual rotor speed reaches 300 rpm at t = 13.28 s. It takes about 2.28 s to reverse the rotational direction from the negative direction to the positive direction. When the motor has already operated in the positive direction, the motor control algorithm is switched to the sensorless control mode again at t = 14.79 s. The *I-f* control mode is implemented in a period of 3.79 s in this direction reversion transition. Then, the command speed is varied by the sequence: 300 → 700 → 1200 → 300 rpm. Thirdly, a similar procedure is applied when the command speed is changed to reverse the rotational direction from +300 to −300 rpm at t = 36 s. The actual rotor speed reaches −300 rpm at t = 38.26 s. It takes about 2.26 s to reverse the rotational direction from the positive direction to the negative direction. The *I-f* control mode is active in a period of 4.02 s. During the operation of the motor, the estimated rotor speed almost tracks the command speed and approximates the actual rotor speed for both the sensorless control mode and the *I-f* control mode. Moreover, the current response in Figure 18b indicates that the reference current i_q^* is regulated differently for the sensorless control mode and the *I-f* control mode in the case of the rotational direction reversion. Therefore, the rising time and settling time are different in the speed transition between the rotational direction reversion and the uni-direction speed variation. It takes a larger time to reverse the rotational direction. In the first direction reversion transition, the current i_q^* is regulated from −0.3652 to 0.6226 A, and then reduced to 0.3362 A before switching to the sensorless control mode. In the second transition, those corresponded values are 0.3694, −0.6558, and −0.3237 A, respectively.

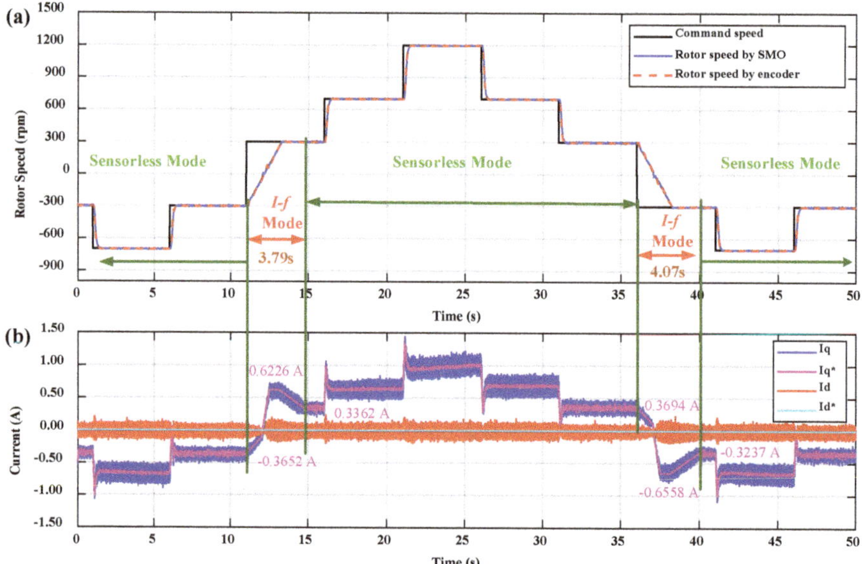

Figure 18. Experimental results in the case of the varied rotational direction for: (**a**) speed response; and (**b**) current response.

Figure 19a–c illustrates the rotor position when the motor changes the rotational direction to the positive direction. When the actual speed crosses the zero value, there is a large estimation error because the modified PLL changes the parameter θ_{offset} from 180° to 0°. However, this error is reduced significantly when the system is switched to the sensorless control mode. In Figure 19e–f, a similar result is analyzed for rotor position in the case of changing the rotor speed from the positive value to negative value. The parameter θ_{offset} is transferred from 0° to 180°. Finally, Figures 18 and 19 verify that the *I-f* control mode makes the motor change the rotational direction stably and the modified conventional PLL is worked effectively, which is not affected by noise, inaccurate back-EMF estimation or large estimation error in the low-speed range. The estimated position still approaches the actual value after the direction reversion transition has occurred.

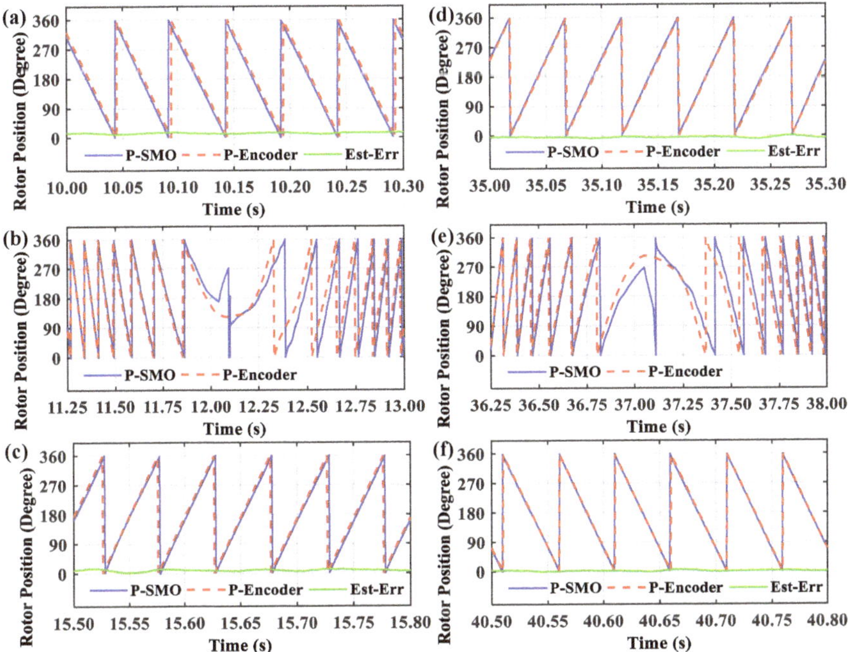

Figure 19. Experimental results of rotor position in the case of the varied rotational direction: from negative to positive (**a–c**); and from positive to negative (**d–f**).

Figure 20 presents the system performance for the PI speed controller and the NFC-based speed control when the motor startups with the initial resistance load of 100 Ω and 400 W. The motor startups at t = 2.821 s and switches to the sensorless control mode at t = 10.915 s. The rotor speed follows the sequence of 0 → 300 → 500 → 1000 rpm. Then, the command speed is regulated as a square wave with a period of 5s and the speed variation from 1000 to 1500 rpm. In the PI controller, at t = 28.94 s, more power resistors are added to increase the external load, and the total resistance load transfers to the new value of 50 Ω and 800 W. At t = 29.06 s, the rotor speed reduces to the low peak, with the estimated value (blue line) of 878 rpm and the actual value (red line) of 867 rpm. The rotor speed steadies at 1000 rpm again at t = 29.84 s. The recovery time is 0.78 s with the actual speed reduction of 133 rpm. In the NFC controller, the external load is enhanced at t = 30.00 s. The rotor speed reduces to the low peak, with the estimated value of 930 rpm and the actual value of 914 rpm at t = 30.09 s. The rotor speed steadies at 1000 rpm again at t = 30.67 s. The recovery time is 0.58 s with the actual speed reduction of 86 rpm.

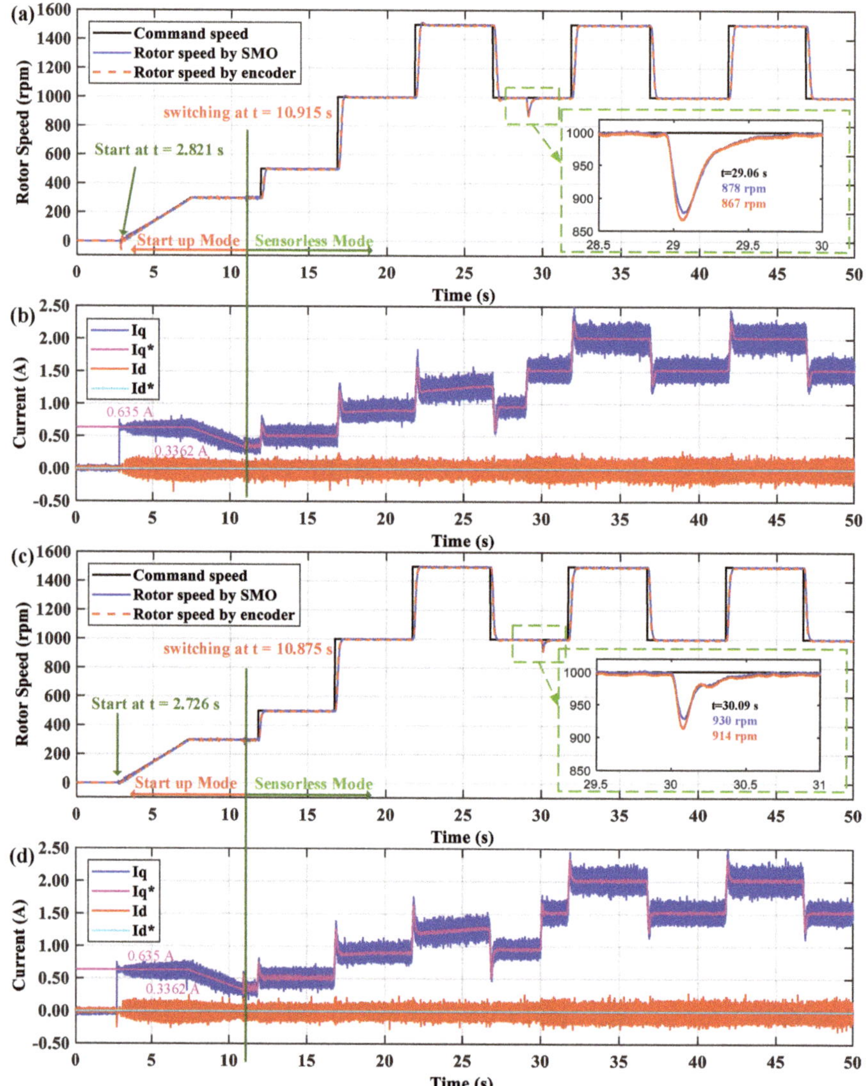

Figure 20. Experimental results for PI controller: (**a**) speed response; (**b**) current response; and for NFC controller: (**c**) speed response; and (**d**) current response in the case of $R_L = 100 \rightarrow 50\ \Omega$.

By the initial resistance load of 50 Ω and 800 W, the system performance for the PI speed controller and the NFC-based speed controller is shown in Figure 21. The speed command is the same as its waveform presented in Figure 20. Comparing to the results in Figure 20, because a larger external load is applied in the startup mode, the reference current i_q^* is ramped down to 0.5063 A, and then the system switches to the sensorless control mode in a shorter time. It takes a period of 6.087 s (for NFC) in the *I-f* control mode. The corresponding values in Figure 20 are 0.3362 A and 8.094 s, respectively. In the PI controller, at t = 28.08 s, some power resistors are removed to reduce the external load, and the total resistance load varies to the new value of 100 Ω and 400 W. At t = 28.19 s, the rotor speed increases to the up peak, with the estimated value of 1132 rpm and the actual value of 1141 rpm. The rotor speed steadies at 1000 rpm again at t = 28.89 s. The recovery time is 0.70 s with the actual speed increment of

141 rpm. In the NFC controller, the external load is reduced at t = 28.92 s. The rotor speed goes to the up peak with the estimated value of 1072 rpm and the actual value of 1083 rpm at t = 29.00 s. The rotor speed stabilizes at 1000 rpm again at t = 29.50 s. The recovery time is 0.50 s with the actual speed increment of 83 rpm. Lastly, Figures 20 and 21 indicate that the sensorless motor control algorithm still works stably and successfully, robust to the disturbance of the external load.

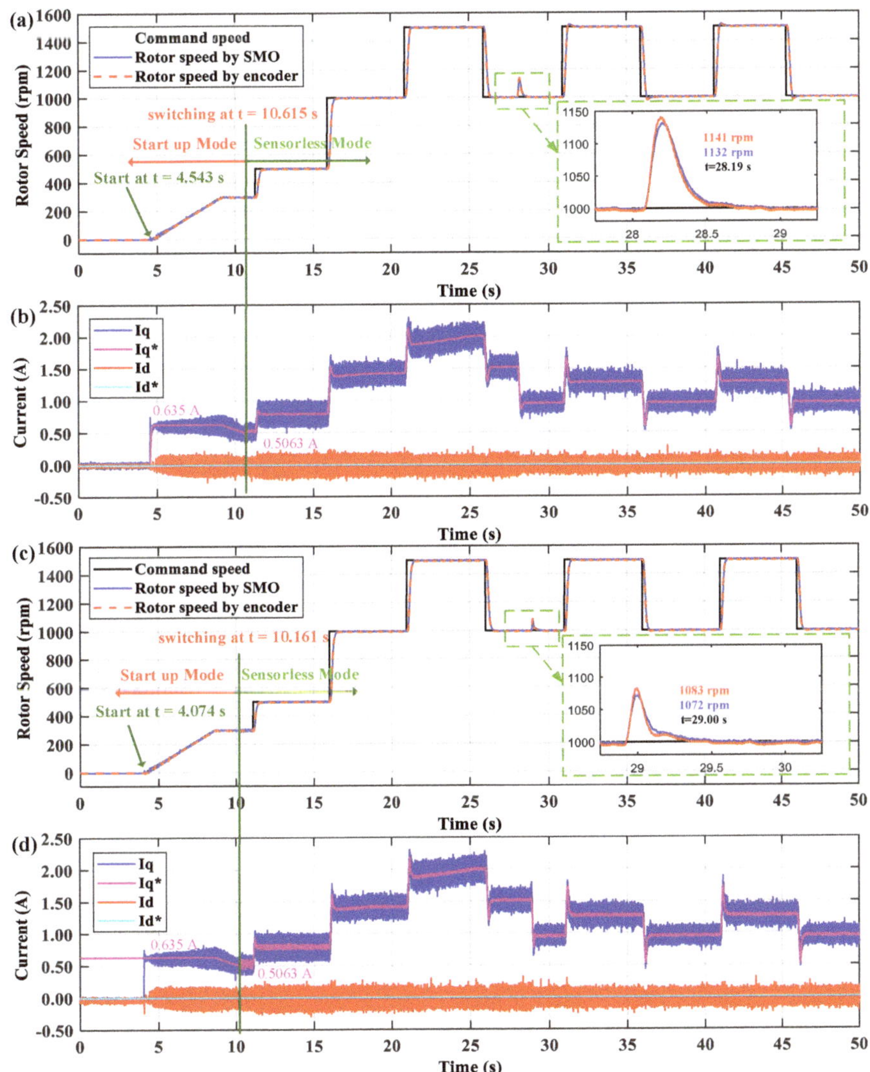

Figure 21. Experimental results for PI controller: (**a**) speed response; (**b**) current response; and for NFC controller: (**c**) speed response; and (**d**) current response in the case of the $R_L = 50 \rightarrow 100\ \Omega$.

The comparison of the detailed speed performances for the PI speed controller and the NFC-based speed controller is presented in Figures 22 and 23. This demonstrates that the NFC-based speed controller creates better performance. The motor speed tracks the command speed perfectly, without the overshoots or undershoots for two cases of external load and the settling time of 0.50 s. However, there is a little difference in the speed performance for the PI speed controller. In the case of $R_L = 100\ \Omega$ and 400 W, the system has an overshoot and an undershoot, at about +17 rpm, and the settling time of 0.78 s. While the system has a good performance, without overshoots or undershoots, and the settling time of 0.53 s in the case of $R_L = 50\ \Omega$ and 800 W. Figures 22 and 23 prove again that the PI speed controller can only work properly at the specific condition ($R_L = 50\ \Omega$ and 800 W), while the proposed NFC-based speed controller works effectively for both cases to obtain a good performance because it has a mechanism to adjust its parameters, adapting to the dynamic system characteristic.

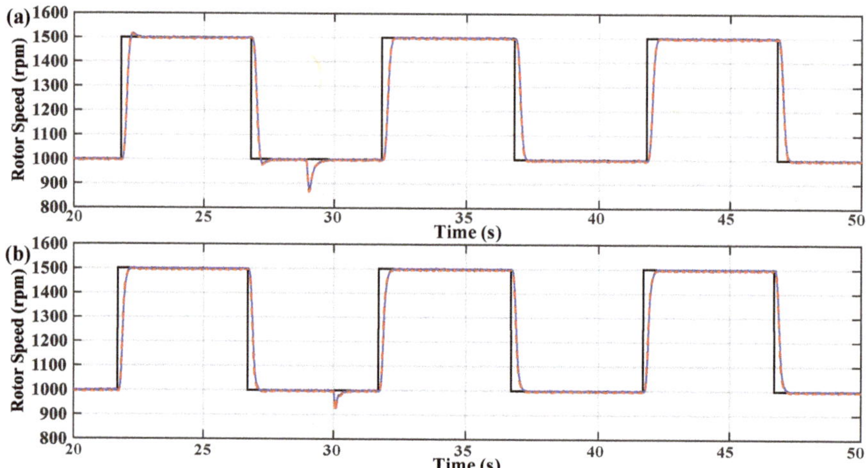

Figure 22. Comparison of speed response of experimental results in the case of the varied external load for $R_L = 100 \rightarrow 50\ \Omega$: (**a**) PI controller; and (**b**) NFC controller.

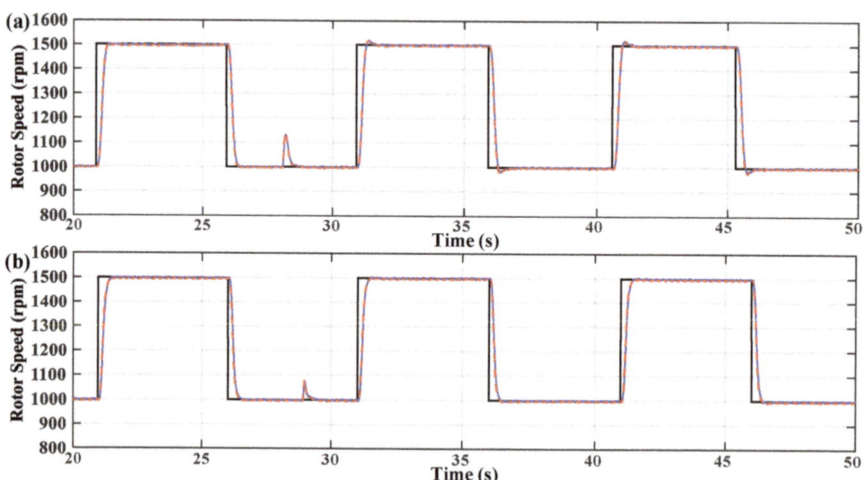

Figure 23. Comparison of speed response of experimental results in the case of the varied external load for $R_L = 50 \rightarrow 100\ \Omega$: (**a**) PI controller; and (**b**) NFC controller.

In summary, the experimental results in Figures 16–23 verify that the proposed control algorithm for the sensorless PMSM drive control system is realized effectively. The motor can startup with different initial external load and switch to the sensorless control mode smoothly. Furthermore, the motor operates in two rotational directions and the direction reversion transition is executed stably. It confirms that the estimator is implemented successfully in combination with the *I-f* control strategy. The estimated values approach the actual value for both rotor speed and rotor position. Moreover, compared to the PI speed controller, the NFC-based speed controller improves the system performance excellently and presents robustness against disturbance. The rotor speed tracks the command properly, without the overshoot or undershoot. The steady-state error almost comes close to zero (within ±5 rpm in the tolerance). Additionally, the DSP application for the PMSM drive control system is properly designed in MATLAB Simulink, deployed to Code Composer Studio software (Version 8.0.0.00016, Texas Instruments, Inc., Dallas, TX, USA) and successfully realized in the real-time system. Therefore, the development time for DSP application is substantially shortened by this deployment method.

6. Conclusions

In this work, the NFC-based speed controller and SMO-PLL estimator for the sensorless PMSM drive control system are presented and realized effectively. The proposed control algorithm was designed in MATLAB Simulink and deployed to the real-time platform, based on a DSP F28379D. Different experimental conditions were executed to evaluate system performance. The combination of the SMO-PLL estimator and the *I-f* control strategy eliminates the initial rotor position estimation and overcomes the reversal problem. The motor can startup with a diverse external load, operate in two directions in a wide speed range, and switch the rotational direction stably. The analyzed results demonstrate that the estimator works correctly. The estimated values come close to the actual value for both rotor position and speed so that the estimation error is almost insignificant. Furthermore, the NFC-based speed controller improves the system performance outstandingly and presents robustness against disturbance. The rotor speed almost tracks the command perfectly, with a zero steady-state error, without overshoot or undershoot, and having shorter settling time in comparison to the PI speed controller. Therefore, the system performance demonstrates that the proposed control algorithm for the sensorless PMSM drive control system is correct and effective.

Author Contributions: H.-K.H. wrote this article, designed the control method, implemented the hardware platform, drew the figures, and performed the simulations as well as the experiments. S.-C.C. supervised and coordinated the investigations, reviewed the manuscript, and checked its logical structure. C.-F.C. supplied the materials and reviewed the article draft. All authors have read and agreed to the published version of the manuscript.

Funding: This research was funded by Fukuta Electric & Machinery Co., Ltd, Taiwan, under contract No. 14001080308.

Conflicts of Interest: The authors declare no conflict of interest.

References

1. Hafez, B.; Abdel-Khalik, A.S.; Massoud, A.; Ahmed, S.; Lorenz, R.D. Single-Sensor-Based Three-Phase Permanent-Magnet Synchronous Motor Drive System with Luenberger Observers for Motor Line Current Reconstruction. *IEEE Trans. Ind. Appl.* **2014**, *50*, 2602–2613. [CrossRef]
2. Andersson, A.; Thiringer, T. Motion Sensorless IPMSM Control Using Linear Moving Horizon Estimation With Luenberger Observer State Feedback. *IEEE Trans. Transp. Electrif.* **2018**, *4*, 464–473. [CrossRef]
3. Risfendra; Kung, Y.-S.; Wang, M.-S.; Huang, L.-C. Realization of a sensorless speed controller for permanent magnet synchronous motor drives based on field programmable gate array technology. *Adv. Mech. Eng.* **2017**, *9*, 1–19. [CrossRef]
4. Park, J.B.; Wang, X. Sensorless Direct Torque Control of Surface-Mounted Permanent Magnet Synchronous Motors with Nonlinear Kalman Filtering. *Energies* **2018**, *11*, 969. [CrossRef]
5. Than, H.; Kung, Y.-S. FPGA-Realization of an RBF-NN tuning PI controller for sensorless PMSM drives. *Microsyst. Technol.* **2019**, 1–14. [CrossRef]
6. Khlaief, A.; Boussak, M.; Chaari, A. A MRAS-Based stator resistance and speed estimation for sensorless vector controlled IPMSM drive. *Electr. Power Syst. Res.* **2014**, *108*, 1–15. [CrossRef]

7.   Kivanc, O.C.; Ozturk, S.B. Sensorless PMSM Drive Based on Stator Feedforward Voltage Estimation Improved With MRAS Multiparameter Estimation. *IEEE/ASME Trans. Mechatron.* **2018**, *23*, 1326–1337. [CrossRef]

8.   Ma, Z.; Zhang, X. FPGA Implementation of Sensorless Sliding Mode Observer With a Novel Rotation Direction Detection for PMSM Drives. *IEEE Access* **2018**, *6*, 55528–55536. [CrossRef]

9.   Wang, Y.; Wang, X.; Xie, W.; Dou, M. Full-Speed Range Encoderless Control for Salient-Pole PMSM with a Novel Full-Order SMO. *Energies* **2018**, *11*, 2423. [CrossRef]

10.   Ye, S. A novel fuzzy flux sliding-mode observer for the sensorless speed and position tracking of PMSMs. *Optik* **2018**, *171*, 319–325. [CrossRef]

11.   Wang, Y.; Xu, Y.; Zou, J. Sliding-Mode Sensorless Control of PMSM With Inverter Nonlinearity Compensation. *IEEE Trans. Power Electron.* **2019**, *34*, 10206–10220. [CrossRef]

12.   Chen, S.; Zhang, H.; Wu, X.; Tan, G.; Chen, X. Sensorless Control for IPMSM Based on Adaptive Super-Twisting Sliding-Mode Observer and Improved Phase-Locked Loop. *Energies* **2019**, *12*, 1225. [CrossRef]

13.   Chen, Y.; Li, M.; Gao, Y.-W.; Chen, Z.-Y. A sliding mode speed and position observer for a surface-mounted PMSM. *ISA Trans.* **2019**, *87*, 17–27. [CrossRef] [PubMed]

14.   Gong, C.; Hu, Y.; Gao, J.; Wang, Y.; Yan, L. An Improved Delay-Suppressed Sliding-Mode Observer for Sensorless Vector-Controlled PMSM. *IEEE Trans. Ind. Electron.* **2020**, *67*, 5913–5923. [CrossRef]

15.   Hoai, H.-K.; Chen, S.-C.; Than, H. Realization of the Sensorless Permanent Magnet Synchronous Motor Drive Control System with an Intelligent Controller. *Electronics* **2020**, *9*, 365. [CrossRef]

16.   Lin, S.; Zhang, W. An adaptive sliding-mode observer with a tangent function-based PLL structure for position sensorless PMSM drives. *Int. J. Electr. Power Energy Syst.* **2017**, *88*, 63–74. [CrossRef]

17.   Lu, Q.; Quan, L.; Zhu, X.; Zuo, Y.; Wu, W. Improved sliding mode observer for position sensorless open-winding permanent magnet brushless motor drives. *Prog. Electromagn. Res. M* **2019**, *77*, 147–156. [CrossRef]

18.   Zhan, Y.; Guan, J.; Zhao, Y. An adaptive second-order sliding-mode observer for permanent magnet synchronous motor with an improved phase-locked loop structure considering speed reverse. *Trans. Inst. Meas. Control.* **2019**, *42*, 1008–1021. [CrossRef]

19.   Qian, J.; Ji, C.; Pan, N.; Wu, J. Improved Sliding Mode Control for Permanent Magnet Synchronous Motor Speed Regulation System. *Appl. Sci.* **2018**, *8*, 2491. [CrossRef]

20.   Rubaai, A.; Young, P. Hardware/Software Implementation of Fuzzy-Neural-Network Self-Learning Control Methods for Brushless DC Motor Drives. *IEEE Trans. Ind. Appl.* **2015**, *52*, 414–424. [CrossRef]

21.   Kung, Y.-S.; Tsai, M.-H. FPGA-Based Speed Control IC for PMSM Drive with Adaptive Fuzzy Control. *IEEE Trans. Power Electron.* **2007**, *22*, 2476–2486. [CrossRef]

22.   Chou, H.-H.; Kung, Y.-S.; Quynh, N.V.; Cheng, S.; Quynh, N.V. Optimized FPGA design, verification and implementation of a neuro-fuzzy controller for PMSM drives. *Math. Comput. Simul.* **2013**, *90*, 28–44. [CrossRef]

23.   Wang, Z.; Lu, K.; Blaabjerg, F. A Simple Startup Strategy Based on Current Regulation for Back-EMF-Based Sensorless Control of PMSM. *IEEE Trans. Power Electron.* **2012**, *27*, 3817–3825. [CrossRef]

24.   Liu, J. *Radial Basis Function (RBF) Neural Network Control for Mechanical Systems: Design, Analysis and Matlab Simulation*; Springer Science & Business Media: Berlin/Heidelberg, Germany, 2013.

Article

Sensorless Fractional Order Control of PMSM Based on Synergetic and Sliding Mode Controllers

Marcel Nicola [1,*] and Claudiu-Ionel Nicola [1,2,*]

[1] Research and Development Department, National Institute for Research, Development and Testing in Electrical Engineering—ICMET Craiova, 200746 Craiova, Romania
[2] Department of Automatic Control and Electronics, University of Craiova, 200585 Craiova, Romania
* Correspondence: marcel_nicola@yahoo.com (M.N.); claudiu@automation.ucv.ro (C.-I.N.)

Received: 30 July 2020; Accepted: 6 September 2020; Published: 11 September 2020

Abstract: The field oriented control (FOC) strategy of the permanent magnet synchronous motor (PMSM) includes all the advantages deriving from the simplicity of using PI-type controllers, but inherently the control performances are limited due to the nonlinear model of the PMSM, the need for wide-range and high-dynamics speed and load torque control, but also due to the parametric uncertainties which occur especially as a result of the variation of the combined rotor-load moment of inertia, and of the load resistance. Based on the fractional calculus for the integration and differentiation operators, this article presents a number of fractional order (FO) controllers for the PMSM rotor speed control loops, and i_d and i_q current control loops in the FOC-type control strategy. The main contribution consists of proposing a PMSM control structure, where the controller of the outer rotor speed control loop is of FO-sliding mode control (FO-SMC) type, and the controllers for the inner control loops of i_d and i_q currents are of FO-synergetic type. Superior performances are obtained by using the control system proposed, even in the case of parametric variations. The performances of the proposed control system are validated both by numerical simulations and experimentally, through the real-time implementation in embedded systems.

Keywords: permanent magnet synchronous motor; fractional order control; synergetic control; sliding mode control

---

1. Introduction

The permanent magnet synchronous motor (PMSM) is widely used in industrial applications, the aerospace industry, electric vehicles, robotics, electric drives and computer peripherals. The popularity of using the PMSM for a very wide range of applications is due to a set of advantages such as efficiency, small size, high power and high torque density. Naturally, for the control of the PMSM, a number of algorithms and control strategies have been developed, both in the range of the classic type of control, and also as through modern and unconventional approaches. The field oriented control (FOC) and direct torque control (DTC) [1–7] can be distinguished among the control strategies of the PMSM. The DTC strategy is characterized by a simpler structure in terms of controllers which are generally ON-OFF, but inherently the performance of the control system is affected by the occurrence of oscillations. The FOC strategy contains a cascade control structure, where the outer loop controls the PMSM rotor speed, and the inner control loops control currents i_d and i_q. In the classical approach, the FOC strategy controllers are PI type. This approach includes all the advantages provided by the simplicity of using such controllers, but inherently the control performances are limited due to the nonlinear model of the PMSM, the need for wide-range and high-dynamics speed and load torque control, but also due to the parametric uncertainties which occur especially as a result of the variation of the combined rotor-load moment of inertia, and of the load resistance.

Among the more complex control systems used to obtain superior performances we can mention the adaptive control [8–10], the predictive control [11–13], the robust control [14,15], the backstepping control [16], the sliding mode control (SMC) [17,18] and the synergetic control [19,20]. These types of control systems provide superior performance in terms of the response time, the overshoot and the parametric robustness, and the real-time implementation in embedded systems can be achieved with digital signal processors (DSP) with common performance, with a very good performance/price ratio.

Among these control systems, due to their robustness to parametric variations, the SMC-type control systems have a special role, as well as the development of low-order controllers, with obvious advantages in the real-time implementation in embedded systems. To counter SMC's main disadvantage due to the occurrence of the chattering phenomenon, a series of techniques have been developed, among which we mention the use of a proportional-integral type of sliding surface plus integrator or a second-order SMC. Further, we recall that the synergetic control can be considered as a generalization of the SMC-type control, retaining the decoupling and model order reduction properties of this type of control and the advantages provided by this approach for the synthesis of the controller.

We can mention the following intelligent control systems: fuzzy [21], neuro-fuzzy [22], artificial neural network (ANN) [23–25], particle swarm optimization (PSO) or genetic algorithms [26]. These types of control also provide superior performance, but in the real-time implementation in embedded systems it is necessary to use very fast DSPs, with relatively high costs, in addition to a number of specific software libraries of the control application development environments. Thus, the performance/price ratio does not recommend the widespread use of these types of controllers.

Regarding the elimination of the speed sensors, in order to increase the reliability of the system, Luenberger [27], model reference adaptive system (MRAS) [28], and sliding mode observer (SMO) speed observers [29,30] are used for the deterministically described systems, and Kalman type observers [31] are used for the stochastic description of the system. Evidently, according to the performance-cost criterion, the deterministic observers are the most commonly used. Furthermore, a range of observers have been developed for the detection of faults, one of the most useful observers being used for the detection of faults in current sensors on the supply phases of the PMSM.

One of the special applications of using the fractional calculus for integration and differentiation operators consists of obtaining the fractional order (FO) controllers [32–35], in order to obtain superior control performance. In this sense, the first approaches were obviously aimed at obtaining FO-PI-type controllers. Although the development of the FO-type controllers is very attractive due to the possibility for finer tuning of certain tuning parameters, which are traditionally integer and fixed parameters (for example the power of operator s in the structure of the PI or PID-type controller), the study of these controllers was greatly accelerated with the development of specialized toolboxes such as the fractional order modeling and control (FOMCON) integrated into the MATLAB/Simulink environment.

Among the usual applications of the PMSM control systems we mention: maintaining the speed according to a profile set by a speed reference generator, but also master/slave type multi-motor applications where the coupling is rigid or flexible and it is necessary to maintain the same speed or maintain the torque developed by each engine in the narrowest range possible [36]. Furthermore, the electric vehicles drive control applications raise the problem of multi-motor speed control [37]. Applications such as the automatic control of the hydropower dam spillway require that the error accumulated in each drive chain corresponding to each engine be less than the set value [38]. These applications are generally achieved using the controllers described above, but of the integer order type. In this article we will focus on the fractional order controllers which provide superior control performance, but also on the increased difficulties regarding the implementation in embedded systems.

This article compares the performances obtained using FO-PI, tilt integral derivatives (TID), FO-lead lag controller, and SMC speed controllers against the classic PI-type speed controller in an FOC-type control structure of the PMSM, under the conditions where the controller of the current loops is of PI type. It also presents the performances obtained by using the synergetic control for the control of currents i_d and i_q, within an FOC-type control structure of the PMSM with PI speed controller.

The main contribution consists of proposing a PMSM control structure, where the controller of the outer rotor speed control loop is of FO-SMC type, and the controllers for the inner control loops of i_d and i_q currents are of FO-synergetic type. Superior performances are obtained by using the control system proposed, even in the case of parametric variations. The FO-SMC controller outputs the current reference i_{qref}, while $i_{dref} = 0$ according to the FOC control strategy. The FO-synergetic-type controllers directly provide the control inputs u_d and u_q, and the control of the inverter is performed through the inverse Park and Clarke transformations from d-q reference frame system to abc reference frame system. The validation of the results presented is achieved by numerical simulations, but also by real-time implementation in embedded systems.

The rest of the paper is organized as follows: the basic concepts of the fractional calculus for integration and differentiation operators are presented in Section 2, the FOC-type control strategy and the transfer function of the PMSM are presented in Section 3. The fractional order speed controllers for the PMSM are presented in Section 4, Section 5 presents the fractional order synergetic current controllers for the PMSM and Section 6 presents the observers for rotor speed estimation and fault detection. Sections 7 and 8 present the numerical simulations and the experimental results, respectively. Some conclusions are presented in the last section.

2. Fractional Order Calculus

Let us note that the non-integer order operator for integration and differentiation as aD_t^α, where α represents the fractional order, a and t represent the limits of the interval at which the operator is applied [27,28].

$$aD_t^\alpha = \begin{cases} \frac{d^\alpha}{dt^\alpha} & \mathrm{Re}(\alpha) > 0 \\ 1 & \mathrm{Re}(\alpha) = 0 \\ \int_a^t (dt)^{-\alpha} & \mathrm{Re}(\alpha) < 0 \end{cases} \tag{1}$$

Although there are several ways to define this operator, not all have a generally accepted meaning. The most widely used definition is the Riemann–Liouville differintegral [32,33]:

$$aD_t^\alpha f(t) = \frac{1}{\Gamma(m-\alpha)} \left(\frac{d}{dt}\right)^m \int_\alpha^t \frac{f(\tau)}{(t-\tau)^{\alpha-m+1}} d\tau \tag{2}$$

where $m - 1 < \alpha < m$, $m \in N$, and $\Gamma(\cdot)$ represents Euler's gamma function. Another useful definition in practical applications is given by Grünwald–Letnikov [32,33]:

$$aD_t^\alpha f(t) = \lim_{h\to 0}\frac{1}{h^\alpha} \sum_{j=0}^{\left(\frac{t-a}{h}\right)} (-1)^j \binom{\alpha}{j} f(t-jh) \tag{3}$$

where $(\cdot)$ represents the integer part.

Similarly to the case of the integer order of the operator defined in (1), the Laplace transform is applied and the transfer function for signals and with fractional derivative is defined. For example, if the orders of the fractional operator s are integer multiples in relation to the commensurate order q, $(q \in R^+, 0 < q < 1, \alpha_k = kq)$, the transfer function $H(\lambda)$ can be expressed as follows:

$$H(\lambda) = \frac{\sum\limits_{k=0}^{m} b_k \lambda^k}{\sum\limits_{k=0}^{n} a_k \lambda^k} \tag{4}$$

where $\lambda = s^q$.

However, in the case of linear and time-invariant systems, in the fractional case, the state space representation becomes:

$$D^q x(t) = Ax(t) + Bu(t)$$
$$y(t) = Cx(t) + Du(t)$$

(5)

Furthermore, the stability of the system (5) can be verified by fulfilling the following relationship:

$$|arg(eig(A))| > q\frac{\pi}{2}$$

(6)

where $0 < q < 1$ represents the commensurate order, and $eig(A)$ is the eigenvalue of the associated matrix A.

To obtain a good approximation of a transfer function with fractional order in a specified frequency range (ω_b, ω_h) and of order N, Oustaloup's recursive filter for s^γ and $0 < \gamma < 1$ can be used as follows [32,33]:

$$G_f(s) = K \prod_{k=-N}^{N} \frac{s + \omega'_k}{s + \omega_k}$$

(7)

where ω'_k, ω_k and K are given by:

$$\omega'_k = \omega_b \left(\frac{\omega_h}{\omega_b}\right)^{\frac{k+N+\frac{1}{2}(1-\gamma)}{2N+1}} ; \; \omega_k = \omega_b \left(\frac{\omega_h}{\omega_b}\right)^{\frac{k+N+\frac{1}{2}(1+\gamma)}{2N+1}} ; \; k = \omega_h^\gamma$$

(8)

Furthermore, a refined Oustaloup filter is given by [32]:

$$s^\alpha \approx \left(\frac{d\omega_h}{b}\right)^\alpha \left(\frac{ds^2 + b\omega_h s}{d(1-\alpha)s^2 + b\omega_h s + d\alpha}\right) G_p$$

(9)

$$G_p = K \prod_{k=-N}^{N} \frac{s + \omega'_k}{s + \omega_k}; \; \omega_k = \left(\frac{b\omega_h}{d}\right)^{\frac{\alpha+2k}{2N+1}}; \; \omega'_k = \left(\frac{d\omega_b}{b}\right)^{\frac{\alpha-2k}{2N+1}}$$

(10)

In Equations (9) and (10), usually $b = 10$ and $d = 9$.

3. Mathematical Model of PMSM. Transfer Function Representation. FOC Strategy of PMSM

The mathematical model of the PMSM in the rotor reference frame (d-q frame) by applying the Park transform and according to [1,2] is obtained in the following form:

$$\begin{bmatrix} u_q \\ u_d \end{bmatrix} = \begin{bmatrix} R_q + \rho L_q & \omega_e L_d \\ -\omega_e L_q & R_d + \rho L_d \end{bmatrix} \begin{bmatrix} i_q \\ i_d \end{bmatrix} + \begin{bmatrix} \omega_e \lambda_0 \\ \rho \lambda_0 \end{bmatrix}$$

(11)

where u_d, u_q and i_d, i_q are the stator voltages and currents in the d-q reference frame of the PMSM, L_q, L_d and R_q, R_d are the stator inductances and resistances in the d-q reference frame, ω_e is the electrical angular velocity of the rotor, λ_0 is the flux linkage, and ρ is the differential operator.

The flux on the d-q axes is expressed as:

$$\lambda_q = L_q i_q$$
$$\lambda_d = L_d i_d + \lambda_0$$

(12)

By indicating the electromagnetic torque developed by the PMSM as T_e, the following relations on the PMSM dynamics can be expressed:

$$T_e = \frac{3}{2} n_p \left(\lambda_d i_q - \lambda_q i_d\right); \; T_e = K_t i_q$$
$$T_e = T_L + B\omega + J\frac{d\omega}{dt}$$

(13)

where $K_t = \frac{3}{2}n_p\lambda_0$ represents the torque constant, B represents the viscous friction coefficient, J represents the rotor inertia, n_p represents the number of pole pairs, and T_L represents the load torque.

By assuming the following simplifications $L_d = L_q = L$, $R_d = R_q = R_s$, and $\omega_e = n_p \cdot \omega$, where ω is the angular velocity of the rotor, the following PMSM model can be obtained:

$$
\begin{pmatrix} \dot{i_d} \\ \dot{i_q} \\ \dot{\omega} \end{pmatrix} =
\begin{pmatrix}
-\frac{R_s}{L} & n_p\omega & 0 \\
-n_p\omega & -\frac{R_s}{L} & -\frac{n_p\lambda_0}{L} \\
0 & \frac{K_t}{J} & -\frac{B}{J}
\end{pmatrix}
\begin{pmatrix} i_d \\ i_q \\ \omega \end{pmatrix} +
\begin{pmatrix} \frac{u_d}{L} \\ \frac{u_q}{L} \\ -\frac{T_L}{J} \end{pmatrix}
\tag{14}
$$

Using the Equations (11)–(14) describing the PMSM, Figure 1 shows the block diagram by reduced transfer functions of the PMSM rotor speed control system. In addition to the notations presented above, we denote the transfer functions of current and speed sensors as $H_c(s)$ and $H_\omega(s)$ in Figure 1. Usually these are 1st order transfer functions where the time constant is in the order of milliseconds, thus allowing a simplified approach to the reduction of these transfer functions to constants.

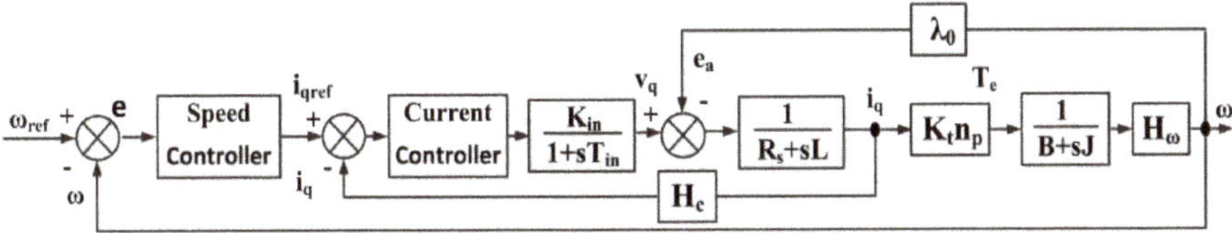

Figure 1. The block diagram of the speed loop of the permanent magnet synchronous motor (PMSM) drive.

The transfer function of the inverter is as follows:

$$
H_{in} = \frac{K_{in}}{1 + sT_{in}}
\tag{15}
$$

where: $K_{in} = 0.65 \cdot (V_{dc}/V_{cm})$ and $T_{in} = 1/(2 \cdot f_c)$, V_{dc} represents the dc link voltage (input of the inverter), V_{cm} represents the maximum control voltage, and f_c represents the switching frequency of the inverter.

In Figure 1 by shifting the point of intersection of the back-electromotive force (back EMF) loop with the speed loop to the point of intersection with the current loop, an equivalent form of defining the fixed part (the current control loop of i_q) as transfer functions is shown in Figure 2 [39].

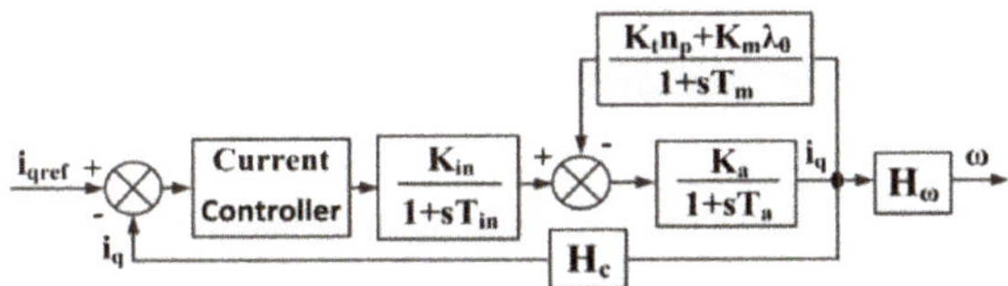

Figure 2. The block diagram of the current control loop of the PMSM drive.

In Figure 2, using the following notations:

$$
\begin{cases}
K_a = \frac{1}{R}; \ T_a = \frac{L}{R}; \ K_m = \frac{1}{B}; \ T_m = \frac{J}{B}; \ K_b = K_t K_m \lambda_0; \\
T_i = 0.1T_m; \ K_{in} = 20; \ K_i = (T_m K_{in})/(T_2 K_b)
\end{cases}
\tag{16}
$$

is the transfer function of the inner current loop which represents the fixed part of the speed control system (outer control loop) which becomes as follows:

$$H_f = \frac{K_t K_i}{(as^3 + bs^2 + cs + d)} \tag{17}$$

where

$$\begin{cases} a = T_i L J; \; b = L J + T_i (L B + R_s J) \\ c = L B + R_s J + T_i (R_s B + \lambda_0 K_t); \; d = R_s B + \lambda_0 K_t \end{cases} \tag{18}$$

Figure 3 shows the proposed general block diagram of the enhanced FOC control type strategy for PMSM. In this paper, the speed controller of the outer control loop is a classic PI-type controller, but, as shown in Section 4, it can be replaced with FO-PI, TID, FO-lead-lag, and FO-SMC controllers. Furthermore, compared to the classic approach, the PI-type current controllers in the inner control loop can be replaced with synergetic and FO-synergetic controllers, as presented in Section 5. The output of these controllers provides the reference i_{qref} for the inner current control loop, where $i_{dref} = 0$. Section 6 presents improvements that can be made to the classic FOC-type scheme by using an SMO-type observer to estimate the rotor position and speed, but also an FDO observer to detect the faults on the PMSM supply phases.

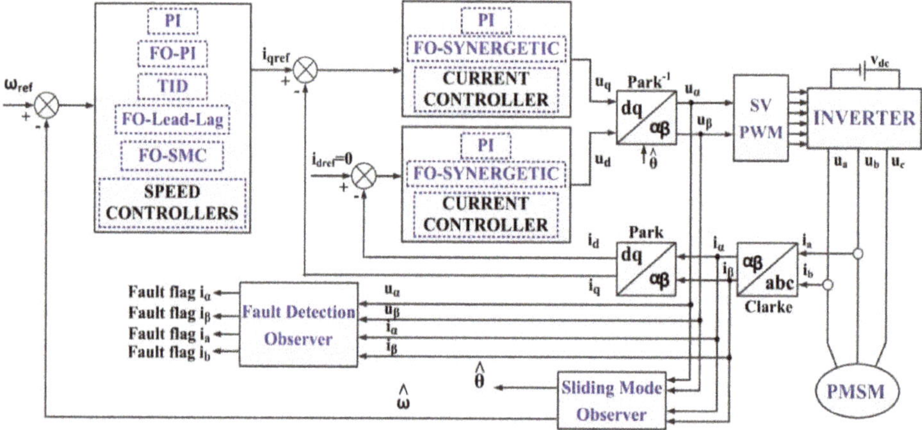

Figure 3. Enhanced field oriented control (FOC) control strategy for the PMSM—general block diagram.

4. Fractional Order Speed Controllers for PMSM

By using the fractional-type control, whose basic elements were described in Section 2, this section presents the equations of certain fractional-type controllers that will replace the PI-type speed controller in the FOC-type control strategy of PMSM.

4.1. FO-PI Speed Controller

Furthermore, in terms of the fractional controllers, the most commonly used are the $PI^\lambda D^\mu$ controllers, which can be expressed as follows [33]:

$$u(t) = K_p e(t) + K_i D^{-\lambda} e(t) + K_d D^\mu e(t) \tag{19}$$

where $e(t)$ represents the error signal.

After applying the Laplace transform to Equation (19), by assuming zero initial conditions, the following equation is obtained:

$$G_c(s) = K_p + \frac{K_i}{s^\lambda} + K_d s^\mu \tag{20}$$

where K_p represents the proportional gain, K_i represents the integrator gain, λ represents the integrator order (positive), K_d represents the differentiator gain, and μ represents the differentiator order. For $\lambda = \mu = 1$, the result is the usual integer-order PID controller.

4.2. TID Speed Controller

Another fractional controller used in applications is the TID controller, which can be described by the next transfer function [33]:

$$G_c(s) = \frac{K_t}{s^{1/n}} + \frac{K_i}{s} + K_d s \tag{21}$$

where K_t represents the tilt gain, n represents the tilt integrator order, K_i represents the integrator gain, and K_d represents the differentiator gain.

4.3. Lead-Lag Speed Controller

The general form of the transfer function of an FO-lead-lag controller is given by [33]:

$$G_c(s) = K_c\left(\frac{s + \frac{1}{\lambda}}{s + \frac{1}{x\lambda}}\right)^{\alpha} = K_c x^{\alpha}\left(\frac{\lambda s + 1}{x\lambda s + 1}\right)^{\alpha}, \ 0 < x < 1 \tag{22}$$

where λ represents the fractional order of the FO-lead-lag controller.

It is noted that, for $\alpha > 0$, a lead effect of the FO-lead-lag controller is obtained, while for $\alpha < 0$, a lag effect of the FO-lead-lag controller is obtained.

For $k' = K_c x^{\alpha}$, the common form of the FO-lead-lag controller is obtained:

$$G_c(s) = k'\left(\frac{\lambda s + 1}{x\lambda s + 1}\right)^{\alpha} \tag{23}$$

For $k' = \alpha = 1$, $\lambda = \frac{K_p}{K_i}$, and x has a very high value (for example $x > 10{,}000$), the transfer function of the FO-lead-lag controller becomes the transfer function of the FO-PI controller. It can therefore be concluded that there is an increased flexibility in the use of the FO-lead-lag controller in a control loop.

4.4. FO-SMC Speed Controller

To achieve an SMC-type controller for the control of a PMSM motor described by Equation (14), the state variables x_1 and x_2 described below are selected [1]:

$$x_1 = \omega_{ref} - \omega \tag{24}$$

where x_1 represents the tracking error of the speed.

$$x_2 = \dot{x}_1 = \frac{\omega_{ref} - \omega}{dt} = -\dot{\omega} \tag{25}$$

Equation (26) defines the sliding surface S of the zero-error manifold. Through differentiation, the following equation is obtained (27):

$$S = cx_1 + x_2 \tag{26}$$

$$\dot{S} = cx_2 + \dot{x}_2 = cx_2 - D\dot{i}_q \tag{27}$$

where c represents the positive adjustable parameter and $D = \frac{3n_p\lambda_0}{2J}$ is obtained from Equation (13).

In order to control the response time of the PMSM control system, the condition of time evolution of the surface S is imposed like in Equation (28):

$$\dot{S} = -\varepsilon \mathrm{sgn}(S) - qS, \quad \varepsilon, q > 0 \tag{28}$$

where ε and q represent the positive adjustable parameters; $sgn()$ represents the *signum* function.

To reduce the chattering effect (characteristic of the SMC design), the *sgn* function is replaced with the sigmoid function defined as follows:

$$H(x) = \frac{2}{1 + e^{-a(x-c)}} - 1 \tag{29}$$

For $a = 4$ and $c = 0$, $H \in [-1\ 1]$, the transition of the function H from -1 to 1 is smoothed and ensures the reduction of the chattering effect. Based on these, after some calculations, the SMC-type controller output value is obtained in the following form:

$$i_{qref}(t) = \frac{1}{D} \int_0^t [cx_2 + \varepsilon H(S) + qS]dt \tag{30}$$

It can be specified that i_{qref} represents the current reference for the control loop on q axis, while i_{dref} is set to zero according to the FOC-type control strategy [23]. To demonstrate the stability of the PMSM control system under the action of the control law given by Equation (30), the Lyapunov function candidate is selected in the following form $V = \frac{1}{2}S^2$ [1].

$$\dot{V} = S\dot{S} = S[-\varepsilon H(S) - qS] = -\varepsilon H(S) - qS^2 \tag{31}$$

After some calculations, $\dot{V} \leq 0$ is obtained, where $\dot{V}$ is given by the relation (31). To achieve the FO-SMC controllers, the sliding surface S is selected as follows:

$$S = k_p x_1 + k_d D^\mu x_1 = k_p x_1 + k_d D^{\mu-1} x_2 \tag{32}$$

After differentiation, the following relation is obtained:

$$\dot{S} = k_p \dot{x}_1 + k_d D^{\mu+1} x_1 = k_p x_2 + k_d D^{\mu-1} \dot{x}_2 \tag{33}$$

Based on the mathematical model of the PMSM described in Section 3, the following relation is obtained:

$$\dot{x}_2 = \ddot{\omega}_{ref} - \frac{3n_p \lambda_0}{2J} \dot{i}_q + \frac{1}{J}\dot{T}_L + \frac{B}{J}\dot{\omega} \tag{34}$$

By inserting Equation (34) into Equation (33), the following relation is obtained:

$$\dot{S} - k_p x_2 + k_d D^{\mu-1}\left(\ddot{\omega}_{ref} - \frac{3n_p \lambda_0}{2J} \dot{i}_q + \frac{1}{J}\dot{T}_L + \frac{B}{J}\dot{\omega}\right) \tag{35}$$

For $\dot{S} = 0$, the following relation is obtained:

$$-\varepsilon H(S) - qS - k_p x_2 = k_d D^{\mu-1}\left(\ddot{\omega}_{ref} - \frac{3n_p \lambda_0}{2J} \dot{i}_q + \frac{1}{J}\dot{T}_L + \frac{B}{J}\dot{\omega}\right) \tag{36}$$

By applying operator $D^{1-\mu}$ (described in Section 2) to both members of Equation (36), the following relation is obtained:

$$D^{1-\mu}\left(-\varepsilon H(S) - qS - k_p x_2\right) = k_d\left(\ddot{\omega}_{ref} - \frac{3n_p \lambda_0}{2J} \dot{i}_q + \frac{1}{J}\dot{T}_L + \frac{B}{J}\dot{\omega}\right) \tag{37}$$

This results in the following relation:

$$\frac{1}{k_d}D^{1-\mu}\left(-\varepsilon H(S) - qS - k_p x_2\right) = \ddot{\omega}_{ref} - \frac{3n_p \lambda_0}{2J} \dot{i}_q + \frac{1}{J}\dot{T}_L + \frac{B}{J}\dot{\omega} \tag{38}$$

Equation (38) can be rewritten as follows:

$$\frac{1}{J}\frac{3}{2}n_p\lambda_0 i_q = \ddot{\omega}_{ref} + \frac{1}{J}\dot{T}_L + \frac{B}{J}\dot{\omega} - \frac{1}{k_d}D^{1-\mu}\left(-\varepsilon H(S) - qS - k_p x_2\right) \tag{39}$$

The current reference i_{qref} is obtained from Equation (39) as follows:

$$i_{qref}(t) = \frac{1}{\frac{1}{J}\frac{3}{2}n_p\lambda_0}\int_0^t\left[\ddot{\omega}_{ref} + \frac{1}{J}\dot{T}_L + \frac{B}{J}\dot{\omega} - \frac{1}{k_d}D^{1-\mu}\left(-\varepsilon H(S) - qS - k_p x_2\right)\right]dt \tag{40}$$

5. Fractional Order Synergetic Current Controllers for PMSM

The synergetic control can be considered as a generalization of the sliding mode control (SMC), retaining the decoupling design procedure and model order reduction properties of this type of control and the advantages provided by this approach for the synthesis of the control. Thus, in this section, the PI-type current controllers will be replaced for the inner current control loops in the FOC-type control strategy, with synergetic and FO-synergetic type controllers to obtain superior performances.

For the synergetic control, a macro-variable is defined as a function of the states of the system, as follows [19]:

$$\Psi = \Psi(x, t) \tag{41}$$

The synthesized control inputs will force the system to operate on the manifold $\Psi = 0$, in a similar manner to the SMC. A number of macro-variables equal to the number of control inputs are defined. The dynamic evolution of each macro-variable is defined according to the following equation:

$$T\dot{\Psi} + \Psi = 0,\ T > 0 \tag{42}$$

where T is selected so as to achieve the rate of convergence of the system evolution towards the desired manifold.

By differentiating the macro-variable Ψ:

$$\dot{\Psi} = \frac{d\Psi}{dx}\dot{x}, \tag{43}$$

and by inserting Equation (43) into (42) using the explicit description of the states $\dot{x}$ from the mathematical model, in the case of the PMSM expressed by Equation (14), the control law is obtained as follows:

$$u = u(x, \Psi(x, t), T, t) \tag{44}$$

In case of applying the FOC type strategy for the PMSM control (see Figure 4), the outer speed control loop supplies the reference i_{qref} at the PI-type speed controller output for the inner control loop whose controller proposed in this paper is synergetic. According to Equations (42)–(44), the synergetic controller provides the controls u_d and u_q. Furthermore, according to the FOC control strategy of the PMSM $i_{dref} = 0$.

According to [20], in order to achieve superior control performance under static and dynamic regime, ω_{acc} and ω_{dec} are defined as the angular velocity of the rotor by selecting the current limit mode of operation for accelerating and decelerating transients, respectively, of the following form:

$$\begin{aligned}
\omega_{acc} &= \omega_{ref} - k_q\left(\left|i_{qmax}\right| - i_{qref}\right)\\
\omega_{dec} &= \omega_{ref} - k_q\left(-\left|i_{qmax}\right| - i_{qref}\right)
\end{aligned} \tag{45}$$

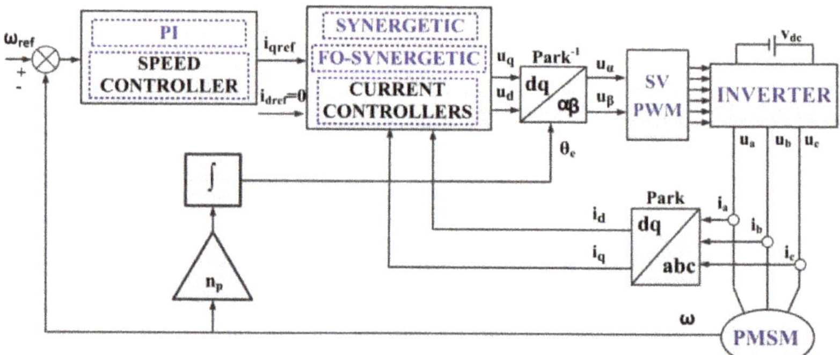

Figure 4. FO-synergetic current control for the PMSM—general block diagram.

For $\omega > \omega_{acc}$ and $\omega < \omega_{dec}$ the macro-variable on q axis Ψ_q can be represented as in the following equation:

$$\Psi_q = \left(\omega(t) - \omega_{ref}\right) + k_q\left(i_q(t) - i_{qref}\right) \tag{46}$$

and for $\omega \leq \omega_{acc}$ the macro-variable on q axis Ψ_q is represented in Equation (47), and for $\omega \geq \omega_{dec}$ the macro-variable on q axis Ψ_q is represented in Equation (48).

$$\Psi_q = i_q(t) - |i_{qmax}| + k_{iq}\int_0^t \left(i_q(t) - |i_{qmax}|\right)dt \tag{47}$$

$$\Psi_q = i_q(t) + |i_{qmax}| + k_{iq}\int_0^t \left(i_q(t) + |i_{qmax}|\right)dt \tag{48}$$

where i_{qmax} is the maximum admissible current on q-axis, and k_q is a value which is dynamically adjusted as a function of the angular velocity of the rotor error. The control law for q axis can be expressed as follows: for $\omega_{acc} < \omega < \omega_{dec}$, u_q is given by Equation (49), for $\omega \leq \omega_{acc}$, u_q is given by Equation (50), and for $\omega \geq \omega_{dec}$, u_q is given by Equation (51):

$$u_q(t) = R_s i_q + n_p\omega(L i_d + \lambda_0) + \frac{L}{T_q}\left(i_{qref} - i_q\right) + \frac{L}{T_q k_q}\left(\omega_{ref} - \omega\right) + \frac{L}{J k_q}\left(-K_t i_q + B\omega + T_L\right) \tag{49}$$

$$u_q(t) = R_s i_q + n_p\omega(L i_d + \lambda_0) + \frac{L}{T_q}\left(|i_{qmax}| - i_q\right) + k_{iq}L\left(|i_{qmax}| - i_q\right) - \frac{k_{iq}L}{T_q}\int_0^t \left(i_q - |i_{qmax}|\right)dt \tag{50}$$

$$u_q(t) = R_s i_q + n_p\omega(L i_d + \lambda_0) + \frac{L}{T_q}\left(-|i_{qmax}| - i_q\right) + k_{iq}L\left(-|i_{qmax}| - i_q\right) - \frac{k_{iq}L}{T_q}\int_0^t \left(i_q + |i_{qmax}|\right)dt \tag{51}$$

The macro-variable on d axis Ψ_d can be represented as in the following equation:

$$\Psi_d = \left(i_d(t) - i_{dref}\right) + k_{id}\int_0^t \left(i_d(t) - i_{dref}\right)dt \tag{52}$$

After some calculus, the control law for d axis can be expressed as follows [20]:

$$u_d(t) = R_s i_d - n_p \omega L i_q - \frac{L}{T_d}\left(i_d - i_{dref}\right) - k_{id}L\left(i_d - i_{dref}\right) - \frac{k_{id}L}{T_d}\int_0^t \left(i_d - i_{dref}\right)dt \tag{53}$$

Following the calculations made for the synergetic control of the PMSM, using definitions (1) and (2) of the fractional calculus for integration and differentiation operators, in this section we will obtain the values of the controls $u_d(t)$ and $u_q(t)$.

We define the macro-variable on q axis Ψ_q similarly to (46) for $\omega > \omega_{acc}$ and $\omega < \omega_{dec}$, as follows:

$$\Psi_q(x) = D^\mu x_3 + k_q x_1 \tag{54}$$

where: $x_1 = i_q - i_{qref}$, $x_3 = \omega - \omega_{ref}$, and $\mu > 0$. Next, $\dot{\Psi}_q$ is calculated, and the following relation is obtained:

$$\dot{\Psi}_q(x) = D^\mu \dot{x}_3 + k_q \dot{x}_1 = D^\mu \dot{\omega} + k_q \dot{i}_q \tag{55}$$

By inserting (55) into the Equation (42) of dynamic evolution of macro-variable Ψ_q, for $T = T_q$, the following relation is obtained:

$$T_q\left[D^\mu\left(\frac{K_t i_q}{J} - \frac{B\omega}{J} - \frac{T_L}{J}\right) + k_q\left(-\frac{R_s i_q}{L} - n_p \omega i_d - \frac{n_p \lambda_0 \omega}{L} + \frac{u_q}{L}\right)\right] +$$
$$+ D^\mu\left(\omega - \omega_{ref}\right) + k_q\left(i_q - i_{qref}\right) = 0 \tag{56}$$

By rearranging the terms in Equation (56) the following relation is obtained:

$$T_q D^\mu\left(\frac{K_t i_q}{J} - \frac{B\omega}{J} - \frac{T_L}{J}\right) + T_q k_q\left(-\frac{R_s i_q}{L} - n_p \omega i_d - \frac{n_p \lambda_0 \omega}{L}\right) +$$
$$+ \frac{T_q k_q u_q}{L} + D^\mu\left(\omega - \omega_{ref}\right) + k_q\left(i_q - i_{qref}\right) = 0 \tag{57}$$

After some calculations, Equation (56) becomes:

$$\frac{T_q k_q u_q}{L} - T_q D^\mu\left(-\frac{K_t i_q}{J} + \frac{B\omega}{J} + \frac{T_L}{J}\right) + \frac{T_q k_q R_s i_q}{L} +$$
$$+ n_p \omega(L i_d + \lambda_0)\frac{T_q k_q}{L} + D^\mu\left(\omega - \omega_{ref}\right) + k_q\left(i_q - i_{qref}\right) = 0 \tag{58}$$

Based on this, the control u_q of the PMSM is obtained:

$$u_q(t) = \frac{L}{J k_q}D^\mu\left(-K_t i_q + B\omega + T_L\right) + R_s i_q + n_p \omega(L i_d + \lambda_0)$$
$$+ \frac{L}{T_q k_q}D^\mu\left(\omega - \omega_{ref}\right) + \frac{L}{T_q}\left(i_q - i_{qref}\right) \tag{59}$$

Similarly to relations (47) and (48), control u_q is obtained for $\omega \leq \omega_{acc}$ and $\omega \geq \omega_{dec}$, respectively. Next, the macro-variable on d axis Ψ_d is defined:

$$\Psi_d(x,t) = D^\mu x_2 + k_d \int_0^t x_2(t)dt \tag{60}$$

where $x_2 = i_d - i_{dref}$ and $\mu > 0$.

Next, $\dot{\Psi}_d$ is calculated, and the following relation is obtained:

$$\dot{\Psi}_d(x) = D^\mu \dot{x}_2 + k_d x_2 = D^\mu \dot{i}_d + k_d\left(i_d - i_{dref}\right) \tag{61}$$

By inserting (61) into the Equation (42) of dynamic evolution of macro-variable Ψ_d, for $T = T_d$, the following relation is obtained:

$$T_d\left\{\left[D^\mu\left(n_p\omega i_q - \frac{R_s i_d}{L} + \frac{u_d}{L}\right)\right] + k_d\left(i_d - i_{dref}\right)\right\} + D^\mu\left(i_d - i_{dref}\right) + k_d\int_0^t \left(i_d - i_{dref}\right)dt = 0 \tag{62}$$

By applying the operator defined in (1), $D^{-\mu}$ (which becomes I_μ) to both members of Equation (62) the following relation is obtained:

$$T_d\left(n_p\omega i_q - \frac{R_s i_d}{L} + \frac{u_d}{L}\right) + T_d k_d I_\mu\left(i_d - i_{dref}\right) + i_d - i_{dref} + k_d I_{\mu+1}\left(i_d - i_{dref}\right) = 0 \tag{63}$$

After some calculations, Equation (63) becomes:

$$\frac{T_d u_d}{L} = \frac{R_s i_d T_d}{L} - n_p\omega i_q T_d - T_d k_d I_\mu\left(i_d - i_{dref}\right) - \left(i_d - i_{dref}\right) - k_d I_{\mu+1}\left(i_d - i_{dref}\right) = 0 \tag{64}$$

Based on this, the control u_d of the PMSM is obtained:

$$u_d(t) = R_s i_d - n_p\omega i_q L - L k_d I_\mu\left(i_d - i_{dref}\right) - \frac{L}{T_d}\left(i_d - i_{dref}\right) - \frac{L k_d}{T_d}I_{\mu+1}\left(i_d - i_{dref}\right) \tag{65}$$

6. Rotor Speed Estimation and Fault Detection

This section presents two observers, which complete the classic FOC-type control structure. Thus, the sensorless characteristic of the control is ensured by an SMO-type observer which estimates the PMSM speed. Furthermore, the use of an FDO-type observer enables the fault detection of the current sensors on the PMSM supply lines.

6.1. Rotor Speed and Position Estimations Based on SMO-Type Observer

By using the PMSM operating equations given by the relations (11)–(14), and by applying the inverse Park transform, the equations of the currents i_α and i_β and the back-EMF e_α and e_β are obtained in α-β frame [2]:

$$\begin{aligned} i_\alpha &= i_d \cos(\theta_e) - i_q \sin(\theta_e) \\ i_\beta &= i_d \sin(\theta_e) + i_q \cos(\theta_e) \end{aligned} \tag{66}$$

$$\begin{aligned} e_\alpha &= \frac{d\lambda_\alpha}{dt} = -\lambda_0 \omega_e \sin(\theta_e) \\ e_\beta &= \frac{d\lambda_\beta}{dt} = -\lambda_0 \omega_e \cos(\theta_e) \end{aligned} \tag{67}$$

Based on these, the PMSM operating equations can be rewritten as:

$$\begin{aligned} \frac{di_\alpha}{dt} &= -\frac{R_s}{L}i_\alpha - \frac{1}{L}e_\alpha + \frac{1}{L}u_\alpha \\ \frac{di_\beta}{dt} &= -\frac{R_s}{L}i_\beta - \frac{1}{L}e_\beta + \frac{1}{L}u_\beta \end{aligned} \tag{68}$$

The equations of the SMO-type observer according to which the rotor speed and position can be estimated are given by the equations [23,25]:

$$\begin{aligned} \frac{d\hat{i}_\alpha}{dt} &= -\frac{R_s}{L}\hat{i}_\alpha + \frac{1}{L}u_\alpha - \frac{1}{L}kH(\hat{i}_\alpha - i_\alpha) \\ \frac{d\hat{i}_\beta}{dt} &= -\frac{R_s}{L}\hat{i}_\beta + \frac{1}{L}u_\beta - \frac{1}{L}kH(\hat{i}_\beta - i_\beta) \end{aligned} \tag{69}$$

where: k represents the observer gain, and H is a sigmoid type function described in Equation (29).

The sliding vector is selected as follows:

$$S_n = [S_\alpha\ S_\beta]^T = [\hat{i}_\alpha - i_\alpha\ \hat{i}_\beta - i_\beta]^T = [\bar{i}_\alpha\ \bar{i}_\beta]^T \tag{70}$$

To demonstrate the stability of the proposed observer, a Lyapunov function is selected of the form [25]:

$$V = \frac{1}{2}S_n^T S_n = \frac{1}{2}\left(S_\alpha^2 + S_\beta^2\right) \tag{71}$$

The current error system is defined in the form:

$$\begin{aligned}
\dot{\bar{i}}_\alpha &= \dot{\hat{i}}_\alpha - \dot{i}_\alpha = -\frac{R_s}{L}\bar{i}_\alpha + \frac{1}{L}e_\alpha - \frac{1}{L}kH(\bar{i}_\alpha) \\
\dot{\bar{i}}_\beta &= \dot{\hat{i}}_\beta - \dot{i}_\beta = -\frac{R_s}{L}\bar{i}_\beta + \frac{1}{L}e_\beta - \frac{1}{L}kH(\bar{i}_\beta)
\end{aligned} \tag{72}$$

According to these, $\dot{V}$ is calculated, and the following relation is obtained:

$$\dot{V} = -\frac{R_s}{L}\left(\bar{i}_\alpha^2 + \bar{i}_\beta^2\right) + \frac{1}{L}\left[\left(e_\alpha - k\right)\bar{i}_\alpha H(\bar{i}_\alpha) + \left(e_\beta - k\right)\bar{i}_\beta H(\bar{i}_\beta)\right] < 0 \tag{73}$$

By selecting the observer gain as $k \geq \max\left(|e_\alpha|, |e_\beta|\right)$, the stability condition of the observer is obtained: $\dot{V} < 0$.

Based on this, on the sliding surface, the following relation is obtained:

$$\left[\dot{S}_\alpha \ \dot{S}_\beta\right]^T = \left[S_\alpha \ S_\beta\right]^T \approx \left[0\ 0\right] \tag{74}$$

Based on relations (73) and (74), are obtained the estimations for e_α and e_β:

$$\begin{aligned}
\hat{e}_\alpha &= kH(\bar{i}_\alpha) = -\lambda_0 \hat{\omega}_e \sin\theta_e \\
\hat{e}_\beta &= kH(\bar{i}_\beta) = \lambda_0 \hat{\omega}_e \cos\theta_e
\end{aligned} \tag{75}$$

Finally, the estimates for the speed and position of the PMSM rotor can be obtained as below:

$$\begin{cases}
\hat{\omega}_e = \dfrac{\sqrt{\hat{e}_\alpha^2 + \hat{e}_\beta^2}}{\lambda_0} \\
\hat{\theta}_e(t) = \int\limits_{t_0}^{t} \hat{\omega}_e(t)dt + \theta_0
\end{cases} \tag{76}$$

where: θ_0 represents the initial electrical position of the rotor.

Figure 5 shows the implementation in the MATLAB/Simulink environment in order to perform the numerical simulations of the SMO-type observer for the estimation of the position and rotor speed of the PMSM.

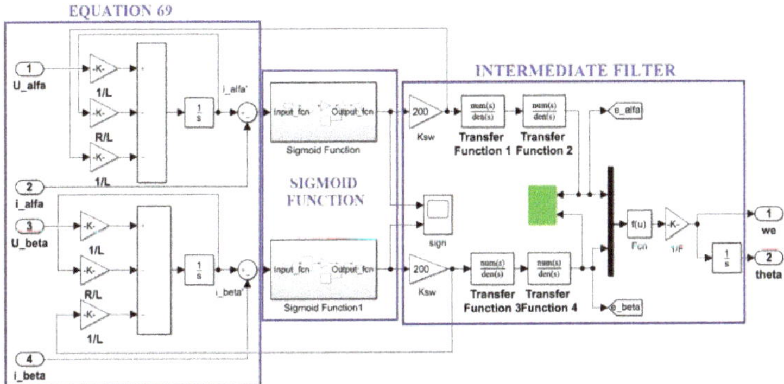

Figure 5. Sliding mode observer (SMO)-type observer—MATLAB/Simulink implementation block diagram.

6.2. Fault Detection Based on FDO-Type Observer

To detect the faults of the current sensors on the PMSM supply phases, an FDO-type observer will be used. In the α-β frame, the equations analogous to those described in Equation (11) for a PMSM are as follows:

$$\begin{bmatrix} u_\alpha \\ u_\beta \end{bmatrix} = R_s \begin{bmatrix} i_\alpha \\ i_\beta \end{bmatrix} + \rho L \begin{bmatrix} i_\alpha \\ i_\beta \end{bmatrix} + \omega_e \begin{bmatrix} \lambda_0 & 0 \\ 0 & \lambda_0 \end{bmatrix} \begin{bmatrix} -\sin\theta_e \\ \cos\theta_e \end{bmatrix} \tag{77}$$

According to Equation (77), the following relation can be written:

$$\rho \begin{bmatrix} i_\alpha \\ i_\beta \end{bmatrix} = \frac{R_s}{L} \begin{bmatrix} i_\alpha \\ i_\beta \end{bmatrix} + \frac{1}{L} \begin{bmatrix} u_\alpha \\ u_\beta \end{bmatrix} + \frac{\omega_e}{L} \begin{bmatrix} \lambda_0 & 0 \\ 0 & \lambda_0 \end{bmatrix} \begin{bmatrix} -\sin\theta_e \\ \cos\theta_e \end{bmatrix} \tag{78}$$

Considering the flux linkage $\lambda_{ext} = \lambda_0$ in α-β frame is obtained the following relation:

$$\lambda_{ext,\alpha\beta} = \begin{bmatrix} \lambda_{ext,\alpha} \\ \lambda_{ext,\beta} \end{bmatrix} = \begin{bmatrix} \lambda_0 & 0 \\ 0 & \lambda_0 \end{bmatrix} \begin{bmatrix} \cos\theta \\ \sin\theta \end{bmatrix} \tag{79}$$

Let us note: $x = \begin{bmatrix} i_\alpha & i_\beta \end{bmatrix}^T$, $u = \begin{bmatrix} u_\alpha & u_\beta \end{bmatrix}^T$, $A = \begin{bmatrix} -\frac{R_s}{L} & 0 \\ 0 & -\frac{R_s}{L} \end{bmatrix}$, $B = \begin{bmatrix} \frac{1}{L} & 0 \\ 0 & \frac{1}{L} \end{bmatrix}$, $C = \begin{bmatrix} 1 & 0 \\ 0 & 1 \end{bmatrix}$,

$F = \begin{bmatrix} 0 & \frac{\omega_e}{L} \\ \frac{\omega_e}{L} & 0 \end{bmatrix} = -\frac{\omega_e}{L} J$, $J = \begin{bmatrix} 0 & -1 \\ 1 & 0 \end{bmatrix}$.

According to Equation (78), the equation of the PMSM which has an embedded FDO-type observer can be written as [40]:

$$\begin{cases} \dot{x}(t) = Ax(t) + Bu(t) + F\lambda_{ext,\alpha\beta} + Ed \\ y(t) = Cx(t) + Gf_s \end{cases} \tag{80}$$

where $f_s = \begin{bmatrix} f_{s\alpha} & f_{s\beta} \end{bmatrix}^T$ is the stator current sensor fault vector, $d = \begin{bmatrix} d_1 & d_2 \end{bmatrix}^T$ is the unknown but bounded disturbance vector, x is the state vector, u and y represents the input and output vector, respectively, $G = \begin{bmatrix} 1 & 0 \\ 0 & 1 \end{bmatrix}$, and $E = \begin{bmatrix} 1 & 0 \\ 0 & 1 \end{bmatrix}$.

Consider a new state variable z, which is just variable y, but filtered, with a, b constants:

$$\dot{z} = -az + by \tag{81}$$

By selecting $a = 0$, $b = 1$, Equation (81) becomes:

$$\dot{z} = y = Cx(t) + Gf_s \tag{82}$$

Based on relations (80)–(82), the system of equations of the PMSM with embedded FDO observer becomes:

$$\begin{cases} \dot{x}(t) = Ax(t) + Bu(t) + F\lambda_{ext,\alpha\beta} + Ed \\ \dot{z}(t) = Cx(t) + Gf_s \\ w = z \end{cases} \tag{83}$$

Consider f_s the actuator fault of the system described by (83). The FDO-type observer has the following form [40]:

$$\begin{cases} \dot{\hat{x}}(t) = A\hat{x}(t) + Bu(t) + F\lambda_{ext,\alpha\beta} + Ev_1 \\ \dot{\hat{z}}(t) = C\hat{x}(t) + G\hat{f}_s + v_2 \end{cases} \tag{84}$$

v_1 and v_2 are selected, as control signals for the correction of the sliding mode, as follows:

$$\begin{cases} v_1 = L_1 H(e_x) \\ v_2 = L_2 H(e_z) \end{cases} \tag{85}$$

where $e_x = x - \hat{x}$ and $e_z = z - \hat{z}$, and L_1 and L_2 are positive design constants. Let us note $e_s = f_s - \hat{f}_s$.

Based on this, the equations of the errors can be written as follows:

$$\begin{cases} \dot{e}_x = \dot{x} - \dot{\hat{x}} = A e_x + E(d - v_1) \\ \dot{e}_z = \dot{z} - \dot{\hat{z}} = C e_x + G e_s - v_2 \end{cases} \tag{86}$$

To demonstrate the stability of FDO-type observers, the Lyapunov function is selected as follows:

$$\dot{V} = e_x^T e_x + e_z^T e_z + e_s^T Q e_s \tag{87}$$

where $Q > 0$ is a constant matrix with appropriate dimensions.

By calculating $\dot{V}$, the following relation is obtained:

$$\dot{V} \le 2\|e_x\| \, \|E\| (\|d\| - L_1) + 2\|e_z\| (\|C\| \, \|e_x\| - L_2) + 2e_s^T \left(G e_z - Q\dot{\hat{f}}_s \right) \tag{88}$$

To ensure stability, by selecting $L_1 > \|d\|$ and $L_2 > C\|e_x\|$, the law of adaptation for faults vector f_s is obtained:

$$\dot{\hat{f}}_s = Q^{-1} G e_z \tag{89}$$

Since i_α and i_β can be derived from $i_{a,b,c}$ in the form:

$$\begin{cases} i_\alpha = i_a \\ i_\beta = \dfrac{i_b - i_c}{\sqrt{3}} = \dfrac{2i_b + i_a}{\sqrt{3}} \end{cases} , \tag{90}$$

Then, the effect of the occurrence of faults $f_{s\alpha}$ and $f_{s\beta}$, propagate in the form of faults f_a and f_b in phases a and b of the PMSM power windings in the form of equations and laws of adaptation described by Equations (91)–(93):

$$\begin{cases} f_{s\alpha} = f_a \\ f_{s\beta} = \dfrac{2f_b + f_a}{\sqrt{3}} \end{cases} \tag{91}$$

$$\dot{\hat{f}}_a = \dot{\hat{f}}_{s\alpha} = Q^{-1} e_{z1} \tag{92}$$

$$\dot{\hat{f}}_b = \frac{\sqrt{3}\dot{\hat{f}}_{s\beta} - \dot{\hat{f}}_{s\alpha}}{2} = Q^{-1} \frac{\sqrt{3}e_{z2} - e_{z1}}{2} \tag{93}$$

In order to achieve the implementation of the FDO-type observer in the MATLAB/Simulink environment, it is necessary to define the equations presented in this section explicitly, by components. For implementation, $Q = G = I_2$. Thus, Equation (83) is defined explicitly in the form of Equations (94)–(96).

$$\begin{cases} \dot{i}_\alpha = -\dfrac{R_s}{L} i_\alpha + \dfrac{u_\alpha}{L} + \dfrac{\omega_e \lambda_0}{L} + d_1 \\ \dot{i}_\beta = -\dfrac{R_s}{L} i_\beta + \dfrac{u_\beta}{L} - \dfrac{\omega_e \lambda_0}{L} + d_2 \end{cases} \tag{94}$$

$$\begin{cases} \dot{z}_\alpha = y_\alpha = i_\alpha + f_\alpha \\ \dot{z}_\beta = y_\beta = i_\beta + f_\beta \end{cases} \tag{95}$$

$$\begin{cases} z_\alpha = \frac{1}{s}(i_\alpha + f_\alpha) \\ z_\beta = \frac{1}{s}(i_\beta + f_\beta) \end{cases} \tag{96}$$

Equations (84) and (85) are defined explicitly by components in the form of Equations (97)–(99).

$$\begin{cases} \dot{\hat{i}}_\alpha = -\dfrac{R_s}{L}\hat{i}_\alpha + \dfrac{u_\alpha}{L} + \dfrac{\omega_e \lambda_0}{L} + L_1 H\left(i_\alpha - \hat{i}_\alpha\right) \\ \dot{\hat{i}}_\beta = -\dfrac{R_s}{L}\hat{i}_\beta + \dfrac{u_\beta}{L} - \dfrac{\omega_e \lambda_0}{L} + L_1 H\left(i_\beta - \hat{i}_\beta\right) \end{cases} \tag{97}$$

$$
\begin{cases}
\dot{\hat{z}}_\alpha = \hat{i}_\alpha + \hat{f}_\alpha + L_2 H(z_\alpha - \hat{z}_\alpha) \\
\dot{\hat{z}}_\beta = \hat{i}_\beta + \hat{f}_\beta + L_2 H\left(z_\beta - \hat{z}_\beta\right)
\end{cases}
\tag{98}
$$

$$
\begin{cases}
e_{z\alpha} = z_\alpha - \hat{z}_\alpha \\
e_{z\beta} = z_\beta - \hat{z}_\beta
\end{cases}
\tag{99}
$$

Equation (89) is implemented as follows:

$$
\begin{bmatrix} \hat{f}_\alpha \\ \hat{f}_\beta \end{bmatrix} = \begin{bmatrix} z_\alpha - \hat{z}_\alpha \\ z_\beta - \hat{z}_\beta \end{bmatrix};
\quad
\begin{cases}
\hat{f}_\alpha = \frac{1}{s}(z_\alpha - \hat{z}_\alpha) \\
\hat{f}_\beta = \frac{1}{s}\left(z_\beta - \hat{z}_\beta\right)
\end{cases}
\tag{100}
$$

We specify that s represents the complex variable in Equations (70)–(73). Figure 6 shows the implementation in MATLAB/Simulink of the FDO-type observer using the equations defined explicitly by components (94)–(100).

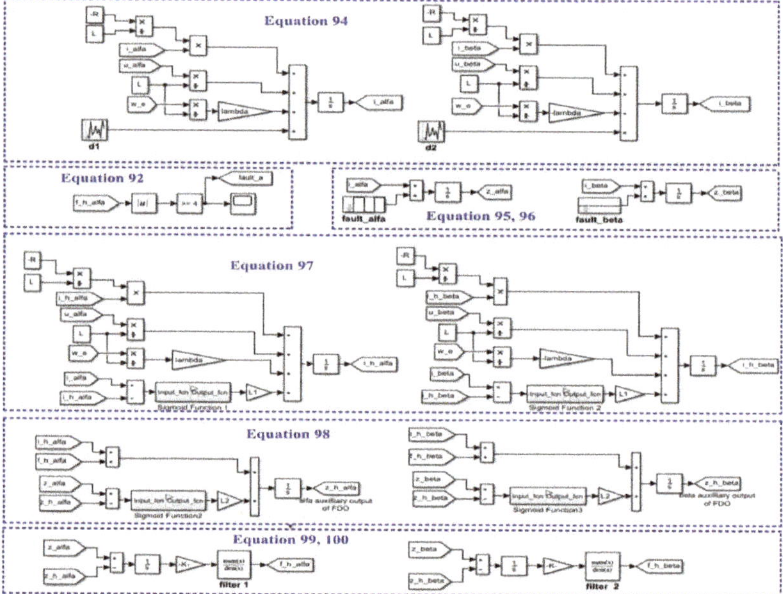

Figure 6. FDO-type observer—MATLAB/Simulink implementation block diagram.

7. Numerical Simulations

For the numerical simulations performed in MATLAB/Simulink and presented in this section, the PMSM parameters are shown in Table 1.

Table 1. Nominal parameters of PMSM.

Motor Parameter	Symbol	Value	Unit
Stator resistance	R_s	2.875	Ω
d axes inductance	L_d	0.0085	H
q axes inductance	L_q	0.0085	H
Combined inertia of rotor and load	J	0.0008	kg·m^2
Combined viscous friction of rotor and load	B	0.005	N·m·s/rad
Flux induced by permanent magnets of rotor in stator phases	λ_0	0.175	Wb
Pole pairs number	n_p	4	-

7.1. Numerical Simulations—Fractional Order Speed Controllers for PMSM

The block diagram for the implementation in MATLAB/Simulink of the PMSM control system based on the FOC strategy proposed in this article is presented in Figure 7. The results of the behavior of the PMSM control system are presented comparatively, using the classic PI controller, FO-PI controller, TID controller, FO-lead-lag controller and FO-SMC controller for the control of the outer PMSM rotor speed control loop. The other blocks in Figure 7 implement the speed reference generator, the load torque generator, the time evolution I/O signals and the model of the PMSM drive. The sensorless function of the control system of the rotor speed is provided by the use of an SMO-type observer described in Section 6. Additionally, to detect the faults of the current sensors on the supply phases of the PMSM, an FDO-type observer as described in Section 6 is embedded in the general control structure in Figure 7.

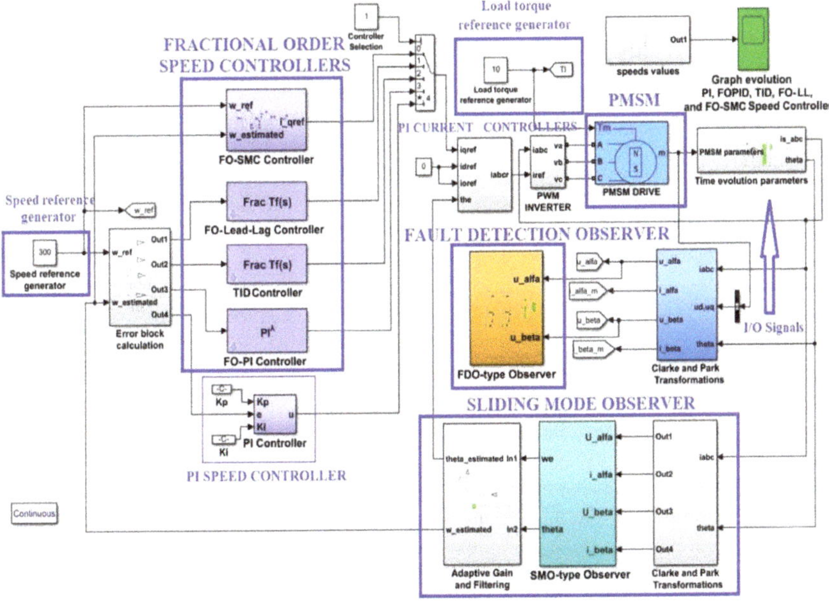

Figure 7. MATLAB/Simulink implementation block diagram for sensorless control of PMSM based on fractional order speed controllers, PI current controllers, SMO speed observer and FDO.

Starting from Equations (17) and (18), which represent the transfer function of the fixed part of the PMSM rotor speed control system, for the nominal parameters presented in Table 1, the theoretical transfer function of the fixed part is obtained:

$$H_f(s) = \frac{2.857}{5.44 \cdot 10^{-4} s^3 + 2.588 \cdot 10^{-5} s^2 + 0.03761\,s + 0.7637} \tag{101}$$

The discrete form obtained for the transfer function based on Equation (101) according to the Tustin method for a sampling period of 0.1 ms is:

$$H_f(z) = \frac{0.004648 z^3 + 0.01394 z^2 + 0.01394 z + 0.004648}{z^3 - 2.166 z^2 + 1.837 z - 0.6607} \tag{102}$$

The transfer function of the fixed part of the PMSM rotor speed control system is obtained by using the MATLAB System Identification Toolbox, through identification in continuous and discrete form in Equations (103) and (104), respectively:

$$H_{f\_ident}(s) = \frac{0.002373}{s^3 + 0.0663s^2 + 0.01484s + 8.163 \cdot 10^{-6}} \tag{103}$$

$$H_{f\_ident}(z) = \frac{2.96 \cdot 10^{-3} z^3 + 8.89 \cdot 10^{-3} z^2 + 8.89 \cdot 10^{-3} z + 2.96 \cdot 10^{-3}}{z^3 - 3z^2 + 3z - 0.99} \tag{104}$$

The tuning of PI controllers by using Ziegler–Nichols methods is a well-known technique. In the fractional case, the FOMCON toolbox for the MATLAB utility program is used for the tuning of FO-PI controllers. In order to obtain the optimal tuning parameters in the fractional case, a number of optimization methods are incorporated in the FOMCON toolbox, both in the frequency range and in the time domain. In the frequency range, the goal of optimizing the parameters is achieved by obtaining optimal performance in terms of the sensitivity function $S(j\omega)$ for disturbance rejection for the low and middle frequency range and the rejection of the high frequency noise using the complementary sensitivity function $T(j\omega)$. In the time domain, the tuning of fractional controllers is carried out by minimizing optimal criteria, such as the integral absolute error (IAE) [41].

Using the FOMCON toolbox for MATLAB, an FO-PI controller described by Equation (20) can be tuned for the control of the PMSM rotor speed. For $K_p = 1.2$, $K_i = 12$, $\lambda = 1.1$, and $K_d = \mu = 0$, the transfer function of the FO-PI controller is obtained:

$$H_{PI} = \frac{1.2s^{1.1} + 12}{s^{1.1}} \tag{105}$$

The closed loop transfer function of the PMSM rotor speed control, where the controller is given by Equation (20) and the fixed part is given by Equation (103), is expressed in the following form:

$$H_{CL\_PI} = \frac{4.3272 \cdot 10^8 s^{1.1} + 4.3272 \cdot 10^9}{s^{3.2} + 73730s^{2.2} + 4.3746 \cdot 10^8 s^{1.1} + 4.3272 \cdot 10^9} \tag{106}$$

Using the FOMCON toolbox for MATLAB, the stability of the closed loop system of the PMSM rotor speed control, in other words, the fulfillment of the condition given by the relation (6) is graphically presented in Figure 8, where it is noted that the system is stable.

The step response for the FO-PI controller given by the relation (105), considering the fixed part is represented by both the theoretical transfer function and by the transfer function obtained through identification, in other words, the relations (101) and (103), respectively, is presented in Figure 9.

A similar response is noted in the two cases, with a good behavior under both the dynamic and stationary regimes and which additionally proves the similarity between the transfer function of the fixed part obtained theoretically and by identification.

Figure 10 shows the closed loop system consisting of the FO-PI controller given by relation (105) and theoretical transfer function given by relation (101), Bode diagram, Nyquist diagram and Nichols diagram. The stability of the system can be inferred from the specific interpretation of these diagrams and, in addition (to the conclusions in Figure 8), an amplitude stability margin of 12 dB can be noted.

For the implementation in DSP, it is necessary to obtain the transfer function of the FO-PI controller given in Equation (105) in an equivalent form as discrete variable z, but of integer order. For this, according to those presented in Section 2, an approximation of the fractional order transfer function can be obtained with an integer-order continuous transfer function, by using the Oustaloup filter.

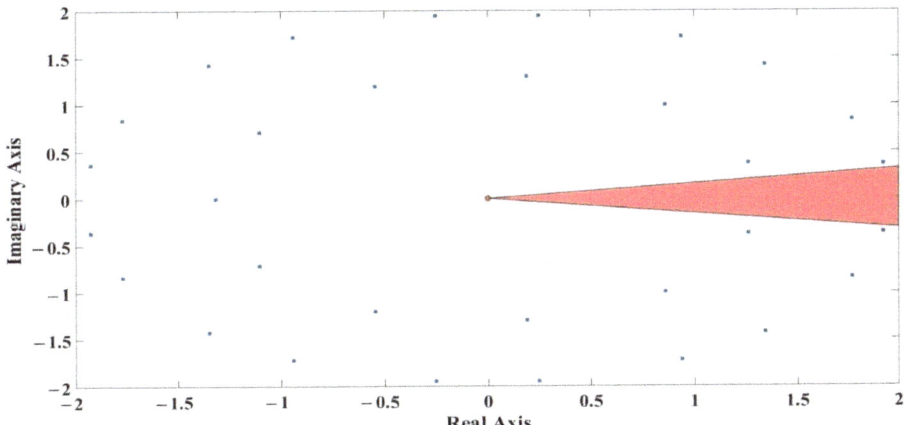

Figure 8. Graphical representation for the stability of the PMSM rotor speed control system—closed loop.

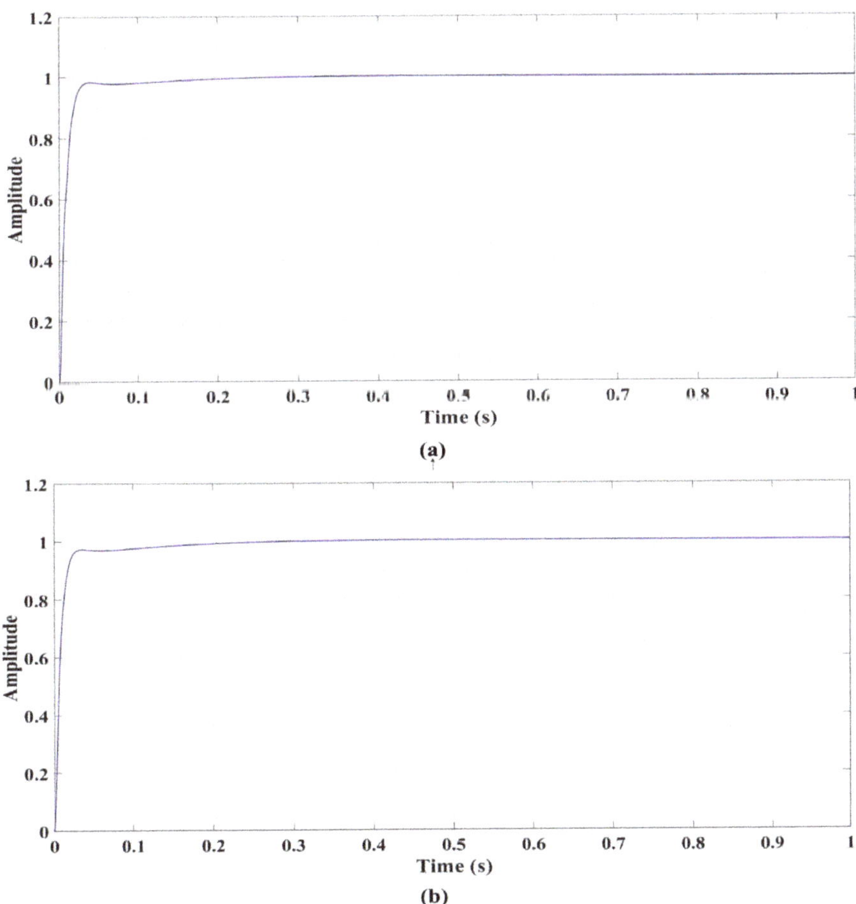

(a)

(b)

Figure 9. Unit step response of the PMSM rotor speed control system with FO-PI controller and fixed part with: (**a**) theoretical transfer function; (**b**) transfer function obtained through identification.

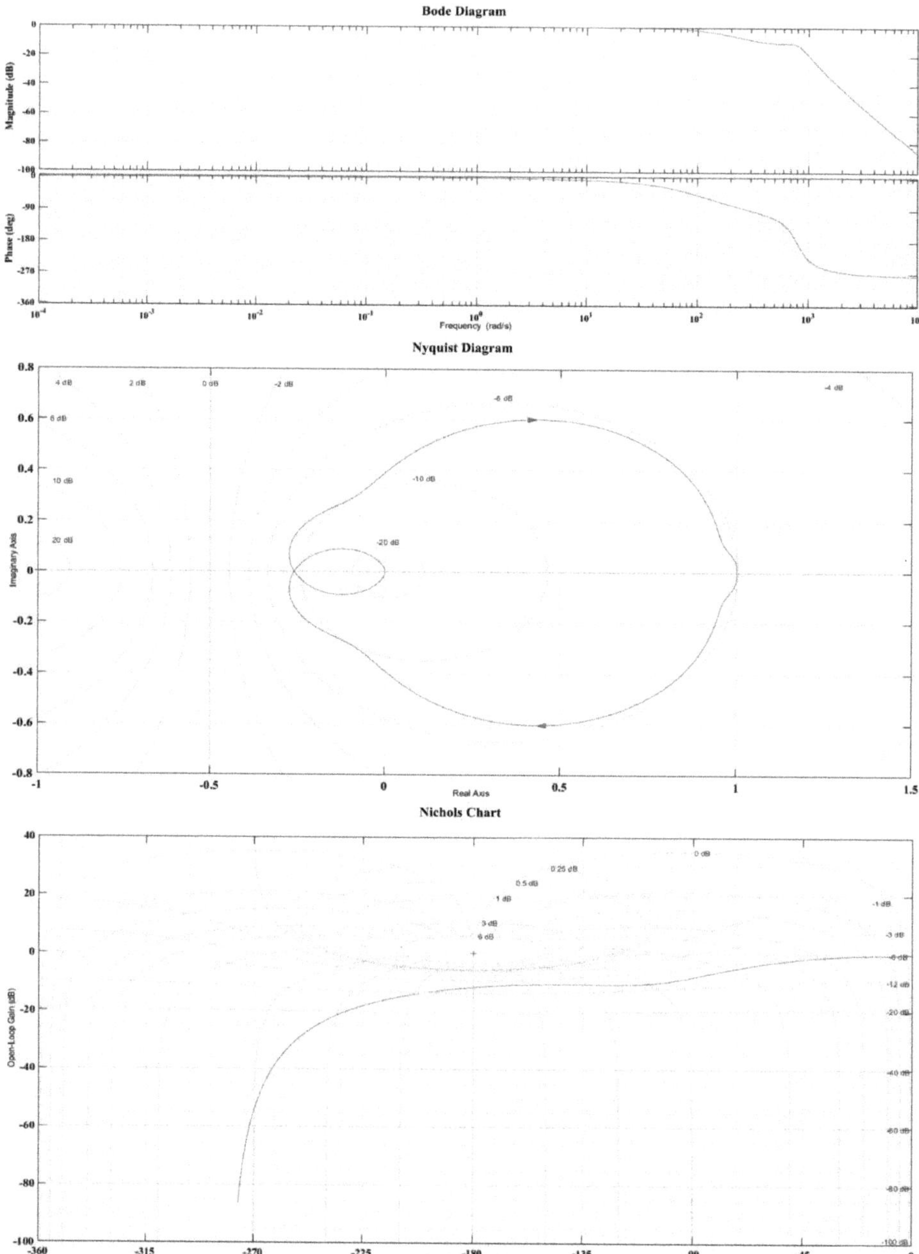

Figure 10. Bode, Nyquist and Nichols diagram representations for closed-loop system composed from FO-PI controller and theoretical transfer function of PMSM given by Equations (101) and (105).

In the usual frequency range for the presented application $\omega = (10^{-2};\, 10^{3})$ rad/s, in Equation (107) is expressed and the equivalent transfer function obtained. In order to obtain the discrete form of this equivalent transfer function, the Tustin substitution is used, and the obtained discrete transfer function is expressed in Equation (108).

$$H_{PL\_INT}(s)= \frac{1.2(s+280.6)(s+354.8)(s+35.48)(s+25.88)(s+6.918)(s+4.422)}{s(s+354.8)(s+218.8)(s+35.48)(s+28.18)(s+3.548)} \cdot \frac{(s+3.548)(s+0.3575)(s+0.3548)(s+0.0355)(s+0.03548)}{(s+0.3548)(s+0.2818)(s+0.03548)(s+0.02818)}$$

(107)

$$H_{PL\_INT}(z)= \frac{1.2031(z-0.7554)(z-0.7013)(z-0.9651)(z-0.9745)}{(z-1)^5(z-0.9972)(z-0.9965)(z-0.9722)} \cdot \frac{1.2031(z-0.7554)(z-0.7013)(z-0.9651)(z-0.9745)}{(z-1)^5(z-0.9972)(z-0.9965)(z-0.9722)}$$

(108)

Another fractional order controller is presented in Equation (21) and for $K_t = 1.2$, $K_i = 12$, $n = 10$, and $K_d = 0$, the following transfer function of the fractional order TID controller is obtained:

$$H_{TID} = \frac{1.2 + 12s^{0.1}}{s^{1.1}}$$

(109)

The closed loop transfer function for the fixed part presented above in Equation (103) and the TID controller given by Equation (109) is the following:

$$H_{CL\_TID} = \frac{4.3272 \cdot 10^8 s^{0.91} + 4.3272 \cdot 10^9}{s^{3.1} + 7373 s^{2.1} + 4.3746 \cdot 10^8 s^{0.91} + 4.3272 \cdot 10^9}$$

(110)

The response to the unit step of the PMSM rotor speed control loop with fractional TID controller is shown in Figure 11. A good performance is noted in dynamic and stationary regime.

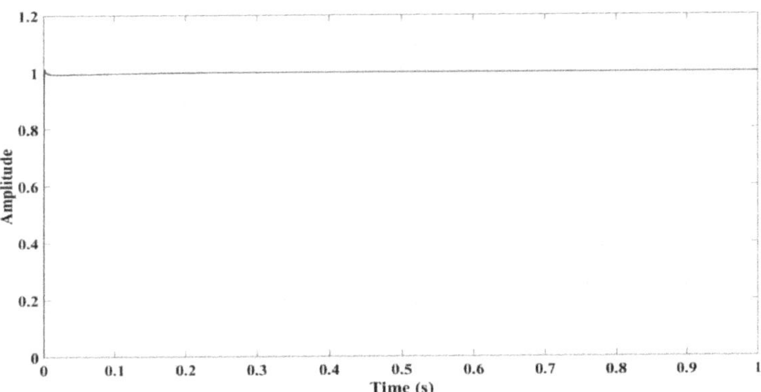

Figure 11. Unit step response of the PMSM rotor speed control system with transfer function obtained through identification for the fixed part and fractional TID controller.

Proceeding as above in the case of the FO-PI controller, using the Oustaloup filter zpk, in the TID controller case the results for integer-order will be as follows:

$$H_{TID\_INT}(s) = \frac{656.44(s+114.1)(s+17.33)(s+6.938)(s+0.8839)(s+0.08997)}{s(s+901.6)(s+90.16)(s+9.016)(s+0.9016)(s+0.09016)}$$

(111)

$$H_{TID\_INT}(z) = \frac{0.25176(z+0.7516)(z-0.8922)(z-9828)(z-0.9931)(z-1)}{(z-1)^3(z-0.991)(z-0.9138)(z-0.4059)}$$

(112)

In case of another fractional order controller which is presented in Equation (23), and for $k' = 300$, $x = 50$, $\lambda = 1.4$ and $\alpha = 0.11$, the following transfer function of the FO-lead-lag controller is obtained:

$$H_{LL} = \frac{3.606 \cdot 10^8}{s^{2.1} + 7.373 \cdot 10^4 s^{1.1} + 4.74 \cdot 10^6} \tag{113}$$

The closed loop transfer function for the fixed part presented above in Equation (103) and the FO-lead-lag controller given by Equation (113) is the following:

$$H_{CL\_LL} = \frac{abs\left(1.0561 \cdot 10^{11} - 6.942 \cdot 10^9 \cdot i\right)}{s^{2.1} + 73730 s^{1.1} + abs\left(1.0561 \cdot 10^{11} - 6.942 \cdot 10^9 \cdot i\right)} \tag{114}$$

The response to the unit step of the PMSM rotor speed control loop with the FO-lead-lag controller is shown in Figure 12. Furthermore, for this type of fractional order controller a good performance is noted in the dynamic and stationary regimes.

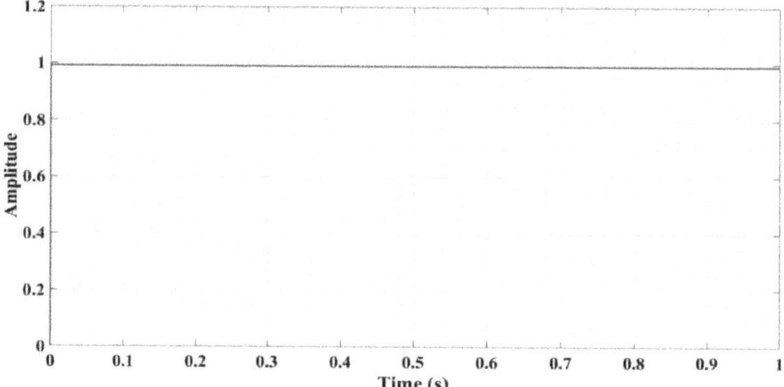

Figure 12. Unit step response of the PMSM rotor speed control system with transfer function obtained through identification for the fixed part and FO-lead-lag controller.

By using the Oustaloup filter zpk in the FO-lead-lag controller case the results for integer-order will be as follows:

$$
\begin{aligned}
H_{LL\_INT}(s) \quad &- \frac{1.8073 \cdot 10^8 (s + 354.8)^2 (s + 35.48)^2}{(s + 7.37 \cdot 10^4)(s + 354.8)(s + 272.4)(s + 35.48)(s + 3.603)(s + 3.548)} \cdot \\
&\frac{(s + 3.548)^2 (s + 0.3548)^2 (s + 0.03548)^2}{(s + 0.3552)(s + 0.3548)(s + 0.03548)(s + 0.03548)(s^2 + 68.99 s + 1465)}
\end{aligned} \tag{115}
$$

$$
\begin{aligned}
H_{LL\_INT}(z) \quad &= \frac{1.2103(z + 1.071)(z - 1)^4 (z - 0.9965)^2}{z(z - 1)^4 (z - 0.9965)(z - 0.9964)(z - 0.9651)} \cdot \\
&\frac{(z - 0.9651)^2 (z - 0.7013)^2 (z + 0.0003655)}{(z - 0.7616)(z - 0.7013)(z^2 - 1.932 z + 0.9333)}
\end{aligned} \tag{116}
$$

Figure 13 presents the results of the simulation of the PMSM rotor speed control, where the speed controller is FO-PI-type and PI-type, respectively. For a speed reference of 2000 rpm, with no load torque, good dynamic and static results are obtained, but with an obvious advantage of the FO-PI controller. The variations of the stator currents $i_{a,b,c}$, and of currents i_d and i_q are presented, and compliance with reference $i_{dref} = 0$ is noted according to the FOC strategy.

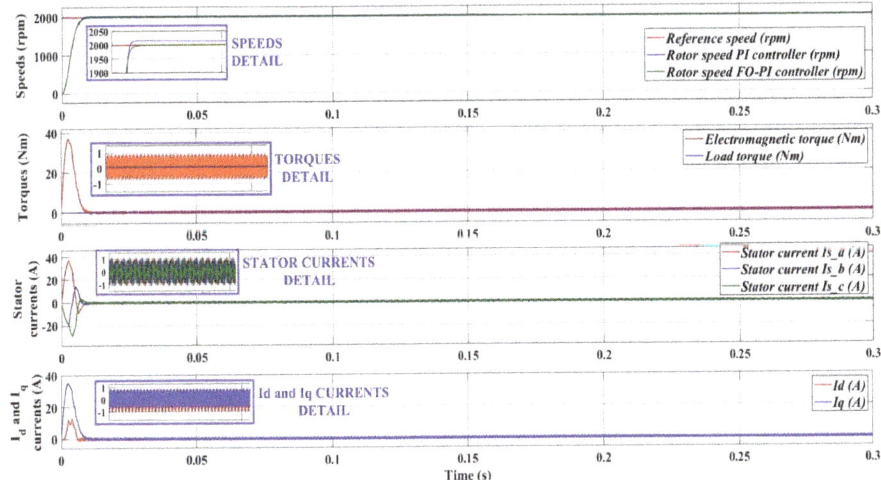

Figure 13. Comparative simulation of the PMSM time evolution with FO-PI speed controller and PI speed controller—ω_{ref} = 2000 rpm, T_L = 0 Nm, K_p = 1.2, K_i = 12, λ = 1.1 and K_d = μ = 0.

In Figure 14, the FO-PI controller is compared to the PI controller for the PMSM rotor speed control system, for a speed reference of 2000 rpm and a load torque of 5 Nm. The overshooting by the speed PI-type controller and the good static and dynamic performance provided by the speed FO-PI-type controller can be noted in the detail in the said figure.

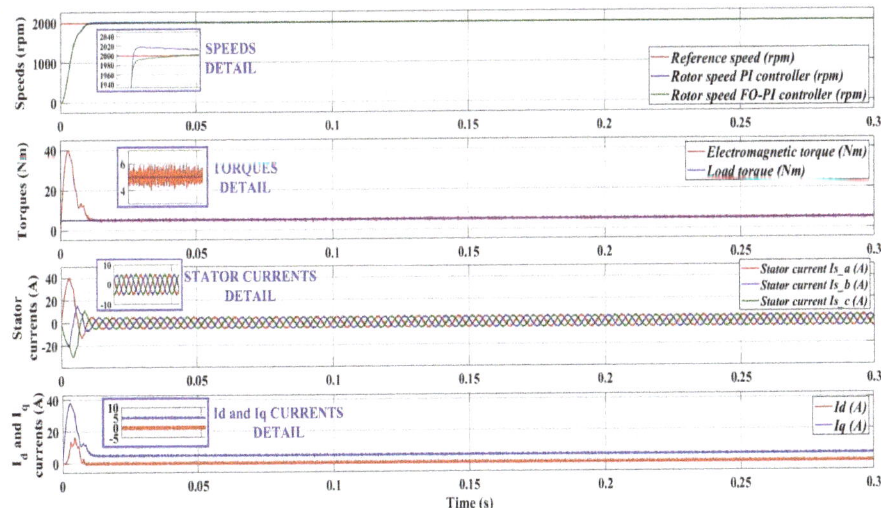

Figure 14. Comparative simulation of the PMSM time evolution with FO-PI speed controller and PI speed controller—ω_{ref} = 2000 rpm, T_L = 5 Nm, K_p = 1.2, K_i = 12, λ = 1.1 and K_d = μ = 0.

The parametric robustness provided by the FO-PI controller for the PMSM rotor speed control loop is presented in Figure 15, where, for the speed and load torque references presented in Figure 14 plus a 50% increase of the J parameter (combined inertia of rotor and load), it is presented by maintaining a response with good dynamic and stationary performance.

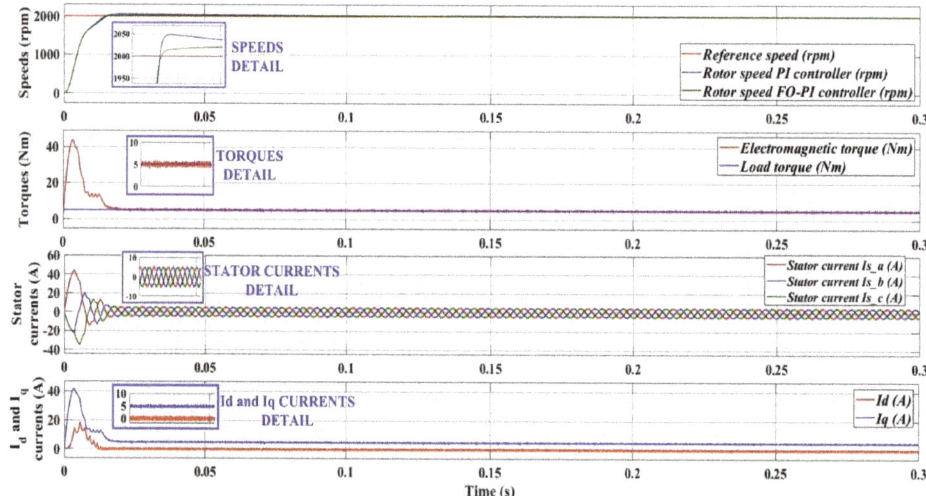

Figure 15. Comparative simulation of the PMSM time evolution with FO-PI speed controller and PI speed controller—ω_{ref} = 2000 rpm, T_L = 5 Nm, K_p = 1.2, K_i = 12, λ = 1.1, $K_d = \mu = 0$ and 100% increase of J parameter.

Figure 16 shows the response of the control system in the case of ω_{ref} = 300 rpm and double the nominal load torque, T_L = 10 Nm. Figure 16 also shows the time evolution of the electromagnetic torque, of the stator currents, as well as currents i_d and i_q. A very good response time (12 ms) is noted, given the lack of overshooting and the evolution of the i_d current around zero. For the implementation of the FO-SMC-type control described in Section 4, the parameters ε = 300, q = 200, c = 100 and μ = 0.55 were selected.

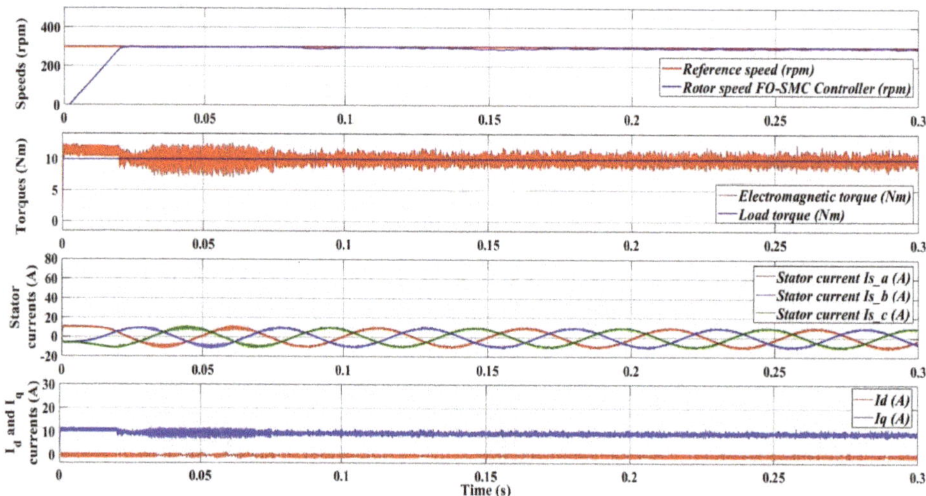

Figure 16. Simulation of the PMSM time evolution with FO-SMC speed controller for ω_{ref} = 300 rpm, T_L = 10 Nm, ε = 300, q = 200, c = 100 and μ = 0.55.

Figure 17 also shows the good performance of the PMSM control system using the FO-SMC-type controller for the outer rotor speed control loop if the speed reference ω_{ref} = 2200 rpm and the load torque T_L = 2 Nm.

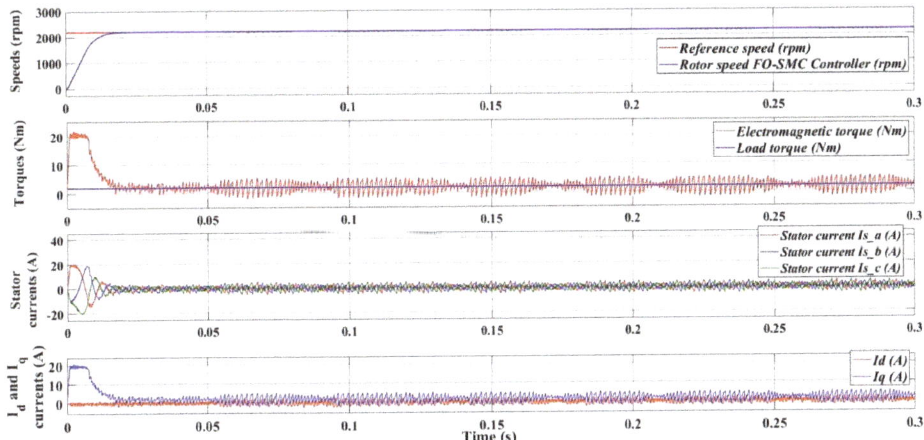

Figure 17. Simulation of the PMSM time evolution with FO-SMC speed controller for ω_{ref} = 2200 rpm, T_L = 2 Nm, ε = 300, q = 200, c = 100 and μ = 0.55.

The parametric robustness of the PMSM control system is noted in Figure 18, where, for the same speed and load torque references in Figure 17, however, with a 100% increase of J parameter and an added uniformly distributed noise, good static and dynamic performances are noted, while there is an override of less than 2%.

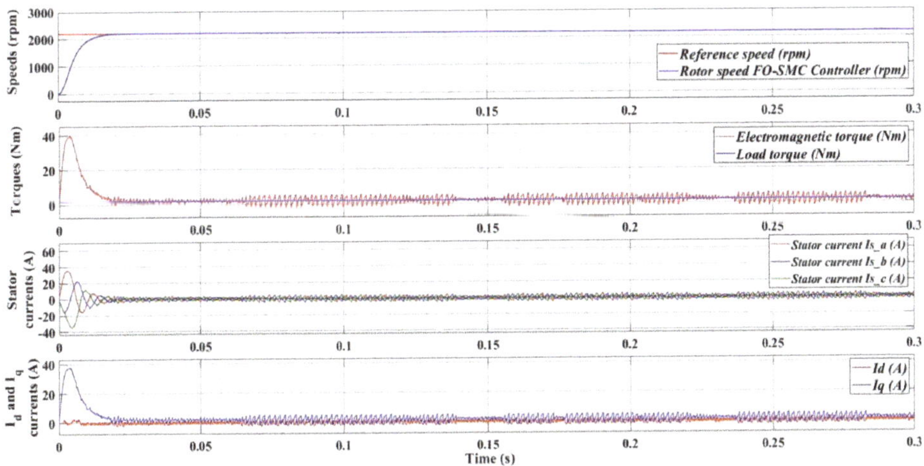

Figure 18. Simulation of the PMSM time evolution with FO-SMC speed controller for ω_{ref} = 2200 rpm, T_L = 2 Nm and uniformly distributed noise, ε = 300, q = 200, c = 100, μ = 0.55 and 100% increase of J parameter.

For a comparison between the performance obtained for the PMSM rotor speed control by using the speed PI controller, the FO-PI speed controller, the TID speed controller, the FO-lead-lag speed controller and the FO-SMC speed controller, Figure 19 shows the system response in closed loop for a reference of 300 rpm and a load torque of 10 Nm. The superiority of the speed FO-PI-type speed controller, TID-type speed controller, FO-lead-lag speed controller and FO-SMC speed controller over the speed PI-type controller is noted. This can be intuited by the fact that the first two controllers mentioned have a higher number of tuning parameters than the PI-type controller, and their mathematical model can be considered more accurate due to the use of the fractional calculus for integration and differentiation operators. Of the four speed controllers compared, the superiority is apparent for the speed FO-SMC-type controller which does not feed any overshooting, and the response time is below 15 ms for the nominal load torque.

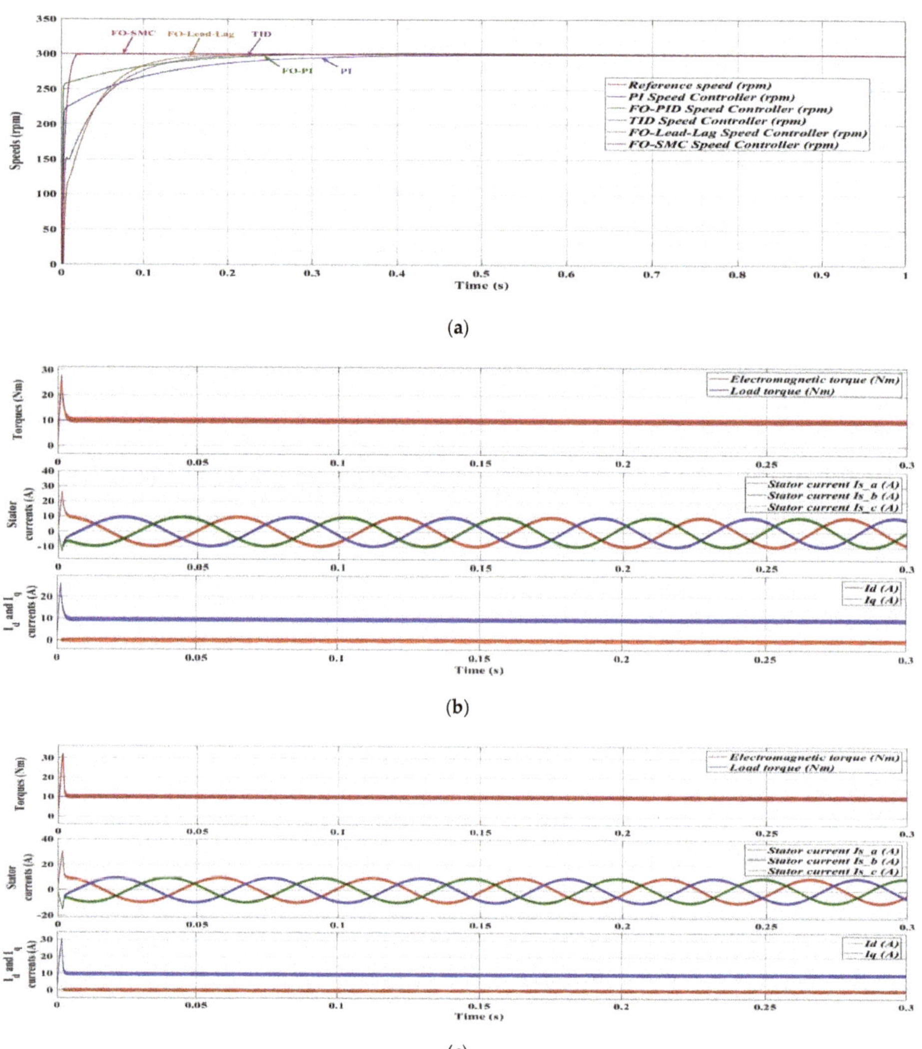

(a)

(b)

(c)

Figure 19. *Cont.*

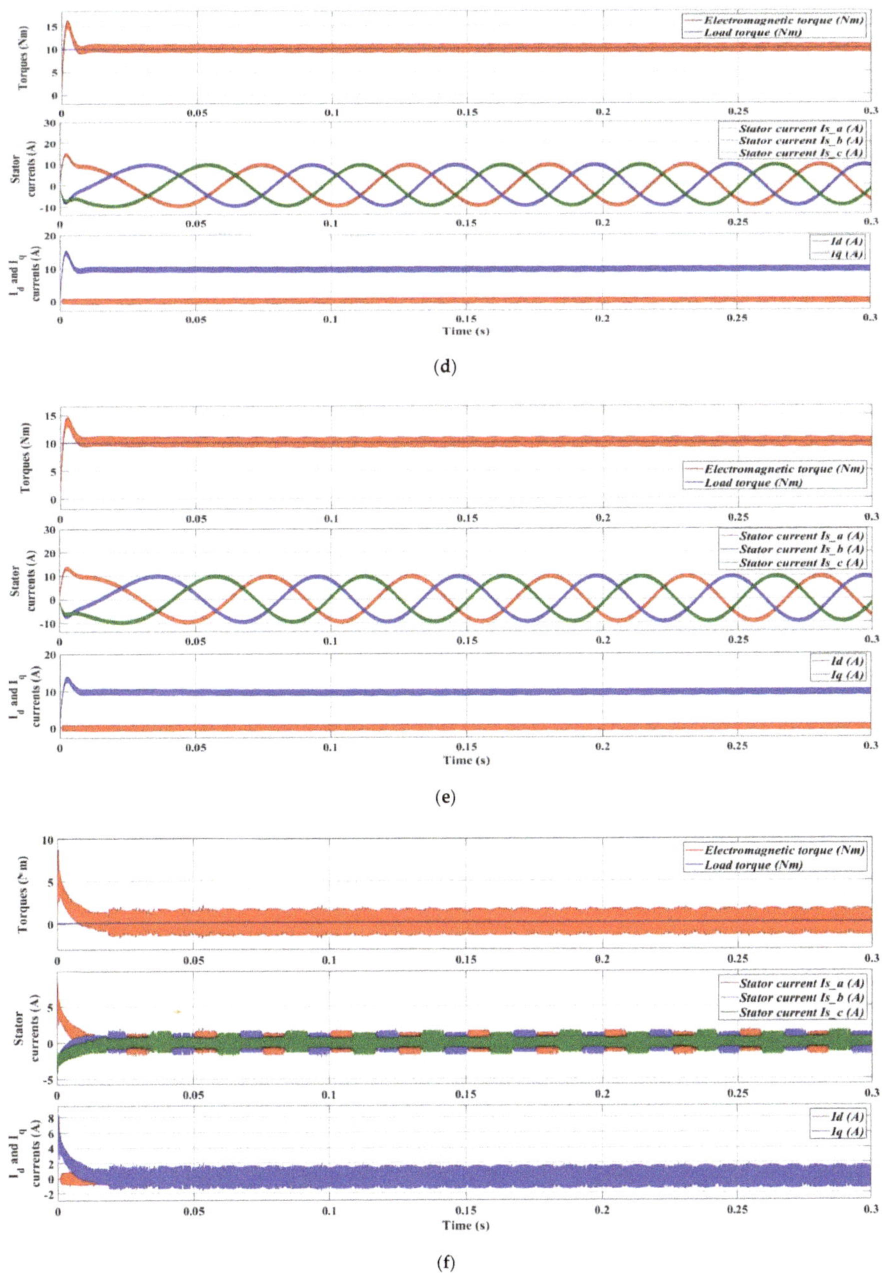

Figure 19. Comparative simulation of the PMSM time evolution with FO-SMC speed controller, FO-lead-lag speed controller, TID speed controller, FO-PI speed controller and PI speed controller—ω_{ref} = 300 rpm, T_L = 10 Nm: (**a**) speed comparison; (**b**) torques, stator currents and i_d and i_q currents for PI speed controller; (**c**) torques, stator currents and i_d and i_q currents for FO-PI speed controller; (**d**) torques, stator currents and i_d and i_q currents for TID speed controller; (**e**) torques, stator currents and i_d and i_q currents for FO-lead-lag speed controller; and (**f**) torques, stator currents and i_d and i_q currents for FO-SMC speed controller.

The value of the speed/torque ripple is defined as follows:

$$x_{rip} = \sqrt{\frac{1}{N} \sum_{i=1}^{N} \left(x(i) - x_{ref}(i)\right)^2} \tag{117}$$

where N represents the number of samples, x and x_{ref} represent the rotor speed/torque and the prescribed reference of speed/torque, respectively.

Table 2 compares a number of performance indices obtained by using fractional speed controllers for a reference speed $\omega_{ref} = 300$ rpm and a load torque reference $T_L = 10$ Nm. The performances of these types of controllers are presented both in the case of nominal parameters, and in the case of J parameter doubling. It can be noted that the fractional controllers and especially the FO-SMC speed controller have superior performance.

Table 2. Comparison of performance indices of the fractional order proposed speed controllers.

Performance Indices	PI Speed Controller	FO-PI Speed Controller	TID Speed Controller	FO-Lead-Lag Speed Controller	FO-SMC Speed Controller
Overshoot (%) nominal J	0	0	0	0	0
Overshoot (%) double J	0	0	0	0	0
Settling time (ms) nominal J	300	240	220	180	16
Settling time (ms) double J	450	300	280	220	40
Steady state error (%) nominal J	0.11	0.11	0.1	0.1	0.09
Steady state error (%) double J	0.12	0.12	0.11	0.1	0.09
Speed ripple (rpm) nominal J	121.78	81.14	142.24	112.94	102.81
Speed ripple (rpm) double J	289.28	192.35	216.14	204.91	131.15
Torque ripple (Nm) nominal J	13.68	17.16	10.08	9.23	18.91
Torque ripple (Nm) double J	11.8	15.45	12.91	12.1	16.01

7.2. Numerical Simulations for Rotor Speed Estimation and Fault Detection

Figure 20 shows the evolution of the SMO-type observer for the estimation of the PMSM rotor speed. Very good stationary results are noted, except for the first 100 ms, when, normally, at the start of PMSM, it is controlled in the open loop, according to a predefined sequence.

The numerical simulation parameters $L_1 = 150{,}000$ and $L_2 = 50$ (amplification factors of the FDO-type observer) are selected to test the efficiency of the FDO-type observer described in Section 6. Thus, Figure 21 shows the efficiency of this type of observer due to the fact that it very precisely reconstructed the currents i_α and i_β together with the outputs z_α and z_β provided by Equation (84), which describes the implementation of the FDO-type observer under the conditions where faults of the current sensors may occur.

Based on Equations (91)–(93), which express the relationships between the fault flags on phases α and β in the α-β reference frame and the real supply phases a, b, c of the PMSM, Figure 22 shows that, for the occurrence of a fault on phase a, it is detected, and a specific fault flag is set to logic 1, during the occurrence of such a fault of the current sensor. The detection threshold used is 4 A, and a fast response of the FDO-type observer of 60 ms is noted.

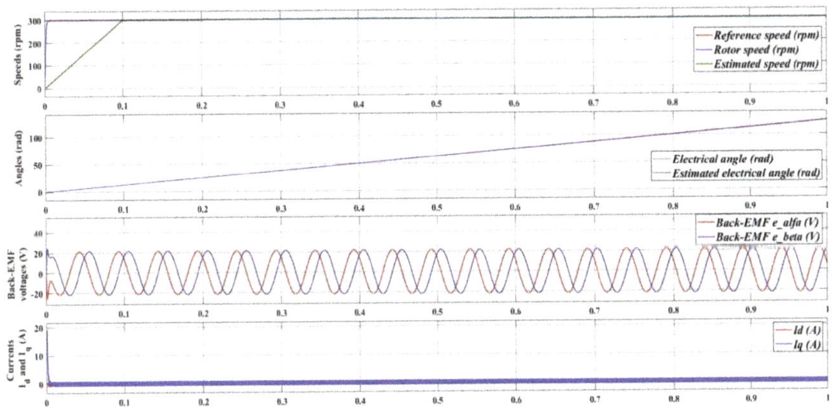

Figure 20. Simulation of the back-EMF, rotor position and rotor speed time evolution from the SMO-type observer.

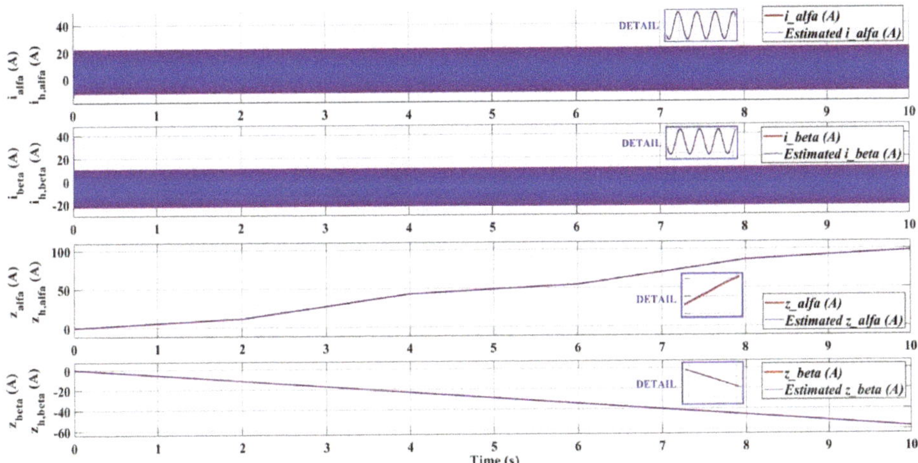

Figure 21. Estimated signals time evolution based on FDO-type observer.

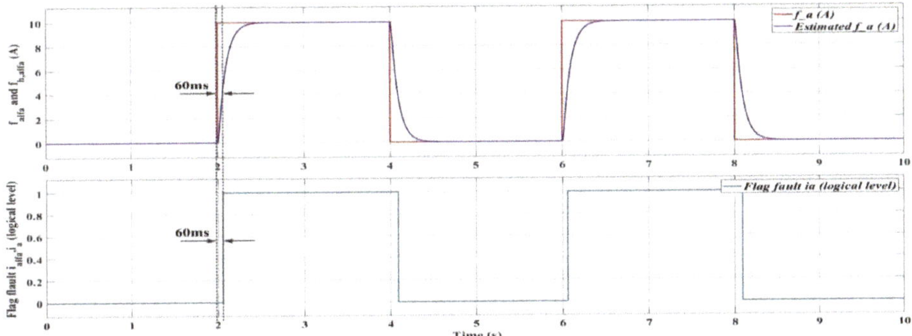

Figure 22. Estimated fault on phase "a" based on FDO-type observer.

The response time of the FDO-type observer is the time between the moment of occurrence of a fault of the sensor and the setting of the fault flag, under the conditions where the occurrence of false fault detections is eliminated by setting of the 4 A threshold, and the disturbance on the system (d_1 and d_2, described in Equation (80) used the in this numerical simulation is a uniform random variable with amplitude in the range -10 to 10 A.

7.3. Numerical Simulations—Fractional Order Synergetic Current Controllers and PI Speed Controller for PMSM

Figure 23 presents the block diagram of the MATLAB/Simulink implementation of the PMSM sensorless control system based on the synergetic current control or the FO-synergetic current control for the inner current loops control.

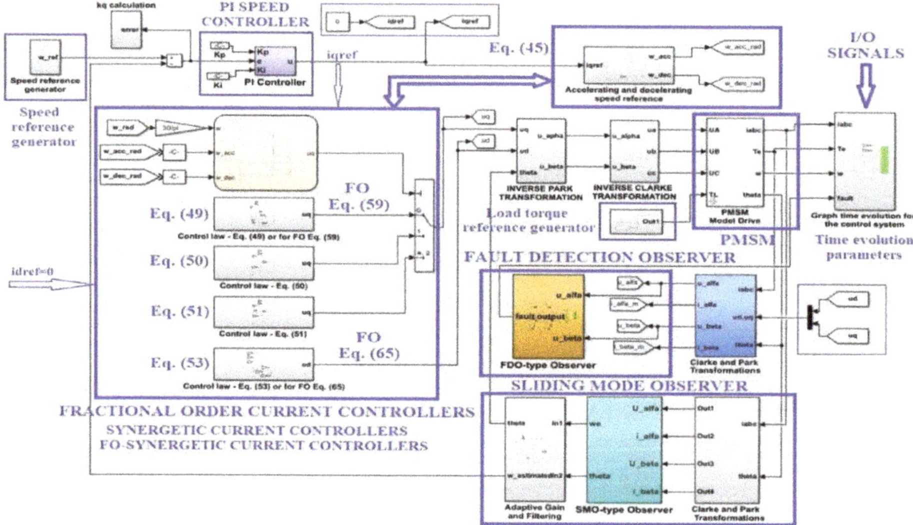

Figure 23. Simulink block diagram for the PMSM sensorless control based on PI speed controller, FO-synergetic current controllers, SMO speed observer and FDO.

The main functional blocks represented in the block diagram of the control system are: the PI speed controller which supplies current i_{qref} ($i_{dref} = 0$ according to the FOC control strategy), the speed reference generator, the synergetic controller or the FO-synergetic controller, the load torque generator, the PMSM motor drive, the SMO-type observer, the acceleration and deceleration of the rotor angular velocity and the Clarke and Park transformation.

The synergetic current controllers or the FO-synergetic current controller block provides the controls u_d and u_q by implementing Equations (49)–(51) and (53) for the synergetic current control, and Equations (59) and (65) for the FO-synergetic current control, respectively.

For a speed reference of 500 rpm and a torque load of 1 Nm when using a synergetic controller, Figure 24 presents the qualitatively superior response of the system with an override below 8%, but with a notable performance response time of 2 ms.

For the numerical simulations in which the FO-synergetic type controller is used, the parameters described in Section 5 are: $k_{iq} = 10,000$, $k_q = 10,000$, $i_{qmax} = 50$, $T_d = 3$, $T_q = 3$, $k_{id} = 10,000$ and $\mu = 0.5$ (for Equations (54) and (60)).

In Figure 25, by replacing the synergetic controller with the FO-synergetic controller, for the same conditions presented in Figure 24, the reduction of the override to 4% and an excellent performance determined by the 1 ms response time are noted.

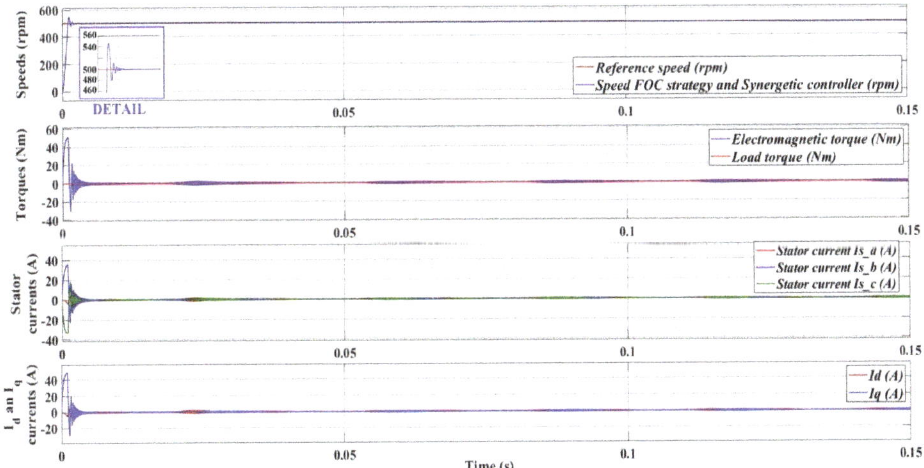

Figure 24. Simulation of the PMSM time evolution with synergetic controller and FOC strategy—ω_{ref} = 500 rpm, T_L = 1 Nm, k_{iq} = 10,000, k_q = 10,000, i_{qmax} = 50, T_d = 3, T_q = 3 and k_{id} = 10,000.

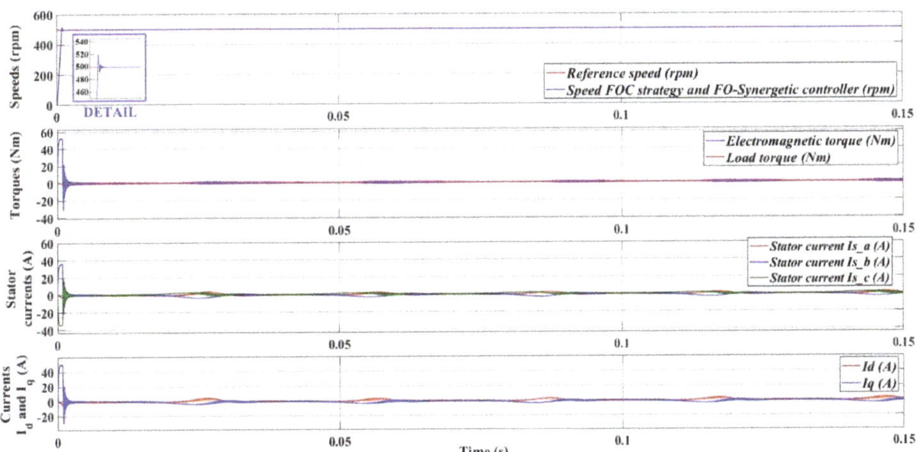

Figure 25. Simulation of the PMSM time evolution with FO-synergetic controller, and FOC strategy—ω_{ref} = 500 rpm, T_L = 1 Nm, k_{iq} = 10,000, k_q = 10,000, i_{qmax} = 50, T_d = 3, T_q = 3, k_{id} = 10,000 and μ = 0.5.

Figures 24 and 25 additionally present the evolution of the electromagnetic torques, load torques and stator currents i_a, i_b, i_c and currents i_d and i_q.

Figure 26 presents the comparative time evolution of the numerical simulation for the FOC strategy with PI current controllers, synergetic current controllers and FO-synergetic current controllers of the PMSM. Figure 26 shows that very good performance achieved by using the synergetic control for the inner current loops i_d and i_q. Obviously, due to the additional control parameters, between the two types of synergetic control, the best performance is obtained by using the FO-synergetic current controllers.

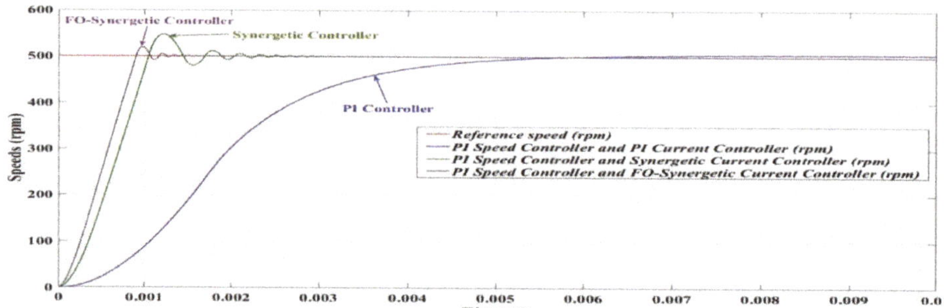

Figure 26. Comparative time evolution of the numerical simulation for FOC strategy with PI controller, synergetic controller and FO-synergetic controller of the PMSM—ωref = 500 rpm and T_L = 1 Nm.

Table 3 compares a number of performance indices obtained by using fractional current controllers for a reference speed ω_{ref} = 500 rpm and a load torque reference T_L = 1 Nm. The performances of these types of controllers are presented both in the case of nominal parameters, and in the case of J parameter doubling. It can be noted that the synergetic controllers, and especially the FO-synergetic current controller, have superior performance.

Table 3. Comparison of performance indices of the fractional order proposed current controller.

Performance Indices	PI Current Controller	Synergetic Current Controller	FO-Synergetic Current Controller
Overshoot (%) nominal J	0	8	2
Overshoot (%) double J	0	14	3.5
Settling time (ms) nominal J	6	1.2	1
Settling time (ms) double J	11	1.8	1.6
Steady state error (%) nominal J	0.1	0.08	0.07
Steady state error (%) double J	0.1	0.08	0.07
Speed ripple (rpm) nominal J	182.16	112.91	102.45
Speed ripple (rpm) double J	214.91	129.54	107.63
Torque ripple (Nm) nominal J	15.21	17.95	15.82
Torque ripple (Nm) double J	19.44	21.02	18.73

7.4. Numerical Simulations—Fractional Order Speed Controllers and Fractional Order Synergetic Current Controller for PMSM

This subsection presents the numerical simulations for the structure proposed in this article, namely, an FOC control strategy of the PMSM where the speed controller is of the SMC or FO-SMC types, and the controllers for the current control loops are of synergetic and FO-synergetic type. Figure 27 shows the block diagram of the MATLAB/Simulink implementation.

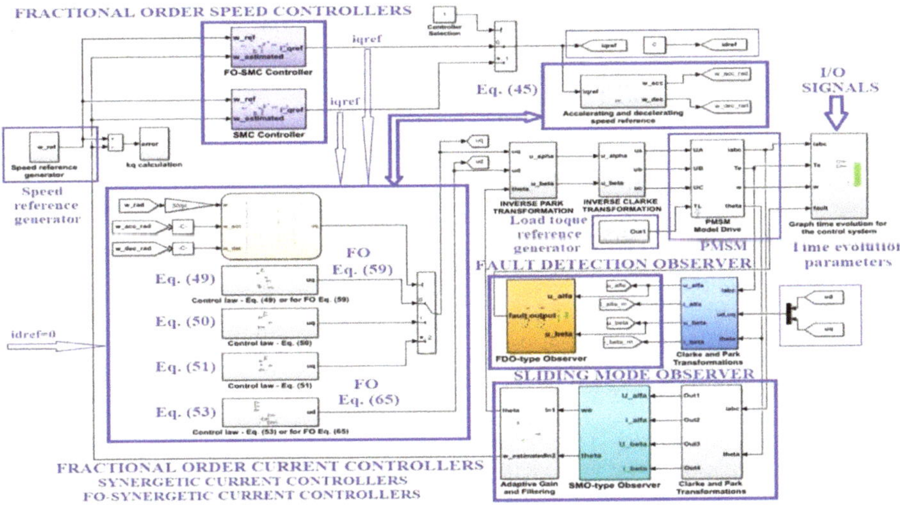

Figure 27. Simulink block diagram for the PMSM sensorless control based FO-SMC speed controller, FO-synergetic current controller, SMO speed observer and FDO.

For the implementation of the FO-SMC-type control described in Section 4, the parameters $\varepsilon = 300$, $q = 200$, $c = 100$ and $\mu = 0.55$ were selected, and for the FO-synergetic type controller described in Section 5, the parameters $k_{iq} = 10{,}000$, $k_q = 10{,}000$, $i_{qmax} = 50$, $T_d = 3$, $T_q = 3$, $k_{id} = 10{,}000$ and $\mu = 0.5$, for Equations (54) and (60), were selected.

The evolution of the time response for the PMSM control system with FO-SMC speed controller and FO-synergetic current controller for $\omega_{ref} = 500$ rpm, $T_L = 1$ Nm is presented in Figure 28. Very good static and dynamic performances can be noted, of which we mention the response time of 0.92 ms and an overshoot of less than 1.2%.

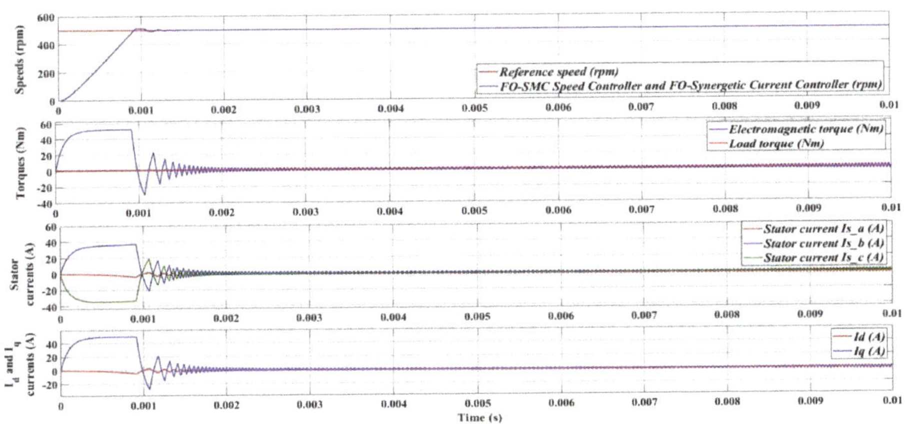

Figure 28. Simulation of the PMSM time evolution with FO-SMC speed controller and FO-synergetic current controller—$\omega_{ref} = 500$ rpm, $T_L = 1$ Nm, ($\varepsilon = 300$, $q = 200$, $c = 100$ and $\mu = 0.55$ for FO-SMC speed controller), ($k_{iq} = 10{,}000$, $k_q = 10{,}000$, $i_{qmax} = 50$, $T_d = 3$, $T_q = 3$, $k_{id} = 10{,}000$ and $\mu = 0.5$ for FO-synergetic current controllers).

In Figure 29, under the same conditions, but for a load torque of 10 Nm, very good performances of the PMSM control system with a response time of 1.2 ms are also obtained. The parametric robustness of the proposed control system is demonstrated in Figure 30 by obtaining a response time of 1.6 ms under the conditions where a uniformly distributed noise with 0.2 Nm magnitude additionally acts on the load torque, J parameter has a 100% increase, and also the stator resistance R_s has double the nominal value.

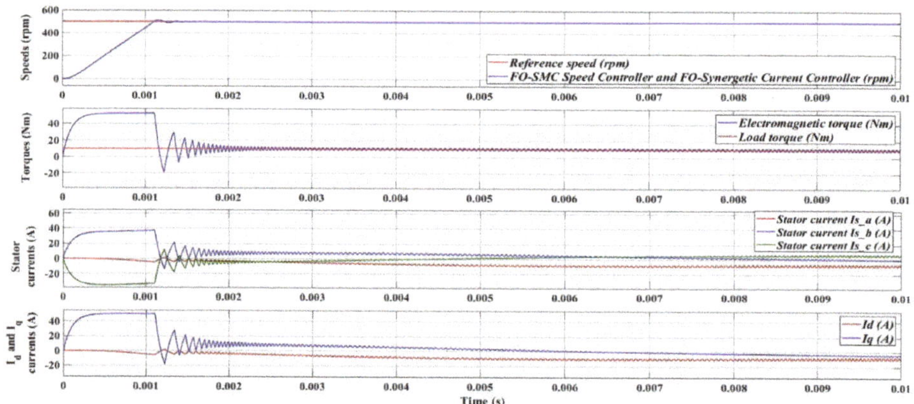

Figure 29. Simulation of the PMSM time evolution with FO-SMC speed controller and FO-synergetic current controller—ω_{ref} = 500 rpm, T_L = 10 Nm, (ε = 300, q = 200, c = 100 and μ = 0.55 for FO-SMC speed controller), (k_{iq} = 10,000, k_q = 10,000, i_{qmax} = 50, T_d = 3, T_q = 3, k_{id} = 10,000 and μ = 0.5 for FO-synergetic current controllers).

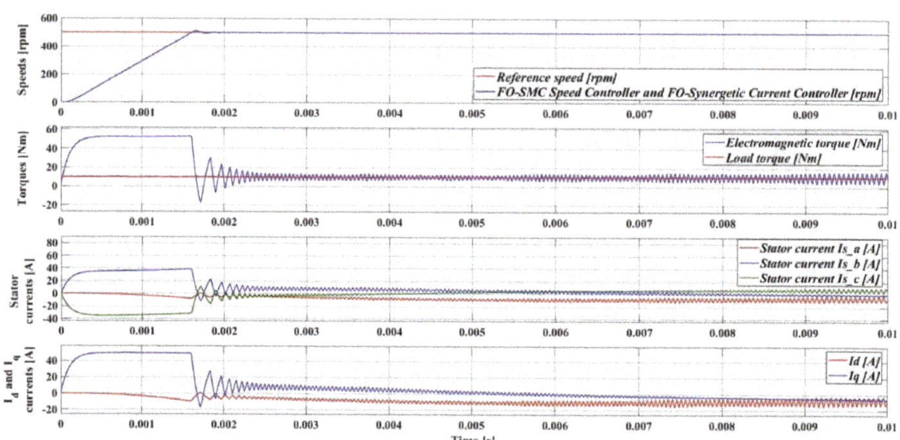

Figure 30. Simulation of the PMSM time evolution with FO-SMC speed controller and FO-synergetic current controller—ω_{ref} = 500 rpm, T_L = 10 Nm and uniformly distributed noise, (ε = 300, q = 200, c = 100 and μ = 0.55 for FO-SMC speed controller), (k_{iq} = 10,000, k_q = 10,000, i_{qmax} = 50, T_d = 3, T_q = 3, k_{id} = 10,000 and μ = 0.5 for FO-synergetic current controllers), 100% increase of J parameter and 100% increase of stator resistance R_s.

Figure 31 compares the response of four PMSM control systems obtained by combinations of SMC and FO-SMC speed control systems, and the current controllers are of synergetic and FO-synergetic type. Table 4 presents the comparative results of the performance of these control systems according to: overshoot, settling time, steady state error and speed ripple defined in relation (117). It is obvious that the control system proposed in this article based on FO-SMC speed controller and FO-synergetic current controllers has the best performance.

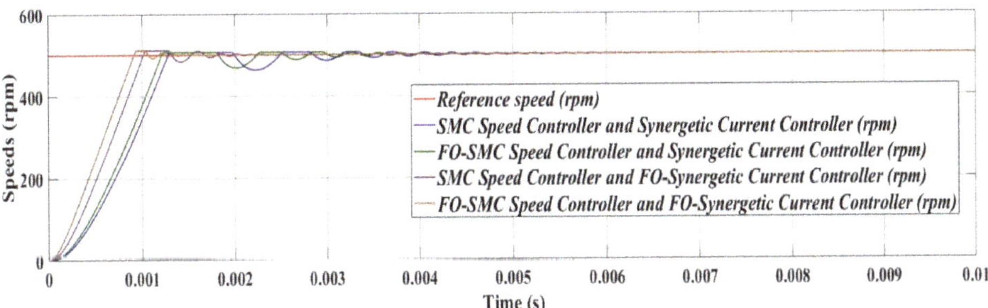

Figure 31. Simulation of the PMSM time evolution comparative fractional order controllers—ω_{ref} = 500 rpm and T_L = 1 Nm.

Table 4. Comparison of performance indices of the fractional order proposed controllers.

Performance Indices	SMC Speed Controller and Synergetic Currents Controller	FO-SMC Speed Controller and Synergetic Currents Controller	SMC Speed Controller and FO-Synergetic Currents Controller	FO-SMC Speed Controller and FO-Synergetic Currents Controller
Overshoot (%) nominal J	1.15	1.15	1.18	1.15
Overshoot (%) double J	1.18	1.9	1.8	1.2
Settling time (ms) nominal J	1.4	1.3	1	0.92
Settling time (ms) double J	1.5	1.7	1.22	1.8
Steady state error (%) nominal J	0.07	0.07	0.06	0.06
Steady state error (%) double J	0.08	0.07	0.06	0.06
Speed ripple (rpm) nominal J	123.03	118.73	104.50	95.34
Speed ripple (rpm) double J	149.25	148.16	120.85	83.09
Torque ripple (Nm) nominal J	14.92	14.74	126.29	14.71
Torque ripple (Nm) double J	20.96	20.82	112.13	14.33

It is obviously noticeable that the values obtained for the settling time are 0.92 ms under nominal parameters of the PMSM, along with the other performance indices which can be considered as very good for the FO-SMC speed controller and the FO-synergetic current controller. The PMSM used in these numerical simulations is implemented in Power Systems/Simscape Electrical toolbox from Simulink, and in many scientific papers it is used as benchmark, and, to our best knowledge, the settling time of 0.92 ms obtained when using the FO-SMC speed controller and the FO-synergetic current controller is the best settling time obtained for a usual range of the speed reference and load torque, and for any other proposed controllers.

In addition, under the same conditions as a classic FOC-type control system of a PMSM, Figure 32 also shows an improvement in the THD of the currents in the PMSM supply phases. Thus, THD is reduced from 50% to 22%. Obviously, for further reduction of the THD, additional specialized systems as those presented in [42] can also be added.

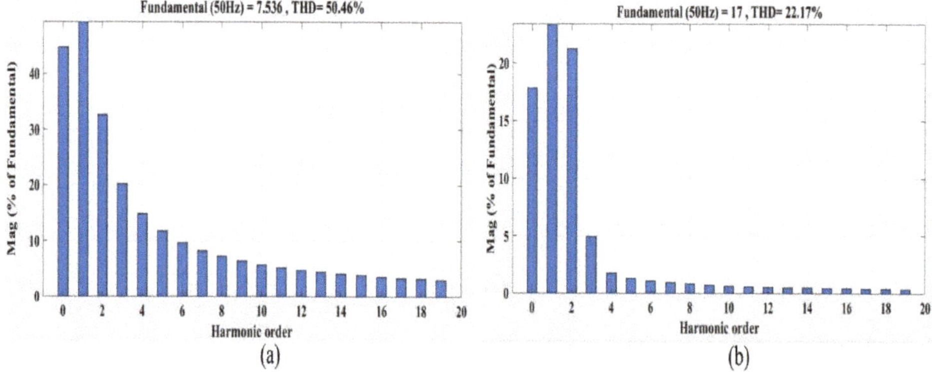

Figure 32. THD current analysis: (**a**) PMSM control system with PI speed controller and PI current controllers; (**b**) PMSM control system with FO-SMC speed controller and FO-synergetic current controllers.

8. Experimental Results

For the experimental validation of the simulations of the proposed PMSM control algorithms, which were presented and validated by numerical simulations in the previous section, this section presents the development platform used for their real time implementation in embedded systems.

The developed algorithms are implemented in MATLAB/Simulink and use dedicated functions from specialized libraries in the Model-Based Design Toolbox S32K1xx Series which contains the Automotive Math and Motor Control Library Set for NXP S32K14x devices, a toolbox which is dedicated to the PMSM control.

The hardware platform dedicated for the PMSM control for the experimental testing of the proposed algorithms is an S32K144 development kit which contains an S32K144 evaluation board (S32K144EVB-Q100), DEVKIT-MOTORGD board based on SMARTMOS GD3000 pre-driver and Linix 45ZWN24-40 PMSM type. The controller of the development platform is S32K144 MCU which is of 32bit Cortex M4F type, which has a time base of 112 MHz with 512 KB of flash memory and 54 KB of RAM.

There are also a number of dedicated hardware peripherals (FTM, ADC, PDB, PWM, timers) for the PMSM control, common analog and digital I/O processing blocks, but also a wide range of communication blocks which are common for the industrial environment. Among the communication interfaces we mention OpenSDA serial debug interface, CAN controller with CAN-FD protocol.

Figure 33 shows an image of the experimental platform.

The software application block diagram of the implementation in MATLAB/Simulink and Model-Based Design Toolbox S32K1xx Series NXP for the embedded system of the PMSM control system is presented in Figure 34. The main blocks presented are: data acquisition and commands, which is supervised by the dispatcher software interrupters, current controllers from inner loop and speed controller from outer loop.

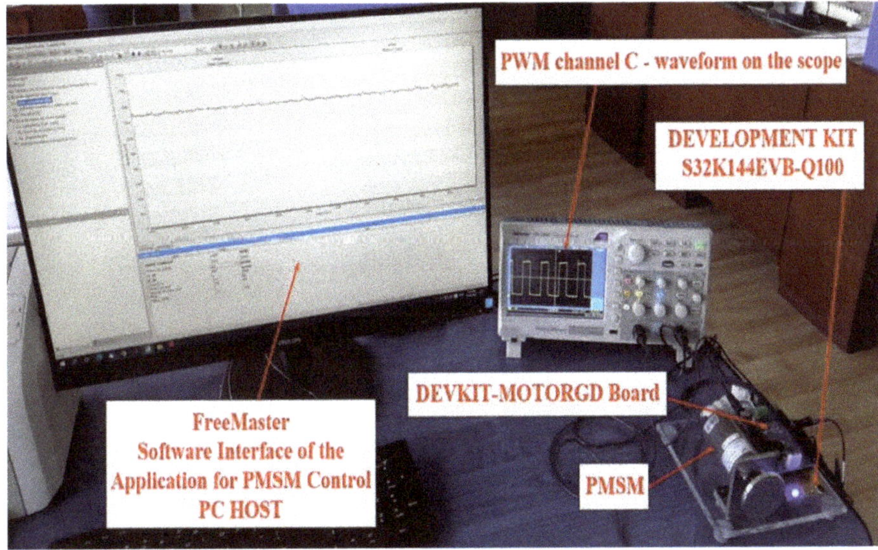

Figure 33. Experimental platform image.

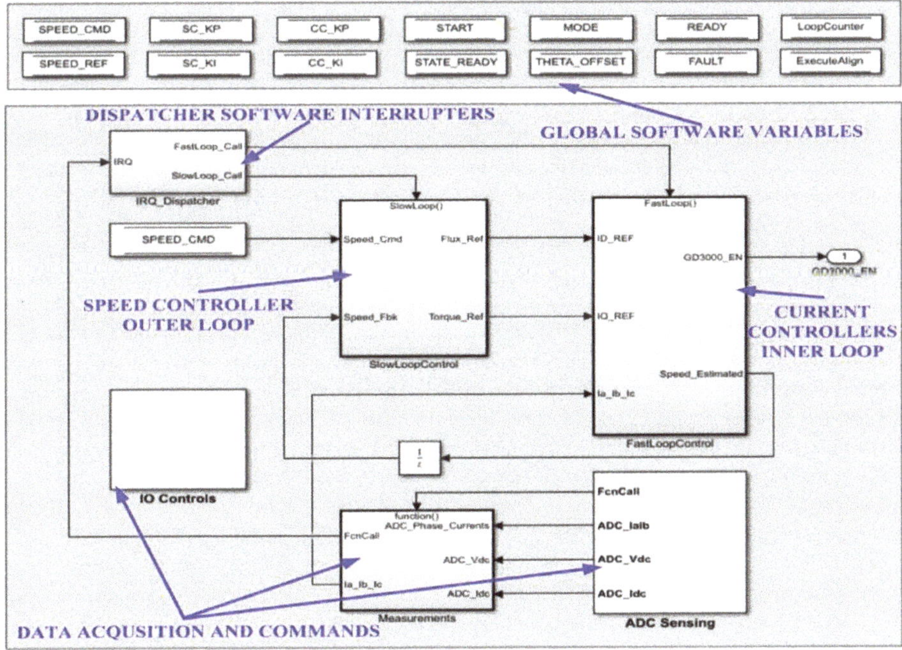

Figure 34. Software block diagram of the implementation in MATLAB/Simulink and Model-Based Design Toolbox S32K1xx Series NXP embedded system for PMSM control.

Figure 35 presents the block diagram of the software for the outer speed control loop MATLAB/Simulink model utilizing the bit accurate models for the Automotive Math and Motor Control Library Set for NXP S32K14x devices. The main blocks are: speed reference, initialization speed loop, switching block output command, classical PI speed controller and FO-PI speed controller.

The Discrete Zero-Pole function from Simulink [43] is used for the implementation of the FO-PI speed controller described in Section 4. Equation (108), which represents the discrete form of the FO-PI speed controller, is implemented through this function.

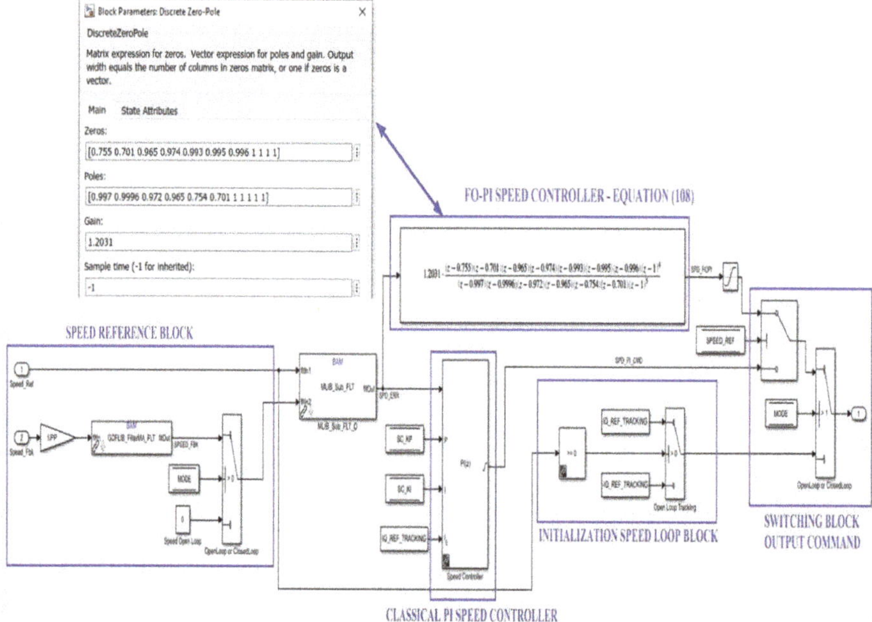

Figure 35. Software block diagram of the outer loop speed control MATLAB/Simulink model utilizing the bit accurate models for Automotive Math and Motor Control Library Set for NXP S32K14x devices.

The inner loop for the current control runs every 0.1 ms and the outer loop for the speed control runs every 1 ms. By comparison, each program implemented for numerical simulations, presented in Section 7, runs at each 0.001 ms.

Based on FreeMaster, which is a real-time debugging monitoring software interface for data visualization, configuration and tuning of embedded software applications, the next figure presents the real time evolution of the parameters of the PMSM control system. For reasons of communication between the FreeMaster software interface from the host PC and the controller of the PMSM, the sampling time for the evolution of parameters in Figures 36–44 is 10 ms. Figures 36–38 present the real-time evolution of the PMSM rotor speed with classical PI speed controller.

Figures 39–41 present the real-time evolution of the PMSM rotor speed with FO-PI speed controller. The superior performance of the FO-PI speed controller is clearly noticeable.

The following figures show the real-time evolution of the main PMSM control parameters of interest using the FO-PI speed controller. The real-time evolution of the stator currents are presented in Figure 42, the real-time evolution of i_d and i_q currents are presented in Figures 43 and 44 presents the real-time evolution of the electromagnetic torque.

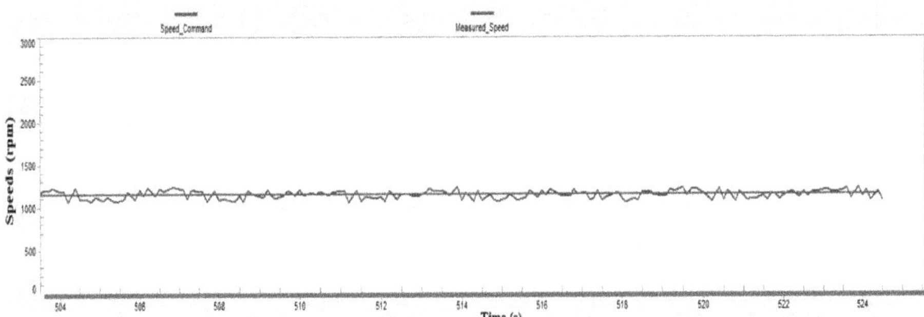

Figure 36. Real time evolution of the rotor speed of PMSM with classical PI speed controller for ω_{ref} = 1150 rpm and no load torque.

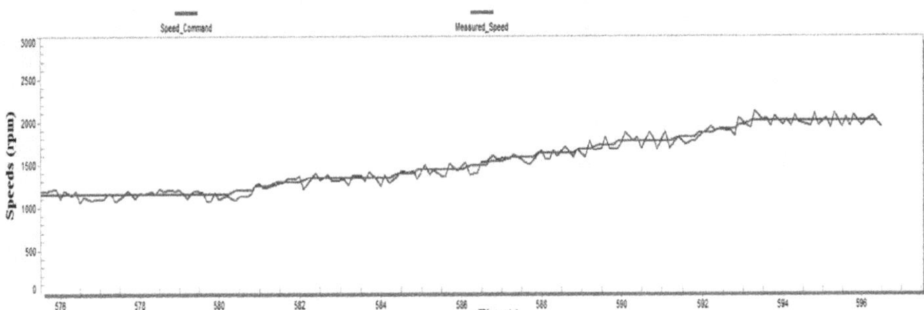

Figure 37. Real time evolution of the rotor speed of PMSM with classical PI speed controller for ω_{ref} between 1150 rpm and 2000 rpm and no load torque.

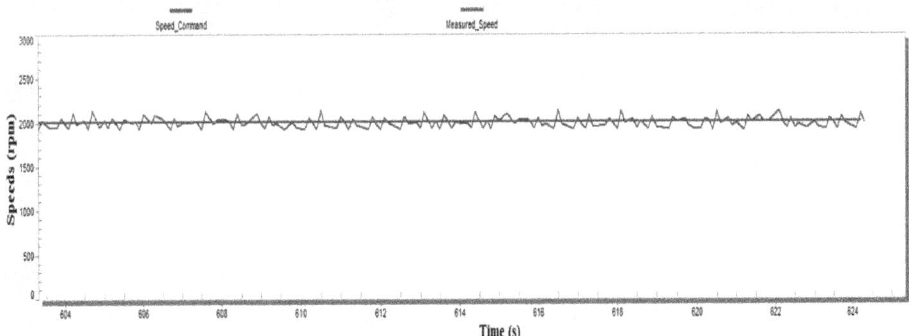

Figure 38. Real time evolution of the rotor speed of PMSM with classical PI speed controller for ω_{ref} = 2000 rpm and no load toque.

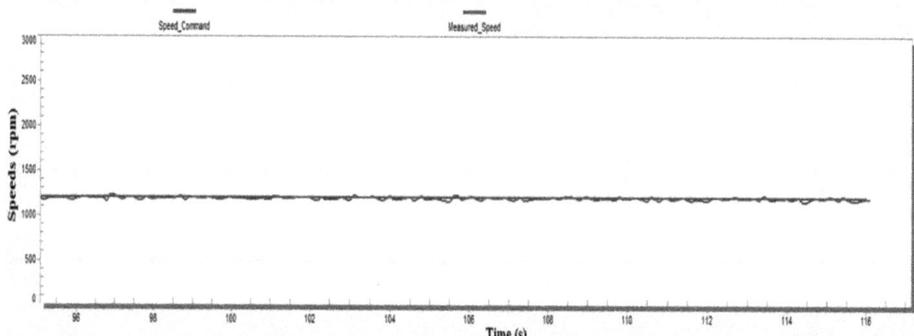

Figure 39. Real time evolution of the rotor speed of PMSM with FO-PI speed controller for ω_{ref} = 1150 rpm and no load toque.

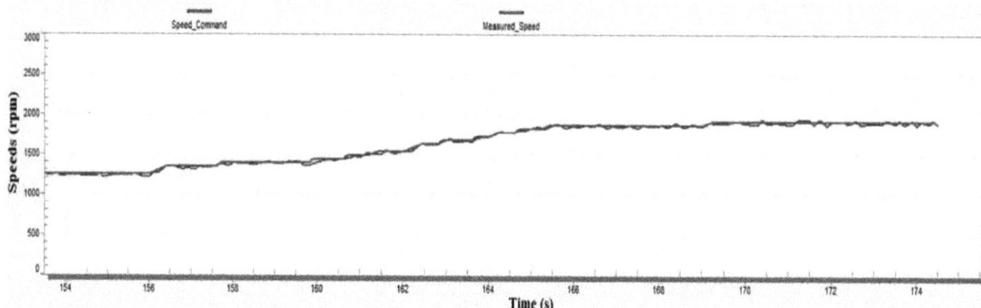

Figure 40. Real time evolution of the rotor speed of PMSM with classical FO-PI speed controller for ω_{ref} between 1150 rpm and 2000 rpm and no load toque.

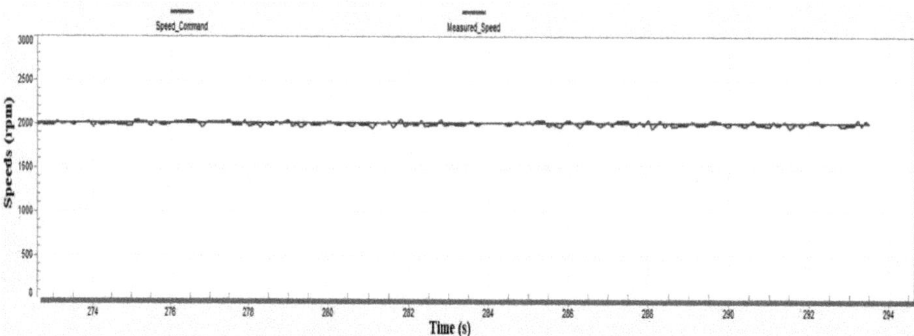

Figure 41. Real time evolution of the rotor speed of PMSM with FO-PI speed controller for ω_{ref} = 2000 rpm and no load toque.

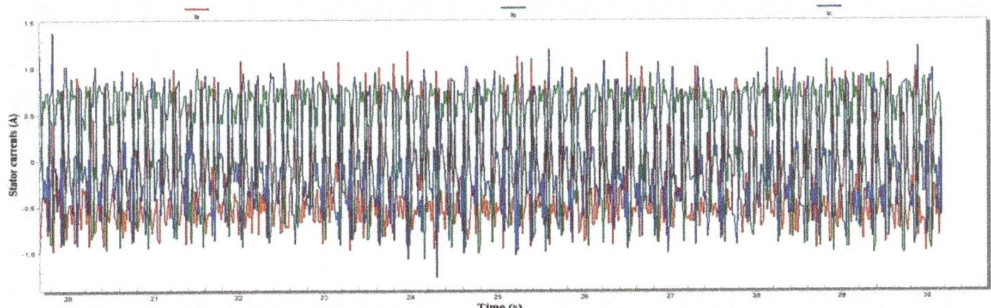

Figure 42. Real time evolution of the stator currents of PMSM with FO-PI speed controller.

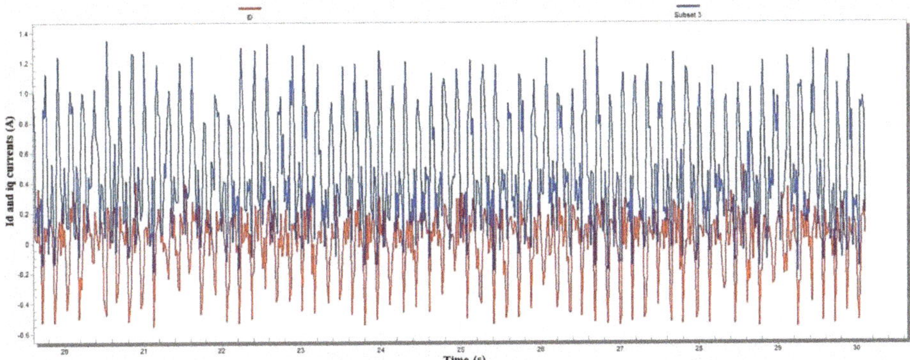

Figure 43. Real time evolution of the i_d and i_q currents of PMSM with FO-PI speed controller.

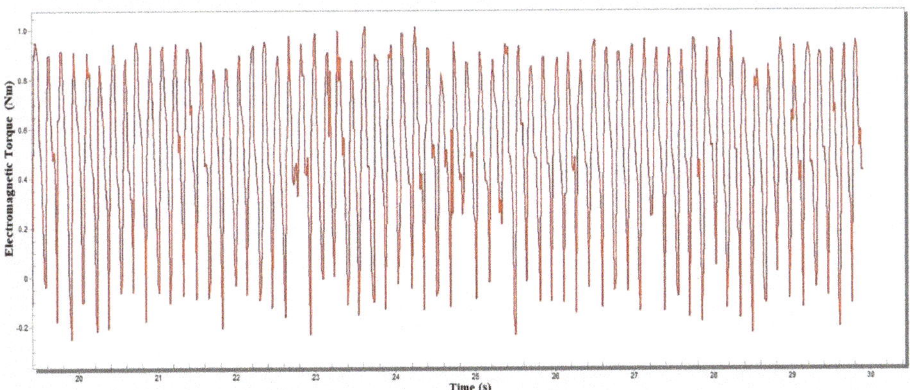

Figure 44. Real time evolution of the electromagnetic torque of PMSM with FO-PI speed controller.

It can be noted that there is a similarity between the results obtained by numerical simulation in Figure 13 for the control of the PMSM using an FO-PI speed controller, ω_{ref} = 2000 rpm and no load torque, and the experimental results presented in Figures 42–44 concerning the stator currents, the electromagnetic torque and i_d and i_q currents. Furthermore, Figures 36–41 show the superiority of the FO-PI speed controller over the classic PI speed controller, in the case of implementation in an embedded system.

9. Conclusions

Based on fractional calculus, this article presents a number of fractional controllers for the PMSM rotor speed control loops and i_d and i_q current control loops in the FOC-type control strategy. The proposed system for the PMSM control is based on an FO-SMC speed controller and FO-synergetic current controllers. Due to the additional control parameters generated in the structure of the fractional order controllers, superior performances are obtained for the PMSM rotor speed control. In addition, the sensorless-type PMSM control structure detects the faults of the current sensors and performs a significant reduction in the THD. The parametric robustness of the proposed control system is demonstrated by very good control performances achieved even when the uniformly distributed noise is present in the load torque T_L, and under variations by 100% of the load torque T_L, of the moment of inertia of J rotor and of the stator resistance R_s. The performances of the proposed control system are validated both by numerical simulations and experimentally, through real-time implementation in embedded systems.

Author Contributions: Conceptualization, M.N.; data curation, M.N. and C.-I.N.; formal analysis, M.N. and C.-I.N.; funding acquisition, M.N.; investigation, M.N. and C.-I.N.; methodology, M.N. and C.-I.N.; project administration, M.N.; resources, M.N. and C.-I.N.; software, M.N. and C.-I.N.; supervision, M.N. and C.-I.N.; validation, M.N. and C.-I.N.; visualization, M.N. and C.-I.N.; writing—original draft, M.N. and C.-I.N.; writing—review and editing, M.N. and C.-I.N. All authors have read and agreed to the published version of the manuscript.

Funding: This paper was developed with funds from the Ministry of Research and Innovation as part of the NUCLEU Program: PN 19 38 01 03.

Conflicts of Interest: The authors declare no conflict of interest.

References

1. Utkin, V.; Guldner, J.; Shi, J. Automation and Control Engineering. In *Sliding Mode Control in Electromechanical Systems*, 2nd ed.; CRC Press: Boca Raton, FL, USA, 2009; pp. 223–271.
2. Bose, B.K. *Modern Power Electronics and AC Drives*; Prentice Hall: Upper Saddle River, NJ, USA, 2002; pp. 439–534.
3. Alkorta, P.; Barambones, O.; Cortajarena, J.A.; Martija, I.; Maseda, F.J. Effective Position Control for a Three-Phase Motor. *Electronics* **2020**, *9*, 241. [CrossRef]
4. Yuan, T.; Wang, D. Performance Improvement for PMSM DTC System through Composite Active Vectors Modulation. *Electronics* **2018**, *7*, 263. [CrossRef]
5. Wang, H.; Leng, J. Summary on development of permanent magnet synchronous motor. In Proceedings of the Chinese Control and Decision Conference (CCDC), Shenyang, China, 9–11 June 2018; pp. 689–693.
6. Bao, G.; Qi, W.; He, T. Direct Torque Control of PMSM with Modified Finite Set Model Predictive Control. *Energies* **2020**, *13*, 234. [CrossRef]
7. Wu, Z.; Gu, W.; Zhu, Y.; Lu, K. Current Control Methods for an Asymmetric Six-Phase Permanent Magnet Synchronous Motor. *Electronics* **2020**, *9*, 172. [CrossRef]
8. Wang, W.; Tan, F.; Wu, J.; Ge, H.; Wei, H.; Zhang, Y. Adaptive Integral Backstepping Controller for PMSM with AWPSO Parameters Optimization. *Energies* **2019**, *12*, 2596. [CrossRef]
9. Wang, J. Fuzzy Adaptive Repetitive Control for Periodic Disturbance with Its Application to High Performance Permanent Magnet Synchronous Motor Speed Servo Systems. *Entropy* **2016**, *18*, 261. [CrossRef]
10. Lu, S.; Tang, X.; Song, B. Adaptive PIF Control for Permanent Magnet Synchronous Motors Based on GPC. *Sensors* **2013**, *13*, 175–192. [CrossRef]
11. Zhao, Y.; Liu, X.; Zhang, Q. Predictive Speed-Control Algorithm Based on a Novel Extended-State Observer for PMSM Drives. *Appl. Sci.* **2019**, *9*, 2575. [CrossRef]
12. Tang, M.; Zhuang, S. On Speed Control of a Permanent Magnet Synchronous Motor with Current Predictive Compensation. *Energies* **2019**, *12*, 65. [CrossRef]
13. Zhang, G.; Chen, C.; Gu, X.; Wang, Z.; Li, X. An Improved Model Predictive Torque Control for a Two-Level Inverter Fed Interior Permanent Magnet Synchronous Motor. *Electronics* **2019**, *8*, 769. [CrossRef]
14. Liu, X.; Zhang, Q. Robust Current Predictive Control-Based Equivalent Input Disturbance Approach for PMSM Drive. *Electronics* **2019**, *8*, 1034. [CrossRef]

15. Ma, Y.; Li, Y. Active Disturbance Compensation Based Robust Control for Speed Regulation System of Permanent Magnet Synchronous Motor. *Appl. Sci.* **2020**, *10*, 709. [CrossRef]

16. Larbaoui, A.; Belabbes, B.; Meroufel, A.; Tahour, A.; Bouguenna, D. Backstepping Control with Integral Action of PMSM Integrated According to the MRAS Observer. *IOSR J. Electr. Electron. Eng.* **2014**, *9*, 59–68. [CrossRef]

17. Merabet, A. Cascade Second Order Sliding Mode Control for Permanent Magnet Synchronous Motor Drive. *Electronics* **2019**, *8*, 1508. [CrossRef]

18. Qian, J.; Ji, C.; Pan, N.; Wu, J. Improved Sliding Mode Control for Permanent Magnet Synchronous Motor Speed Regulation System. *Appl. Sci.* **2018**, *8*, 2491. [CrossRef]

19. Bounasla, N.; Hemsas, K.E.; Mellah, H. Synergetic and sliding mode controls of a PMSM: A comparative study. *J. Electr. Electron. Eng.* **2015**, *3*, 22–26. [CrossRef]

20. Bogani, T.; Lidozzi, A.; Solero, L.; Di Napoli, A. Synergetic Control of PMSM Drives for High Dynamic Applications. In Proceedings of the IEEE International Conference on Electric Machines and Drives, San Antonio, TX, USA, 15 May 2005; pp. 710–717.

21. Khettab, K.; Ladaci, S.; Bensafia, Y. Fuzzy adaptive control of fractional order chaotic systems with unknown control gain sign using a fractional order Nussbaum gain. *IEEE/CAA J. Autom. Sin.* **2019**, *6*, 816–823. [CrossRef]

22. Wang, M.-S.; Hsieh, M.-F.; Lin, H.-Y. Operational Improvement of Interior Permanent Magnet Synchronous Motor Using Fuzzy Field-Weakening Control. *Electronics* **2018**, *7*, 452. [CrossRef]

23. Nicola, M.; Nicola, C.-I.; Duţă, M. Sensorless Control of PMSM using FOC Strategy Based on Multiple ANN and Load Torque Observer. In Proceedings of the 2020 International Conference on Development and Application Systems (DAS), Suceava, Romania, 21–23 May 2020; pp. 32–37.

24. Hoai, H.-K.; Chen, S.-C.; Than, H. Realization of the Sensorless Permanent Magnet Synchronous Motor Drive Control System with an Intelligent Controller. *Electronics* **2020**, *9*, 365. [CrossRef]

25. Wang, M.-S.; Tsai, T.-M. Sliding Mode and Neural Network Control of Sensorless PMSM Controlled System for Power Consumption and Performance Improvement. *Energies* **2017**, *10*, 1780. [CrossRef]

26. Liu, X.; Du, J.; Liang, D. Analysis and Speed Ripple Mitigation of a Space Vector Pulse Width Modulation-Based Permanent Magnet Synchronous Motor with a Particle Swarm Optimization Algorithm. *Energies* **2019**, *9*, 923. [CrossRef]

27. Zhu, Y.; Tao, B.; Xiao, M.; Yang, G.; Zhang, X.; Lu, K. Luenberger Position Observer Based on Deadbeat-Current Predictive Control for Sensorless PMSM. *Electronics* **2020**, *9*, 1325. [CrossRef]

28. Nicola, M.; Nicola, C.-I.; Sacerdotianu, D. Sensorless Control of PMSM using DTC Strategy Based on PI-ILC Law and MRAS Observer. In Proceedings of the 2020 International Conference on Development and Application Systems (DAS), Suceava, Romania, 21–23 May 2020; pp. 38–43.

29. Ye, M.; Shi, T.; Wang, H.; Li, X.; Xiai, C. Sensorless-MTPA Control of Permanent Magnet Synchronous Motor Based on an Adaptive Sliding Mode Observer. *Energies* **2019**, *12*, 3773. [CrossRef]

30. Urbanski, K.; Janiszewski, D. Sensorless Control of the Permanent Magnet Synchronous Motor. *Sensors* **2019**, *19*, 3546. [CrossRef]

31. Kamel, T.; Abdelkader, D.; Said, B.; Padmanaban, S.; Iqbal, A. Extended Kalman Filter Based Sliding Mode Control of Parallel-Connected Two Five-Phase PMSM Drive System. *Electronics* **2018**, *7*, 14. [CrossRef]

32. Tepljakov, A. Fractional-Order Calculus Based Identification and Control of Linear Dynamic Systems. Master's Thesis, Tallinn University of Technology—Department of Computer Control, Tallinn, Estonia, 2011.

33. Tepljakov, A.; Petlenkov, E.; Belikov, J. FOMCON: Fractional-order modeling and control toolbox for MATLAB. In Proceedings of the 18th International Conference Mixed Design of Integrated Circuits and Systems—MIXDES, Gliwice, Poland, 16–18 June 2011; pp. 684–689.

34. Mohd Zaihidee, F.; Mekhilef, S.; Mubin, M. Robust Speed Control of PMSM Using Sliding Mode Control (SMC)—A Review. *Energies* **2019**, *12*, 1669. [CrossRef]

35. Wang, D.; Song, B.; Kang, C.; Xu, J. Design of Fractional Order PI Controller for Permanent Magnet Synchronous Motor. In Proceedings of the 2nd IEEE Advanced Information Management, Communicates, Electronic and Automation Control Conference (IMCEC), Xi'an, China, 25–27 May 2018; pp. 780–784.

36. Nicola, M.; Sacerdotianu, D.; Nicola, C.-I.; Hurezeanu, A. Simulation and Implementation of Sensorless Control in Multi-Motors Electric Drives with High Dynamics. *Adv. Sci. Technol. Eng. Syst. J.* **2017**, *2*, 59–67. [CrossRef]

Electronics **2020**, *9*, 1494

37. Nicola, M.; Nicola, C.-I.; Vintila, A. Sensorless Control of Multi-Motors PMSM using Back-EMF Sliding Mode Observer. In Proceedings of the Electric Vehicles International Conference (EV2019), Bucharest, Romania, 3–4 October 2019; pp. 1–6.

38. Nicola, M.; Velea, F. Automatic Control of a Hidropower Dam Spillway. *Ann. Univ. Craiova Electr. Eng. Ser.* **2010**, *34*, 1–4.

39. Nicola, M.; Nicola, C.I.; Duță, M. Delay Compensation in the PMSM Control by using a Smith Predictor. In Proceedings of the 2019 8th International Conference on Modern Power Systems (MPS), Cluj Napoca, Romania, 21–23 May 2019; pp. 1–6.

40. Huang, G.; Luo, Y.-P.; Zhang, C.-F.; He, J.; Huang, Y.-S. Current Sensor Fault Reconstruction for PMSM Drives. *Sensors* **2016**, *16*, 178. [CrossRef]

41. Biris, I.; Muresan, C.; Nascu, I.; Ionescu, C. A Survey of Recent Advances in Fractional Order Control for Time Delay Systems. *IEEE Access* **2019**, *7*, 30951–30965. [CrossRef]

42. Hwang, J.-C.; Wei, H.-T. The Current Harmonics Elimination Control Strategy for Six-Leg Three-Phase Permanent Magnet Synchronous Motor Drives. *IEEE Trans. Power Electron.* **2015**, *6*, 3032–3040. [CrossRef]

43. Ponce, P.; Molina, A.; Mata, O.; Ibarra, L.; MacCleery, B. *Power System Fundamentals*; CRC Press: Boca Raton, FL, USA, 2018; pp. 55–120.

MDPI

St. Alban-Anlage 66

4052 Basel

Switzerland

Tel. +41 61 683 77 34

Fax +41 61 302 89 18

www.mdpi.com

Electronics Editorial Office

E-mail: electronics@mdpi.com

www.mdpi.com/journal/electronics